HANDBOOK OF ELECTRICAL AND ELECTRONICS TECHNOLOGY

Edited by Curtis D. Johnson
University of Houston

Prentice Hall
Englewood Cliffs, New Jersey | Columbus, Ohio

Library of Congress Cataloging-in-Publication Data

Johnson, Curtis D.
    Handbook of electrical and electronics technology/Curtis D. Johnson.
        p.   cm.
    Includes index.
    ISBN 0-13-210618-3
    1. Electric engineering—Handbooks, manuals, etc. I. Title.
TK151.J64   1996
621.3—dc20   95-42875
                                                                              CIP

Editor: Charles Stewart
Production Editor: Alexandrina Benedicto Wolf
Cover photo: © T. Skrivan/H. Armstrong Roberts, Inc.
Production Manager: Deidra M. Schwartz
Marketing Manager: Debbie Yarnell

This book was set in Times Roman by Bi-Comp, Inc. and was printed and bound by
Quebecor Printing/Book Press. The cover was printed by Phoenix Color Corp.

Printed in the United States of America

10 9 8 7 6 5 4 3 2 1

ISBN: 0-13-210618-3

Prentice-Hall International (UK) Limited, *London*
Prentice-Hall of Australia Pty. Limited, *Sydney*
Prentice-Hall of Canada, Inc., *Toronto*
Prentice-Hall Hispanoamericana, S. A., *Mexico*
Prentice-Hall of India Private Limited, *New Delhi*
Prentice-Hall of Japan, Inc., Tokyo
Simon & Schuster Asia Pte. Ltd., *Singapore*
Editora Prentice-Hall do Brasil, Ltda., *Rio de Janeiro*

PREFACE

The objective of this handbook is to provide a quick reference for technicians, technologists, and engineers on practical, applications-oriented issues in electrical and electronics technology. It is intended to serve as a reminder of things already learned and to present practical considerations about applications that may not have been covered in the classroom. It is more informational than instructional in presentation.

The handbook should be turned to when a practitioner

- has trouble recalling some concept or principle once learned,
- needs practical, applications-oriented information about an electrical or electronics topic, or
- needs a quick review of a topic as a precursor to detailed study.

Each of the 18 chapters of the handbook addresses a major topic in electrical and electronics technology, with practical worked examples where appropriate. The first three chapters cover general circuit concepts and sources of electrical power. Chapters 4 through 10 cover the common circuit components from resistors through transistors as well as transformers and motors. Chapters 11 through 15 are devoted to integrated circuit systems of both an analog and digital nature. The last three chapters present the specialized topics of filters, measurement, and communication. The appendices contain tables and other general reference material.

Although not intended as a text for instruction, the handbook can be introduced to students in the early years of their study as a reference they can use throughout much of their education in electrical and electronics technology and on into their professional careers.

Many professionals have been involved in the writing of this handbook. Where a particular author has been responsible for a major portion of a chapter, they have been recognized by name. The handbook took many years to complete and I wish to thank all those involved in writing, reviewing, and production for their understanding and forbearance.

Curtis D. Johnson

BRIEF CONTENTS

CONTENTS

1

NETWORK CONCEPTS AND ANALYSIS*

This chapter reviews the basic concepts of electrical networks and the techniques used to analyze them.

1.1   NETWORK FUNDAMENTALS

An electrical network is an interconnection of elements called sources and components. As illustrated in Figure 1.1, each element has a voltage across its terminals and a current passing through its terminals. The objective of network analysis is to determine the magnitude and polarity of the voltages and magnitude and direction of the currents as a function of time.

Units   The set of units used in electrical network analysis is the Système International d'Unites (SI units). This is a metric system based on seven fundamental quantities including the meter, kilogram, and second. It is more completely defined in Appendix 1.

1.1.1   Charge

The most basic electrical quantity is a property of atomic particles called *charge*. Charge comes in two forms, positive and negative. There is a force between charged particles, called the *Coulomb force,* which causes like charged particles to be repelled from each other and unlike charges to be attracted. This force is the basis for electrical behavior.

Matter is composed of units called *atoms,* which may be pictured as being composed of a nucleus containing positively charged particles called *protons,* sur-

---

* This chapter was written by Dr. Farrokh Attarzadeh, College of Technology, University of Houston.

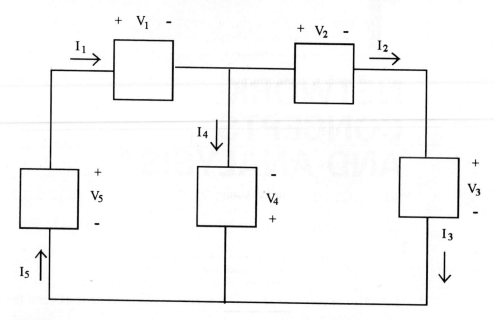

FIGURE 1.1
A network is a connection of elements, each with a voltage and current.

rounded by an equal number of negatively charged particles, called *electrons*. Normally the number of positive and negative charges in a given quantity of matter is equal and we say that it is *uncharged* or *neutral*. If electrons are removed from a neutral quantity of matter, the result is *positively* charged matter. Similarly, if electrons are added to the matter, the result is *negatively* charged matter.

Although not one of the basic seven, the SI defined unit of charge is the *coulomb* (C). This is the amount of charge that results from an electrical current flow of one ampere for one second. It is also the charge possessed by 6.24×10^{18} electrons. As a variable, the charge is described by the variable label Q for fixed charge or $q(t)$ for time changing charge.

1.1.2 Current

An electrical current exists whenever charge, $q(t)$, is transferred from one point in the conductor to another. The SI system defines the basic unit of current as the *ampere* (A). This represents a current for which charge is being transferred at the rate of one coulomb in one second. In equation form, we have the current defined as the rate of change of charge, i.e., the derivative,

$$i(t) = \frac{dq}{dt}$$

(1.1)

where $i(t)$ = current in amperes (A)
         $q(t)$ = charge in coulombs (C)
         t = time in seconds (s).

If the flow is uniform, i.e., a constant current, then Eq. (1.1) reduces to the form

$$I = Q/t$$

Here, I is the steady current and Q is the amount of charge transferred in a time t. Note that the capital letter I is used for constant current and lowercase letter $i(t)$ for varying current.

The polarity of the current really indicates direction of flow and depends on the polarity of the charges being transferred. Two standards are used to describe the direction. The most common, called *conventional current,* defines positive current direction as the motion of positive charges, thus opposite to the direction in which

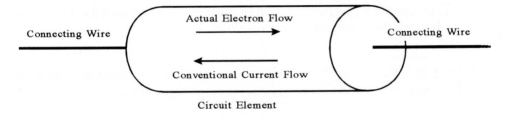

Connecting Wire   Connecting Wire

Actual Electron Flow

Conventional Current Flow

Circuit Element

FIGURE 1.2
Conventional current flow direction is opposite that of actual electron flow.

electrons flow, as shown in Figure 1.2. Some treatments use a standard called *electron flow* wherein current direction is defined with a positive sign for the direction of electron flow. This handbook uses conventional current.

EXAMPLE 1.1   Thirty coulombs of charge pass a given point in a wire in 5 s. How many amperes of current are flowing?

Solution
From Eq. (1.1), although for constant current, we have

$$I = Q/t = (30 \text{ C})/(5 \text{ s}) = 6 \text{ C/s} = 6 \text{ A}$$

1.1.3 Energy

The law of conservation of energy states that energy cannot be created or destroyed, only transformed. The electrical form of energy may be produced from many other types of energy, such as chemical (as in a battery), mechanical (as in a hydroelectric generator) and atomic (as in a nuclear reactor). The SI unit of energy is the *joule* (*J*). The letter symbol *W* or *w(t)* is often used for energy.

In electricity and electronics we are concerned with doing *work,* which is the expenditure of energy. If an electrical motor lifts a heavy load, that is a form of work. If a TV presents an image for viewing, that too is a form of work. Electrical work is done by the movement of charges, i.e., by an electrical current.

1.1.4 Voltage

If energy is expended (as work) on a quantity of charge, then the ratio of the energy expended to the quantity of charge is called the *voltage.* Voltage is the force that makes the charges move, i.e., voltage produces the current, which does work. For example, a battery uses chemical processes to do work on charged particles to make them move through a conductor. Thus, a voltage exists across the battery terminals to force the charges to move. The unit of voltage in the SI system is the *volt* (*V*). The letter symbol *V* (for fixed), or *v(t)* (for varying), is often used for the voltage. Unfortunately this often causes confusion with the unit symbol, which is also V.

EXAMPLE 1.2

If 100 J of energy is used to move 20 C of charge through an element, what is the voltage across the element?

Solution
Because voltage is the ratio of energy to charge, we have $V = W/Q = (100 \text{ J})/(20 \text{ C}) = 5 \text{ V}$

1.1.5 Power

When work is done over a period of time a definition of the *rate at which work is done* is more useful than the amount of work. This is called the *power* and is described by the SI unit of *watt* (*W*), which is work done at a rate of one joule in one second. The letter symbol *P* (for fixed) or *p(t)* (for varying) is often used for the power label. Since it is a rate, power is defined by the derivative

$$p(t) = \frac{dw}{dt} \qquad (1.2)$$

or, for fixed power, $P = W/t$.

If the power is known the energy can be found, in general, by the relation

$$w(t) = \int_{-\infty}^{t} p(\tau)\, d\tau \qquad (1.3)$$

or, for constant power, $W = Pt$. The variable of integration is τ.

The specification of conditions at $t = -\infty$ is not always convenient so an alternative to Eq. (1.3) makes use of known conditions at some initial time, t_0.

$$w(t) = \int_{t_0}^{t} p(\tau)\, d\tau + w(t_0) \tag{1.4}$$

Since current is the time rate of change of charge and voltage is the rate of change of work with charge, we see that Eq. (1.2) may be generalized for electrical variables as

$$p(t) = i(t)v(t) \tag{1.5}$$

or, when the voltage and current do not vary in time, as

$$P = IV$$

EXAMPLE 1.3

An amount of energy equal to 200 J is used in 10 s by an electrical motor. What is the power?

Solution
From Eq. (1.2), $P = W/t = (200\ \text{J})/(10\ \text{s}) = 20\ \text{W}$

EXAMPLE 1.4

A 60-W light bulb operates on 120 V. How much current does it require?

Solution
From Eq. (1.5), $I = P/V = (60\ \text{W})/(120\ \text{V}) = 0.5\ \text{A}$

1.1.6 Network Classifications

Electrical networks may be defined by the following four classifications.

Linear or Nonlinear   A linear network can be described by a set of linear equations; otherwise, it is nonlinear. A network is linear if the network variables of voltage and current occur only to the positive first power. Most network analysis is restricted to finding the solution to linear networks. When nonlinear elements are present, such as bipolar transistors, approximate linear models must be used in order to solve the network.

Time Invariant and Time Varying   A time-invariant network implies that none of the network element values varies with time; otherwise, it is time varying. If it is time invariant, then the solutions will be voltages and currents that do not change in time.

Passive and Active   If the total energy delivered into a given network is non-negative at any instant of time, the network is passive; otherwise, it is active. In a passive network the elements can only consume energy, whereas in an active network elements can produce electrical energy by transformation of energy from one form to another.

Lumped and Distributed If the dimensions of the network elements are small compared to the wavelength of the highest signal frequency applied to the network, it is lumped; otherwise, it is distributed. When the dimensions of the elements and signal wavelengths become comparable, radiation effects require significant modifications to be made to the network analysis problem.

Another classification involves the pattern which the elements make when connected. Common terms for patterns are ladders, lattices, and bridged-T networks. Figure 1.3 illustrates some common patterns.

1.1.7 Structure of Linear Networks

The *elements* of a linear network are of two types, passive and active. Passive elements, which inject no energy into the network, are *components* such as resistors, capacitors, inductors, and related elements. Active elements, which may inject energy into the network, are real and controlled *sources* of voltage and current.

Figure 1.4 shows a network that helps define the following terms.

A *node* is a point in a network at which two or more elements are joined, as shown in Figure 1.4. If there are three or more elements connected at a node, that node is also called a *junction*.

A *branch* of a network extends from one junction to another and may consist of a single element or a series of elements. Thus, there is a node at the ends of each element and a junction at the end of each branch.

A *loop* is a closed path for current in a network. A *mesh* is a combination of loops.

The term *bilateral* is used to describe coupling between circuits due to mutual elements. If current in one circuit produces voltage in another and if the same current in the second circuit would produce the same voltage in the first, the coupling is called symmetrically bilateral. Ordinary electrical elements, including transformers, are symmetrically bilateral. The transistor is an example of an asymmetrical bilateral coupling.

A passive element receives power from the network. An active element delivers power to the network. Linear networks use ideal sources that either maintain a specified voltage between their terminals, regardless of current, or maintain a specified current regardless of voltage.

1.1.8 Sources

Ideal Voltage Source An ideal voltage source is an element that produces a voltage or potential difference of V volts across its terminals, regardless of what load is connected. Figure 1.5(a) illustrates the ideal voltage source. It has zero internal resistance, inductance, and capacity. A real voltage source, such as a battery, is modeled as an ideal voltage source in series with an internal resistance or impedance, as shown in Figure 1.5(b).

Ideal Current Source An ideal current source is an element that delivers a constant current of I amperes into a load regardless of what load is connected.

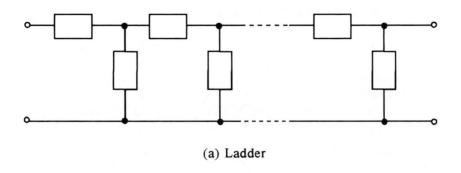

(a) Ladder

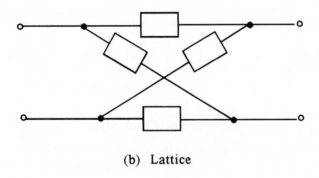

(b) Lattice

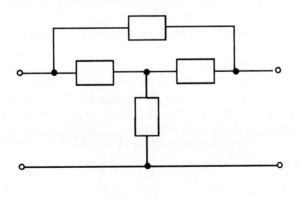

(c) Bridged-T

FIGURE 1.3
Networks form many patterns including the ladder, lattice, and bridged-T patterns.

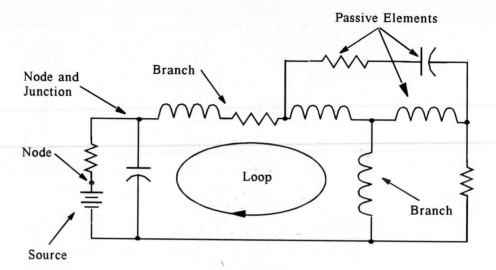

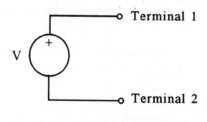

FIGURE 1.4
This general schematic serves to define network terms.

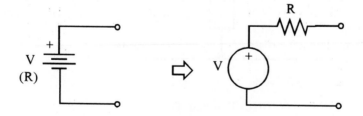

(a) Ideal Voltage Source

(b) Real Voltage Source

FIGURE 1.5
An ideal voltage source is used to model real voltage sources such as batteries.

8

FIGURE 1.6
An ideal current source delivers the same current into any load.

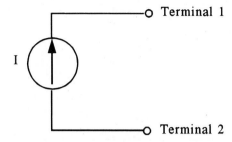

Terminal 1

I

Terminal 2

Figure 1.6 illustrates the ideal current source. It has an infinite internal resistance and zero inductance and capacity.

Controlled Sources   Controlled sources are ideal voltage or current sources whose magnitude and polarity depend on, or are controlled by, some other voltage or current in the network (see Section 1.4.1).

1.2   FUNDAMENTAL NETWORK CIRCUIT LAWS

This section presents a brief summary of the fundamental network circuit laws that are used to analyze electrical networks.

Wires   The elements of an electrical network are connected by metal wires to carry the current. For network analysis purposes these wires are ideal in that they possess no resistance, inductance, or capacity.

1.2.1   Ohm's Law

A *resistor* is a two-terminal component with the property that the ratio of voltage across its terminals to current through it is a constant. That constant is called its *resistance*.

Ohm's law defines this fact in equation form, as

$$R = \frac{V}{I} \tag{1.6}$$

where R = the resistance in ohms (Ω)
V = the voltage drop in volts (V)
I = the current passed in amperes (A).

Note that the voltage drop polarity is always such that the voltage is positive on the terminal of the resistor into which the conventional current enters.

FIGURE 1.7
Circuit for Example 1.5.

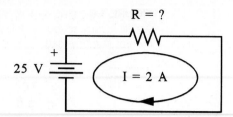

EXAMPLE 1.5 Determine the resistance of the resistor shown in Figure 1.7.

Solution
Using Ohm's law, $R = (25 \text{ V})/(2 \text{ A}) = 12.5 \ \Omega$

1.2.2 Kirchhoff's Current Law

Kirchhoff's current law (KCL) states that the algebraic sum of all current at a node or junction is zero at every instant of time. This simply means that if the sum of all currents directed into a node is subtracted from the sum of all current directed out of the node, the result will always be zero.

To apply KCL, we first note the current direction for each element connected to the node and then apply the law. Thus for the situation shown in Figure 1.8 we have the following equation:

$$I_1 - I_2 + I_3 - I_4 = 0 \qquad\qquad (1.7)$$

FIGURE 1.8
KCL specifies that the sum of currents at a node will be zero.

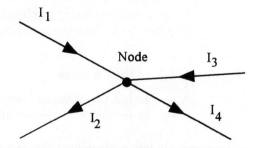

EXAMPLE 1.6 In Figure 1.9(a) all currents are given in both magnitude and direction except for that through R_3. Find the current through R_3.

Solution
To apply KCL, we first note that the interconnection of resistors at the top forms one junction of four elements as illustrated in Figure 1.9(b). KCL states that the sum of currents entering this junction will equal the sum of currents leaving the junction. Using the given current values and directions, we have

$$I_1 + I_2 + I_3 = I_T$$

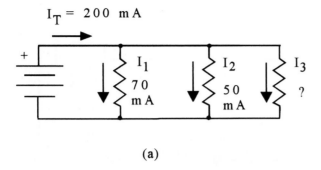

(a)

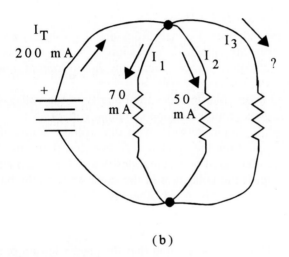

(b)

FIGURE 1.9
Circuit for Example 1.6 showing how KCL is applied.

or, solving for the unknown,

$$I_3 = I_T - I_1 - I_2$$
$$I_3 = 200 - 70 - 50$$
$$I_3 = 80 \, \text{mA}$$

Since it turns out positive we know it has the direction assumed in the schematic.

1.2.3 Kirchhoff's Voltage Law

Kirchhoff's voltage law (KVL) states that the algebraic sum of all voltages around any closed loop of a network is zero at every instant of time. This is also stated in

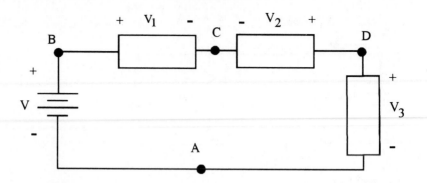

FIGURE 1.10
KVL specifies that the sum of voltages around a loop will be zero.

terms of voltage drops and rises. Voltage drops are the voltages across passive components. Voltage rises are voltages across sources. In this case the law states that the algebraic sum of voltage drops around any closed loop of a network will equal the algebraic sum of the voltage rises.

An algebraic sum is one that includes the polarity of the voltages as well as the magnitudes as one goes around the loop summing voltages.

Figure 1.10 shows a simple network with one voltage source and voltage reference polarities assigned for the elements. Notice that there are four nodes. Starting with the source connected between nodes A and B we sum all remaining voltages around the loop. If the polarity is in the same sense as the battery as we go around the loop, the magnitude is added; otherwise, it is subtracted. Thus we find,

$$E - V_1 + V_2 - V_3 = 0 \tag{1.8}$$

This equation follows KVL because we find that the algebraic sum of all voltages around the loop is zero. In the alternate form we would sum the drops and set them equal to the rises:

$$V_1 - V_2 + V_3 = E \tag{1.9}$$

This is the same as Eq. (1.8).

EXAMPLE 1.7   For the network shown in Figure 1.11, determine the unknown voltage drop, V_2.

Solution
The polarities and magnitudes of the known voltages are given. We arbitrarily assign a polarity V_2 as positive on the right, for example. Then, from KVL, we can write,

$$V_1 - V_2 + V_3 = E$$

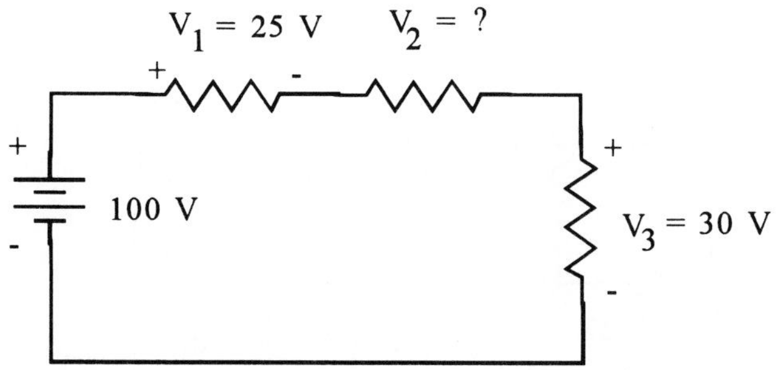

FIGURE 1.11
Circuit for Example 1.7 showing how KVL is applied.

or

$$V_2 = V_1 + V_3 - E$$
$$V = 25 + 30 - 100$$
$$V_2 = -45 \text{ V}$$

Since the assigned polarity resulted in a negative value, the voltage is actually positive on the left.

1.2.4 Thévenin's Theorem

Any two points in an electrical network containing only linear passive elements and ideal sources can be replaced by a single voltage source in series with an impedance.

This means that all electrical characteristics of the original network, between those two points, are replicated by the equivalent source, called the *Thévenin voltage,* and the equivalent impedance, called the *Thévenin impedance,* or, if only resistors are present, the *Thévenin resistance.*

The voltage source in the equivalent circuit is the open-circuit voltage appearing between the selected points of the original network. The term *open circuit* means that no external element is placed between the points.

The Thévenin impedance is the impedance of the original network between the selected points, with all real voltage sources replaced by short circuits and all real current sources replaced by open circuits.

EXAMPLE 1.8   Use Thévenin's theorem to find the equivalent circuit between points a and b in Figure 1.12(a). Use this circuit to find the load voltage across the external load R_L.

Solution
The Thévenin impedance in this case is a resistance. It is found by replacing the 25-V source with a short circuit and finding the resistance between points a and b.

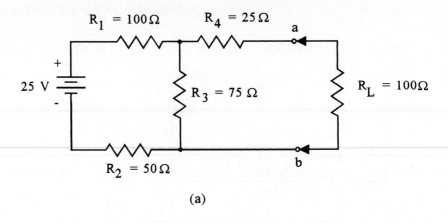

(a)

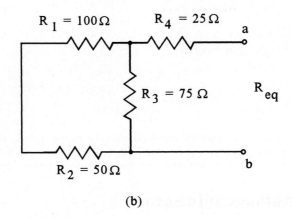

(b)

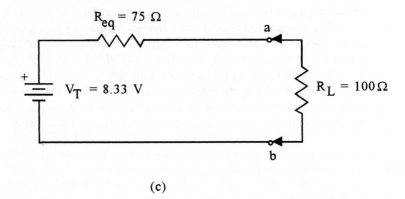

(c)

FIGURE 1.12
Circuit for Example 1.8 and Thévenin's theorem.

Figure 1.12(b) shows the network for finding the resistance.

$$R_{eq} = R_4 + \frac{R_3(R_1 + R_2)}{R_1 + R_2 + R_3}$$

$$= 25\,\Omega + \frac{(150\,\Omega)(75\,\Omega)}{150\,\Omega + 75\,\Omega} = 75\,\Omega$$

The Thévenin voltage is simply the voltage dropped across the 75-Ω resistor in Figure 1.12(a) since the 25-Ω resistor carries no current (open circuit from a to b). Thus the Thévenin voltage is,

$$V_T = \frac{R_3}{R_1 + R_2 + R_3}\,V = \frac{75\,\Omega}{100\,\Omega + 50\,\Omega + 75\,\Omega}\,25 = 8.33\,\text{V}$$

Figure 1.12(c) shows the equivalent circuit with the load attached. The load voltage is simple to find now because

$$I_L = \frac{V_T}{R_{eq} + R_L} = \frac{8.33\,\text{V}}{175\,\Omega} = 47.6\,\text{mA}$$

so

$$V_L = I_L R_L = (47.6\,\text{mA})(100\,\Omega) = 4.76\,\text{V}$$

1.2.5 Norton's Theorem

Any two points in a network consisting of passive components and ideal sources can be replaced by an equivalent current source in parallel with an impedance.

The single current source, called the *Norton current source,* and the parallel impedance, called the *Norton impedance,* demonstrate all the electrical characteristics of the original network.

The current source has the magnitude and direction of current through a short circuit placed between the selected points. The Norton impedance is the impedance between the selected terminals with all real voltage sources replaced by a short circuit and all real current sources replaced by an open circuit.

EXAMPLE 1.9

The model of a physical battery is an ideal voltage source in series with a resistor, as shown in Figure 1.13(a). Find the Norton equivalent circuit to show how it can be modeled as a current source.

Solution
The Norton current will be the current through a short between output terminals a and b.

$$I_N = I_{sc} = V/R = 12\,\text{V}/0.01\,\Omega = 1200\,\text{A}$$

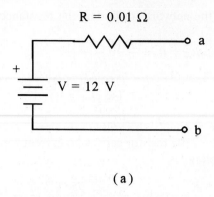

(a)

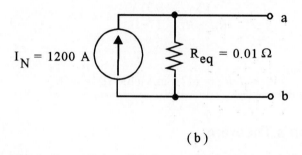

(b)

FIGURE 1.13
Circuit for Example 1.9 and Norton's theorem.

If the ideal voltage source is replaced by a short to find the Norton resistance, the value will simply be the series resistance, $R_N = R = 0.01\ \Omega$. Thus the Norton equivalent circuit is shown in Figure 1.13(b).

EXAMPLE 1.10   Determine the current through the 5-Ω resistor for the circuit in Figure 1.14(a) using Norton's theorem.

Solution
Let us find the Norton equivalent circuit between points a and b of the circuit, without the 5-Ω resistor. The short-circuit current will be the sum of the current due to the 10-V source and the current due to the 20-V source, as shown in Figure 1.14(b):

$$I_N = (10\ \text{V}/10\ \Omega) + (20\ \text{V}/50\ \Omega) = 1.4\ \text{A}$$

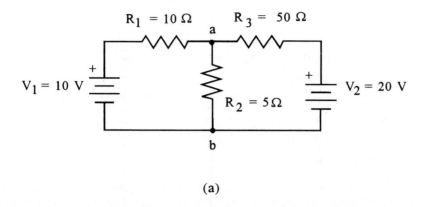

(a)

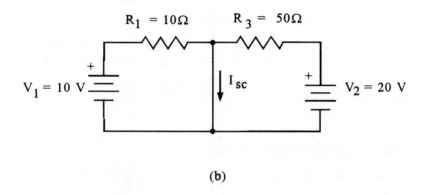

(b)

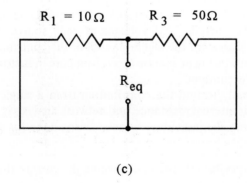

(c)

FIGURE 1.14
Circuit for Example 1.10 and Norton's theorem.

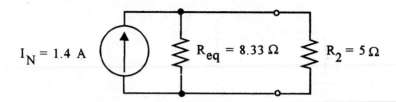

FIGURE 1.15
Solution to Example 1.10.

The Norton resistance between points a and b, with the two voltage sources replaced by short circuits, is simply the parallel combination of the 10- and 50-Ω resistors, as shown in Figure 1.14(c):

$$R_N = (10\ \Omega)(50\ \Omega)/(10\ \Omega + 50\ \Omega) = 8.33\ \Omega$$

The Norton equivalent circuit is shown in Figure 1.15 along with the 5-Ω resistor to which it is to be connected. From the current divider rule we now find

$$I_2 = [R_N/(R_N + R_2)]I_N = 0.875\ \text{A}$$

1.2.6 Superposition Theorem

The superposition principle states that if a system is affected by two or more causes acting jointly, it is permissible, if the relations are linear, to consider that each cause acts *independently* and then to superimpose the two or more related effects.

In networks with two or more sources, the principle of superposition says that we may find the currents and voltages of each source independently. The net currents and voltages are the algebraic sum of the contributions from each source. All elements must be linear so transistors, iron-core inductors, etc., must be modeled in a linear representation.

Operationally to find the contribution from a single source, the others are replaced by their internal resistance, and network analysis is used to find the currents and voltages from the individual source. This is done for each source.

EXAMPLE 1.11
Using the superposition theorem, determine the current through R_2 in the network of Figure 1.16(a).

Solution
First to find the contribution from V_1 we replace V_2 by a short circuit, since it is an ideal source. The resulting network, shown in Figure 1.16(b), is easy to solve. First the total resistance presented to V_1 is

$$R_T = R_1 + R_2\|R_3 = 14.55\ \Omega$$

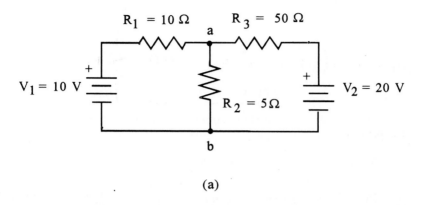

(a)

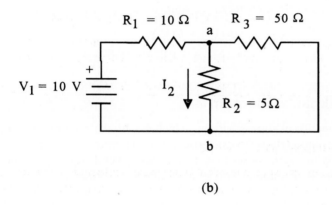

(b)

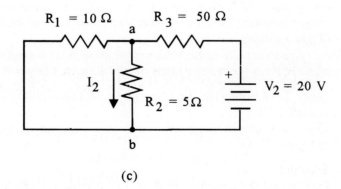

(c)

FIGURE 1.16
Circuit for Example 1.11 and the superposition theorem.

so the total current is

$$I_T = V_1/R_T = 10 \text{ V}/14.55 \ \Omega = 0.6875 \text{ A}$$

Then the current through R_2 is

$$I_2(V_1) = [R_3/(R_2 + R_3)]I_T = 0.625 \text{ A}$$

Now V_1 is replaced by a short circuit and the current from V_2 is found from the network in Figure 1.16(c):

$$R_T = R_3 + R_1 \| R_2 = 53.33 \ \Omega$$
$$I_T = V_2/R_T = 20 \text{ V}/53.33 \ \Omega = 0.375 \text{ A}$$
$$I_2(V_2) = [R_1/(R_1 + R_2)]I_T = 0.25 \text{ A}$$

The final solution is the algebraic combination of the two currents. They both have the same direction through R_2 so we simply add them:

$$I_2 = I_2(V_1) + I_2(V_2)$$
$$= 0.625 \text{ A} + 0.25 \text{ A} = 0.875 \text{ A}$$

1.2.7 Millman's Theorem

The inconvenience of superpositions, which essentially requires that a problem be worked out completely for each source, can sometimes be avoided by Millman's theorem. This theorem shows that if there are several sources in a network, they can sometimes be combined into a single effective source.

Figure 1.17 illustrates how this can be done for a network with four ideal voltage sources. In this case all four sources and resistances are replaced by a single voltage source and resistor.

EXAMPLE 1.12

Use Millman's theorem to find the voltage across and current through R_L in Figure 1.18(a).

Solution
We deal with the network to the left of points a and b, i.e., we do not include the load. Then, following the prescription of Figure 1.17 [note that one Sieman (S) equals $1/\Omega$],

$$G_T = G_1 + G_2 = 0.1 \text{ S} + 0.05 \text{ S} = 0.15 \text{ S}$$
$$R_T = 1/G_T = 6.67 \ \Omega$$
$$I_1 = V_1 G_1 = (10 \text{ V})(0.1 \text{ S}) = 1 \text{ A}$$
$$I_2 = V_2 G_2 = (20 \text{ V})(0.05 \text{ S}) = 1 \text{ A}$$
$$I_T = I_1 + I_2 = 2 \text{ A}$$
$$E_T = I_T/G_T = 2 \text{ A}/0.15 \text{ S} = 13.33 \text{ V}$$

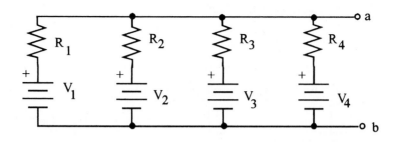

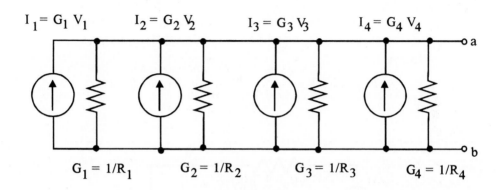

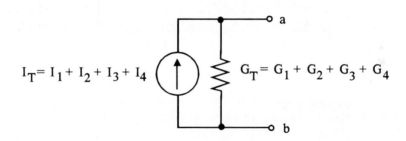

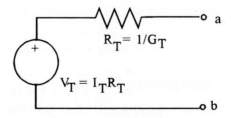

FIGURE 1.17
Millman's theorem shows how sources can be combined.

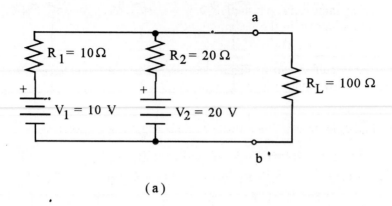

(a)

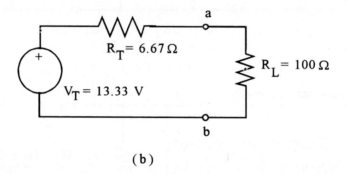

(b)

FIGURE 1.18
Circuit for Example 1.12.

Thus, the equivalent circuit is shown in Figure 1.18(b), from which we can easily find the load voltage and current:

$$V_L = [R_L/(R_T + R_L)]V_T = 12.5 \text{ V}$$
$$I_L = V_L/R_L = 0.125 \text{ A}$$

1.2.8 Reciprocity Theorem

The reciprocity theorem states that if a voltage applied in one branch of a linear, bilateral, passive network produces a certain current in a second branch, the same voltage applied in the second branch will produce the same current in the first branch.

The same principle applies if it is a current applied in the first branch to produce a voltage in the second branch.

EXAMPLE 1.13   For the circuit shown in Figure 1.19(a), determine the current through the 5-Ω resistor. Prove that reciprocity is satisfied.

Solution
We find I_4 in branch 2 due to the voltage source, V_1, in branch 1.

$$R_T = R_1 + R_2 \| (R_3 + R_4) = 20 \, \Omega$$
$$I_T = V_1/R_T = 20 \, \text{V}/20 \, \Omega = 1 \, \text{A}$$
$$I_4 = [V_1 - R_1 I_T]/(R_3 + R_4)$$
$$= [20 \, \text{V} - (10 \, \Omega)(1 \, \text{A})]/(15 \, \Omega + 5 \, \Omega) = 0.5 \, \text{A}$$

Now, to test reciprocity, we place a 20-V source in branch 2 and take out the voltage source in branch 1, as shown in Figure 1.19(b), and solve for I_1:

$$R_T = R_4 + R_3 + R_1 \| R_2 = 26.67 \, \Omega$$
$$I_T = V_4/R_T = 20 \, \text{V}/26.67 \, \Omega = 0.75 \, \text{A}$$

Then

$$I_1 = [R_2/(R_1 + R_2)] = 0.5 \, \text{A}$$

This shows the reciprocity.

1.2.9   Substitution Theorem

The substitution theorem states that if a network branch has a current through it of I_{ab} or a voltage across it of V_{ab}, then it may be replaced by a current source of the same value or voltage source of the same value with no effect on the rest of the network.

EXAMPLE 1.14   For the circuit shown in Figure 1.20, replace branch a–b with an appropriate voltage source and current source.

Solution
It is necessary merely to find the voltage across the terminals of R_4:

$$R_T = R_1 + R_2 \| (R_3 + R_4) = 20 \, \Omega$$
$$I_T = V_1/R_T = 1 \, \text{A}$$

Now, the current through the 5-Ω resistor is

$$I_{ab} = [V_1 - (I_T R_1)]/(R_3 + R_4) = 0.5 \, \text{A}$$

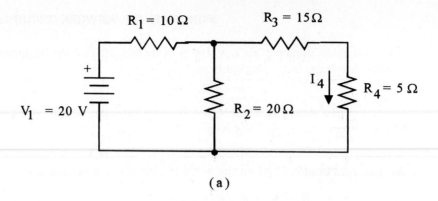

(a)

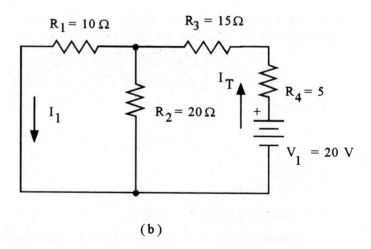

(b)

FIGURE 1.19
Circuit for Example 1.13 illustrating reciprocity.

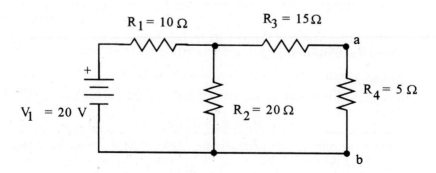

FIGURE 1.20
Circuit for Example 1.14 and the substitution theorem.

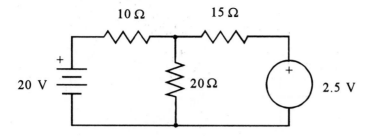

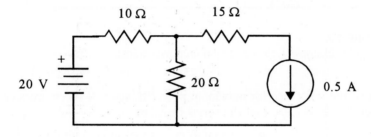

FIGURE 1.21
Solution for Example 1.14.

so the voltage is

$$V_{ab} = I_{ab}R_4 = (0.5 \text{ A})(5 \text{ V}) = 2.5 \text{ V}$$

Figure 1.21 shows the equivalent networks with voltage source substitution and current source substitution. Note that the 5-Ω resistor is now out of the circuit.

1.2.10 Maximum Power Transfer Theorem

When a source of voltage V supplies a load of resistance R_L through a series resistance R_s, maximum power is transferred to the load if its resistance is equal to R_s.

EXAMPLE 1.15   Four solar cells are connected in the series parallel arrangement shown in Figure 1.22. Find the load resistance for maximum power transfer.

Solution
Thévenin's theorem shows that the network may be replaced by a single source in series with a single resistor. The Thévenin resistance is the series resistance, so it

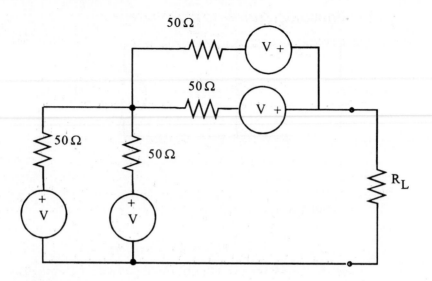

FIGURE 1.22
Circuit for Example 1.15 and maximum power transfer calculations.

will be the resistance for maximum power transfer. If the sources are replaced by short circuits, we see that the series resistance is

$$R_s = 50 \ \Omega \| 50 \ \Omega + 50 \ \Omega \| 50 \ \Omega = 50 \ \Omega$$

Therefore, the load resistance for maximum power transfer is 50 Ω.

1.3 AC CONCEPTS

Many applications of electricity and electronics involve voltages and currents that are constant in time. These networks are called *direct current* (*dc*), although the abbreviation dc is used with voltage as well as current. Time-varying voltage and current sources are also of great importance in electricity and electronics. The nature of the time variation can take on a great many forms, from the chaotic variations induced by noise and the complex patterns of voice signals to the regular 60-Hz variations in an electrical power system.

A *periodic* time variation is one for which a particular signal shape versus time repeats itself continuously. One *cycle* of a periodic signal is represented by the repeated shape.

A special class of periodic time dependence is that in which the variations are described mathematically by trigonometric functions of time. This is called *alternating current* (*ac*), although again, the abbreviation ac is used for both voltage and current. Strictly speaking ac refers only to these trigonometric signals, although the term is often used for any time variation.

1.3.1 Sinusoidal Currents and Voltages

A sinusoidal voltage is described by a mathematical function of time having the form,

$$v(t) = A \cos(\omega t + \phi) \qquad \textbf{(1.10)}$$

where $v(t)$ = a function that changes value with time t
   A = the constant amplitude in volts
   ω = the angular frequency in radians per second
   ϕ = the phase in radians.

A sinusoidal function is completely defined for all time when the amplitude, angular frequency, and phase are known. Equation (1.10) has been written in terms of the trigonometric cosine function. It could just as well be specified in terms of the sine function because the only difference between the sine and cosine is a phase shift of 90° or $\pi/2$ radians. Thus we could write

$$v(t) = A \sin(\omega t + \theta) \qquad \textbf{(1.11)}$$

where $v(t)$ = the same function as Eq. (1.10), i.e., it would look the same
      if plotted versus time
   A = the same amplitude
   ω = the same angular frequency
   θ = the phase, where $\phi = \theta + \pi/2$.

The physical meanings of A, ω, and ϕ can be explained in terms of a graph of Eq. (1.10) as shown in Figure 1.23.

Amplitude   The amplitude is the maximum positive value, A, of the signal, also called the peak value. The minimum value is $-A$ and the peak-to-peak (pp) value is $2A$. The peak value is often designated by V_m for voltage and I_m for current.

Angular Frequency   The angular frequency, ω, has the units of radians per second (rad/s). It specifies the number of radians of angle by which the argument

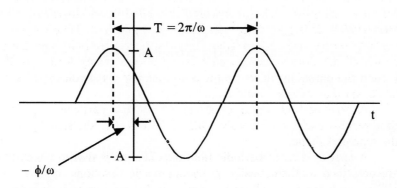

FIGURE 1.23
An ac signal characterized by amplitude, period, and phase.

of the function varies or sweeps out in 1 s. Although the proper unit of angle is the radian, the degree is also used where 1 radian = $(180/\pi)° \approx 57.3°$.

It is more common to express ω in terms of the *frequency, f*, of the sinusoid. The frequency is the number of cycles completed by the function in 1 s. A frequency of one cycle per second is an SI unit named *hertz* (Hz). Angular frequency and the frequency in hertz are related by

$$\omega = 2\pi f \qquad\qquad (1.12)$$

The *period* of the signal is the reciprocal of the frequency, $T = 1/f = \omega/2\pi$. This is shown in Figure 1.23 as the time between maxima, but it is the time between any equivalent two points from one cycle to the next.

Phase Angle   A pure cosine function is at a maximum at $t = 0$ and every T seconds after and before that. A pure sine has its maxima shifted by $\pi/2$ radians from the cosine. The phase shift specifies how the maxima of the cosine (or sine) are shifted from their values with no phase shift. Figure 1.23 shows the shifted maxima of Eq. (1.10) for a phase shift ϕ. Again the proper unit for phase is the radian but the degree of angle is often used.

EXAMPLE 1.16

Determine the angular frequency and period of an ac signal with a frequency of 120 Hz. If the amplitude is 5 and the value at $t = 0$ is 4.6, as described in Figure 1.23, write an equation for the sinusoid in terms of the cosine function.

Solution

The angular frequency is computed from Eq. (1.12) as

$$\omega = 2\pi f = 2\pi(120 \text{ Hz}) = 754 \text{ rad/s}$$

and the period is

$$T = 1/f = 1/120 \text{ Hz} = 8.33 \text{ ms}$$

Since the amplitude is $A = 5$, we know the function can be written in the form of Eq. (1.10) as

$$v(t) = 5 \cos(754t + \phi)$$

To find the phase angle, we simply note that at $t = 0$ we have

$$4.6 = 5 \cos(\phi)$$

or

$$\phi = \cos^{-1}(4.6/5) = 23° \text{ or } 0.40 \text{ rad}$$

Thus,

$$v(t) = 5 \cos(754t + 0.4)$$

1.3.2   Phasor Representation

A special representation of voltage and current is used for ac network analysis. This is based on the Euler relation between the complex exponential function and the sine and cosine functions, where $j = \sqrt{-1}$,

$$e^{j\theta} = \cos(\theta) + j\sin(\theta)$$

From this you can see that Eq. (1.10) can be written in the form

$$v(t) = \text{Re}(Ae^{j\phi}e^{j\omega t})$$

where Re() means the "real part" of the complex number in the brackets and Im() will represent the imaginary part. A *phasor* is defined as the complex quantity

$$\mathbf{F} = F_m e^{j\omega t} \tag{1.13}$$

where $F_m = Ae^{j\phi}$ = amplitude at $t = 0$.

The phasor is often presented in a graph, or *phasor diagram*, with the imaginary part plotted vertically and the real part plotted horizontally as shown in Figure 1.24. We say the phasor rotates its amplitude A counterclockwise about the origin as a function of time. The phase ϕ specifies the starting angle of the rotation at $t = 0$.

Note that in many cases the quantity F_m alone is called the phasor and is used for network analysis. This can be done in most cases because the ac time dependence, $e^{j\omega t}$, is common throughout the network. Strictly speaking, however, the phasor includes the $e^{j\omega t}$ term and resulting time variation.

Phasors are often represented in the notation of amplitude and phase (in degrees) separated by an angle symbol, $\angle$. Thus, the result of Example 1.16,

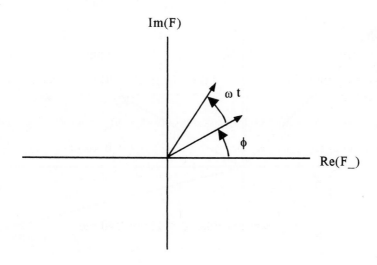

FIGURE 1.24
A phasor plot showing rotation in time.

$f(t) = 5 \cos(754t + 0.4)$ would be written

$$\mathbf{F} = 5\angle 23° \qquad \text{since } 0.4 \text{ rad} = 23°$$

Addition and Subtraction of Phasors Because of the phase shift, phasors must be added and subtracted like vectors, i.e., by the horizontal and vertical components. This can be done graphically or using complex algebra.

EXAMPLE 1.17 Given two phasors, $\mathbf{F1} = 3\angle 30°$ and $\mathbf{F2} = 4\angle 135°$, find the sum and difference using both graphical and complex algebra techniques.

Solution

Figure 1.25 shows both phasors plotted on a phasor diagram. The components of the two phasors are shown as projections onto the horizontal and vertical axes, respectively. The components are added (or subtracted) to obtain the resultant component. The results become the horizontal and vertical components of the answer. Figure 1.25 shows this for the sum and the difference. Using graphical estimation and a protractor for the angle we find

$$\mathbf{F1} + \mathbf{F2} = 4.3\angle 94°$$
$$\mathbf{F1} - \mathbf{F2} = 5.5\angle -14°$$

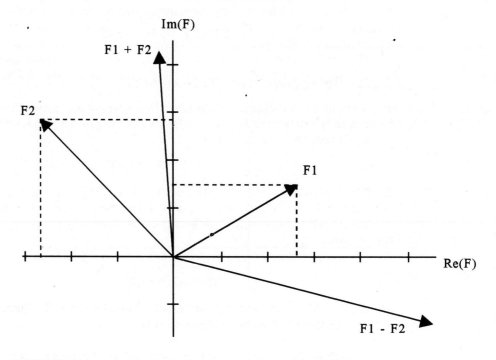

FIGURE 1.25
Example 1.17 shows that phasors add and subtract like vectors.

For the mathematical treatment, we write the two phasors in complex form using the Euler relation:

$$\mathbf{F1} = 3\angle 30° = 3\cos(30) + 3\sin(30)j = 2.60 + 1.50j$$

and

$$\mathbf{F2} = 4\angle 135° = 4\cos(135) + 4\sin(135)j = -2.83 + 2.83j$$

Now the sum and differences can be formed as the sum and differences of the real and imaginary parts:

$$\mathbf{F1} + \mathbf{F2} = (2.60 - 2.83) + (1.50 + 2.83)j$$
$$= -0.23 + 4.33j$$
$$\mathbf{F1} - \mathbf{F2} = (2.60 + 2.83) + (1.50 - 2.83)j$$
$$= 5.43 - 1.33j$$

These can be converted back to the phasor notation by using the expression for a complex number in polar form:

$$x + jy = \mathrm{R}e^{j\theta}$$

where $R = \sqrt{x^2 + y^2}$ and $\theta = \tan^{-1}(y/x)$. This gives

$$\mathbf{F1} + \mathbf{F2} = 4.32\angle 93°$$
$$\mathbf{F1} - \mathbf{F2} = 5.59\angle -13.8°$$

1.3.3 Components in AC Networks

The current-versus-voltage relationship for two-terminal components when the input is ac is defined by the quotient of phasor voltage across the device and the phasor current through its terminals:

$$Z = V/I = R + jX \qquad \textbf{(1.14a)}$$

Since both phasors have $e^{j\omega t}$ dependence, this will cancel in the quotient so that the ratio, called the *impedance,* will be a time-independent complex number with units of ohms, Ω. The real part is called the *resistance, R*, and the imaginary part is called the *reactance, X*.

The impedance is often written in the polar complex form:

$$Z = |Z|e^{j\theta} = |Z|\angle \theta \qquad \textbf{(1.14b)}$$

where $|Z| = \sqrt{R^2 + X^2}$ and $\theta = \tan^{-1}(R/X)$. Note, however, that impedance is *not* a phasor because it does not depend on time.

Combinations Impedances in series add and impedances in parallel add as the inverses. When forming combinations the complex nature of the impedance must be included. For addition and subtraction, it is easiest to express impedance in rectangular form, as an Eq. (1.14a), and combine the real and imaginary parts separately.

For the product and quotient of two impedances, it is easier to express in polar form, as in Eq. (1.14b). When multiplying, the amplitudes are multiplied and the phases added. When dividing, the magnitudes are divided and the phase of the divisor is subtracted from the phase of the dividend.

Resistor The impedance of a resistor is a pure real number

$$Z_R = R = R\angle0^0$$

where R = resistance in Ω.

Inductor The impedance of an ideal inductor is a pure imaginary number, i.e., reactance only,

$$Z_L = jX_L = j\omega L = \omega L\angle90°$$

where L = inductance in henrys and ω = angular frequency in rad/s.

Capacitor The impedance of an ideal capacitor is a pure imaginary number, i.e., reactance only,

$$Z_C = jX_C = -j(1/\omega C) = (1/\omega C)\angle-90°$$

where C = capacitance in farads.

Combinations of resistors, capacitors, and inductors will have an overall impedance determined by the series and parallel combination of component impedances, taking into account the complex numbers.

EXAMPLE 1.18

What is the impedance of a 100-Ω resistor in series with a 10-mH inductor and a 5-μF capacitor if the frequency is 1000 Hz?

Solution
Impedances in series add, so we have

$$Z_T = Z_R + Z_L + Z_C$$

where $Z_R = 100 \ \Omega$, $Z_L = j[(2\pi)(0.01 \text{ H})(1000 \text{ Hz})] = 62.8j \ \Omega$, and $Z_C = -j\{1/[(2\pi)(0.000005 \text{ F})(1000 \text{ Hz})]\} = -31.8j \ \Omega$. So

$$Z_T = 100 + 62.8j - 31.8j = 100 + 31j \ \Omega$$

EXAMPLE 1.19

What is the impedance of the network of Figure 1.26 at a frequency of 120 Hz?

Solution
Let Z_1 and Z_2 be the impedances of the resistor–capacitor and the resistor–inductor, respectively. Then, for two impedances in parallel we can write

$$1/Z_T = 1/Z_1 + 1/Z_2$$

or, for two,

$$Z_T = Z_1Z_2/(Z_1 + Z_2)$$

Now

$$Z_1 = 50 - 66.3j = 83\angle-53°$$
$$Z_2 = 100 + 754j = 760.6\angle82.4°$$
$$Z_1 + Z_2 = 150 + 694.3j = 703.9\angle77.7°$$
$$Z_1Z_2 = (83)(760.6)\angle(82.4 - 53)° = 63129.8\angle29.4°$$

so

$$Z_T = (63129.8/703.9)\angle(29.4 - 77.7)° = 89.7\angle-48.3°$$

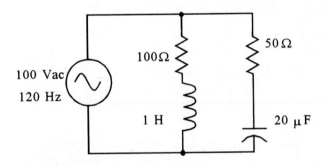

FIGURE 1.26
Circuit for Examples 1.19 and 1.20 showing impedance calculations.

EXAMPLE 1.20 What is the phasor current drawn from the 100-V, 120-Hz source in Figure 1.26?

Solution
The 100-V, 120-Hz source is the reference, so $v = 120\angle0°$. The phasor current is then given by

$$I = V/Z_T = (100\angle0°)/(89.7\angle-48.3°)$$
$$= (100/89.7)\angle(0 + 48.3)° = 1.1\angle48.3° \text{ A}$$

1.3.4 AC Network Analysis

Kirchhoff's voltage and circuit laws (KVL and KCL), defined earlier in this chapter, hold equally well for ac circuits, *provided we use phasors to represent the voltage and current* and impedance for the components. This is also true of the other network laws and principles: Thévenin and Norton equivalent circuits, superposition, and substitution. The following example illustrates the nature of ac network analysis.

EXAMPLE 1.21

Find the Thévenin equivalent circuit for the ac network of Figure 1.27(a).

Solution

The Thévenin voltage is the open-circuit voltage across the terminals. The Thévenin impedance is the impedance across the terminals with the voltage source replaced by its internal resistance. We model the 5-V, 5-kHz source as an ideal source in series with 600 Ω.

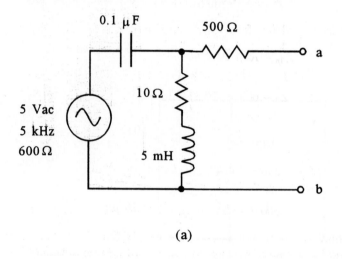

(a)

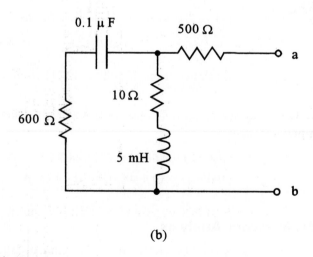

(b)

FIGURE 1.27
Circuit for ac Thévenin circuit calculations in Example 1.21.

Let's solve for the impedance first. Figure 1.27(b) shows the network with the source replaced by a short circuit. The resistor and capacitor in series become

$$Z_1 = R_s + jX_C$$
$$= 600 + jX_C = 600 - j1/[(2\pi)(5000 \text{ Hz})(0.1 \times 10^{-6}\,\text{F})]$$
$$= 600 - 318.3j\,\Omega = 679.2\angle-28°\,\Omega$$

The resistor and inductor become

$$Z_2 = R_1 + jX_L = 10 + j[(2\pi)(5000 \text{ Hz})(0.005 \text{ H})]$$
$$= 10 + 157.1j\,\Omega = 157.4\angle86.4°\,\Omega$$

Thus, the net impedance from a to b, and therefore the Thévenin impedance, is

$$Z_{Th} = R_2 + Z_1Z_2/(Z_1 + Z_2)$$
$$Z_1Z_2 = (679.2\angle-28°)(157.4\angle86.4°) = 106906\angle58.4°$$
$$Z_1 + Z_2 = 600 - 318.3j + 10 + 157.1j = 610 - 161.2j$$
$$= 630.9\angle-14.8°$$

and

$$Z_{Th} = 500 + 106906\angle58.4°/630.9\angle-14.8°$$
$$= 500 + 169.4\angle73.2°$$
$$= 500 + 48.9 + 162j = 548.9 + 162j = 572.3\angle16.4°$$

To find the Thévenin voltage, we simply find the voltage across the 10-Ω resistor and 5-mH inductor, since no current goes through the 500-Ω resistor. From KVL we have

$$V_s = V_{600} + V_C + V_{Th}$$

or

$$V_{Th} = V_s - V_{600} - V_C$$

Now, $V_s = 5\angle0°$, $V_{600} = 600I$, and $V_C = Z_CI$, so we need I. Therefore, $I = V_s/Z_T$ and, using previous results,

$$Z_T = 610 - 161.2j = 630.9\angle-14.8°$$

Hence,

$$I = 5\angle0°/630.9\angle-14.8° = 7.93\angle14.8°\,\text{mA}$$

Thus,

$$V_{600} = 600(7.93\angle14.8°\,\text{mA}) = 4.76\angle14.8°$$
$$V_C = 318.3\angle-90°(7.93\angle14.8°\,\text{mA}) = 2.52\angle-75.2°$$

The answer is

$$V_{\text{Th}} = 5\angle 0° - 4.76\angle 14.8° - 2.52\angle -75.2°$$
$$= 5 - 4.60 - 1.22j - 0.64 + 2.44j$$
$$= -0.24 + 1.22j = 1.24\angle 101°$$

This means that the Thévenin source has a magnitude of 1.24 V at 5 kHz, is shifted in phase from the source by 101°, and is available through a series impedance of $572.3\angle 16.4°$ Ω.

1.3.5 AC Power

The relationship among network voltage, current, and power delivered to or taken from a network depends on the phase relation between the voltage and current. In general, the *instantaneous power* associated with two terminals of a network is related to the voltage across the terminals and current through the terminals by a previous expression, Eq. (1.5):

$$p(t) = i(t)v(t) \tag{1.5}$$

This equation means that at any instant of time t, the power in watts per second delivered to the network or taken from the network through two terminals is given by the product of the voltage and the current at that same instant.

Power is being delivered into the network if the conventional current is directed into the more positive terminal of the network. Power is being taken from the network if the conventional current is directed out of the more positive terminal.

For ac networks certain generalizations of the instantaneous power equation can be made. Since the current and voltage oscillate in value in time, so does the power. In terms of work done, it is more appropriate to consider the *average* power in ac networks to eliminate the oscillation.

Resistance Loads If the load of an ac source is purely resistive, i.e., $Z_L = R\angle 0°$, then the current and voltage are in phase and the power, shown in Figure 1.28(a), oscillates but is always positive (delivered to the load). The average power is given by

$$P_{\text{ave}} = I_m V_m / 2 \tag{1.15a}$$

If we use the root-mean-square (rms) values of voltage and current, defined by

$$I_{\text{rms}} = I_m / \sqrt{2} \quad \text{and} \quad V_{\text{rms}} = V_m / \sqrt{2}$$

then the average power is given by

$$P_{\text{ave}} = I_{\text{rms}} V_{\text{rms}} \tag{1.15b}$$

The interpretation is that a dc source of the same current and voltage as the ac rms values above will deliver the same power to the load, i.e., the ac rms values *act like* equivalent dc values.

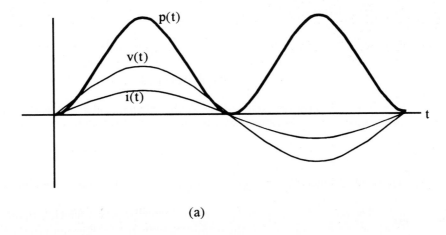

(a)

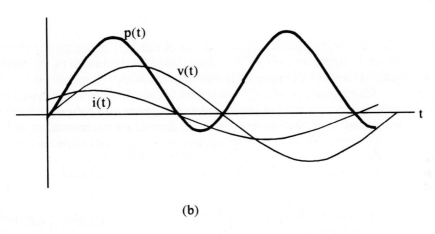

(b)

FIGURE 1.28
AC power can flow into or out of an element depending on the phase.

EXAMPLE 1.22

What is the rms current through an ordinary 60-W light bulb? What is the resistance of the bulb?

Solution
The current can be found from Equation (1.15b), since the rms voltage is known to be 120 V.

$$I_{rms} = P_{ave}/V_{rms} = 60 \text{ W}/120 \text{ V} = 0.5 \text{ A}$$

Now the resistance can be found since the rms values act like the dc equivalent and we can use Ohm's law:

$$R = V_{rms}/I_{rms} = 120 \text{ V}/0.5 \text{ A} = 240 \ \Omega$$

General Impedance Loads If the load is not purely resistance, then a phase shift will occur between ac voltage and current. This will mean that over part of the oscillating power cycle, power is delivered into the load, and over the rest it is taken from the load, as shown in Figure 1.28(b). This is easy to understand for a capacitor that, over a cycle, is charged (power delivered into the load) and discharged (power delivered from the load).

For a general impedance load Z, the *apparent power, S,* is that given by the product of the rms current and voltage:

$$S = I_{rms}V_{rms} \tag{1.16}$$

To avoid confusing this with real power, we use a unit designation of *volt-amperes* (*VA*) to describe this quantity.

The *real power, P,* is that part of the apparent power which actually constitutes work or energy transfer. It is the power taken by the resistive part of the impedance. If the phase difference between the voltage and current is given by ϕ, then the real power is

$$P = S\cos(\phi) \tag{1.17}$$

where the term $\cos(\phi)$, called the *power factor* (PF), is defined as the ratio of real power to apparent power PF $= P/S$. The optimum value of the power factor is unity (1) so that all apparent power is real power.

The *reactive power, Q,* is that part of the apparent power which is associated with the reactive term of the impedance. It represents energy exchanged between source and load, with no work done. It is given by the alternative expression

$$Q = S\sin(\phi) = \pm\sqrt{S^2 - P^2} \tag{1.18}$$

Since this power does no work, it must be distinguished from real power. This is done by means of a special unit designation, the VAR, which comes from volt-ampere-reactive. The sign is determined by the capacitative (+) or inductive (−) nature of the reactance.

EXAMPLE 1.23

A load of $Z = 100 + 50j = 111.8\angle26.6°$ is connected across a 120-Vac, 60-Hz source. Find the apparent power, real power, reactive power, and power factor.

Solution

Since the 120 V is an rms value, the phasor of this source is $120\sqrt{2}\angle0° = 169.7\angle0°$ V, so we can find the phasor current from the expression

$$i = (169.7\angle0°)/(111.8\angle26.6°) = 1.52\angle-26.6° \text{ A}$$

So the rms current is $I_{rms} = 1.52/\sqrt{2} = 1.073$ A. The apparent power is

$$S = (120 \text{ V})(1.073 \text{ A}) = 128.8 \text{ VA}$$

while the real power is given by

$$P = 128.8 \cos(-26.6°) = 115.2 \text{ W}$$

and the reactive power is given by

$$Q = 128.8 \sin(-26.6) = -57.7 \text{ VAR}$$

The power factor is simply PF $= \cos(26.6) = 0.89$.

EXAMPLE 1.24

Suppose a capacitor is placed in parallel with the load of Example 1.23. If the capacitive reactance is 250 how are the results changed?

Solution
The capacitor impedance is $-250j$ or $250\angle-90°$ and this is in parallel with the original load. The new load impedance is given by

$$
\begin{aligned}
Z_{new} &= (111.8\angle26.6°)(250\angle-90°)/(100 + 50j - 250j) \\
&= (27950\angle-63.4°)/(223.6\angle-63.4°) \\
&= 125\angle0° \quad \text{(pure resistive)}
\end{aligned}
$$

Now the current phasor is

$$I = (169.7\angle0°)/(125\angle0°) = 1.36\angle0° \text{ A}$$

and $I_{rms} = 0.96$ A so

$$S = (120 \text{ V})(0.96 \text{ A}) = 115.2 \text{ VA}$$
$$P = 115.2 \cos(0°) = 115.2 \text{ W}$$
$$Q = 115.2 \sin(0°) = 0 \text{ VAR}$$

You can see that the peak and rms current has been reduced yet the power delivered to the resistive part of the load remains the same. The power factor is 1 (or 100%). This is an example of a *power factor correction*, in which a reactance is added in parallel to the load to increase the power factor and reduce the rms and peak current drawn from the source. This reduces, for example, power loss in the feed lines.

1.4 CONTROLLED SOURCES AND MODELING

Most of the network laws used to solve networks are valid only for linear systems, i.e., systems for which the voltage and current appear to only the first power. Yet many special devices in circuits, such as diodes and transistors, are nonlinear. To obtain approximate solutions to networks with these devices, we make linear approximations of their *I-V* characteristics. These approximations involve the use of controlled sources and device modeling.

1.4.1 Controlled Sources

A controlled source is one whose output, either voltage or current, depends on some other voltage or current in the network in which it is used. These are also called *dependent* sources to emphasize this dependency. There are four possible types:

> Voltage-controlled voltage source (VCVS)
>
> Voltage-controlled current source (VCIS)
>
> Current-controlled voltage source (ICVS)
>
> Current-controlled current source (ICIS)

Controlled sources are dependent on both magnitude and polarity or direction of the controlling voltage or current. Figure 1.29 shows the symbols used to represent these special sources.

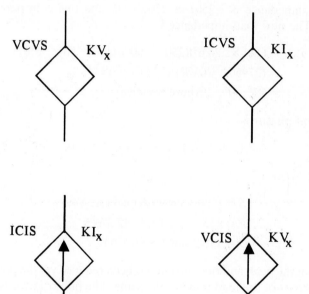

FIGURE 1.29
Circuit symbols for the four types of controlled sources.

EXAMPLE 1.25

For the network shown in Figure 1.30(a), find the voltage across the load resistor, R_L.

Solution

The VCIS indicates that the load current is given by 0.05 times the voltage across R_2. The direction will be into the load if the V_1 polarity is as shown. Let us first find the Thévenin equivalent circuit for V_s, R_1, and R_2. We have

$$R_{Th} = R_1 R_2 / (R_1 + R_2) = 353 \ \Omega$$
$$V_{Th} = R_2 V_s / (R_1 + R_2) = 16.9 \ V$$

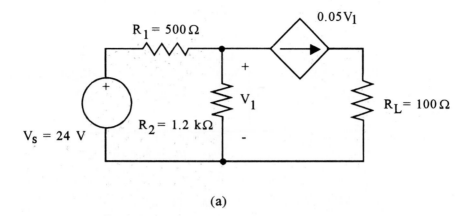

(a)

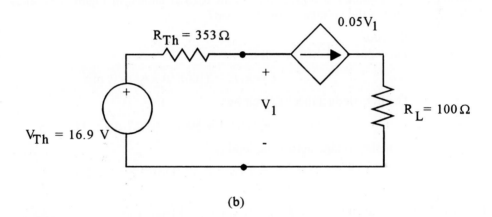

(b)

FIGURE 1.30
Controlled source circuit for Example 1.25.

The new circuit is as shown in Figure 1.30(b). If I_L is the load current, then $V_1 = 16.9 - 353 I_L$ and $I_L = 0.05 V_1$. Combining,

$$V_1 = 16.9 - 353(0.05 V_1) = 16.9 - 17.65 V_1$$
$$= 16.9/18.65 = 0.91 \text{ V}$$

Now the current through the load can be found:

$$I_L = 0.05(0.91) = 45.5 \text{ mA}$$

and the load voltage is

$$V_L = (0.0455 \text{ A})(100 \ \Omega) = 4.55 \text{ V}$$

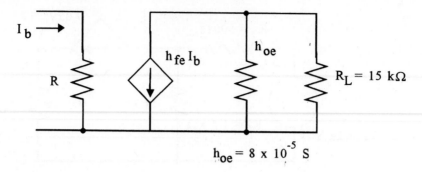

FIGURE 1.31
Controlled source circuit for Example 1.26.

EXAMPLE 1.26

An amplifier is represented as an ICIS as shown in Figure 1.31. If $I_b = 10 \ \mu A$, what is the voltage across the load?

Solution
The value of the ICIS will be

$$I = h_{fe}I_b = (100)(10 \ \mu A) = 1 \ mA$$

The net conductance is given by

$$G = h_{oe} + (1/R_L) = 80 \ \mu S + 66.7 \ \mu S = 146.7 \ \mu S$$

Now the voltage across the load is

$$V_L = I/G = 1 \ mA/146.7 \ \mu S = 6.82 \ V$$

EXAMPLE 1.27

Find currents I_1 and I_2 for the circuit of Figure 1.32.

Solution
This circuit has an ICIS. We apply KCL, where there are only two nodes and the bottom is chosen as the reference. Then we have

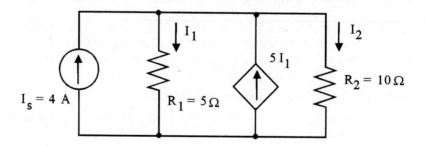

FIGURE 1.32
Controlled source circuit for Example 1.27.

$$I_1 + I_2 = 5I_1 + I_s$$
$$-4I_1 + I_2 = 4$$

but $I_1 = V/5$ and $I_2 = V/10$, so $-4(V/5) + V/10 = 4$, which gives $V = -5.71$ V. Now I_1 and I_2 are

$$I_1 = -5.71 \text{ V}/5 \,\Omega = -1.14 \text{ A}$$
$$I_2 = -5.71 \text{ V}/10 \,\Omega = -0.571 \text{ A}$$

1.4.2 Modeling

Linear models of nonlinear devices are designed to be valid in certain operating regimes with respect to current, voltage, and frequency. Once a model has been selected, traditional network analysis can be used to determine voltages and currents.

Figure 1.33(a) shows the common schematic symbol of a bipolar transistor with appropriate currents and voltages to the base, collector, and emitter. When used in the common-emitter configuration, at low frequencies, forward biased, and with small magnitude signals (compared to the sources) the model of Figure 1.33(b) can be used. Notice the two controlled sources, a VCVS in the base and an ICIS in the collector.

1.5 NETWORK ANALYSIS PROCEDURES

A number of formal procedures have been developed for obtaining a solution of complex networks. Most are now available in computer programs that require only the network parameters as input. The following two procedures are quite efficient for manual solution of networks. These two procedures are valid for setting up the equations for networks that contain only resistors, ideal voltage or current sources of any time dependence, and controlled voltage or current sources.

1.5.1 Nodal Analysis

In this procedure, which is based on KCL, equations are set up for the voltages of all nodes in the network with respect to a reference node. Resistors should be represented as their conductance (inverses). Real and controlled voltage sources should be converted to real and controlled current sources.

The formal procedures are as follows:

1. Identify each node, select one as the reference, and assign to the rest unknown voltage symbols: $V_1, V_2, \ldots, V_i, \ldots, V_n$.
2. Express each resistor in terms of its conductance. Thus a resistor, $R_i = 100 \,\Omega$, becomes a conductance, $G_1 = 1/R_i = 0.01$ S.
3. There are n equations for the n unknown voltages. Each equation is com-

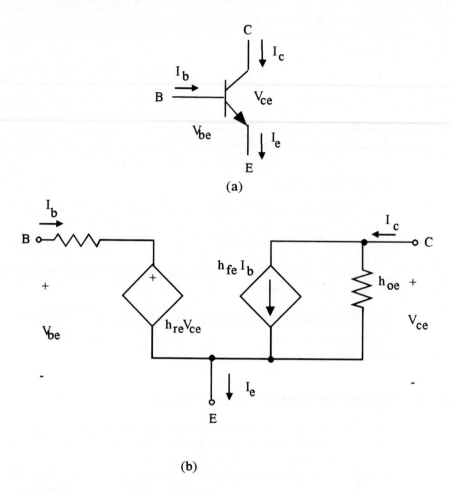

FIGURE 1.33
Controlled source model of a bipolar transistor.

posed of the sum of coefficients times unknown voltages equal to a sum of sources. The coefficients and sum of sources are found from the following rules:

    a. Coefficient g_{ii} of unknown voltage V_i in equation i will be the sum of all conductances connected directly to that node.

    b. Coefficients g_{ij} and g_{ji} of the other $n - 1$ unknown voltages, V_j where $(j \neq i)$, in equation i will be the negative of conductances directly connected between node j and node i.

    c. The sum of sources I_i consists of all current sources directly connected to node i. The source is added if it is directed into the node and subtracted if it is directed out of the node.

The equations can be solved by any of the standard methods of solving simultaneous equations: substitution, Cramer's rule, determinants, matrix algebra, or computer programs.

EXAMPLE 1.28

Find the solution for the output voltage in Figure 1.34(a) using nodal analysis.

Solution

Notice that the network has an ICIS whose value is 10 times the current through the 10-Ω resistor. In Figure 1.34(b) the network has been redrawn with the following changes:

1. The 5-V source in series with a 10-Ω resistor has been transformed into a 0.5-A source shunted by a 0.1-S (10-Ω) conductance.
2. All resistors have been written in terms of conductance.
3. The reference node was selected as the bottom common line and two other nodes were identified and labeled V_1 and V_2, with V_2 as the desired output voltage.

Following the preceding rules, we form two simultaneous equations:

$$g_{11}V_1 + g_{12}V_2 = I_1$$
$$g_{21}V_1 + g_{22}V_2 = I_2$$

The coefficients are

$$g_{11} = 0.1 + 0.2 + 0.05 = 0.35 \text{ S}$$
$$g_{12} = -0.05 \text{ S}$$
$$g_{21} = -0.05 \text{ S}$$
$$g_{22} = 0.5 + 0.05 = 0.55 \text{ S}$$

The sources sum is

$$I_1 = 0.5 + 10I_x$$
$$I_2 = 2 - 10I_x$$

The current i_x can be expressed in terms of the unknown voltages by noting from Figure 1.34(a) and Ohm's law that

$$I_x = (5 - V_1)/10 = 0.5 - 0.1V_1$$

Combining all this gives

$$0.35V_1 - 0.05V_2 = 0.5 + 5 - V_1$$
$$-0.05V_1 + 0.55V_2 = 2 - 5 + V_1$$

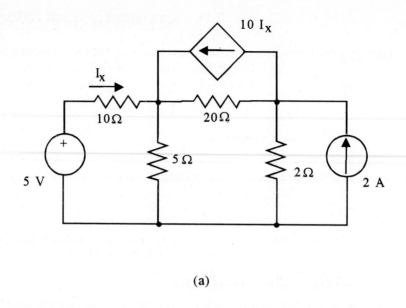

(a)

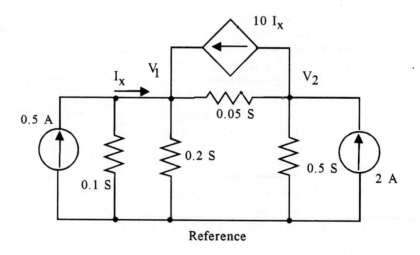

Reference

(b)

FIGURE 1.34
Nodal analysis circuit for Example 1.28.

Rearranging,

$$1.35V_1 - 0.05V_2 = 5.5$$
$$-1.05V_1 + 0.55V_2 = -3$$

We solve for V_2 using Cramer's rule:

$$V_2 = \frac{\begin{vmatrix} 1.35 & 5.5 \\ -1.05 & -3 \end{vmatrix}}{\begin{vmatrix} 1.35 & -0.05 \\ -1.05 & 0.55 \end{vmatrix}} = 2.5 \text{ V}$$

1.5.2 Loop or Mesh Analysis

This approach, which is based on KVL, will set up simultaneous equations for the currents of all elementary loops in the network. The network must consist of resistors and real and controlled voltage sources. The real voltage sources can have any time dependence. Any current sources must be transformed into voltage sources.

The equations are developed according to the following rules of construction:

1. Identify all of the elementary loops of the network and assign to each a clockwise unknown current label: $I_1, I_2, \ldots, I_i, \ldots, I_n$. An elementary loop is one that encloses no other network element.
2. There will be n equations, one for each loop and composed of the sum of coefficients times the unknown currents equal to a sum of sources. The equations are constructed as follows:
   a. The coefficient, r_{ii}, of current I_i in the equation for the i'th loop is found as the sum of all resistors in that loop.
   b. The coefficients, r_{ij} and r_{ji}, of current I_j, where $i \neq j$, in the equation for the i'th loop are given as the negative of resistances through which both current I_i and I_j pass, i.e., which *share* the two loop currents.
   c. The sum of sources is constructed by adding all sources in the i'th loop for which the loop current is directed out of the positive terminal and subtracting sources for which the loop current is directed into the positive terminal of the source.

EXAMPLE 1.29 Find an equation for the output voltage for the network of Figure 1.35(a).

Solution

Note that the network contains a time-dependent voltage source, $2 \sin(5t)$, and a ICIS whose current is 10 times the current through the input 10-Ω resistor. Figure 1.35(b) shows the network with the current source converted to a voltage source and with three elementary loops identified and labeled. The output will be $V_{\text{out}} = 200I_3$.

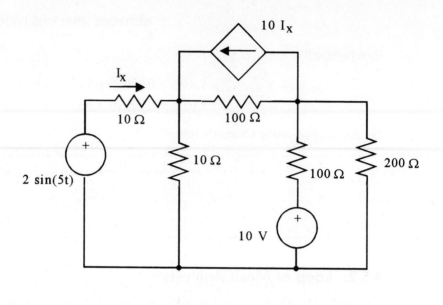

(a)

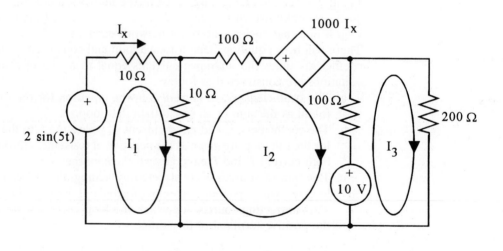

(b)

FIGURE 1.35
Mesh analysis circuit for Example 1.29.

The equations will have the form

$$r_{11}I_1 + r_{12}I_2 + r_{13}I_3 = V_1$$
$$r_{21}I_1 + r_{22}I_2 + r_{23}I_3 = V_2$$
$$r_{31}I_1 + r_{32}I_2 + r_{33}I_3 = V_3$$

The coefficients are

$$r_{11} = 10 + 10 = 20, \quad r_{12} = -10, \quad r_{13} = 0 \text{ (no shared resistor)}$$
$$r_{21} = -10, \quad r_{22} = 10 + 100 + 100 = 210, \quad r_{23} = -100$$
$$r_{31} = 0, \quad r_{32} = -100, \quad r_{33} = 100 + 200 = 300$$

The voltage sums are

$$V_1 = 2 \sin(5t), \quad V_2 = -1000I_x - 10, \quad V_3 = 10$$

We can express the dependency in terms of the unknown currents since clearly, from Figure 1.35(b), $I_x = I_1$.

The set of equations becomes

$$20I_1 - 10I_2 \qquad = 2 \sin(5t)$$
$$-10I_1 + 210I_2 - 100I_3 = -1000I_1 - 10$$
$$- 100I_2 + 300I_3 = 10$$

Simplifying gives the final set of equations,

$$20I_1 - 10I_2 \qquad = 2 \sin(5t)$$
$$990I_1 + 210I_2 - 100I_3 = -10$$
$$- 100I_2 + 300I_3 = 10$$

We need I_3 multiplied by 200, so from Cramer's rule the solution is

$$V_{out} = 200I_3 = 200 \frac{\begin{vmatrix} 20 & -10 & 2\sin(5t) \\ 990 & 210 & -10 \\ 0 & -100 & 10 \end{vmatrix}}{\begin{vmatrix} 20 & -10 & 0 \\ 990 & 210 & -100 \\ 0 & -100 & 300 \end{vmatrix}}$$

Evaluation of these determinants gives

$$V_{out} = 200[0.03 - 0.047 \sin(5t)] = 6 - 9.4 \sin(5t)$$

So the output is inverted and the gain is 4.7.

1.5.3 Capacitors and Inductors

When capacitors and inductors are introduced into the networks, the equations that result are generally integrodifferential. This means that the equations contain the integrals and derivatives of the unknowns. Of course, in the case of steady-state sinusoidal sources, the methods of ac analysis can be used. In all other cases the integrodifferential equations must be solved.

In simple networks the simultaneous set of equations can be solved by conversion to differential equations and using classical methods of solving such equations. In more complicated networks, the methods of Laplace transforms are used to obtain a solution.

2

SOURCES OF ELECTRICAL ENERGY*

The purpose of this chapter is to review the various types of sources used to provide electrical energy to circuits and devices. The focus of the chapter is on small-scale sources such as those required for stationary or portable electrical and electronic equipment. This includes chemical batteries, fuel cells, and exotic sources such as photovoltaic cells and thermocouples. This chapter does not include rotating machinery and other large-scale electrical energy sources such as those employed by electric utility companies.

2.1 GENERAL CHARACTERISTICS OF SOURCES

The Thévenin equivalent circuit of a general source consists of a voltage source, V_s, in series with an internal impedance, Z_s, as shown in Figure 2.1(a). If it is an ac source such as an ac generator then the voltage will vary periodically in time. For the steady-state or dc source, the voltage will be constant in time and the internal impedance is often represented by a resistance, as shown in Figure 2.1(b). When connected to a load the voltage appearing across the actual terminals of the source, V_t, may differ from the open-circuit source voltage because of voltage dropped across the internal resistance.

The objective of the source is to deliver energy to an electrical circuit or device so some function can be performed. For example, an electric motor consumes energy and converts it to rotational motion, and a TV consumes energy in order to convert electrical signals to visual and audio information.

Energy Energy is measured with the unit of *joules* (*J*). One joule is defined as the energy possessed by a 1-kg mass moving at a speed of 1 m/s, i.e., 1 J = 1

---

* This chapter was written by Dr. Curtis D. Johnson, College of Technology, University of Houston.

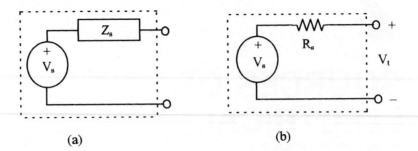

(a)   (b)

FIGURE 2.1
Source models.

kg-m/s. For example a 2-ounce tennis ball served at 80 mph has an energy of about 36 J. The battery is an electrical source that possesses a fixed amount of energy which can be delivered over time to a consuming electrical or electronic system. A generator, on the other hand, converts one form of energy into another (mechanical into electrical) and so does not possess any energy itself. In either case, in electrical applications, the net energy is not usually of primary interest because the consuming equipment operates continuously. It is the flow of energy over time, i.e., the *power,* that is of primary interest.

Power   The flow of energy in time is measured in joules per second. This is described by the unit *watt* (W), which is defined as a flow of one joule per second (1 J/s). In general the power, $p(t)$, and energy, $e(t)$, are related by the following derivative equation:

$$p(t) = \frac{de(t)}{dt} \tag{2.1}$$

When power is constant this relation reduces to a simple ratio,

$$P = \frac{E}{t} \tag{2.2}$$

where by convention capital letters have been used for the power, P, and energy, E, to indicate that their values are not varying in time. Equation (2.2) says that if a total energy E is delivered over a time t then the rate of that energy delivery, i.e., the power, is given by the energy divided by the time.

Electrical Power   In electrical circuits the power either *consumed by* a load or *delivered by* a source is found from a product of the voltage across the element times the current flowing through the element:

$$p(t) = i(t)v(t) \tag{2.3}$$

If the current and voltage are constant, i.e., when there is no time variation, this equation is written as

$$P = IV \qquad \text{(2.4)}$$

where capital letters denote time-invariant variables.

Energy is delivered from the source to the load when the (conventional) current flows from the positive terminal of the source to the load. Figure 2.2 shows a dc source connected to a load. Notice that the upper connecting wire is positive in voltage with respect to the lower. Since the current flows out of the positive terminal of the source, it is delivering energy (power) to the load. Since the current is entering into the positive terminal of the load, it is consuming power.

FIGURE 2.2

Relation between a source and a load.

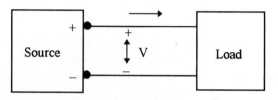

EXAMPLE 2.1

A 6.0-V battery has an internal resistance of 0.2 Ω. What is the terminal voltage when connected to resistive loads of 1000, 10, 1, 0.2, and 0.1 Ω? What is the power delivered to each load?

Solution

In Figure 2.3 the battery has been modeled as a 6.0-V source in series with a 0.2-Ω resistor. The battery is then connected to a load resistance, R_L. The total current is found from Ohm's law to be

$$I = V_s/(R_s + R_L) = 6.0/(0.2 + R_L)$$

The voltage across the terminals, V_T, is found by subtracting the voltage dropped across the internal resistance:

$$V_T = V_s - IR_s = 6.0 - 0.2\frac{6.0}{(0.2 + R_L)}$$

FIGURE 2.3

Circuit model for Example 2.1.

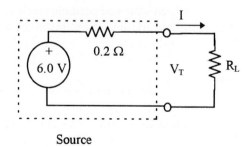

Source

The power delivered to each load is simply found by the familiar expression, $P = I^2R_L$. Putting this all together, the terminal voltage and power for each load is found by simply using the three values of load resistance in the preceding equations. The result is:

R_L (Ω)	V_T (V)	P (W)
1000	5.999	0.036
10	5.882	3.46
1	5.000	25
0.2	3.000	45
0.1	2.000	40

Notice in Example 2.1 how the source terminal voltage and power delivery from the source depend on the relation between the source resistance and load resistance. This example also illustrates the fact that maximum power delivery occurs when the load resistance equals the source internal resistance.

AC Power Relations   In general, if the source voltage and current delivered from an ac source are in phase then Eq. (2.3) can be used to determine the power. In many cases, however, a phase shift occurs between the voltage and current. Power and, therefore, energy is only delivered to the load for that part of the period when the current flows from the source terminal, which has a more positive voltage. Figure 2.4 contrasts the two cases. In Figure 2.4 the voltage, $V(t)$, and current $I1(t)$ from a source are in phase, and Eq. (2.3) can be used to find the power delivered to the load. In the case of $V(t)$ and current $I2(t)$, however, there is a phase shift between current and voltage. Therefore, power is only delivered to the load over that part of the cycle when the current is directed out of the positive terminal of the source, i.e., when both voltage and current in Figure 2.4 are positive. Over the rest of the cycle, energy actually flows back from the load into the source!

Issues such as this in the area of ac power generation and delivery are beyond the scope of this chapter and will not be considered. Please see a text on ac power systems for information on this topic.

In the following sections of this chapter various types of conventional sources of small-scale and portable electrical power are described. In each case the characteristics of the source are given along with typical applications.

FIGURE 2.4
Phase shift between current and voltage affects energy delivery.

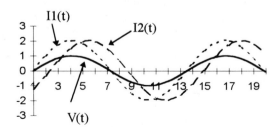

2.2   BATTERIES

A battery is a source of electrical energy that derives its energy from reactions between chemicals in the device. It therefore has a fixed amount of energy, since, when the reacting chemicals are used up, no more electrical energy can be produced. In this section the operational characteristics of batteries are defined and then the properties of specific types of batteries are presented.

Battery and Cell   A single system for the production of electrical energy from chemical processes is called a *cell*. When these cells are packaged—alone or arranged in a series or parallel system as an integral unit—the system is called a *battery*. Thus, the common C size flashlight battery consists of a single 1.5-V cell, and the common 9-V battery consists of a series of six 1.5-V cells.

2.2.1   Basic Properties of Batteries

As later sections will show, a great number of chemical mixes are employed to produce batteries. Nevertheless, certain common characteristics can be defined.

Generic Battery Operation   Figure 2.5 will help demonstrate the basic operating principles of battery cells. The exact nature of the various chemicals is not critical to understanding basically what happens. The various chemical materials are denoted by capital letters. Two metallic electrodes, A (the anode) and C (the cathode), are inserted into a chemical electrolyte, B. The nature of an electrolyte is that it disassociates into positive and negative ions,

$$B \leftrightarrow D^+ + E^-$$

where the double arrow indicates that both disassociation and recombination are occurring, i.e., the reaction is normally occurring in both directions. In equilibrium the reaction is occurring in both directions at the same rate so that the concentrations of each material are constant.

At the anode, a chemical reaction occurs between material A and the electrolyte negative ion, which forms a different compound, F, and causes electrons to be released into the electrode. The anode thus becomes a source of electrons,

FIGURE 2.5
Generic model of a chemical battery.

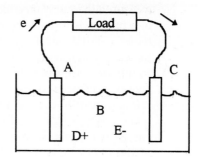

$$A + E^- \leftrightarrow F + e^- \text{ (in anode)}$$

At the cathode, a chemical reaction occurs between material C, the electrolyte positive ion, and an electron from the metallic cathode. This forms a new compound, G, and leaves the cathode depleted of electrons. The cathode thus becomes a sink for electrons,

$$C + D^+ + e^- \text{ (from cathode)} \leftrightarrow G \text{ (in cathode)}$$

In some cell types these reactions are reversible (secondary cells) and in others they occur only to the right (primary cells). In either case they at first operate primarily to the right until the buildup of electrons at the anode and deficit of electrons at the cathode produce a difference of potential sufficient to make reactions to the right and left equal, i.e., equilibrium is reached. This then produces the *open-circuit voltage* of the cell.

If the anode and cathode are connected through a load via electrical wires, the excess electrons of the anode will travel through the wires and load, forming a current and combining with positive ions in the cathode. This process upsets the chemical equilibrium so that the reactions just discussed again continue primarily to the right. In this way a continuous supply of electrons is provided from the chemical reactions. Eventually the supply of active chemical materials, A, C, D^+ and E^-, is exhausted by the preceding reactions and the cell fails to operate. The battery is then *discharged* or *dead.*

Anode and Cathode Definition   In a battery care must be taken to understand the relationship between the terms *anode* and *cathode* and the polarity of the voltage produced by the cells. From the preceding definitions you can see that the "anode" of the cell has an excess of electrons while the "cathode" is depleted of electrons. Therefore, the "cathode" terminal has a positive voltage with respect to the "anode." Conventional current flows from this positive terminal to the negative while actual electron flow is from the negative or anode terminal to the cathode.

Energy Content—Capacity   In general, battery energy content would be measured in joules since that is the correct unit of energy. The total energy content of a battery is a function of the specific type of materials used and the amount of those materials available. To compare batteries, it is more appropriate to specify the energy density of a battery as the joules per unit volume of the chemical mix. In this view, for example, a common carbon-zinc battery would be said to have an energy density of about 180 mWh/cm^3, where m stands for milli- (10^{-3}) and *Wh* refers to watt-hours and thus represents an energy of one watt delivered for one hour, 1 Wh = (1 J/s)(1 h)(3600 s/h) = 3600 J. Thus for this carbon-zinc cell the energy density is 648 J/cm^3. In principle, this means that every cubic centimeter of the battery could deliver 648 J over whatever time period was desired. This representation of battery energy content is used in conjunction with another expression more appropriate for comparison of batteries. The following paragraphs explain the alternative description.

The energy content of a battery is also defined as its *capacity*. This is not to be confused with the traditional capacity of the electrical component called a capacitor. Battery capacity relates to the battery voltage and current delivery over a time period. If the battery terminal voltage is given by V_T then the power delivered by the battery when a current I is drawn is given from Eq. (2.4) as $P = IV_T$. The total energy delivered over a period of time t is then found from Eq. (2.2) to be

$$E = Pt$$

or

$$E = tIV_T \qquad\qquad (2.5)$$

It has become common practice to describe the capacity by giving the product of the time it would take to discharge the battery at a given current assuming a terminal voltage V_T. The unit then is that of ampere-seconds (As) or, as more usually given, ampere-hours (Ah). The actual energy can then be found by multiplication of this quantity by the battery voltage.

EXAMPLE 2.2
It is found that a 1.5-V battery can deliver a current of 600 mA for a time of 2.5 h before becoming discharged. What is the capacity of the battery? How much actual energy is this in joules (assuming the voltage remains constant over the entire time)? How long could the battery deliver a current of 20 mA?

Solution
The capacity would be described then by simply giving the product of current and time, or

$$\text{Capacity} = (600 \text{ mA})(2.5 \text{ h}) = 1.5 \text{ Ah}$$

where "Ah" is the unit designation for ampere-hours. The actual amount of energy in joules is then given by $E = (1.5 \text{ V})(1.5 \text{ Ah})(3600 \text{ s/h}) = 8100$ J. Note that the hours had to be converted to seconds. From the 1.5-Ah rating we see that at a current of 20 mA the battery would last for $t = (1.5 \text{ Ah}/20 \text{ mA})$ or 75 h.

Wet Cell   In the early development of batteries, the electrolyte was always a liquid into which the metallic anode and cathodes were immersed. Many batteries in common use today still use liquid electrolytes, such as the common automotive lead-acid battery. For small size and portable activity, however, such cells are difficult to use because they contain a liquid and must usually be operated in an upright orientation or sealed in some fashion.

Dry Cell   In 1868 a Frenchman named Georges Leclanche devised a way to construct a battery in which the electrolyte impregnated another material to make a paste instead of a liquid. Cell electrical action was maintained however. In other variations, the electrolyte is impregnated into paper or other absorbent material. These are referred to as *dry cells,* although they are not truly dry. These cells can be operated in any physical orientation and form the backbone of modern portable battery applications.

Primary Batteries   When the cells of a battery have depleted the reactant materials, as noted in the generic description given earlier, the battery is said to be discharged. *Primary batteries* are those for which the cell reactants cannot be replenished, i.e., the cell reactions go only to the right and thus the cells cannot be recharged with chemical energy and used again.

Secondary Batteries   Some types of batteries employ cells in which the reactants can be restored and therefore can be employed again. Such units are called *secondary batteries* and are said to be *rechargeable.* The recharging action is usually accomplished by connecting the battery to another source of electrical energy and essentially reversing the chemical reactions by driving a current and therefore energy back into the cells. This restores the original reactants so that the battery can be used again.

2.2.2   Battery Specifications

There are a number of practical issues associated with the operation and use of batteries. The following paragraphs provide descriptions of typical battery specifications as an aid to determination of how they should be used in specific applications.

Voltage   The terminal voltage produced by a battery is primarily determined by the type of chemicals used for the electrodes and the electrolyte. Ideally this voltage would be constant until the battery becomes discharged. In reality, however, the voltage decreases as the battery is discharged. The rate of decrease is determined by how much current is being drawn from the battery, by the operating temperature, and by whether the battery is in use continuously or intermittently. Battery specifications generally provide information such as curves of how the terminal voltage decreases in time under various conditions.

There is some question about when a battery is discharged since the terminal voltage drops off with time. The *endpoint voltage* is that voltage at which, under some operating current, the battery is considered to be discharged. The length of time the battery will last is then determined by how long it takes for the terminal voltage to drop to the endpoint value. Thus, if some piece of equipment needs at least 1.1 V at some specified current to operate and a battery with a starting value of 1.5 V is being used, the endpoint voltage would be 1.1 V.

Internal Resistance   The model presented in Figure 2.1(b) shows that the battery has an effective resistance in series with a voltage source. This source or *internal* resistance means that there is a maximum current which the battery can deliver. Essentially if the battery terminals are shorted a maximum current of $I_{max} = V_s/R_s$ would result. This internal resistance also means that the actual terminal voltage of the battery can be loaded down by voltage dropped across the internal resistance, as Example 2.1 showed.

The internal resistance is quite distinct from the voltage and capacity of the battery. It is quite possible to have a battery with some open-circuit terminal voltage and a large capacity (energy) but from which only very small current can be drawn because of inherently high internal resistance. If an attempt is made to draw a large current, the terminal voltage drops to a small value because of voltage dropped across the internal resistance.

EXAMPLE 2.3

A battery measures 8.98 V on a very high input resistance digital voltmeter (DVM). When it is connected to a 9-V radio, the radio does not work. The radio normally draws 20 mA but now is found to draw only 0.5 mA and the battery voltage drops to 0.225 V when it is connected to the radio. What is wrong?

Solution

Well, if the radio normally draws 20 mA, then it represents an effective load resistance of $R_L = 9$ V/0.02 A $= 450$ Ω. The battery effective internal resistance, R_s, has increased to the point where even at 0.5 mA most of the 9.0 V is dropped across the internal resistance. In fact, the internal resistance is

$$R_s = (8.98 \text{ V} - 0.225 \text{ V})/0.0005 \text{ A} = 17.51 \text{ k}\Omega$$

Since the DVM has a very high input resistance and draws nearly zero current, there was minimal voltage drop across the internal resistance when the measurement was made disconnected from the radio.

Capacity   The capacity as described in the previous section was appropriate for an ideal battery. The practical question is not how much capacity (stored energy) a battery has, but how much of this energy can be extracted. The *working capacity* depends on a number of other factors. In addition, the actual energy content is not as easily found as in Example 2.2 because the terminal voltage is not constant during discharge. Calculations such as that in the example provide a rough estimate of capacity. In many cases the average terminal voltage over the discharge period is used instead of the starting voltage.

An important factor in determination of working capacity is whether the battery is used in a condition of continuous or intermittent discharge. For continuous use the working capacity can be significantly less. To provide a design criterion for this, the effective capacity in Ah of a battery is often specified for several different *rates* of discharge and for continuous or intermittent duty. Intermittent duty is often defined as four hours per day of discharge at some specified rate.

The rate of discharge may be used to specify battery capacity in one of two ways. In some cases the rate is given as the current which is to be drawn. Thus, a specification may say that a battery will have a capacity of 1.5 Ah for a discharge rate of 100 mA. Then it would be expected to last for a time of (1.5 Ah/100 mA) = 15 h. If this same battery is discharged at 200 mA it may be *expected* to last (1.5 Ah/200 mA) = 7.5 h. However, that same battery may have its capacity reduced to only 1.0 Ah if discharged at 200 mA and thus last only 5 h. So, the discharge current affects the capacity.

Another common expression of capacity is via a common time required to fully discharge the battery. The term discharge *rate* refers to the length of time to fully discharge the battery. A common specification for large secondary batteries is the "20-h rate." In this view, for example, a certain battery may be said to have a capacity of 30 Ah at a 20-h rate. This would mean that if discharged at a current of $I = (30 \text{ Ah}/20 \text{ h}) = 1.5$ A, the battery would be discharged in 20 hours. However it must be noted that the capacity of this battery for a 5-h rate would most likely be less than 30 Ah. In some cases specifications will give the Ah capacity of a battery as a function of the discharge rate in hours, where again this is the number of hours to discharge the battery.

Storage Life   When disconnected from any load a battery reaches a condition of equilibrium, as noted earlier, such that the reactions stop or, if reversible, are occurring equally in both directions. In principle then there should be no net loss of capacity (stored energy). In reality, however, other factors lead to a gradual depletion of the active materials in the reactions and thus a discharge of the battery. The shelf or storage life is a measure of the loss of capacity over time. This is generally given as the percentage of remaining capacity as a function of storage time or as the percent loss of capacity per unit time.

Temperature Dependence   Since the battery cell action depends on chemical reactions, which in turn depend on temperature, it is expected that battery characteristics will depend on temperature. Generally the battery working capacity will be reduced when operated at reduced temperatures. Thus, a battery with a capacity of 500 mAh at 70°F (21°C) may have a capacity of only 50% or 250 mAh at 32°F (0°C). Generally the capacity is recovered when the battery is returned to the higher temperature.

Battery Size Standards   The American National Standards Institute (ANSI), the International Electrochemical Commission (IEC) and others have been working to define certain standards for battery sizes. Manufacturers are urged to cooperate in manufacturing batteries according to these standards so that interchangeability in equipment is assured. Tables 2.1, 2.2, and 2.3 present some of these standards for the most common batteries.

In general, there are round, button, wafer, or coin and rectangular batteries. The round types are those commonly found in flashlights, electronic games, radios and TVs, toys, and many other moderate- to small-size electrical and electronic equipment. Button types are used in watches, calculators, hearing aids, and other

TABLE 2.1
Round cell sizes

Designations			Nominal Size (mm)	
IEC		ANSI	Diameter	Length
Mono	R20	D	34.2	61.5
Baby	R14	C	26.2	50.0
Mignon	R6	AA	14.5	50.5
Lady	R1	N	12	30.0
Micro	R03	AAA	10.5	44.5

TABLE 2.2
Button cell sizes

Designation	Nominal Size (mm)	
IEC	Diameter	Height
R9	16	6.2
R44	11.6	5.4
R43	11.6	4.2

TABLE 2.3
9-Volt rectangular cell size

Designation	Nominal Size (mm)		
IEC	Length	Height	Width
6F22	25	47.5	16

miniature electronic devices. Diameters range from 5.8 to 11.6 mm and many different heights.

The wafer or coin style batteries are very thin and round and are used in calculators and watches. Diameters range from 20 to 24.5 mm with thicknesses from 1.6 to 3 mm. This style is often used for lithium-based cells. Square batteries appear in many sizes and are used in everything from camera film packs to radios. The large-scale batteries such as those used in golf carts and automobiles are not described in these size tables because of the tremendous variety of sizes.

Figure 2.6 shows the dimension definitions for Tables 2.1, 2.2, and 2.3. The dimensions given are nominal values and do not take into account the labels and protective covers that manufacturers may employ.

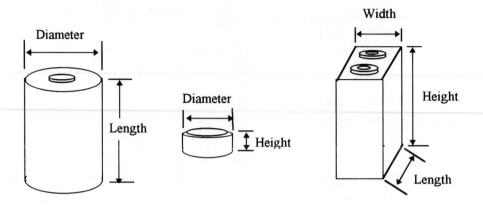

FIGURE 2.6
Dimensions of common battery types.

2.2.3 Primary Battery Types and Characteristics

Primary batteries are of the "dry cell" type in the use of paste or other impregnated materials to hold the electrolyte. In the following sections the most common types of primary batteries are described with typical physical characteristics and common specifications. Two general comparison tables are given at the end of the section to illustrate differences between types.

Polarization Since the electrolyte is in a paste or has been impregnated in a host material, its diffusion is reduced from what would occur in the liquid state. Under high discharge rates it is possible for a condition to occur wherein the distribution of electrolyte ions lags behind reactions at the electrodes. This is called *polarization* of the battery. In this event the battery cells may display a reduction of terminal voltage and/or an apparent discharged state even though capacity still remains in the cells. If removed from the load and allowed to "rest" the cells will rebound. Of course, what actually happens is that diffusion restores the supply of electrolyte and ions at the cell electrodes. This can be speeded up by warming the battery.

Carbon-Zinc: Leclanche Cell One of the first primary or dry cells developed is composed of a zinc (Zn) anode and a cathode of manganese dioxide (MnO_2). The electrolyte is ammonium chloride (NH_4Cl) and/or zinc chloride ($ZnCl_2$) formed as a paste with the manganese dioxide and other materials. The electrolyte also impregnates a chemically porous separator membrane between the zinc and paste. Carbon is inserted into the paste as an electrical contact and current collector for the cathode and serves as the positive voltage terminal of the cell. It is interesting that the carbon plays no active role in the chemical reactions producing the electrical energy, even though this is called a carbon-zinc cell. Figure 2.7 shows a generic cross-section diagram of the physical construction of a round carbon-zinc cell. Note that the zinc serves a dual function as anode and as a vessel to hold the cell materials.

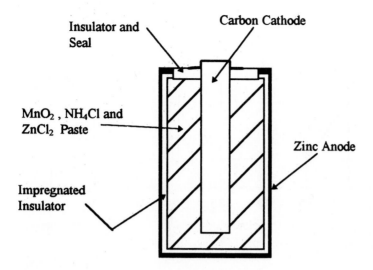

FIGURE 2.7
Cross section of a Leclanche cell.

Voltage Characteristics. The open-circuit voltage of a fresh carbon-zinc cell will be between 1.5 and 1.6 V. Single-cell batteries are common in the AAA, AA, C, and D round sizes. The common 9-V battery for transistor radios and other portable electronic equipment is often made from six small carbon-zinc cells. Figure 2.8 shows a typical reduction of the terminal voltage of a D size cell as a function

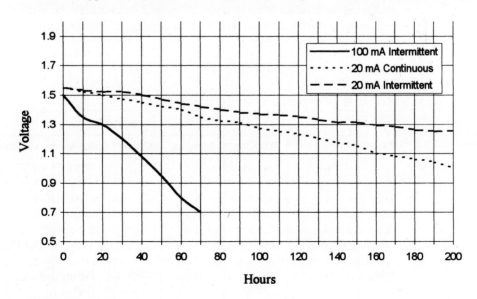

FIGURE 2.8
Leclanche cell voltage variations under various conditions.

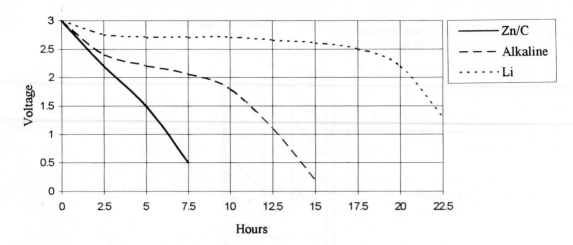

FIGURE 2.9
Comparison of cell voltage variation during discharge.

of time with various discharge currents and for continuous and intermittent (4 h/day) use. This figure shows that the terminal voltage drops quickly at first and then gradually declines as the cell discharges. This voltage drop is *not* due to increased internal resistance since it is measured with a high impedance voltmeter. The curve with the fastest decline is for an intermittent (2 h/day) discharge current of 100 mA. The other two curves are for steady discharge at 20 mA and intermittent (4 h/day) discharge at 20 mA. Leclanche cells have very poor voltage stability during discharge compared with other battery types. (Figure 2.9 compares the voltage drop of the Leclanche cell with alkaline and lithium cells for a constant current of 60 mA and two AA size cells in series.)

TABLE 2.4
Nominal battery capacity by type and size

Size	Nominal Capacity (Ah)						
	Leclanche[a]	Alkaline	Mercury	Silver	Zinc/Air	Lithium	Ni/Cad
D	6.5	14.0	14.0			14.0 (3 V)	1.2–4
C	3.0	7.0				5.0 (3 V)	1–1.8
AA	1.1	2.1	2.4			1.9 (3 V)	0.5
N	0.3	0.7	0.8				0.015
AAA	0.4	1.0	1.0				0.18
9V	0.3 (9 V)	0.6 (9 V)	0.6 (9 V)			1.2 (9 V)	0.01
R9			0.35				0.06
R44			0.22	0.18	0.36		
R43		0.15	0.15	0.13	0.25		

[a] The Leclanche exhibits great variation in working capacity depending on usage conditions.

Capacity.   The capacity of carbon-zinc cells depends strongly on the current drain, the extent of daily use, the selected endpoint voltage, and the temperature of operation. This cell has the lowest energy density of all the commercially available portable primary cells, but it is also the cheapest. The basic energy density ranges from 110 to 190 mWh/cm^3, depending on quality of materials and construction methods. Table 2.4 gives the nominal Ah rating for these batteries in the common sizes in comparison with other types. Remember that at higher currents and/or continuous use the actual working capacity can be significantly less.

EXAMPLE 2.4   What is the longest time that a 9-V carbon-zinc battery could be expected to operate a radio that draws 20 mA? How long would six AA cells operate the radio?

Solution
Caution!!! It is only possible to compute an *upper limit* estimate from data normally available for carbon-zinc cells and the radio. It is not known, for example, what the endpoint voltage for the radio is, i.e., at what voltage it will fail to operate. Since the cell voltage drops off gradually with discharge there will come a point where, even though the cell still has capacity, it cannot operate the radio! So, to calculate the upper limit, the data of Table 2.4 can be used. The 9-V Leclanche (carbon-zinc) has a nominal capacity of 0.3 Ah so it would last for a time $t = (0.3$ Ah$)/(0.02$ A$) = 15$ h, as an upper limit. If AA cells were used the time would be $t = (1.1$ Ah$)/(0.02$ A$) = 55$ h, as an upper limit. Notice that since the AA cells are in series it is the capacity of any one cell that determines the time.

Other Characteristics.   This cell has a storage life of 12 to 24 months depending on the quality of manufacture. It is not a good cell to use in low-temperature environments. At a temperature of only $-10°C$ this cell will have only about 50% of capacity available. This is approximately a linear decrease of effective capacity of $-2.5\%/°C$.

The carbon-zinc Leclanche cell is commonly used for standard cylindrical batteries and the familiar rectangular 9-V "transistor" battery, which is composed of six flat cells. It is also used for large 6-V lantern batteries and a number of other nonstandard configurations.

Carbon-Zinc: Alkaline   This cell is an improvement on the Leclanche carbon-zinc system just described. The alkaline cell still uses manganese dioxide (MnO_2) and zinc for the cathode and anode, respectively. However, the electrolyte used is the highly alkaline potassium hydroxide (KOH). Although in outward appearance these cells seem the same as the Leclanche, they are structured quite differently, as Figure 2.10 shows. In this cell the zinc anode is formed as a paste with the electrolyte, which provides a much more distributed reaction surface. The electrical connection is made with a metal electrode inserted into the paste. The MnO_2 cathode is formed as a compressed paste with electrolyte and graphite for electrical

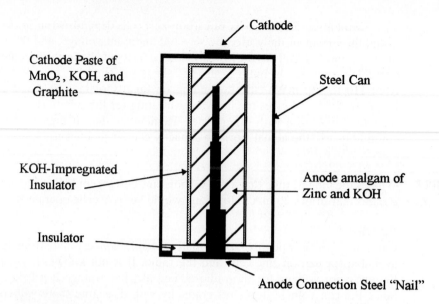

FIGURE 2.10
Cross section of an alkaline carbon-zinc cell.

conductivity. An outer metal case holds the materials together. An electrolyte-impregnated insulator separates the two materials.

The alkaline cell is replacing the Leclanche as the most common cell for popular portable use. It has increased capacity but is somewhat more expensive and hostile to the environment.

Voltage Characteristics. Alkaline cell voltages are slightly lower than the Leclanche but are still nominally 1.5 V in an open-circuit condition. The cells are used individually in the common AAA, AA, C, N, and D sizes as well as the familiar 9-V electronic battery by using six flat cells in series. The voltage does not decrease during use with as great a slope as the Leclanche as illustrated in Figure 2.9, which shows the voltage decline of two AA size cells in series for a current of 60 mA.

Capacity. The nominal capacity of the alkaline cell is greater than the Leclanche but suffers from the same problems of decreased capacity with increased current and continuous usage. The nominal energy density of the chemical mix of this cell ranges from 200 to 300 mWh/cm³ or almost twice that of the Leclanche. Table 2.4 shows the approximate Ah of capacity of this battery in the common configurations.

EXAMPLE 2.5 A printing calculator requires two AA size cells to operate and draws 30 mA when in use. The endpoint voltage is 2.0 V. It will be used approximately 2 hours per day. What is the difference between using cheap carbon-zinc cells or more expensive alkaline cells in terms of days?

Solution

This can be estimated using the curves of Figure 2.9, given toward the front of this section. Since these curves are for 60 mA the times for 30 mA are simply double those on the chart. The carbon-zinc reaches 2.0 V in about 3.0 h so at 30 mA it might be expected to last at least 6.0 h. Actually it would last even longer since the graph is for continuous current draw but this problem is for intermittent draw for which batteries last longer. But this gives an estimate. Anyway, for 2 h per day at 6 h the batteries would last 3 days. The alkaline on the other hand reaches 2.0 V after about 8.5 h at 60 mA or 17 hours at 30 mA. Thus these cells would last for 8.5 days. One would have to buy three sets of carbon-zinc to get the same number of days of usage.

Other Characteristics.   The alkaline cell has a storage life of approximately 24 months. It retains about 50% of capacity at a temperature of $-20°C$, which can be approximated as a linear decrease in working capacity of about $-1\%/°C$.

The carbon-zinc alkaline cell is used in the standard cylindrical cells, in the 9-V rectangular cell, in lantern cells, and in a number of other configurations. Alkaline cells are sometimes used in inexpensive button-type watch and calculator batteries. Some alkaline batteries can be partially recharged but in general they are for a single use.

Zinc-Mercury   The zinc-mercury cell, often simply called a mercury cell, employs zinc as the anode and uses mercury oxide (HgO) for the cathode. The electrolyte is the highly alkaline potassium hydroxide (KOH). Figure 2.11 shows that the basic cell uses an amalgamated zinc anode saturated with the KOH electrolyte, a KOH-impregnated separator and the HgO cathode, which has been mixed with

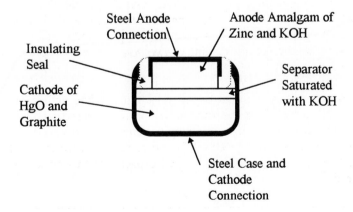

FIGURE 2.11
Cross section of a mercury cell.

graphite to improve conductivity. There are many different configurations for the physical structure. The cell is made in many sizes. It is most commonly made in the small buttons and flat packages for use in cameras, watches, calculators, and other small portable electronic equipment.

The biggest disadvantages of the zinc-mercury cell are its cost, weight, and environmental impact. Disposal becomes a problem because of the dangers associated with mercury. Expenses of the materials, particularly the mercury, make its price higher than comparable sizes of either Leclanche or alkaline batteries. This cell is also significantly heavier than other types. These disadvantages and other factors lead to the result that the principal application of mercury cells is in small button or wafer sizes used in miniature electronic equipment such as watches, hearing aids, and calculators.

Voltage. The unloaded zinc-mercury cell has a voltage of 1.35 to 1.4 V per cell. Although it is often made in the same physical sizes, it is important in applications to consider that the voltage of the cell will be less than that of a fully charged Leclanche or alkaline cell. One of the important characteristics of the mercury cell is that its voltage is quite stable during discharge, particularly at low currents. Figure 2.12 shows a typical graph of cell discharge voltage versus time at a current of 20 mA. From 4 to 44 h, the voltage stayed within 4 to 5% and then dropped rapidly to the discharged state.

Capacity. Mercury cells have one of the highest energy densities of all primary cells, ranging from 400 to 520 mWh/cm^3 or nearly three times the Leclanche. This is one reason for their common use in miniature electronic equipment. Table 2.4 shows the optimum Ah of capacity that can be expected from mercury cells of several sizes.

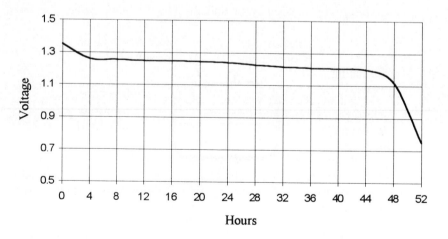

FIGURE 2.12
Variation of mercury cell voltage during discharge.

EXAMPLE 2.6   Two R44 size mercury batteries are to be used to power a small clock. The clock draws an average of 24.5 μA at 2.7 V. Approximately how long would the batteries be expected to last?

Solution

From Table 2.4 the R44 is found to have a nominal capacity of 0.22 Ah. Since the current is given as 24.5 μA, the time is simply $t = 0.22$ Ah$/(24.5 \times 10^{-6}$ A$) \approx 8980$ h or about one year. Since the mercury cell voltage is quite flat up to discharge, it is expected that most of the capacity could actually be used.

Other Characteristics.   Mercury cells do not have good low-temperature performance unless specially manufactured for that type of application. Common cells will retain only about 10% of rated capacity at a temperature of $-10°$C for an approximate linear decrease of $-4.5\%/°$C. Special mercury cells are manufactured to assure improved temperature performance at a cost of maximum current delivery, i.e., higher internal resistance.

Mercury cells are often used as a voltage reference since, under minimal current drain, the cell voltage will remain nearly constant over a long period of time. In one test the open-circuit voltage of a high-quality mercury cell only fell from 1.3570 to 1.3460 V over five years for a decrease of 0.8% or 0.16%/year! Even at 20 mA drain Figure 2.12 shows that the voltage drops by less than 5% over the approximately 40-h life.

The storage characteristic of mercury cells is excellent since their reactions virtually cease when they are not being discharged. Limiting factors are more associated with leakage of the electrolyte than depletion of reactants. Typical storage life exceeds two years.

Silver-Zinc   The silver-zinc or simply silver cell uses a zinc (Zn) anode and a cathode of silver oxide (AgO). The most common electrolyte is potassium hydroxide (KOH). The structure of the cell is essentially the same as that of the mercury cell, as shown in Figure 2.11, with silver oxide in place of the mercury oxide. This primary cell is most commonly used in small, button, or wafer sizes with applications in miniature electronic devices such as watches, hearing aids, and calculators. This cell is more expensive than the alkaline or Leclanche cell because of the silver content. Silver-zinc cells are also rechargeable but often when used in watches, calculators, and small electronic equipment they are employed as a one-time, primary battery.

Voltage.   The unloaded voltage of a fresh silver cell is nearly 1.8 V but drops when discharge starts to a value of about 1.4 to 1.5 V as the internal reactions are started upon loading. The voltage remains quite stable over the discharge time of the battery.

Capacity.   The silver cell has a very high energy density, ranging from 350 to 500 mWh/cm^3 for button types. The typical capacity in mAh for two common sizes is given in Table 2.4.

Other Characteristics.   Capacity of the silver cell drops off approximately linearly with temperature with a slope of about $-1.5\%/°C$. The storage capacity is good with a storage life of over two years.

Zinc-Air   The zinc-air battery uses zinc (Z) as an anode and oxygen (O_2) gas as the cathode. The oxygen comes from air although any other source could be used. The electrolyte is a strong alkaline such as potassium hydroxide (KOH) or sodium hydroxide (NaOH). Of course, oxygen gas by itself is not a conductor so to make the necessary electrical connection the oxygen gas must be allowed to impregnate and diffuse through some solid conducting material that contacts the electrolyte. In most cases this is a granularized form of carbon through which the oxygen gas and electrolyte can diffuse. A special gas-permeable membrane is used to pass the oxygen but block leakage of the electrolyte.
   The basic structure of a miniature zinc-air battery is shown in Figure 2.13. Air is able to enter the cell through small holes and thus diffuse into the carbon cathode. When the cell is manufactured the air holes are sealed so that oxygen cannot enter the structure. The battery is then activated by uncovering the air holes so that oxygen can enter the cell and start the chemical action.

Voltage.   The unloaded terminal voltage of the zinc-air cell is 1.4 V and the internal resistance is quite low. The voltage drop during discharge is much less than that of Leclanche cells.

Capacity.   The zinc-air battery cell has about the highest capacity of any primary type. The energy density for small, button-type cells ranges from 650 to 800 mWh/cm^3. A single button cell can thus have a capacity of 300 to 400 mAh, which is nearly 10 times that available from other types. As the technology improves larger sizes are being manufactured.

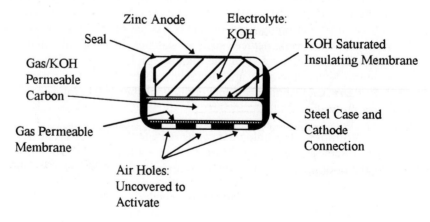

FIGURE 2.13
Cross section of a zinc-air cell.

Other Characteristics. Zinc-air batteries are useful from -10 to $+60°C$, although available capacity is reduced at the lower temperatures. Shelf life of the cells is quite good as long as they remain sealed against atmospheric oxygen. Degradation is typically less than 2% per year as lost capacity for properly sealed units.

The principal difficulty with the zinc-air cell is its sensitivity to atmospheric conditions. Since it is exposed to air for oxygen absorption, it is also exposed to other atmospheric material. Air humidity can affect the cell because of its effect on the electrolyte. If the humidity is low, water will evaporate from the cell, reducing electrolyte; high humidity will result in dilution of the electrolyte through excess water. Cell performance is diminished in either case. Carbon dioxide (CO_2) from the air can react with the electrolyte to form a nonreactive carbonate, which effectively raises the cell internal resistance and therefore reduces terminal voltage under load.

Lithium A number of primary batteries have been developed that use the light and highly reactive metal, lithium (Li), as the anode. A number of different cathode and electrolyte materials are used, each with individual advantages and disadvantages. The basic cell structure is illustrated in Figure 2.14 for small wafer or coin types. The cells must be well sealed because of the high reactivity of lithium with water. Many other configurations are used to make larger round cells. A spiral arrangement of lithium, separator, and cathode material is quite common for larger sizes.

The electrolyte used in lithium cells cannot be water based because lithium reacts violently with water. Thus anhydrous electrolytes are used, which still provide for the transfer of charge via positive and negative ions so that the spirit of the reaction process outlined for primary batteries is maintained.

Some of the basic types of lithium cells are outlined in Table 2.5 along with properties associated with the cells such as voltage and capacity. The second material

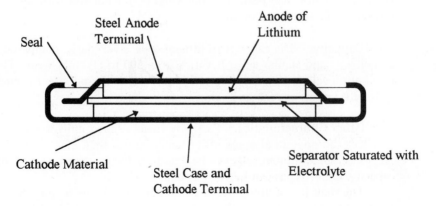

FIGURE 2.14
Cross section of a lithium cell.

TABLE 2.5
Lithium cell characteristics

Lithium-Based Cells	$Li/SOCl_2$	Li/MnO_2	Li/SO_2	Li/I_2	Li/Ag_2CrO_4
Open-circuit voltage	3.65	3.7	3.0	3.1	3.45
Operating voltage	3.5	3.0	2.85	2.8	3.1
Energy density (Wh/cm^3)	800	550	480	550	650
Capacity loss (%/y)	2–3	2–3	2–3	<1	<1
Temperature range (°C)	−55/71	−20/60	−50/70	−40/60	−40/70

given in the table is the cathode composition. The electrolyte includes anhydrous organic and inorganic acids and bases. These are compounds such as propylene carbonate and dimethoxyethane, which are used in the Li/MnO_2 cell.

Lithium cells are being produced in nearly all sizes but will never be a direct replacement since their terminal voltage is nearly twice that of conventional primary cells. They are available for direct replacement in compound systems such as the 9-V rectangular battery. Because of great reliability and lifetime, lithium cells are used in pacemakers as well as watches, calculators, and computer memory backup.

Voltage. The unloaded voltage of most lithium cells is nearly double that of the other primary cells previously described. Thus, a common button cell will have an unloaded voltage of about 3.0 V. One of the most remarkable features of the lithium cell is the stability of its terminal voltage during discharge. Figure 2.9 showed the terminal voltage variation of a typical lithium cell compared with series pairs of carbon-zinc and alkaline so the voltage will be comparable. This is for a current of 60 mA and AA size batteries. We observe the usual primary cell drop of terminal voltage during the onset of discharge, but then the lithium cell maintains a very stable voltage until nearly discharged. The Leclanche cell falls off very rapidly while the alkaline cell performs somewhat better but still with significant drop in voltage as the discharge occurs.

Capacity. The capacity of lithium cells is very large compared with that of carbon-zinc and similar cells. It varies from 500 to 800 mWh/cm^3. This translates into button cells with a capacity of 500 to 1000 mAh and larger round and square cells with ratings as high as 2 Ah.

Other Characteristics. Partly by virtue of their anhydrous electrolyte, lithium cells can operate at much lower temperatures than many other primary cells. At −55°C some lithium cells will be able to deliver as much as 50% of their rated capacity. Other types can be used as high as 150°C.

The shelf life of lithium cells is another remarkable feature. Some units have been shown to retain more than 80% of rated capacity after 10 years of storage. In general, the shelf life can be expected to exceed 5 years.

Clearly there are a number of advantages to lithium-based batteries including high and stable cell voltage, high energy density (capacity), long shelf life, and good operation at low and elevated temperatures. There are also a number of disadvantages including high cost, potential environmental or usage danger, and susceptibility to water. In general, at the present time, lithium batteries can be expected to cost about three to five times as much as carbon-zinc cells.

2.2.4 Battery Selection Data

The decision on what battery to select for a particular system is quite complicated. In fact, it is likely that several of the types available would satisfy the requirements. Table 2.4 and Table 2.6 are presented to provide a comparison between the various types as an aid to battery selection. The nickel/cadmium and lead/acid secondary types of batteries are included in these tables so a composite comparison can be made.

In Table 2.4 the nominal capacity of various types of cell chemistry and sizes was given. The data are presented for cells that are generally commercially available. It is very important to remember that the working capacity may be significantly less than that given depending on use, such as current drain, continuous or intermit-

TABLE 2.6
Comparison of cell types

Type	Operating Voltage (V)	Voltage Flatness (Note 1)	Energy Density (Wh/cm³)	Temperature Limits (50% at °C)	Storage Discharge (%/y)	Cost Factor	Notes
Leclanche	1.25	Poor	0.1–0.2	−10/100	10	1	3
Alkaline	1.20	Fair	0.2–0.3	−20/70	3	2	
Mercury	1.25	Excellent	0.4–0.8	−10/70	2	6	
Silver	1.40	Good	0.35–0.75	−20/70	5	9	4
Zinc/air	1.20	Good	0.8–1.2	−40/60	2	6	5
Lithium	3.0	Excellent	0.4–0.8	−55/150	<1	7	6
NiCd	1.20	Fair	0.04–0.08	−30/50	60	10	7, 8
Lead/acid	2.0	Fair	0.05–0.1	−20/70	60	Note 2	7

Notes:
1. Decrease of terminal voltage during discharge.
2. The lead-acid battery is not made in sizes competitive with the other types.
3. Operation at elevated temperatures for short times.
4. Can be recharged.
5. Storage discharge is without activation.
6. Properties vary greatly among many chemistry types available.
7. Properties given without recharging considered.
8. Other operating voltage levels are available.

tent duty, temperature of the working environment, and even manufacturer. Nevertheless, the table allows some comparison between types.

Table 2.6 provides a comparison between cell chemistry regardless of size. This table includes a cost factor based on the cheapest cell, Leclanche, as unity. Thus you can see that a NiCd cell of comparable size would cost about 10 times the price of the Leclanche cell. Even though the Leclanche does not compare well in any category with other types, it is still half the cost of the closest competitor. This may soon change, however, since many manufacturers are ceasing production of the Leclanche batteries, which will drive up their costs.

For applications where battery replacement is not a problem and where intermittent use is intended, such as flashlights and occasional-use portable electronic equipment, the Leclanche or alkaline cells are the appropriate choice. In applications where low-maintenance and long life are critical, the lithium family of cells is the best choice.

2.2.5 Secondary Battery Types and Characteristics

As previously noted, secondary batteries can be recharged after depletion of the original reactants. This has the obvious advantages of avoiding the necessity of throwing a battery away and installing a new one when a primary battery becomes discharged. The recharging action is accomplished, in a general sense, by making the battery a load to some other electrical source. The electrochemical reactions are then essentially reversed as current is forced into the battery thus restoring the original reactants. Upon completion of the recharging operation, the battery can be put back into service until discharged again. This cycle of charging and discharging can be repeated many times but the battery recharge capacity gradually diminishes with such cycling. One of the specifications of secondary cells relates to the degradation from charge/discharge cycling.

Charging   Essentially charging a secondary battery amounts to driving a current through the battery from some other source. This can be accomplished in a number of ways but there are certain essential factors to consider. Figure 2.15 shows a typical elementary charging setup. On the left we have a discharged battery modeled by some residual source left over after discharge, V_d, and a series resistance, R_d. The charging source is indicated by a source of voltage V_c and series resistance R_c.

Clearly V_c must be greater than V_d for charging to occur, e.g., so that current is directed into the discharged battery. The charging current will then be

$$I_c = \frac{(V_c - V_d)}{(R_d + R_c)} \qquad\qquad (2.6)$$

As the battery charges V_d will increase, which will have the effect of decreasing the current, and R_d will decrease, which will have the effect of increasing the current. It is also true that as the battery becomes charged reversal of the reactions necessary to produce recharging proceeds more slowly. Increasing the current does not help

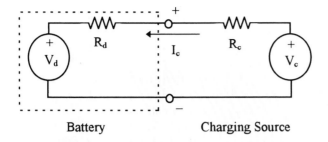

FIGURE 2.15
Circuit model for charging a secondary battery.

but simply causes the battery to overheat due to dissipation by the internal resistance. Whereas this simple method of recharging will work it is not the most efficient in recharging the battery in either the least time or with the least damage to the cells.

Optimum recharging of most secondary cells is accomplished by performing the initial stages of recharging at a constant current, letting the supply source voltage adopt whatever value will provide the constant current. This is the *fast charge* mode when the electrochemical reactions can proceed at a fast rate. At some stage of the process the charger switches to a constant charging voltage, V_c, which is slightly higher than the fully charged battery. In this way the charging current adopts whatever value is determined from the equation above as the battery voltage increases and resistance decreases. The charging current will eventually drop to zero or nearly zero when the battery becomes fully charged.

Overcharging a battery can damage it by causing deterioration of the electrodes. Even more serious, however, is that overcharging often causes hydrolysis of water in the electrolyte and the consequent generation of hydrogen and oxygen gas. This can cause pressure buildup in the battery and even the danger of explosion.

Lead-Acid The lead-acid battery has been the workhorse of the secondary battery applications for many years and will probably remain so into the near future. A fully charged lead-acid battery cell has an anode of lead (Pb) and a cathode of lead dioxide (PbO_2). The electrolyte is a solution of dilute sulfuric acid (H_2SO_4) which disassociates readily into ions of H^+ and SO_4^{2-}.

As the cell is discharged, reactions at the anode produce lead sulfate (PbSO4), which is deposited on the anode, while reactions at the cathode produce water. The result of this is that the sulfuric acid becomes diluted since sulfate ions are taken out of solution and water is added. Battery operation is maintained by diffusion of the H^+ and SO_4^{2-} through the electrolyte solution.

Figure 2.16 shows a typical cell structure. The anode is composed of many plates of "spongy" lead so that a large surface area is presented to the electrolyte. The lead dioxide cathode is not a good conductor and not physically strong so it is deposited on or otherwise arranged with a carrier that is a good conductor.

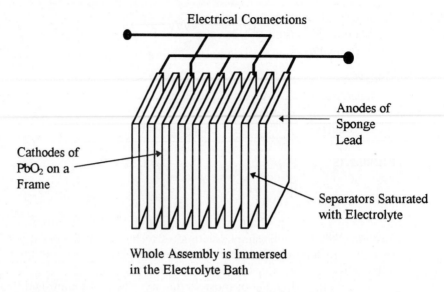

Electrical Connections

Anodes of
Sponge
Lead

Cathodes of
PbO₂ on a
Frame

Separators Saturated
with Electrolyte

Whole Assembly is Immersed
in the Electrolyte Bath

FIGURE 2.16
Basic structure of a lead-acid secondary battery.

The separator that is used between the plates to prevent physical contact must allow passage of the electrolyte. The whole assembly is immersed in the liquid electrolyte.

Since the common lead-acid battery uses a liquid electrolyte it must be used in an upright orientation. It can be used in vehicles and other portable applications where the orientation is relatively constant but not in applications where significant changes in orientation occur. Some lead-acid batteries have been developed that use a "gelled" electrolyte so that ion migration is allowed but the problems of liquid do not occur.

Specific Gravity.   The weight of liquids per unit volume are often described in comparison to water. The ratio of the weight of a volume of liquid to that of water is called the *specific gravity* (SG) of the liquid. Obviously the specific gravity of pure water is one (1.0). The specific gravity of the electrolyte used in lead-acid batteries turns out to be a good indicator of the state of discharge of the battery. The reason is that as a cell discharges its electrolyte becomes diluted with water as noted earlier and therefore its specific gravity comes closer to unity. The electrolyte is sulfuric acid whose SG depends on the degree of dilution with water. A fully charged cell will have an electrolyte specific gravity of 1.210 to 1.275, depending on the size and intended application. The common automobile battery, for example, is typically 1.260 at full charge and drops by about 0.125 to 1.120 when discharged. The state of charge of a battery is often described by the fraction of reduction of specific gravity.

EXAMPLE 2.7 A battery at room temperature (72°F) is fully charged at an SG of 1.245 and considered discharged at 1.120. Suppose the SG is found to be 1.208. What is the percent discharge?

Solution
The change from charged to discharged is about 0.125 and the change is now 1.245 − 1.208 = 0.037. Thus the battery is 0.037/0.125 = 0.296 below full charge or about 30% discharged.

Whereas specific gravity is a convenient way to determine the state of charge of a lead-acid battery, care must be taken in the interpretation of the readings. The value of specific gravity is a function of temperature, dropping by about 0.001 for every 3°F (1.7°C) temperature drop. This can cause an erroneous indication of discharge.

EXAMPLE 2.8 Suppose the battery in Example 2.7 has an SG measured at 32°F of 1.195. What is the state of discharge of the battery?

Solution
This is a drop of 1.245 − 1.195 = 0.050 which would *imply* a discharge state given by 0.050/0.125 = 0.4 or 40% discharged. However, the temperature difference also causes a drop in SG of 0.001(72 − 32)/3 = 0.013 so the actual discharge drop is 0.050 − 0.013 = 0.037, which is about 30% discharged as in the previous example.

Another factor affecting specific gravity is evaporation of water from the electrolyte. This certainly results in an increase in SG because the density of the acid is increased, but it is NOT an indication of battery charge. It is important to keep the level of the electrolyte at the proper value to assure valid SG readings and proper battery operation. Note that it is never necessary to add *acid* to a lead-acid cell since the sulfate is not lost to evaporation, only water.

Voltage. The open circuit of the lead-acid battery is about 2.1 V. The drop in cell voltage in time is very dependent on the discharge rate but even in storage conditions can drop by as much as 2% per day at higher temperatures! Generally lead-acid batteries are preferred for conditions of high discharge rate over short periods, with automatic recharge in between, as in vehicle starting, where stability of voltage over time is not important.

Capacity. The energy density of the lead-acid battery is quite low compared to primary types, at about 0.05 to 0.1 Wh/cm^3, but the battery can is made in sizes with large volume. Therefore, it is not uncommon to have batteries with capacities of 1 to 100 Ah but in industrial applications this can go to 500 Ah. As usual the actual working capacity is less at higher currents and lower temperatures.

Lead-acid battery capacity is often described using the concept of "rate" described earlier. Here a specification of 20 Ah at the 8-h rate means that if the battery is discharged over an 8-h period then 20 Ah of capacity will be available. This implies that the discharge current would be 20 Ah/8 h = 2.5 A. The same battery at a 20-h rate and thus 1 A would be expected to deliver more capacity, while less capacity would be expected at a 4-h rate and 5 A of current.

This effect is illustrated in Figure 2.17. Figure 2.17(a) shows the fraction of capacity that can be obtained from a lead-acid battery specified at the 8-h rate if it is discharged at other rates. Note that at the 8-h rate the capacity available is 100%, while at the 4-h rate it is about 82% and at a 10-h rate one could get about 105% of the 8-h capacity.

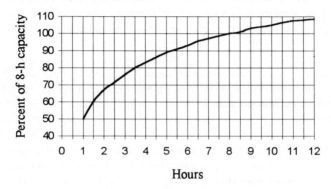

(a) Percent of Capacity

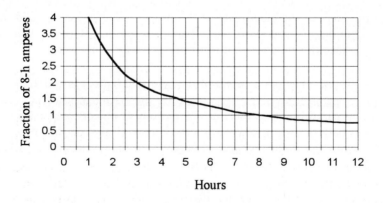

(b) Fraction of Current

FIGURE 2.17
Variation of lead-acid battery characteristics during discharge.

Figure 2.17(b) shows the fraction of current compared to the 8-h current for a range of discharge rates. This shows that if discharged at a 4-h rate the current drawn would be about 1.7 times that at the 8-h rate, while for a 10-h rate it would be about 0.8 times.

EXAMPLE 2.9

A small gasoline-powered vehicle uses a lead-acid battery with a specification of 30 Ah at the 8-h rate to supply electrical needs. If the nominal operating current draw is 2.7 A, approximately how many hours of operation are possible between charges?

Solution

At the 8-h rate the current would be 30 Ah/8 h = 3.75 A. This is a fraction of 2.7/3.75 = 0.72 between usage current and the 8-h rate current. From Figure 2.17(b), the battery will last about 11 h.

Charging. Charging lead-acid batteries can be accomplished using the process suggested earlier of constant current in the early stages and then at a constant voltage as full charge is approached. It is very important not to overcharge a lead-acid battery because when a condition of full charge is reached any further driving of current into the cell will result in hydrolysis of water in the electrolyte and the resulting release of hydrogen and oxygen gas. Buildup of these gases can cause pressure problems but also result in explosions if the gases are ignited. Lead-acid batteries are often *vented* so that gas generated during charging can escape from the cells.

Nickel-Cadmium The charged nickel-cadmium cell has metallic cadmium (Cd) for the anode and a form of nickel hydroxide (NiOOH) for the cathode. The electrolyte is the highly alkaline potassium hydroxide (KOH). During discharge the cadmium of the anode reacts to produce cadmium hydroxide, $Cd(OH)_2$, while the cathode is transformed into another form of nickel hydroxide, $Ni(OH)_2$. The recharging process reverses this chemical process, restoring the original materials as needed.

There are many forms of the nickel-cadmium (NiCd) cell structure. Some versions employ a liquid electrolyte and must be operated in an upright orientation, like the lead-acid system. These are used in high-power applications in a fashion similar to the lead-acid battery. The structure is similar to that given in Figure 2.16.

Of more common interest is the development and use of sealed NiCd rechargeable batteries, which use a gelled electrolyte and can therefore be operated in any orientation. Development of sealed cells represented a great advance in battery technology and brought about the proliferation of NiCd batteries in common use in small-scale, portable electronic equipment. These cells are designed in special ways to reduce or eliminate hydrogen gas production during recharging and to allow for chemical reabsorption of oxygen gas generated during charging. Even so, these cells are manufactured with a vent valve to allow gas pressure release in the event of a malfunction of the gas absorption process. Figure 2.18 shows the basic

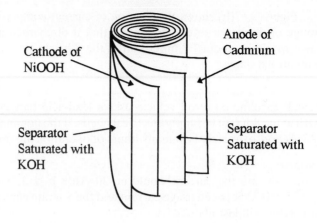

Cathode of NiOOH

Anode of Cadmium

Separator Saturated with KOH

Separator Saturated with KOH

(a) Cylindrical NiCd Cell (Jelly roll)

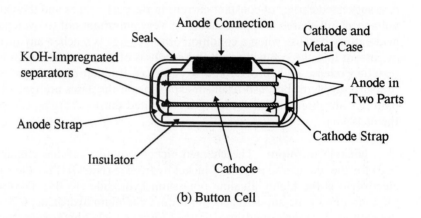

Anode Connection

Seal

Cathode and Metal Case

KOH-Impregnated separators

Anode in Two Parts

Anode Strap

Cathode Strap

Insulator

Cathode

(b) Button Cell

FIGURE 2.18
Nickel-cadmium cylindrical and button cell design.

structure of round and button NiCd batteries. Notice in the button cell that the cathode wafer is sandwiched between two anode wafers, which increases the surface area contact between the electrodes. Internal straps connect the two anode pieces and the cathode to the case.

   Voltage. The open-circuit voltage of a NiCd cell is about 1.4 V but this quickly drops to an operating value of about 1.2 to 1.3 V per cell. During the main part of the discharge period, the cell voltage will drop by only about 5%. For a 10-h rate the voltage can be expected to remain within 5% from about 1 to about 9 h. During the first hour the voltage will drop from about 1.4 to 1.3 and during the next 8 h down to about 1.25. In the next 2 h the cell voltage will quickly drop to the endpoint value of 1.0 to 1.1 V. Of course with higher rates the period of

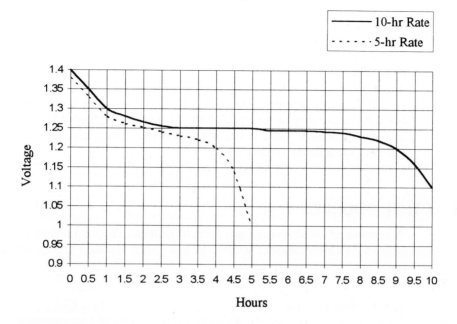

FIGURE 2.19
Variation of NiCd voltage during discharge.

stable voltage will be less. Figure 2.19 illustrates the drop of cell voltage for a NiCd unit when operated at a 10-h rate and a 5-h rate.

Capacity. It is possible to construct NiCd cells in a great range of sizes and shapes. Units are available from as little as 10 mAh to as great as 50 Ah. In general their energy density, 0.01 to 0.08 Wh/cm³, is much lower than available primary cells, as shown in Table 2.6; however, the fact that they can be recharged makes that a less serious consideration.

Charging. The charging process of NiCd batteries follows essentially the same process as described in the beginning of this section. These are quite rugged systems, which can take up to a thousand cycles before serious degradation of performance begins. Some cells exhibit a characteristic called "memory" wherein they will fail to take a full charge if not fully discharged in prior usage. In other words if a cell is only 50% discharged it cannot be fully recharged. This condition can be alleviated by forcing a complete discharge of the battery before recharging.

Other Characteristics. Nickel-cadmium batteries are now made in nearly all popular sizes for portable electrical and electronic applications as shown in Table 2.4. They are commonly used for backup systems in computers and communication equipment where they are kept charged by traditional power. In the event of a power failure they supply electrical power until traditional power is restored, or they become discharged.

EXAMPLE 2.10   A toy requires two AA size batteries to operate. It draws about 45 mA during usage. If it is to be used about a half-hour per day, contrast using alkaline versus NiCd rechargeable batteries.

Solution
Table 2.4 shows that the alkaline AA has about a 2.1-Ah capacity. Thus, it would be expected to last approximately 2.1 Ah/0.045 A = 46.6 h or about 93 days. The NiCd has a capacity of about 0.5 Ah so it would last about 0.5 Ah/0.045 = 11.1 h or about 22.2 days. Thus, it would need to be recharged about four times during the time the alkaline would last.

2.3   OTHER SOURCES

A number of other sources are used to provide electrical power for circuits and devices. The following sections present just a brief overview of some of these sources and their basic characteristics.

2.3.1   Fuel Cells

A fuel cell is a chemical source of electrical energy like batteries but is sufficiently different in operation to be placed in this "other" category. A fuel cell uses gases for both the anodic and cathodic reactions required to produce electricity. To provide for electrical conduction, the gases diffuse into metal electrodes into which the electrolyte has also been impregnated. Reactions then occur in these electrodes and at the interface with the electrolyte. The results are much like chemical batteries in the release of electrons that can pass through a load.

Figure 2.20 shows a generic diagram of the essential features of a fuel cell. The fuel cell itself has no reactive components, other than the electrolyte, so it can be dormant for extended periods without degradation. It is only upon introduction of gases that reactions occur and electricity is produced. The process can be started and stopped via control of the gas flow.

Many materials are employed in fuel cells but the most common are hydrogen and oxygen. Such a system gives energy densities that are hundreds of times that of conventional batteries. Although fuel cells have been known and studied for more than 60 years it is only in the past 20 that technology has allowed practical versions. They are common now on spacecraft and in remote locations.

Although simple in concept the fuel cell has many complications when actually implemented. For example, the product of the hydrogen-oxygen fuel cell is water. This water dilutes the electrolyte, which reduces cell operation. Therefore, a system must be used that extracts water from the electrolyte to maintain its concentration during operation.

Fuel cells operate in essentially three temperature modes. Low-temperature fuel cells, like the hydrogen-oxygen, use liquid electrolytes such as KOH and operate at temperatures under 250°C. Medium-temperature fuel cells use molten salts (car-

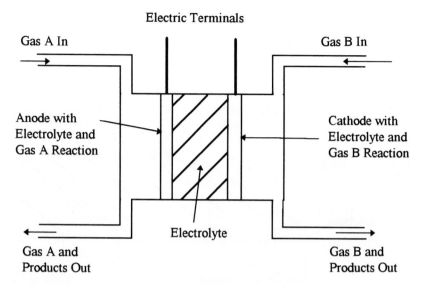

FIGURE 2.20
Basic structure of the fuel cell.

bonates) as the electrolyte and operate up to 500°C. High-temperature fuel cells use a solid ceramic in place of an electrolyte. Ionic gases are able to diffuse directly through these materials. They operate at and above 900°C.

2.3.2   Photovoltaic Cells

Photovoltaic cells are familiar in the form of the "solar cells" used with spacecraft to supply either main or auxiliary electrical power. They are now in very common use throughout the world in land-based applications. Such applications include everything from complex remote sensing stations to home yard lights where photovoltaic cells are used to charge secondary batteries during the day.

This cell converts light energy into electrical energy. The photovoltaic cell is essentially a huge *pn* junction diode that is exposed to light. The most common material used to prepare the *p*-doped and *n*-doped sections is silicon. Figure 2.21 shows a diagram of such a cell. The thicknesses of the various parts of the cell have been exaggerated so the principles can be illustrated.

Voltage   The idea is that when photons of light strike the cell most will penetrate the very thin *p*-layer and pn junction and are absorbed in the *n*-layer. If the photon has sufficient energy an electron–hole pair will be formed and the hole will move across the junction to the *p*-layer. Thus, there is a separation of charges and a difference of potential, i.e., a voltage, forms across the cell. The greater the intensity, the greater the number of electron–hole pairs and thus the voltage increases. The voltage varies approximately as the logarithm of the light intensity.

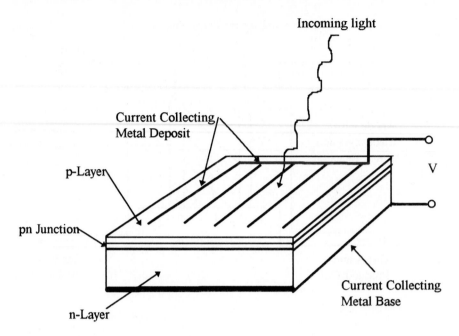

FIGURE 2.21
The photovoltaic cell is a distributed junction diode.

The silicon photovoltaic cell has an internal resistance that depends on the doping, the surface area, and how the cell is fabricated. For use as a power source, groups of cells are placed in series to increase the voltage but with increased net internal resistance. These groups are then placed in parallel to keep the voltage the same but decrease the net internal resistance.

When used as a power source the appropriate load is one for which the power delivery is a maximum. For a traditional source, load resistance is equal to the source internal resistance. In the case of the photovoltaic cell this concept cannot be used because the effective internal resistance depends on how much current is drawn from the cell. Instead one uses that load resistance for which the product of terminal voltage and delivered current is a maximum at some intensity. Figure 2.22a shows curves of voltage versus current for light intensities of 20, 50, and 100 mW/cm^2. By computing the product of current and voltage at every point on a particular curve, the maximum power delivery current and voltage can be determined. Figure 2.22(b) shows power versus voltage.

By dividing the voltage by the current it can easily be computed that the loads for maximum power delivery are approximately 46 Ω for 20 mW/cm^2, 25 Ω for 50 mW/cm^2, and 20 Ω for 100 mW/cm^2. Thus if you provide a load that is optimum at some intensity, it will *not* be the optimum for other light intensities.

2.3.3 Thermoelectric Sources

Many years ago it was found that if two dissimilar wires are connected at each end and the endpoint junctions are at different temperatures, an electromagnetic force

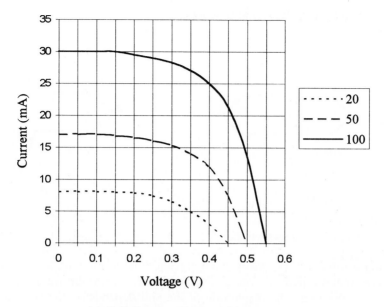

(a) Cell Current Versus Voltage

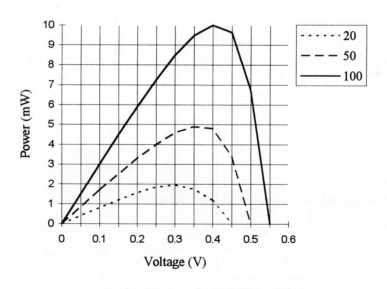

(b) Cell Power Versus Voltage

FIGURE 2.22
Variation of silicon photovoltaic cell voltage and current.

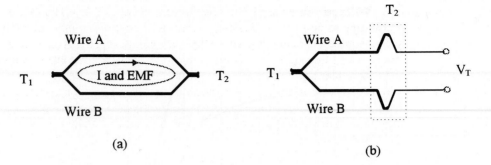

FIGURE 2.23
Basic concept of a thermocouple.

(emf) is formed in the wires and a current flows in the loop formed by the wires. This system is called a thermocouple and Figure 2.23(a) illustrates the concept with a simple diagram. The magnitude of the emf depends on the type of metal used for the wires and the difference in endpoint temperature. Figure 2.23(b) shows that if one of the junctions is separated then the emf is expressed as a voltage between the wires. If a load is connected across the wires, a current will flow and work will be done. Thus in this simple way a battery can be formed that converts thermal energy into electrical energy.

Such a device is actually in common use in water heaters. In this case, as illustrated in Figure 2.24, a thermocouple is placed in the water to be heated. When the water temperature reaches a sufficient value, the voltage and current produced by the thermocouple are sufficient to close an electric solenoid gas valve and stop the heating.

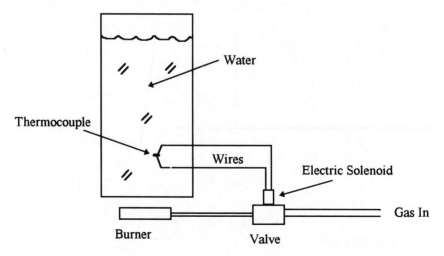

FIGURE 2.24
Use of a thermocouple to provide electric power for a water heater.

Voltage   In general, the voltage generated by thermocouples is very small, amounting to only 10 to 70 μV/°C difference between the junctions. The model of such a source looks like Figure 2.1(b) with the internal resistance composed of the resistance of the wires. This is generally quite small. If a sufficient number of such thermocouples are placed in series, practical amounts of voltage can be generated with an increase in internal resistance, but since the wire resistance is quite small, the resistance does not become very large. Combinations of thermocouples in series and parallel are used to increase the voltage and the current delivery.

3

POWER SUPPLIES AND REGULATORS*

Most power supplies are either ac-to-dc converters operated from a 120-V 60-Hz line or dc-to-dc converters used to convert a dc voltage at one level to a dc voltage at another level. A third type of power supply is called an inverter, which converts a dc voltage to an ac voltage. Inverter applications are rather specialized and limited compared to those of the other two types of supplies.

AC-to-dc converters make use of diodes to rectify the ac line voltage.

3.1 RECTIFIER POWER SUPPLIES

Rectifier power supplies are almost always fed from the secondary of a power transformer. The two general types of single-phase rectifiers are the half-wave and full-wave rectifiers. Rectifier supplies represent an old technology that is still very useful. Numerous improvements have been made over the years in transformers, diodes, and filter capacitors. Even though the circuits are fairly simple, one cannot just throw together some diodes and capacitors and expect to have a power supply that works properly. Diodes for rectifier supplies must have adequate peak inverse voltage ratings and forward current capacity. The transformer must have the proper turns ratio and volt-ampere rating to supply the intended load. The peak inverse voltage (PIV) rating of the diode needed depends on the value of the ac voltage and to some extent on the configuration of the rectifier circuit.

3.1.1 Half-Wave Rectifiers

The half-wave rectifier circuit shown in Figure 3.1 has the following characteristics. The input to the rectifier comes from the secondary of a power transformer. The

---

* This chapter was written by Prof. Orville R. Detraz, Purdue University.

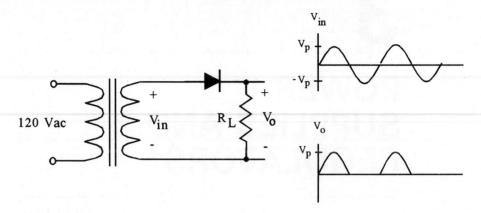

FIGURE 3.1
A half-wave rectifier.

voltage is assumed to be a sine wave whose rms value is V_{rms} volts and whose peak value is V_p, which is equal to $\sqrt{2}V_{rms}$. The output voltage appearing across the load resistor consists of a series of sinusoidally shaped pulses as shown in Figure 3.1.

The maximum value of the load voltage is equal to the peak value of the input sine wave minus the forward drop of the diode. The voltage drop across the diode is usually ignored except for low voltage supplies. The average dc value of the load voltage can be expressed in terms of either the peak or rms voltage. The average dc load voltage is $0.3183 \, V_p$ or $0.45 \, V_{rms}$. The PIV rating for the diode must be at least equal to the peak value of the input voltage.

3.1.2 Full-Wave Rectifiers

Single-phase full-wave rectifier circuits use either a center tapped transformer and two diodes or an untapped transformer and four diodes arranged in a bridge circuit. For either circuit the output voltage is a series of half sine waves having a peak value equal to the peak value of the transformer secondary voltage and an average value equal to $0.6366 \, V_p$ or $0.9 \, V_{rms}$. In the two-diode full-wave circuit the diodes must have a PIV rating equal to twice the peak value of the input voltage, whereas in the bridge circuit the PIV requirement is equal to the peak value of the input voltage. Full-wave rectifier circuits and their outputs are shown in Figure 3.2.

The bridge rectifier circuit does not require a center tapped transformer. Either end of the load resistor could be grounded so the load voltage can be made either positive or negative relative to ground. A bridge circuit that will produce two different voltages is shown in Figure 3.3. If the two load resistors are equal, the two voltages will each be equal to one-half the transformer secondary voltage and will be the negative of each other. That circuit is referred to as a *split supply* and can be used to supply both positive and negative voltages for operational amplifiers.

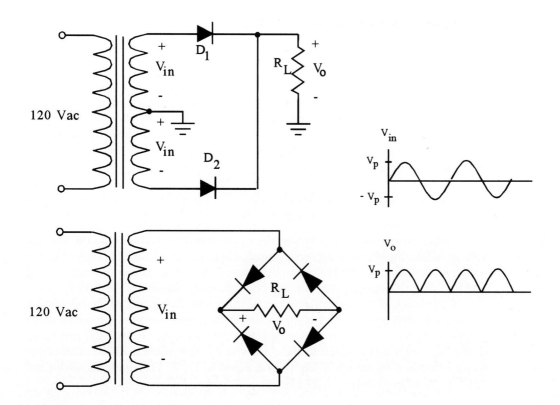

FIGURE 3.2
Full-wave rectifiers.

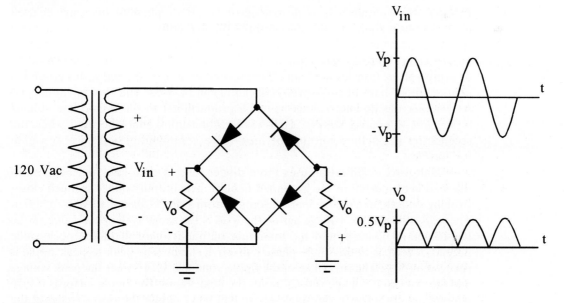

FIGURE 3.3
A split supply.

3.2 RIPPLE VOLTAGES

The variation in dc output voltage shown in Figures 3.1, 3.2, and 3.3 is referred to as the *ripple voltage.* It is often expressed as a percent of the dc voltage. The peak-to-peak variation of the output voltage is one very descriptive way of expressing ripple voltage. The ratio of the rms value of the ac component of the ripple voltage to the dc value of the output voltage is often used to define the ripple factor, in percent:

$$r = \frac{\text{rms ripple voltage}}{\text{dc output voltage}} \times 100\% \tag{3.1}$$

For rectifier supplies with no filtering, the ripple voltage is greater than the average voltage so filtering is almost universally used. There are some simple battery chargers on the market that use the battery to be charged as the load and do not really need filtering since the charger supplies pulses of charging current to the battery.

Calculation of the ripple factor can be quite tedious because of the calculation necessary to determine the ac components of the ripple voltage. Approximations are used whenever they can be justified. For the half-wave rectifier the rms value of the ac component is equal to the peak voltage multiplied by 0.385. The ripple factor calculated by means of Eq. (3.1) is equal to 121%. The ripple factor for the full-wave unfiltered output is 48.4%.

3.3 THREE-PHASE RECTIFIERS

In applications requiring large dc load currents, a three-phase rectifier may be used if a three-phase source is available. A three-phase rectifier offers certain advantages over a single-phase rectifier. Three-phase half-wave and full-wave rectifier circuits are shown in Figure 3.4. The input to the rectifier circuit is taken from the secondary of a three-phase transformer bank. If the secondary is Y connected with a grounded neutral, either half- or full-wave rectification can be used. The choice will depend on whether the dc load voltage is to be proportional to the line-to-line voltage or the line-to-neutral voltage. Half-wave rectification cannot be used when the secondaries are delta connected because there is no neutral and no return path for current.

Half-wave rectification uses three diodes, one in each line, and one end of the load is connected to the grounded neutral. The output voltage is much closer to being smooth than the comparable output from a single-phase half-wave rectifier. The peak value of the voltage across the load is the same as the peak value of the line-to-neutral voltage, and the minimum output is one-half of the peak value instead of zero as in the single-phase half-wave circuit. The peak-to-peak ripple is reduced by 50%. The rms value of the ripple voltage is 0.148 times the peak voltage and the dc output voltage is 0.827 times the peak value. The ripple factor is 0.1976 without any filtering. In the single-phase half-wave circuit the ripple factor of the output voltage without filtering would be 1.21. In the three-phase rectifier the

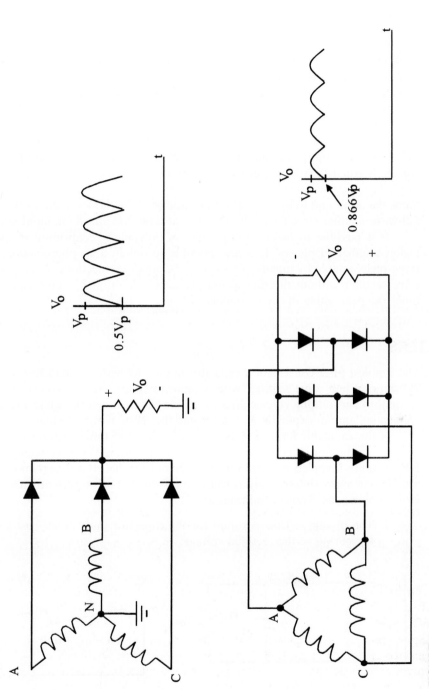

FIGURE 3.4
Three-phase rectifiers.

frequency of the output voltage is three times the line frequency. That makes the task of filtering easier.

In the half-wave rectifier circuit, the transformer windings do not carry current at all times, and as the diodes turn on and off repetitively, transient voltages and currents are reflected back to the primary. These transient voltages can induce substantial noise voltages in telephone lines and other audio equipment in the vicinity.

The full-wave rectifier circuit is always used with a delta connected secondary. Six diodes are required and the rectified output voltage will have a dc value equal to 0.9546 times the peak value of the input voltage. The rms value of the ripple voltage is equal to 0.004 times the peak value of the line voltage. This makes the ripple factor equal to 0.00422. The frequency of the output voltage is equal to six times the input line frequency. The PIV ratings of the diodes in either the half- or full-wave rectifier circuit must be equal to at least 2.45 times the input voltage.

It is possible to extend the process of polyphase rectification to include a higher number of phases. It is also possible to obtain a six-phase source from a three-phase source by means of special transformer connections. A six-phase full-wave rectifier without filtering would produce a dc output voltage equal to 0.98 times the peak value of the secondary voltage.

3.4 FILTERS

The simplest filter consists of a capacitor in parallel with the load resistor. Figure 3.5 shows a half-wave rectifier with a capacitor filter and the voltage across the capacitor. The output voltage follows a sinusoidal waveform while the diode is conducting and an exponential decay while the diode is reverse biased.

If the capacitor filter is to be effective, the $R_L C$ time constant needs to be larger than the period of the input voltage to the filter. The exact analysis of the filter is quite involved and the effort is seldom justified but if certain conditions are met, an approximate analysis will give results that are sufficiently accurate for most applications. Those conditions are:

1. The discharge time constant for the capacitor must be at least ten times the period of the rectified voltage.

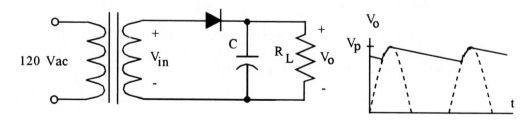

FIGURE 3.5
A capacitor filter.

2. The load resistor should be much smaller than the reverse resistance of the diode.

The capacitor voltage waveform is represented by two straight line segments as shown in Figure 3.6. An exponential decay can be represented by a straight line for a fraction of a time constant. The portion of the sine wave from the point where the capacitor starts recharging until the sine wave peaks can also be represented by a straight line segment. The calculations are simplified further by assuming the filter voltage to be a sawtooth waveform as shown in Figure 3.7. The minimum value of the sawtooth wave will not be less than 90% of the maximum as long as the conditions mentioned earlier are met. For cases where the peak-to-peak value of the sawtooth wave is less than 10% of the maximum value, the approximation is even better.

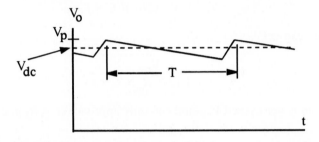

FIGURE 3.6
Approximate output voltage.

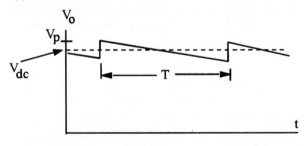

FIGURE 3.7
Sawtooth output voltage.

The straight line approximation implies a constant value of load current being supplied by the capacitive discharge since dv/dt is constant during that interval and the current through the capacitor is given by

$$I = C\frac{dv}{dt} \tag{3.2}$$

where I is in amps when C is in farads and dv/dt is in volts per second. For the sawtooth waveform, the total change in voltage, Δv, is the peak-to-peak load voltage,

V_{pp}, and the total change in time, Δt, is $1/f$ where f is the frequency of the input to the filter. Since the current, I, is constant, the dc load voltage, which is equal to IR_L, is also constant. The value of V_{dc} can be expressed in terms of the peak-to-peak output voltage and the filter components as follows:

$$\frac{V_{dc}}{R_L} = C\frac{dv}{dt} \tag{3.3}$$

$$V_{dc} = R_L \times C \times V_{pp} \times f \tag{3.4}$$

The dc value of the output can also be expressed in terms of the maximum rectified load voltage, V_{max}

$$V_{dc} = V_{max} - V_{pp}/2 \quad \text{or}$$
$$= V_{max} - V_{dc}/(2R_L \times C \times f)$$

from which one can get

$$V_{dc} = \frac{V_{max}}{1 + 1/(2 \times R_L \times C \times f)} \tag{3.5}$$

Note that V_{dc} can never exceed V_{max} and can only approach it as R_L goes to infinity.

EXAMPLE 3.1

Assume that the peak input voltage to a half-wave rectifier operating from a 60-Hz line is 141.42 V, i.e., $(100 \times \sqrt{2})$ and the load resistor is 400 Ω. Using the sawtooth approximation for the output voltage, calculate the minimum value for the filter capacitor needed to make the approximation valid and calculate the dc output voltage for the capacitance chosen.

Solution

We must have $R_L C \geq 10/f$ for the approximation to be valid so C must be at least equal to $10/(400 \times 60)$ or $C \geq 416.7$ μF. We would probably choose a higher value such as 450 or 470 μF. With $C = 470$ μF, the dc output voltage would be

$$V_{dc} = \frac{100\sqrt{2}}{1 + 1/(400 \times 2 \times 0.00047 \times 60)}$$
$$= 135.42 \text{ V}$$

Filter performance can be described either by the ripple rejection ratio given in decibels or the ripple factor or % ripple in the output voltage. The ripple factor, r, was defined in Eq. (3.1).

The important thing to remember in using Eq. (3.1) is that the numerator is the rms value of the ac component of the output voltage, not the rms value of the output voltage. The rms value of a sawtooth wave is $V_{pp}/2\sqrt{3}$. Since the dc value of the output voltage can also be expressed in terms of V_{pp}, the ripple factor, r, can be expressed in terms of R_L, C, and f:

$$r = 1/(2\sqrt{3} \times R_L \times C \times f) \tag{3.6}$$

EXAMPLE 3.2

Calculate the ripple factor for the filter of the previous example.

Solution

$$r = 1/(2\sqrt{3} \times 400 \times 0.00047 \times 60)$$
$$= 0.0256 \text{ or } 2.56\%$$

A word of caution is in order. A 470-μF capacitor has between 5 and 6 Ω of reactance at 60 Hz. At the higher harmonic frequencies its reactance is even less, so it does a fine job of filtering out the harmonics from the output voltage. When the power supply is initially turned on, the filter capacitor is uncharged and looks like a short circuit. The initial current will depend on the impedance of the source and the initial value of the input voltage. It may be necessary to use a small resistor in series with the load and the filter to prevent burning out the rectifier diodes. One should always check the forward surge current rating of diodes used in a rectifier supply.

In addition to the initial surge current, the diodes will also be subject to short high-current pulses during the intervals when they are forward biased. The charge lost by the capacitor during the discharge time is supplied by the source during the short time that the diodes are conducting. Since the charge lost is equal to the charge gained we can put it in equation form:

$$I_D T_D = I_L T_L \tag{3.7}$$

where I_D is the diode current, I_L is the load current, and T_D and T_L are the time intervals during which they flow. In the sawtooth approximation, T_D would be zero, which would make I_D infinite. A realistic approach would be to assume the capacitor discharge to be linear as shown in Figure 3.6 from the peak of the voltage wave to the intersection with the next cycle of the input voltage. We also assume that the capacitor recharges uniformly during the interval T_D. Note that T_D will be greater than zero. If the capacitor voltage drops to 90% of its maximum value before the diode starts conducting, the diode will be forward biased for slightly more than 7% of the period of the input wave. This means that the diode current during that short interval can be almost 13 times as great as the load current. If a better filter is used so that the load voltage drops only to 95% of its maximum, the diode will conduct for about 5% of the time and the diode current will be 19 times as great as the load current. The average diode current in a full-wave rectifier is one-half the load current because the diodes conduct on alternate half cycles.

EXAMPLE 3.3

Design a full-wave rectifier supply to operate from a 120-V, 400-Hz supply. It is to provide an output voltage of 48 Vdc and a load current of 0.5 A. The ripple factor should be no greater than 0.03. Specify the transformer turns ratio, filter capacitor size, and diode ratings.

Solution

Use Eq. (3.6) to see what size filter capacitor is needed to meet the ripple specification. R_L will be 96 Ω and the frequency into the filter will be 800 Hz.

$$C = 1/(0.03 \times 2 \times \sqrt{3} \times 800 \times 96)$$
$$= 125 \ \mu\text{F}$$

The filter time constant is 12 ms and the period of the input is 1/800 or 1.25 ms. The ratio of filter time constant to input period is 9.6, which is high enough to use the sawtooth approximation with only a slight error. The dc load current is to be 0.5 A so the peak load current will be 6 to 6.5 A. The average current per diode is one-half the dc load current or 0.25 A.

Assume a center tapped transformer is to be used. The peak voltage across one-half of the secondary will need to be 48.7 V if you include the forward diode drop. The error would be about 1.5% if the diode drop is neglected. Transformers are rated for rms voltage and current so the secondary needs to be rated at 34.44 V for one-half the secondary winding. The primary rating of the transformer is 120 V. The total power load on the supply is only 24 W so the minimum transformer rating needed is 24 volt-amps. The ratio of total primary to total secondary turns needed is 120/68.88 or 1.742 to 1.

As in the preceding example, the transformer needed to satisfy a specific set of conditions often turns out to be one of a nonstandard size. The solution usually means using one with a higher than needed secondary voltage and then using a more elaborate filter that attenuates the dc voltage to the desired level or using a solid-state regulator on the output.

3.5 COMPLEX FILTERS

3.5.1 RC Filters

To further reduce the ripple in the output voltage, an additional filter stage can be used. One approach is to use a low-pass filter between the load resistor and the filter capacitor as shown in Figure 3.8. An exact analysis of this circuit is very tedious but it can be simplified by assuming that the Thévenin impedance seen looking into the terminals of the input capacitor, C_1, is small compared to the rest of the filter connected to C_1. The voltage, V_i, appearing across C_1 will contain a dc component and an ac component similar to the voltage approximated by the sawtooth wave. That voltage can be considered as a source and superposition can be used to determine the ac and dc components across the load resistor, R_L.

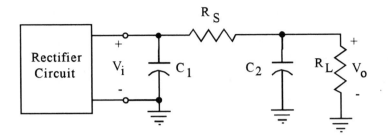

FIGURE 3.8
An RC filter.

The dc component of the voltage across the input to the filter will be attenuated by the voltage divider formed by R_S and R_L. The value of R_S is usually much smaller than R_L in order to minimize attenuation of the dc voltage:

$$V_o = V_i \times R_S/(R_S + R_L) \tag{3.8}$$

The ac analysis is a little more complicated. The ac signal is also attenuated by a voltage divider consisting of a complex impedance in series with R_L. The ratio of output to input ac voltage is the same as for a low-pass filter:

$$V_o = \frac{V_i \times R_L}{\sqrt{(R_S + R_L)^2 + (\omega C R_S R_L)^2}} \tag{3.9}$$

The ac voltage will contain the fundamental plus harmonics. The attenuation of the fundamental is the most important consideration. Since the frequency response of the filter falls off at the rate of 6 dB per octave, harmonics are usually ignored.

EXAMPLE 3.4

Assume the output of the filter of Example 3.3 is to be reduced to 40 V at 0.5 A by inserting an RC filter section between the capacitor and load resistor. Determine the series resistor needed and calculate the ripple in the output if both capacitors are 125 μF.

Solution
The 48 Vdc will be reduced to 40 V by the resistance voltage divider. The load current is to be 0.5 A so the load resistor must be 80 Ω and the series resistor will need to be 16 Ω to drop 8 V at 0.5 A. The voltage across C_1 is assumed to be 48 Vdc with a ripple factor of 0.03, so the rms ripple voltage is 0.03 $\times$ 48 or 1.44 V. The rms ripple voltage at the output is calculated using Eq. (3.9):

$$V_o = \frac{1.44 \times 80}{\sqrt{(16 + 80)^2 + (2\pi\,800 \times 0.000125 \times 16 \times 80)^2}}$$

$$= 0.01432$$

which gives a ripple factor at the output of $r = 3.58 \times 10^{-4}$.

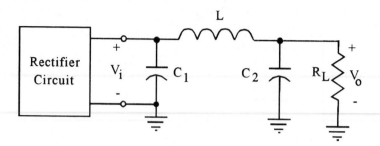

FIGURE 3.9
An LC filter.

3.5.2 LC Filters

For some applications the series resistor is replaced by an inductor as shown in Figure 3.9. The inductor should have a very low dc resistance and will offer substantial opposition to ac current. The analysis of the LC filter is simple if certain conditions are met so that simplifying approximations can be used.

The capacitive reactance of C_2 should be low enough at the fundamental frequency so that the load resistor can be ignored for ac analysis. The reactance of the inductor should be much larger than that of C_2. If those conditions are not met, the filter will not be very effective in reducing ripple in the output. The ratio of output to input ac is given by

$$V_o = V_i X_c / (X_L - X_c) \tag{3.10}$$

and this can be approximated by

$$V_o = V_i(X_c/X_L) \qquad \text{or}$$
$$V_o = V_i/\omega^2 LC \tag{3.11}$$

where ω is the radian frequency of the rectifier output. LC filters are not commonly used for power supplies operated from a 60-Hz line because the inductors tend to be physically large and expensive. For power supplies operated from higher frequency sources, inductors can be used effectively in filters.

3.6 VOLTAGE MULTIPLIERS

Certain applications require dc voltages in the kilovolt range but with load currents in the milliampere range. Voltage multipliers meet those requirements. The basic circuit is an extension of the half-wave rectifier with a capacitor filter. Figure 3.10 shows a voltage doubler circuit with no load resistor. After one or two cycles of the input voltage, C_1 will be charged up to the peak value of the input voltage. It will not discharge on the negative half cycles because D_1 will be reverse biased. On subsequent positive half cycles C_2 will charge through D_2 to twice the value of the input peak voltage. With no load connected to the circuit the capacitors can hold the charge indefinitely.

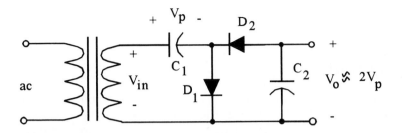

FIGURE 3.10
A voltage doubler.

Additional diodes and capacitors can be added as shown in Figure 3.11. Each capacitor except C_1 will receive a charge of twice the peak input voltage, so it is possible to obtain either an odd or even voltage multiplication. The peak inverse voltage rating required for the diodes is twice the peak value of the input voltage.

It may be necessary to use some current limiting resistance in the secondary of the input transformer because the effective capacitance seen by the transformer can be quite large. This is especially true when a transformer is used to step up the voltage to the multiplier circuit input. The design of a multiplier circuit usually represents a compromise between the number of stages and the input voltage from the transformer secondary. To reduce the possibility of arcing, diodes and capacitors should be derated by 50% of their peak voltage ratings. A 5000-PIV diode would be considered as rated at 2500 V. The peak value of the input voltage would then be limited to 1250 V and the transformer selected accordingly.

The capacitors in series in a multiplier form an equivalent filter capacitor. Since the load on a voltage multiplier is usually in the mega-ohm range, the capacitance value needed is often less than 1 μF. All capacitors in a multiplier string

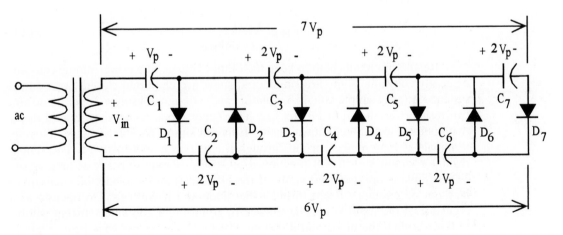

FIGURE 3.11
A voltage multiplier.

should have the same capacitance rating so the voltage distribution will be the same for all the capacitors except the one closest to the transformer secondary.

3.7 REGULATORS

The ideal voltage source is one that maintains a constant terminal voltage regardless of the current taken by the load. The ideal source could be said to have zero internal impedance. All real sources have some internal impedance and consequently the terminal voltage is not constant for all values of load current. Voltage regulation (VR) is a descriptive measure of the way the terminal voltage varies as the load varies. It is defined as the difference between the no load voltage and the full load voltage and is usually expressed as a percent of the full load voltage:

$$VR = \frac{(V_{NL} - V_{FL})}{V_{FL}} \times 100 \qquad (3.12)$$

where V_{NL} is the terminal voltage when the load current is zero and V_{FL} is the terminal voltage when rated load current is being supplied.

The lower the internal impedance of the supply, the lower VR will be. The rectifier type of power supply in general has poor regulation. As the load current is increased, the effect is the same as if the effective load resistance were reduced and this causes a degradation of the filtering. The final stage of filtering requires a large RC time constant in order to be effective. When the effective R is reduced, the peak-to-peak ripple in the output is increased and the dc output voltage is lowered.

A second descriptive measure of power supply performance is the line regulation or LR. It is a measure of the change in output voltage caused by changes in the input line voltage and is usually expressed as the percent change in output voltage per volt change in the input voltage with the load resistance constant. Line regulation is defined by

$$LR = \frac{\Delta V_o / V_o \times 100\%}{\Delta V_{in}} \qquad (3.13)$$

Regulator circuits help to keep the output voltage of a power supply constant or nearly constant. Some are simple and some are elaborate closed-loop control systems. A fairly simple voltage regulator can be designed using a Zener diode and two resistors as shown in Figure 3.12. The Zener diode is a reverse-biased diode that operates in the breakdown region. It has an almost constant voltage across its terminals as long as the current through it is within a certain range. The simple regulator circuit shown will function to keep the output constant as long as the Zener diode is operating normally. If the load resistance is made too small there may not be enough current through the diode and it will cease to operate as a regulator. If the input voltage is allowed to go too high, too much current will try to flow through the diode and it will be destroyed due to excessive heat.

The simple regulator circuit will keep the load voltage constant as long as the diode operates in the breakdown region. The current supplied by the source, I_S, is

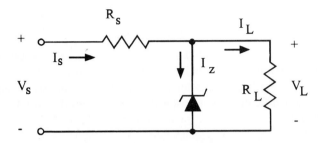

FIGURE 3.12
A Zener diode regulator.

equal to the sum of the Zener current and the load current:

$$I_S = I_Z + I_L \qquad \text{or} \qquad I_Z = I_S - I_L \tag{3.14}$$

The source current depends on the difference between the input and Zener voltages and the series resistance. The calculation of the series resistance is a very important step in the design of a Zener regulator circuit. The maximum value of R_S is the value that just causes the Zener current to drop to zero. The worst case occurs for minimum input voltage and maximum load current:

$$R_{S_{max}} = \frac{V_{S_{min}} - V_L}{I_{L_{max}}} \tag{3.15}$$

This assumes that the Zener will regulate until the current through it drops to zero, which is not really true. Minimum values of Zener currents are not usually given on data sheets but maximum values are given or can be calculated from the maximum power rating. A conservative approach is to assume the minimum current for regulation to be equal to 5% of the maximum.

The Zener diode does have some internal resistance, which is determined by the slope of the volt-amp characteristic curve in the breakdown region. This resistance can cause the terminal voltage to vary as current through the device is varied. The Zener resistance is often ignored but its effect on the performance of a Zener regulator circuit should be calculated to be on the safe side. The Zener can be represented as a voltage source in series with a small resistance.

EXAMPLE 3.5

A 6-V Zener with an internal resistance of 2 Ω is to be used to keep the voltage across a 120-Ω load resistor constant as the input voltage varies between 12 and 18 V. The maximum and minimum currents through the Zener are to be 60 and 5 mA, respectively. (a) Calculate the series resistance needed ignoring the Zener resistance. (b) Consider the effect of the Zener resistance and calculate the load voltage for inputs of 12 and 18 V.

Solution
The series resistance is calculated assuming 5 mA to be the minimum Zener current. The load current is to be 50 mA so

$$R_S = (12 - 6)/0.055 \quad \text{or} \quad 109 \ \Omega$$

The circuit with $R_S = 109 \ \Omega$ is shown in Figure 3.13. Note the model of the Zener as a 6-V source in series with a 2-Ω resistor. If the Zener resistance were not present, the output would be 6 V as long as there was at least 5 mA through the Zener diode. With the Zener resistance included, the analysis is slightly more complicated but not difficult. The use of either mesh or nodal equations will yield the following results:

$$V_S = 12 \ \text{V}, \qquad V_L = 6.0096 \ \text{V}$$
$$V_S = 18 \ \text{V}, \qquad V_L = 6.116 \ \text{V}$$

The variation in output voltage represents a voltage regulation of less than 5%. The Zener current should be checked to see that it is within limits: $I_Z = (V_L - V_Z)/2 \ \Omega$.

For $V_L = 6.0096$ V, $I_Z = 4.80$ mA and for $V_L = 6.116$ V, $I_Z = 58$ mA.

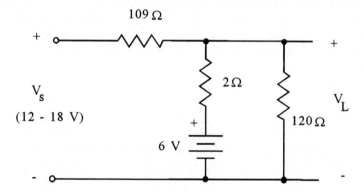

FIGURE 3.13
Regulator example circuit.

The minimum Zener current is a little less than 5 mA, which probably will not hurt anything and the maximum current is within the given limit. The Zener impedance did not affect the device performance appreciably. The Zener resistance can cause problems in some applications because it varies with device voltage rating. Low-voltage Zeners have lower impedances than higher voltage devices. The design of Zener regulator circuits is fairly simple but one must be careful because not every set of conditions given for Zener voltages and current can be realized. A general equation can be derived that relates the load voltage to the input voltage, the Zener voltage, and the resistors in the circuit. Equation (3.16) shows the relationship among those values:

$$V_L = \frac{V_S/R_S + V_Z/R_Z}{(1/R_S + 1/R_Z + 1/R_L)} \qquad (3.16)$$

Also

$$I_Z = \frac{V_L - V_Z}{R_Z}$$

The input voltage must be large enough to make $V_L = V_Z$, otherwise the Zener diode would be operating in the forward-bias mode as a regular diode.

As a general rule, if the Zener resistance value is less than 1% of both the load resistance and the series resistance, the circuit to the left of the load can be replaced by an equivalent circuit consisting of a voltage source whose voltage and equivalent resistance are approximately those of the Zener diode. The approximate equivalent circuit would not be valid over a wide range of input voltages. For circuits whose input voltage varies over a wide range, a single Zener diode will seldom be adequate as a regulator. It is usually necessary in those cases to go to a more elaborate regulator using transistors or operational amplifiers.

Zener diodes usually perform well where the load resistance varies and the input voltage is relatively constant.

3.7.1 Transistor Regulators

Zener diodes by themselves are limited in their capacity to function as voltage regulators due to their limited current capacity. They can serve as a reference voltage source in more complex regulator circuits. Figure 3.14 shows a regulator circuit with an emitter-follower transistor used to supply current to a regulated voltage load. The transistor is referred to as a pass transistor since all the load current passes through it. The Zener diode will keep the transistor base at a constant potential. The base-to-emitter voltage is

$$V_{BE} = V_Z - V_L \tag{3.17}$$

Since V_Z is considered constant, any change in the output voltage will cause a change in V_{BE}. If V_L decreases V_{BE} will increase, causing the transistor to pass

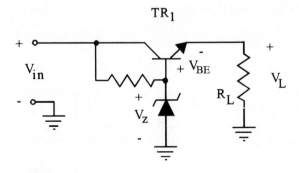

FIGURE 3.14
A transistor regulator.

more load current and V_L to increase. If V_L tries to increase beyond the regulated value, V_{BE} will decrease and reduce the load current so that V_L will decrease.

EXAMPLE 3.6

For the regulator circuit of Figure 3.14 assume $V_Z = 9$ V, $R = 150$ Ω, and $V_{BE} = 0.7$ V. Calculate load voltage and current, the collector-to-emitter voltage drop, and the power dissipated by the pass transistor if (a) $V_{in} = 12$ V and $R_L = 100$ Ω and (b) $V_{in} = 18$ V and $R_L = 75$ Ω.

Solution
As long as $V_{BE} = 0.7$ V and $V_Z = 9$ V, the output voltage should be 8.3 V. For $V_{in} = 12$ V and $R_L = 100$ Ω, the load current will be 83 mA and $V_{CE} = 3.7$ V. The power dissipated is equal to $V_{CE} \times I_L$ and in this case is 307 mW.

For $V_{in} = 18$ V and $R_L = 75$ Ω, $I_L = 111$ mA and $V_{CE} = 9.7$ V. The power dissipated is 9.7×0.111 or 1.08 W. Zener current should also be checked for the anticipated load or input conditions. The current flowing through R is the Zener current plus the base current of the transistor. Assume that the transistor has a dc beta of 100. The base current will vary from 0.83 to 1.11 mA for the two sets of conditions. For an input of 12 V, there will be a 3-V drop across R and 20 mA flowing through it. This makes the Zener current a little over 19 mA. For an input voltage of 18 V, there will be a 9-V drop across R and the current through it will be 60 mA of which almost 59 mA will flow through the Zener.

3.7.2 Op-Amp Regulators

An operational amplifier can be incorporated into a regulator circuit to provide more gain in the control loop and also a higher value of regulated output voltage than can be obtained with a Zener and a pass transistor. Figure 3.15 shows a regulator circuit with an operational amplifier. The Zener voltage is applied to the noninverting input and a voltage proportional to the output, V_L, is applied to the inverting input:

$$V^- = V_L \times R_2/(R_1 + R_2) \qquad \text{and} \qquad V^- = V^+ = V_Z$$

so

$$V_L = (1 + R_1/R_2) \times V_Z$$

Resistor R_3 is picked to maintain the proper current through the Zener to keep V_Z constant. The operational amplifier supplies base current for the pass transistor. If V_L tries to increase, V_{BE} of the pass transistor will decrease, causing the load current to decrease and V_L to decrease. If V_L were to decrease, the process would be reversed.

The op-amp regulator circuit can be designed for a wide range of voltages and can be made almost insensitive to temperature variations. Zener diodes operating at or near 6 V have temperature coefficients that are very close to zero.

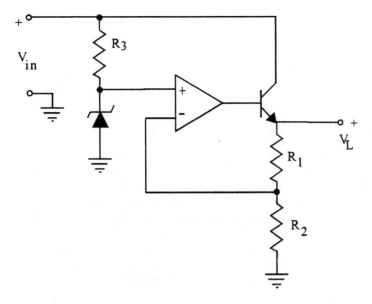

FIGURE 3.15
An op-amp regulator.

EXAMPLE 3.7

We want to provide a regulated output voltage of 15 V for an input voltage from 18 to 30 V. An op-amp regulator circuit is to be used that incorporates a 6-V Zener diode. Assume the Zener current must be at least 15 mA for proper operation.

Solution
From Eq. (3.15), $R_3 = (18 - 6)/0.015 = 800 \ \Omega$. The circuit requires a voltage gain of 2.5 since the output is to be 15 V and the reference is 6 V.

$$1 + R_1/R_2 = 2.5 \quad \text{or} \quad R_1 = 1.5R_2$$

where R_1 and R_2 should be large enough so that the load current is several times larger than the current through R_1 and R_2. Let $R_2 = 10 \ \text{k}\Omega$ and $R_1 = 15 \ \text{k}\Omega$.

3.8 REGULATOR PROTECTION

3.8.1 Current Limiting

It is a good practice to limit the current that can be drawn from a regulated supply. If the output should become shorted, the pass transistor could be destroyed by the heat generated while the transistor is forward biased and the input voltage appears between the collector and the emitter.

In Figure 3.16 Q_2 and R_S are included for current limiting. During normal operation, load current flows through R_S but does not produce enough voltage drop

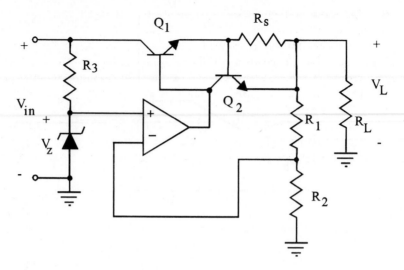

FIGURE 3.16
Current limiting.

to turn on Q_2. If the output became shorted, the maximum current that could flow through R_S would make the drop across R_S equal to about 0.7 V, which is the normal value of V_{BE} when Q_2 is conducting. The value of R_S is chosen to limit I_{SC} (short circuit current).

$$R_S = 0.7//I_{SC} \tag{3.18}$$

Under normal operating conditions Q_1 supplies the load current and the op-amp supplies base current for Q_2. When the load current becomes great enough to cause the base-to-emitter voltage of Q_2 to be about 0.7 V, Q_2 will begin conducting enough collector current so that the base current of Q_1 cannot increase appreciably and there can be no significant increase in load current.

The design of a current-limited regulator is very straightforward. The value of regulated output voltage and the Zener reference voltage are chosen first. The voltage divider resistors are chosen to provide the proper voltage to be fed back and compared to the reference. The current through the divider should be almost negligible compared to the load current. The operational amplifier should have a high open-loop gain and must be able to supply base current for Q_1 and collector current for Q_2. The collector current rating of Q_2 should be about the same as the base current required for Q_1 in order for the turning on of Q_2 to be able to prevent additional current from flowing into the base of Q_1.

3.8.2 Foldback Limiting

Regular current limiting protects the pass transistor by limiting the current that can flow if the load becomes short circuited. Under shorted load conditions, almost all of the input voltage appears across the pass transistor and so it must dissipate a large amount of power.

Foldback limiting functions to reduce both the load voltage and load current if the load current exceeds a certain value. If the load becomes a short circuit, the load current is reduced to a low value. This not only protects the pass transistor but the load as well. Figure 3.17 shows a regulator circuit modified for foldback limiting. This circuit requires two more resistors than the simple limiter circuit. The voltage across the divider consisting of R_4 and R_5 is made almost equal to the output voltage. The emitter of Q_2 is now connected to V_L and the base to the junction of R_4 and R_5. An expression for V_{BE} for Q_2 is given in Eq. (3.19):

$$V_{BE} = \frac{I_L R_S R_5 - V_L R_4}{R_4 + R_5} \tag{3.19}$$

which can be solved for I_L:

$$I_L = \frac{V_{BE}(R_4 + R_5) + V_L R_4}{R_S R_5} \tag{3.20}$$

The load current is dependent on both the load voltage and V_{BE} of Q_2. If V_L is kept constant and the current allowed to increase, a maximum will be reached

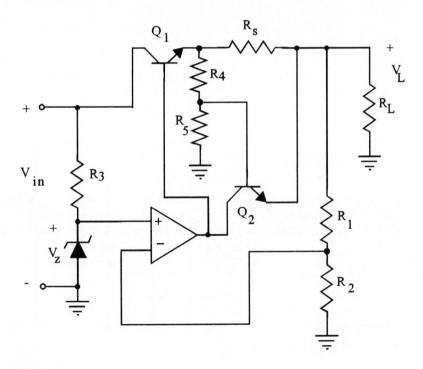

FIGURE 3.17
Foldback limiting.

when V_{BE} is about 0.7 V. If the output voltage is allowed to fall, the load current is reduced also. The load current with the load short circuited is

$$I_{SC} = V_{BE}(R_4 + R_5)/R_S R_5 \tag{3.21}$$

which is less than the maximum given by Eq. (3.20). The value of V_{BE} can never exceed 0.7 V by any significant amount, and under normal operating conditions it will be less than 0.7 V.

3.8.3 Overvoltage Protection

Since excessive heat poses the greatest threat to semiconductor devices, most protective circuits are designed to limit current under abnormal conditions. Many semiconductor loads tend to be voltage sensitive and momentary overvoltages can wipe out an entire load system. The overvoltage protection circuit most often used for power supplies is known as a *crowbar circuit*. A threshold device is used to sense an overvoltage and throw a fast-acting short circuit (crowbar) across the output terminals. A silicon-controlled rectifier (SCR) is used for the short circuit. When the SCR is off it has a high impedance between its anode and cathode and when it is on it has about 1 V between the anode and cathode. Once switched on the SCR will stay on until the anode potential is removed. In most power supply applications this will mean having to turn off the supply.

Figure 3.18 shows a fairly simple single-transistor crowbar circuit. The transistor and the Zener diode provide a well-defined threshold for firing the SCR. Under normal operating conditions Q_1 is biased off. The emitter of Q_1 will always be less than V_L by the amount of the Zener voltage. If V_L increases by ΔV volts, the voltage at the emitter increases by the same amount but the voltage at the base will increase less due to the resistive voltage divider. The net result is that an increase in output

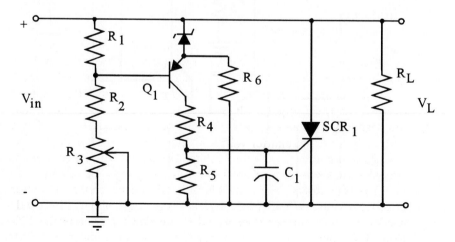

FIGURE 3.18
An SCR crowbar circuit.

voltage causes a decrease in base-to-emitter voltage of Q_1. With the proper values chosen for R_1, R_2, and R_3 and the Zener, Q_1 can be made to turn on whenever V_L increases by a certain value of ΔV. The RC circuit in parallel with the gate-to-cathode terminals serves to integrate out the very narrow noise spikes. The Zener diode must be operated at a current level that will not cause the output voltage to vary appreciably with current since the current that develops the firing voltage at the gate flows through the Zener and Q_1.

3.9 SWITCHING SUPPLIES AND REGULATORS

Series regulators are very useful for many applications, but the power dissipation of the pass transistor may be excessive for some applications. If the pass transistor is turned off, it will dissipate no power. In a switching supply the pass transistor is used as a switch and that switch controls the duty cycle of the supply. The basic principle of the switching supply is shown in Figure 3.19.

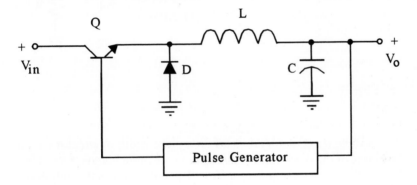

FIGURE 3.19
A basic switching regulator.

The pass transistor is switched on by the train of pulses. When it is on, the emitter voltage is essentially equal to the dc input voltage and when the transistor is off the emitter voltage is zero. The average value of the emitter voltage is determined by the supply voltage and the duty cycle of the pulse train applied to the base. For example, if the duty cycle of the pulse train is 25% and the supply voltage is 20 V, the emitter voltage will have an average value of 5 V.

The output voltage from the filter circuit will be a dc voltage whose value is very close to the average value of the emitter voltage. There may be a small ripple in the output voltage since the filtering will not be perfect. LC filtering is quite effective because the switching frequency is generally about 20 kHz and the inductor needed is much smaller than would be needed for filtering the 120-Hz component of the output of a rectifier power supply operating from a 60-Hz line. The diode is necessary to provide a path for inductor current when the transistor is switched off.

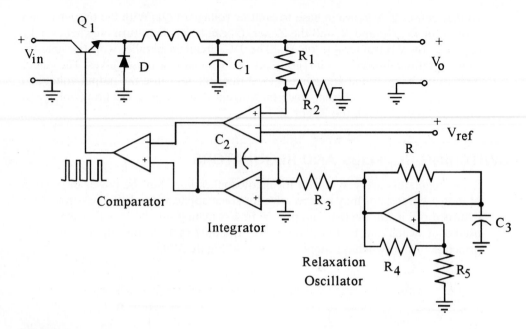

FIGURE 3.20
Switching regulator circuitry.

Switching regulators depend on a pulse width modulator circuit to make the duty cycle of the pulse train dependent on the input voltage level. Figure 3.20 shows the circuitry for a switching regulator. The pulse train is the output of a comparator that has a triangular wave applied to the noninverting input and a dc voltage proportional to the difference between the output and a reference level. If the output increases the pulse duty cycle is lowered, causing the output to decrease.

The triangular wave is the output of an integrator whose input is a square wave produced by a relaxation oscillator. The reference could be made variable if desired. When the regulator is working properly, the output voltage will be equal to the reference voltage multiplied by $(1 + R_1/R_2)$.

Low-power switching regulators are available in integrated circuit form.

3.10 THREE-TERMINAL REGULATORS

Several companies provide three-terminal fixed-voltage regulators in an integrated circuit (IC) package. These are very easy to use. One connects the proper terminals to V_{in}, V_o, and ground.

The LM340 series of regulators manufactured by National Semiconductor is typical of the IC regulators available. The package contains a fixed reference, a pass transistor, and current-limiting and thermal shutdown circuitry. To improve the transient response of the regulated voltage output, an output bypass capacitor of 0.1 μF can be connected between the output terminal and ground. An input bypass capacitor of 0.22 μF is often used to prevent oscillations within the integrated circuit of the regulator. The input bypass capacitor is unnecessary if the regulator package is located within a few inches of the filter capacitor of the unregulated supply.

The LM340 series includes regulators for the following output voltages: 5, 6, 8, 10, 12, 15, 18, and 24 V. The LM340-12 produces an output of 12 V. There is also an LM320 series of regulators that produces the negative of the corresponding voltages produced by the LM340 series. The LM320-12 produces an output of -12 V. The input voltage for devices in the LM340 family must be at least 2 or 3 V greater than the regulated output voltage. The maximum input voltage is determined by the power dissipation limits of the device. An LM340-5 will regulate for an input voltage range from approximately 7 to 20 V.

Ripple rejection of the LM3XX series of regulators is very good and varies from 60 to 80 dB depending on the device. An LM340 regulator is shown in Figure 3.21 as a variable voltage regulator. The common terminal is connected to the top of R_2 and the regulated output voltage appears across R_1. The voltage from pin 2 to ground is given by

$$V_L = V_r(1 + R_2/R_1) + I_Q R_2 \tag{3.22}$$

For the LM340 series I_Q has a maximum value of 8 mA and only varies 1 mA for all load and line changes so V_L is essentially regulated and variable. The same circuit can be used to obtain a controlled current output.

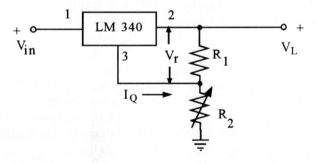

FIGURE 3.21
A variable voltage regulator.

3.11   ADJUSTABLE IC REGULATORS

The 723 IC regulator is an adjustable voltage regulator. It is more complicated than the three-terminal regulators but it is extremely versatile. It can supply a regulated output voltage from 2 to 37 V and if external pass transistors are used it can handle load currents up to 10 A. It can be connected to produce either positive or negative output voltages and can even be used as a switching regulator.

3.12 DC-TO-DC CONVERSION

The fundamental idea is to provide a dc voltage at a higher level than is readily available. Many digital circuits work from a +5-V source but there will be times when other dc voltages are needed and the dc-to-dc conversion technique can be used to good advantage. The fundamental idea is shown in Figure 3.22.

The square-wave oscillator is usually a relaxation oscillator using an IC comparator as shown in Figure 3.23. The comparator has a high open-loop gain and its output voltage will be at either the positive or negative saturation level, which is usually about a volt less than the supply voltage. The voltage at the noninverting input is simply the output voltage multiplied by the resistance voltage divider ratio. That ratio is usually designated by B and is equal to $R_2/(R_1 + R_2)$. The period of the square-wave output of the oscillator depends on B and on the values chosen for R and C. The oscillator shown in Figure 3.23 uses both a positive and a negative supply and is assumed to have equal positive and negative saturation voltages.

The voltage across the capacitor is always going to be an exponential as it tries to charge to V_{sat} or $-V_{sat}$. It never reaches V_{sat} because the output voltage switches every time V_C reaches (or slightly exceeds) BV_{sat}. The time required to complete one cycle is given by Eq. (3.23):

$$T = 2RC \ln \left(\frac{1+B}{1-B} \right) \tag{3.23}$$

and of course the frequency is given by $f = 1/T$. Note that the frequency is independent of the supply voltage or the saturation voltages.

In practice, both positive and negative voltages will probably not be available to power the dc-to-dc converter. The relaxation oscillator circuit can be modified to operate from a single supply. The feedback ratio B will have an effect on the duty cycle as well as the frequency. The duty cycle should be 50% in order to prevent unbalanced currents in the transformer winding. One method used to insure a 50% duty cycle is to feed the output of the relaxation oscillator to a J-K flip-flop whose output will be a square wave with a 50% duty cycle. The frequency of that

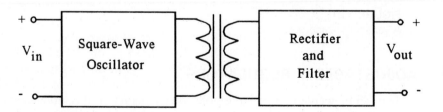

FIGURE 3.22
Basic dc-to-dc conversion scheme.

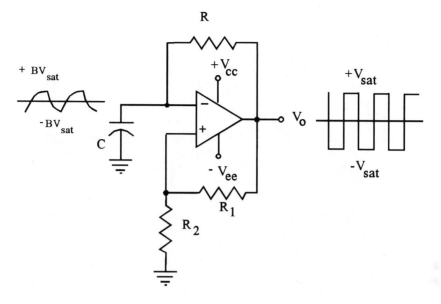

FIGURE 3.23
A relaxation oscillator.

square wave out of the flip-flop will be one-half the frequency of the relaxation oscillator.

The transformer used should have a rectangular hysteresis loop if possible so that the voltage from the transformer secondary is a reasonable square wave. Square-wave voltages are easily rectified and the resultant dc does not require much filtering. The higher the frequency of the square wave, the easier it is to filter the output of the rectifier, but it is difficult to obtain magnetic cores to reproduce a square wave at very high frequencies. A good compromise for the frequency is 20 kHz.

Some sort of buffer or amplifier is needed between the square-wave oscillator and the transformer because the transformer will probably require more current than the oscillator can deliver. A general dc-to-dc converter circuit is shown in Figure 3.24. For a specific application it would be necessary to work out the details such as transformer turns ratio and the values of the emitter and collector resistors for Q_1, which is a phase splitter and will produce a signal at the emitter that is 180° out of phase with the signal at the collector. These two signals feed the bases of the push–pull amplifier made up of Q_2 and Q_3. The output impedance of the two signal sources driving Q_2 and Q_3 are different. If that causes an imbalance, Q_2 and Q_3 could be replaced by Darlingtons.

If the output voltage of the converter is to be held to within a small percent of the nominal value, a three-terminal solid-state regulator can be connected to the output of the rectifier and filter.

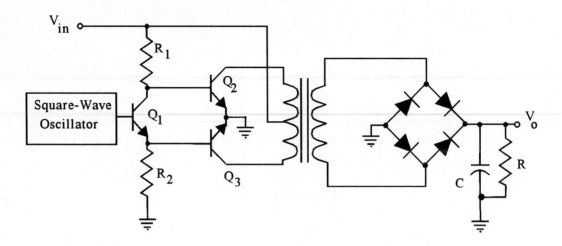

FIGURE 3.24
A dc-to-dc conversion circuit.

3.13 INVERTERS

Inverters convert dc to ac. Rotating machines did that job for years but modern electronic devices and circuits have taken over. Inverters are not in widespread use but there are some important applications. An electronic inverter can produce a variable frequency ac voltage for controlling the speed of induction motors, or an inverter can be used to provide a backup source of three-phase power for emergencies.

The three-phase full-wave bridge rectifier circuit can be modified to function as a three-phase inverter. Replace the diodes with SCRs and the dc load with a dc source as shown in Figure 3.25. The SCRs are turned on and off sequentially by a timing signal, which will determine the frequency of the three-phase ac output. The individual SCRs are turned on and off one at a time so that the current in each

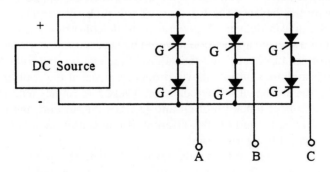

FIGURE 3.25
Basic three-phase inverter.

line reverses during each cycle. There are six distinct steps in each cycle and two different possible timing schemes. One scheme has two SCRs conducting at a time and the other has three SCRs conducting at a time. The details of the switching are not shown. The line currents will not be very sinusoidal in appearance but they will have the proper phase relationship for a three-phase source. Filtering can be used if necessary to obtain output voltages that are more nearly sinusoidal.

The preferred method of turning off the SCRs is to cause the anode current to go to zero by means of forced commutation. Removal of the anode supply could be accomplished by a switching arrangement but that is not very practical because it is likely that a lot of arcing and noise voltages would be generated.

4

RESISTANCE*

The purpose of this chapter is to present the basic definitions of resistance and the resistor. The practical characteristics of resistor types and selection criteria for specific applications are also presented.

4.1 BASIC CONCEPTS

The basic electrical definition of a two-wire element is given by specification of the dynamic (time-dependent) relationship between voltage across and current through the device. Once the electrical behavior definition is understood it is often possible to specify a physical definition, which depends on size, material, etc., and not on current and voltage.

4.1.1 Dynamic Definition of Resistance

Resistance is defined dynamically as follows: "At every instant of time the ratio of voltage across and current through a resistance element will be constant and the voltage and current will be in phase." Furthermore, this definition holds regardless of how the element is connected, i.e., resistance has no inherent polarity.

Ohm's Law Elements that obey this definition are called *resistors*. The constant that represents the ratio of voltage to current is called the *resistance* and is labeled R. The definition is expressed in equation form by Ohm's law:

$$R = \frac{V}{I} \tag{4.1}$$

---

* This chapter was written by Dr. Curtis D. Johnson, College of Technology, University of Houston.

where V = voltage across the element in volts (V)
           I = current through the element in amperes (A)
           R = resistance of the element in ohms (Ω).

Equation (4.1) serves as a definition of the *unit* of resistance, which is called the *ohm,* and is symbolized by a capital Greek omega, Ω. One ohm equals one volt per ampere, i.e., a current of 1 ampere through a resistance of 1 ohm will produce a voltage across the resistor of 1 volt.

Figure 4.1 shows the schematic symbol for the resistor and the voltage across and current through the element. Note the relationship between voltage polarity and current (conventional) direction. The voltage across or *dropped* across the resistor is positive on the end into which the current enters.

FIGURE 4.1

Schematic symbols for a resistor and the IR characteristic curve.

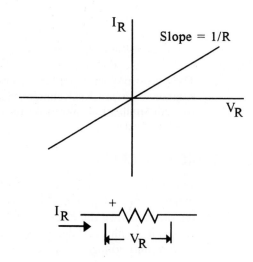

The schematic symbol used in Figure 4.1 is the most common. In some cases, however, a small rectangle is also used for the resistor.

The ideal resistor has zero inductance and zero capacitance. The current versus voltage (*I-V*) curve shown in Figure 4.1 is a straight-line (linear) curve of I versus V with a slope of $(1/R)$.

EXAMPLE 4.1   A resistor in the middle of a TV UHF tuner is found to have a voltage across it of 3.45 V and a current through it of 6.9 mA at a certain instant of time. What is the resistance?

Solution
From Eq. (4.1) and the fact that the resistor *I-V* does not depend on time we find,

$$R = V/I = 3.45 \text{ V}/6.9 \text{ mA} = 500 \ \Omega$$

EXAMPLE 4.2   A 4-kΩ resistor has a voltage dropped across it of 25 V. What is the current?

Solution
From Eq. (4.1) we solve for the current:

$$I = V/R = 25 \text{ V}/4 \text{ k}\Omega = 6.25 \text{ mA}$$

EXAMPLE 4.3   A 2.5-MΩ resistor carries a current of 500 μA. How many volts appear across the element?

Solution
Solving Eq. (4.1) for voltage provides

$$V = IR = (500 \text{ } \mu\text{A})(2.5 \text{ M}\Omega) = 1.25 \text{ kV}$$

Power Dissipation   The fact that the current and dropped voltage of a resistor are in phase indicates that the resistor is taking energy from the circuit at a constant rate. This rate is called the electrical power dissipated and is measured in joules/second (J/s) or watts (W). This power shows up as a *heating* of the resistor as it dissipates the energy to its surroundings. The amount of power is given by the basic relation

$$P = IV \qquad (4.2)$$

where P = power dissipated in watts (W)
I = current through the resistor (A)
V = voltage across the resistor (V).

Equation (4.2) can be expressed in alternative forms by using Ohm's law,

$$P = I^2R \qquad \text{or} \qquad (4.3)$$
$$P = V^2/R \qquad (4.4)$$

EXAMPLE 4.4   Specify the power dissipated in each of the earlier examples.

Solution
For the first example, we use Eq. (4.2) directly:

$$P_1 = IV = (3.45 \text{ V})(6.9 \text{ mA}) = 23.8 \text{ mW}$$

For the second case we will use Eq. (4.3):

$$P_2 = I^2R = (6.25 \text{ mA})^2(4 \text{ k}\Omega) = 156.25 \text{ mW}$$

Finally, for the third example we will use Eq. (4.4):

$$P_3 = V^2/R = (1.25 \text{ kV})^2/2.5 \text{ M}\Omega = 625 \text{ mW}$$

4.1.2 Physical Definition of Resistance

In nearly all cases of current conduction by some element, the current itself is a flow of electrons through the material. In fact, the distinction between a good conductor, semiconductor, and insulator is in the number of electrons in the material free to carry current. In metals there are a great number of such electrons, in semiconductors fewer, and in insulators virtually none.

When these electrons flow through an element they make collisions with atoms of the material, causing atomic vibrations. This amounts to a loss of energy by the electrons and the resulting vibrations of atoms correspond to a rise in the material temperature. The collisions amount to a loss of energy by the current and hence a resistance to current flow.

The electrical resistance of a material is a consequence of the conditions described in the two preceding paragraphs. Even in good conductors with lots of electrons, collisions occur and so some resistance is present.

The *resistivity* of a material is a constant that describes the inherent resistance of the material to current flow. This value expresses the resistance per unit area for a unit length of the material. The units of resistivity are ohm-meters (Ω-m) and Table 4.1 shows the resistivities for an assortment of materials.

The physical definition of resistance is in terms of the material physical size and its resistivity. Figure 4.2 shows a piece of material of constant cross section with area A and of length L. The resistance of this material is given by

TABLE 4.1
Resistivity

Material	Resistivity (Ω-m)
Silver	1.6×10^{-8}
Copper	1.7×10^{-8}
Aluminum	2.8×10^{-8}
Nickel	7.8×10^{-8}
Nichrome	100×10^{-8}
Carbon	$\leq 190 \times 10^{-8}$ (varies)

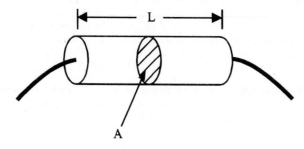

FIGURE 4.2
Physical definition of electrical resistance.

$$R = \rho \frac{L}{A} \tag{4.5}$$

This is the same value that would be obtained if the ratio of voltage across and current through the device were measured and divided, as in Eq. (4.1).

EXAMPLE 4.5

For a meter shunt, it is necessary to provide a resistance of exactly 0.356 Ω. How can this be constructed from copper wire?

Solution
We see from Table 4.1 that copper has a resistivity of 1.7×10^{-8} Ω-m. In the Appendix 2 we find that #30 AWG copper wire has a diameter of 0.010 in. All we have to do is determine the proper length to obtain 0.356 Ω. First we find the cross-sectional area in square meters. Diameter, D, is

$$D = (0.010 \text{ in.})(0.0254 \text{ m/in.}) = 2.54 \times 10^{-4} \text{ m}$$

Then,

$$A = \pi D^2/4 = (3.14159)(2.54 \times 10^{-4})^2/4$$
$$= 5.067 \times 10^{-8} \text{ m}^2$$

Now, from Eq. (4.5),

$$L = RA/\rho = (0.356 \ \Omega)(5.067 \times 10^{-8} \text{ m}^2)/(1.7 \times 10^{-8} \ \Omega\text{-m})$$

or

$$L = 1.061 \text{ m}$$

So, a length of 1.061 m of #30 AWG wire will give the required resistance. Of course, we can wind this into a small coil. Wire of any other AWG could be used by finding the appropriate length.

4.1.3 Combinations of Resistors

In many cases it is desirable to use series and parallel combinations of resistors. This is done to construct special circuits such as voltage or current dividers and to provide nonstandard values of resistance.

Series Combination   The net resistance of a series of resistors, as shown in Figure 4.3, is found as a simple sum of each resistance in the series:

$$R_{net} = R_1 + R_2 + R_3 \tag{4.6}$$

Note that the current through each resistor is the same but the voltage across each is given by a product of its resistance times the common current.

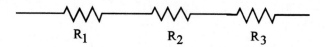

FIGURE 4.3
Series of resistors.

EXAMPLE 4.6

Resistors R_1, R_2, and R_3 in Figure 4.3 have values of 20, 4.7, and 1 kΩ, respectively. The series combination is connected to a 30-V power supply. What is the net resistance, the current, and the voltage across each resistance?

Solution

The net resistance is given by the sum, as in Eq. (4.6),

$$R_{net} = 20\,k\Omega + 4.7\,k\Omega + 1\,k\Omega$$
$$= 25.7\,k\Omega$$

The current is now found from Ohm's law, Eq. (4.1):

$$I = V/R_{net} = 30\,V/25.7\,k\Omega = 1.167\,mA$$

The voltage across each is found from Ohm's law also, but now solving for the voltage,

$$V_1 = IR_1 = (1.167\,mA)(20\,k\Omega)\ = 23.34\,V$$
$$V_2 = IR_2 = (1.167\,mA)(4.7\,k\Omega) =\ \ 5.49\,V$$
$$V_3 = IR_3 = (1.167\,mA)(1\,k\Omega)\ \ \ =\ \ 1.17\,V$$

Parallel Combinations Figure 4.4 shows three resistors in parallel. The *equivalent* resistance of such a parallel combination is given by the inverse relation,

$$\frac{1}{R_{eq}} = \frac{1}{R_1} + \frac{1}{R_2} + \frac{1}{R_3} \tag{4.7}$$

We call this the *equivalent resistance* because the parallel combination acts like a resistor of this value. Note that the equivalent resistance will always have a value less than that of the smallest resistor in the parallel combination.

The voltage across all resistors will be the same but the current through each will be determined by Ohm's law, Eq. (4.1).

EXAMPLE 4.7

The three resistors in Figure 4.4 have values of 82, 47, and 12 kΩ, respectively. A 15-V source is placed across the parallel combination. What is the equivalent resistance, the net current drawn from the source, and the current through each resistor?

FIGURE 4.4
Parallel arrangement of resistors.

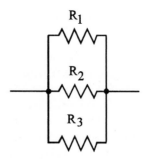

Solution
The equivalent resistance can be found from Eq. (4.7):

$$\frac{1}{R_{eq}} = \frac{1}{82 \text{ k}\Omega} + \frac{1}{47 \text{ k}\Omega} + \frac{1}{12 \text{ k}\Omega}$$

$$= 1.22 \times 10^{-5} + 2.13 \times 10^{-5} + 8.33 \times 10^{-5}$$

$$R_{eq} = 8.56 \text{ k}\Omega \qquad (\textit{Note:} \text{ less than 12 k}\Omega.)$$

The net current is found from Ohm's law, the 15-V source, and the equivalent resistance:

$$I_{net} = V/R_{eq} = 15 \text{ V}/8.56 \text{ k}\Omega = 1.75 \text{ mA}$$

The current through each resistor is found from Ohm's law directly since each has the 15-V source across it:

$$I_1 = 15 \text{ V}/R_1 = 15 \text{ V}/82 \text{ k}\Omega = 0.18 \text{ mA}$$
$$I_2 = 15 \text{ V}/R_2 = 15 \text{ V}/47 \text{ k}\Omega = 0.32 \text{ mA}$$
$$I_3 = 15 \text{ V}/R_3 = 15 \text{ V}/12 \text{ k}\Omega = 1.25 \text{ mA}$$

Note that for the special case of two resistors in parallel, Eq. (4.7) can be shown to reduce to the form:

$$R_{eq} = \frac{R_1 R_2}{R_1 + R_2} \qquad\qquad (4.8)$$

This equation only works for two resistors in parallel, and cannot be extended to three or more resistors.

4.2 PRACTICAL RESISTOR CHARACTERISTICS

Real resistors have properties beyond simply electrical resistance. It is important to understand what these characteristics are and how they affect circuit performance.

4.2.1 Identification

Resistors come in many sizes and shapes but there are two basic forms: axial resistors and power resistors. *Axial resistors* are small cylindrical objects with axial wires, often also with colored bands, as shown in Figure 4.5(a). This is the most common form to be encountered in electronics.

Power resistors are designed to operate at elevated temperatures without failure. They have many forms but are often encapsulated in ceramic or metal cases with lugs for connection as shown in Figure 4.5(b). This form is found in power supplies and high-power electrical and electronic equipment.

There is no uniformity to this rule. Some power resistors may have axial leads and some precision, low-power resistors have lugs instead of axial leads.

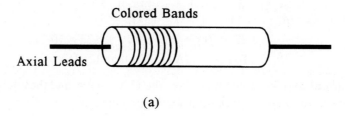

(a)

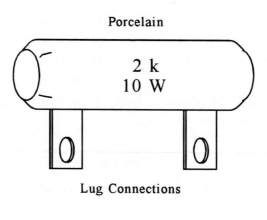

(b)

FIGURE 4.5
(a) Cylindrical resistor with axial leads and (b) ceramic encapsulated resistor with lug lead.

Marking   The resistance of a resistor, as well as other characteristics, must be marked on the device so it can be identified effectively. The resistance is either simply written on the device or coded colored bands are used.

If the value of resistance is written on the resistor, it will be given in ohms or the prefix designation for kilo- (k or K) or mega- (M) will be used. Thus a 1650-ohm resistor might be marked as 1650, 1650 Ω, 1.65 k, 1.65 K, 1.65 kΩ, or 1.65 KΩ. The proper prefix for kilo- is a lowercase letter k, but some manufacturers continue to use the capital letter K.

The most common marking used is a set of coded, colored bands placed closer to one end of the resistor. Table 4.2 gives the digits and multipliers associated with the colors. Figure 4.6 shows a resistor with the colored bands identified. The first two bands give the digits or *significant figures* of the resistance and the third band gives the base 10 multiplier by which the digits will be multiplied. In Figure 4.6 then, the first band (the closest to one end) is yellow (4) the second is violet (7) so that the digits are 47. The third band is again yellow for a multiplier of 10,000 or 10^4. Thus the resistance is

$$R = (47)(10{,}000) = 470{,}000 \ \Omega = 4.7 \times 10^5 \ \Omega$$

Since we use the standard kilo- or mega- prefixes we would write this as either 470 kΩ or 0.47 MΩ.

The fourth band, if present, gives the tolerance of the resistor, which is discussed later.

Note that colored bands are also used for marking capacitors, inductors, and diodes, so the presence of bands does not guarantee that the device is a resistor.

TABLE 4.2

Color code

Color	Band Number			
	1	2	3 (Multiplier)	4 (Tolerance, %)
Black	0	0	1	
Brown	1	1	10	
Red	2	2	100	
Orange	3	3	1000	
Yellow	4	4	10000	
Green	5	5	100000	
Blue	6	6	1000000	
Violet	7	7	10000000	
Grey	8	8		
White	9	9		
None				±20
Silver				±10
Gold				±5

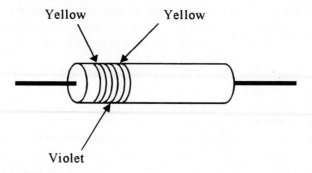

Yellow   Yellow

Violet

FIGURE 4.6
Resistor with color code value identification.

Measurements showing the resistance to be the same in both directions through the device will demonstrate that it is not a diode or capacitor, but it could still be an inductor. If the inductance is measured to be very small it is most likely a resistor.

Standard Values   Resistors, particularly axial resistors, are manufactured in standard sets of values. Thus, if you need a 1350-Ω resistor you are out of luck because that is not a standard value. The closest would be 1200 Ω. For resistances between standard values you can use series/parallel combinations, variable resistors or, when large quantities are needed, arrange for a resistor manufacturer to produce a quantity of the value needed.

The standard values are grouped by digits and multipliers, much like the color code given earlier. Table 4.3 gives the digits and multipliers of the most common set of standard values. A new manufacturing technique has brought about adoption of different standard values by some manufacturers.

TABLE 4.3
Standard values

Digits: 1, 1.1, 1.2, 1.3, 1.5, 1.6, 1.8, 2.0, 2.2
2.4, 2.7, 3.0, 3.3, 3.6, 3.9, 4.3, 4.7, 5.1
5.6, 6.2, 6.8, 7.5, 8.2, 9.1
Multipliers: 1
10
100
1,000
10,000
100,000
1,000,000
10,000,000

4.2.2 Tolerance

A resistor manufacturer produces thousands of a particular resistor in any given production run. Naturally there will be some variation of value across the resistors produced. Thus the manufacturer specifies that a resistor has a *tolerance* about the value indicated by the marking. There are a number of standard tolerances.

The tolerance is expressed as a ± percentage of the marked value. For example, suppose you pick up one resistor from a batch marked 3000 Ω with a specified ±10% tolerance. The *actual* value of the resistor you pick up will lie between 3000 + 0.1(3000) = 3300 Ω and 3000 − 0.1(3000) = 2700 Ω. Of course the resistor you pick up will have a *specific* value but the marking only tells you that it will be within the range given.

Tolerance is either indicated by a number written on the resistor or by a fourth colored band. Table 4.2 shows the tolerances indicated by the various colors. Note that no colored band indicates a ±20% tolerance.

EXAMPLE 4.8

A resistor is found to have the following colored bands: red–violet–orange–gold. What is the value of this resistor? (See Table 4.2.)

Solution
The red–violet–orange bands indicate the nominal value to be 27 × 1000 = 27 kΩ. The gold band indicates a tolerance of ±5% of this value. Therefore, the actual resistance value will lie within the range of 27 kΩ ± 0.05 (27 kΩ) = 27 kΩ ± 1.35 kΩ or between the values of 28.35 and 25.65 kΩ.

4.2.3 Power Rating

When a resistor carries a current there is a voltage drop and thus it is dissipating energy continuously in time. This is expressed as a power dissipation given by Eq. (4.2). The resistor dissipates this energy by heating up to some temperature so that energy is radiated away as heat at the same rate as the input given by Eq. (4.2).

Resistors have a *power rating* that specifies the maximum power which can be dissipated without damage and ultimate failure of the component. This is usually determined by the temperature it can tolerate before failure. External cooling, by a fan or heat sink, for example, can extend the power rating.

For axial resistors the standard power ratings are $\frac{1}{8}$, $\frac{1}{4}$, $\frac{1}{2}$, 1, and 2 W. The rating is determined primarily from the size of the component, as illustrated in Figure 4.7 for carbon composition resistors. For others it is best to consult the manufacturer's specifications.

Power type resistors have power ratings ranging from a few watts to thousands of watts. The power rating is usually written on the component.

FIGURE 4.7
Physical sizes of resistors with different power ratings.

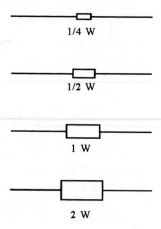

1/4 W

1/2 W

1 W

2 W

EXAMPLE 4.9

A circuit will use a 2-kΩ resistor carrying a maximum current of 30 mA and a 120-kΩ resistor with a voltage across it of 90 V. What power ratings should these resistors have?

Solution

We simply calculate the power to be dissipated by each resistor and then pick a rating that is higher. For the first,

$$P_1 = I^2R = (30 \text{ mA})^2(2 \text{ k}\Omega) = 1.8 \text{ W}$$

The second gives

$$P_2 = V^2/R = (90 \text{ V})^2/120 \text{ k}\Omega = 67.5 \text{ mW}$$

From these results we conclude the 2-kΩ resistor should have a 2-W rating and the 120-kΩ resistor can have either the $\frac{1}{8}$- or $\frac{1}{4}$-W rating. Of course, you can always use a resistor with a higher rating.

Power Derating When a resistor is operated at an elevated temperature, from self-heating or in some heated environment, the power rating must be *derated* from the indicated value. The indicated value is often specified for a "room temperature" of 20 or 25°C (68 to 77°F) but 60 to 70°C is also used. A manufacturer will often provide a derating curve such as that shown in Figure 4.8 for a line of resistors. Note that above 25°C the power rating is decreased by some percentage of the rated value. At 75°C, for example, the rating is only about 60% of the value at 25°C.

4.2.4 Temperature Coefficient

The resistance of a resistor will change with temperature. This dependence on temperature is expressed by a *temperature coefficient* (TC), which may be given as the percentage change of resistance per degree of temperature. The most common

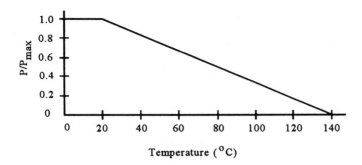

FIGURE 4.8
Resistor power rating is degraded with increasing temperature.

specification is the percent change in resistance per degree Celsius (%/°C) from the resistance at 25°C. The resistance can increase with temperature (positive coefficient) or decrease with temperature (negative coefficient).

Another way of expressing temperature dependence is by the *parts per million* (ppm) change per degree of temperature. Thus a specification of -150 ppm/°C means that the resistance will decrease by a fraction $(150/1,000,000) = 0.00015$ of its value for each rise of one degree Celsius of temperature above 25°C. This is equivalent to 0.015%/°C. Just multiply the ppm by 10^{-4} (or divide by 10,000) to get the percent.

**EXAMPLE
4.10**

A 1500-Ω resistor will be used in an environment that cycles from 0 to 40°C during some tests. What is the range of resistance change between the following choices? (a) carbon composition: -0.1%/°C; (b) metal film: $+200$ ppm/°C.

Solution
For the first resistor, we have a fractional change of -0.001 for every °C. The resistance will increase for temperature from 25 to 0°C and decrease from 25 to 40°C.

$$R_1(0°C) = 1500 + (0.001)(25)(1500) = 1537.5 \ \Omega$$
$$R_1(40°C) = 1500 - (0.001)(40 - 25)(1500) = 1477.5 \ \Omega$$

So the total change was 60 Ω or 4%.
For the metal film, we have $200/10^6 = 0.0002$ fractional change in resistance for every °C from 25°C.

$$R_2(0°C) = 1500 - (0.0002)(25)(1500) = 1492.5 \ \Omega$$
$$R_2(40°C) = 1500 + (0.0002)(40 - 25)(1500) = 1504.5 \ \Omega$$

The total change in this case is only 12 Ω or 0.8%.

4.2.5 Other Characteristics

The following paragraphs cover other characteristics and specifications of resistors.

Voltage Rating The voltage rating of a resistor is based on the electric fields that can exist across a resistor and the effect this may have on its application. Seemingly modest conditions can give rise to significant electric fields.

A 1-MΩ resistor with a current of only 1 mA will have a voltage drop of $V = (1 \text{ mA})(1 \text{ M}\Omega) = 1000$ V! Suppose the resistor is 2 cm long, then the electric field is,

$$E = V/d = 1000/0.02 \text{ m} = 50,000 \text{ V/m}$$

This electric field is sufficient to cause ionization, breakdown of dielectrics, and perhaps fracture of the resistor insulating material.

The voltage rating specifies the maximum working voltage to prevent failure of the unit due to excessive electric fields.

Inductance All resistors have some small self-inductance. In wire-wound components the inductance can become quite large. The inductance limits the applications of resistors in high-frequency applications where this inductance could create a high impedance. Resistors for high-frequency applications are carefully constructed to keep the inductance at a minimum.

Capacitance A resistor will have a small capacity between its leads, as well as the resistance. This will appear as a capacitor in parallel with the resistance. At very high frequencies, this will act as a low-impedance shunt around the resistance.

Noise Because resistance action depends on random collisions between electrons and ions in the resistance material, a certain amount of electrical noise is *generated* by passing a current through the resistor.

4.3 TYPES OF FIXED RESISTORS

Fixed-value resistors are made from many different materials and in many different shapes and sizes. The following sections review the major types of resistors, their special characteristics, and applications.

4.3.1 Carbon Resistors

Carbon has been used to fabricate fixed-value resistors for many years and is still the most common material for resistor construction.

Composition The most common form of carbon resistor is the composition type illustrated in Figure 4.9. The resistance is composed of a combination of carbon

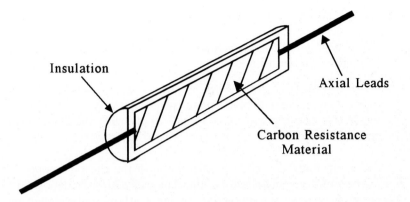

FIGURE 4.9
Physical structure of carbon composition resistors.

and an inert binder, solidified under heat and pressure. Axial leads are attached and the element is encapsulated in an insulating material. The value of resistance for a given size can be varied over many decades of range by changing the mix of carbon and binding agent.

1. *Power:* The resistors are generally available in $\frac{1}{8}$-, $\frac{1}{4}$-, $\frac{1}{2}$-, 1-, and 2-W power ratings. Typical power derating is a reduction of 10 to 15%/°C above 25°C.
2. *Tolerance:* Commonly available tolerances for carbon composition resistors are ±20%, ±10% and ±5%.
3. *Temperature coefficient:* The temperature coefficient of carbon composition resistors changes from negative to positive at about 20°C. Figure 4.10 illustrates the peculiar resistance versus temperature of carbon. In the range of 0 to 60°C, the TC is quite small, varying from about −0.1%/°C at 0°C, passing through zero at about 20°C, to about +0.1%/°C at 60°C. Below 0 and above 60°C the TC becomes quite large.

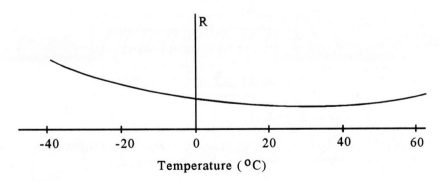

FIGURE 4.10
Resistance versus temperature for carbon resistors.

4. *Applications:* The carbon composition resistor is the workhorse of the electrical/electronics industry. Any application for power less than about 2 W, moderate temperature environments, and moderate temperature variation can be satisfied by this resistor. It is also physically quite rugged and is inexpensive.

The voltage rating depends on the physical length and therefore on the power rating. A $\frac{1}{8}$-W size can withstand about 150 V, while the 2-W version is rated as high as 650 V.

Carbon Film   A modernized version of the carbon composition resistor is the carbon film. In this resistor a thin layer of carbon is deposited on a ceramic cylinder and connected to wires at each end. Carbon can be etched away in a spiral pattern, often by a laser beam, to provide the desired final resistance. Construction is illustrated in Figure 4.11.

This type of resistor is available with an improved tolerance of ±2%. The temperature coefficient is basically the same as the carbon composition.

4.3.2 Metal Resistors

Wire-Wound   Wire-wound resistors are formed by winding a metal wire around an insulating core, often ceramic. An insulating material is then placed over the wire. Basic construction is shown in Figure 4.12. The metal is typically an alloy, such as nichrome, which has a higher resistivity than a pure metal, as Table 4.1 shows. The higher resistivity is used so that high resistances can be achieved without extensive wire lengths and very small wire diameters. In general, however, high resistance values (≈ 1 MΩ and larger) are not available with wire-wound resistors.

1. *Power:* Small precision resistors for use in electronic circuits may have a power rating of less than 1 W. Heavy duty, high-current resistors for electrical power applications are available with power ratings of several hundred

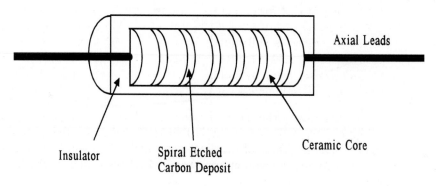

FIGURE 4.11
Structure of a carbon film resistor.

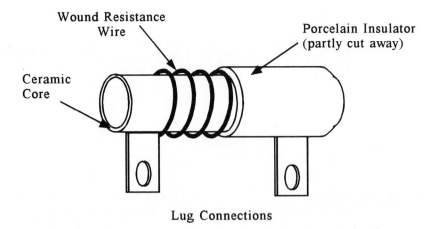

Wound Resistance
Wire

Porcelain Insulator
(partly cut away)

Ceramic
Core

Lug Connections

FIGURE 4.12
Wire-wound resistor.

watts. Power derating is approximately 0.3 W/°C for temperatures above 25°C.

2. *Tolerance:* Wire-wound resistors in the precision category are available with standard tolerances of ±1, ±0.5, and ±0.1%. Those intended for power applications are usually offered with tolerances of ±5 or ±10%.

3. *Temperature coefficient:* Precision wire-wound resistors offer one of the lowest temperature coefficients of all resistor types. The TC will be in the range of 20 to 100 ppm/°C. High power types generally offer larger coefficients, perhaps to 200 ppm/°C.

4. *Applications:* A basic limitation of the wire-wound resistor is that very high resistances (>1 MΩ) are not available. In general, wire-wounds can be used in applications requiring high precision and/or low temperature sensitivity and/or good power-handling capability.

In its simplest form, the wire-wound resistor will have significant inductance since it is just a coil of wire. This limits its usefulness to frequencies low enough that its reactance is still much less than its resistance, which will probably be approximately $f < 50$ kHz. For high-frequency applications, special low-inductance wire-wounds are made by looping the windings in such a way that the inductances of the loops approximately cancel.

Metal Film The most common metal film resistor is constructed as shown in Figure 4.13. A very thin metal film is deposited on the surface of a ceramic cylinder. The thickness of the layer determines the nominal resistance. A laser is then used to etch a spiral cut of the metal off the cylinder. The length of the cut determines the final resistance. Axial metal leads are attached and the assembly is encapsulated in an insulator, which is often a baked-on ceramic. The most common metals are nickel and chromium although alloys and certain oxides are also used.

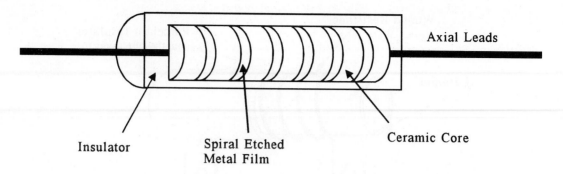

FIGURE 4.13
Structure of a metal film resistor.

1. *Power:* Metal film resistors are available with power ratings from as low as 1/20 W to hundreds of watts. The very low power units are also very small in size for applications where size is an important factor. Derating ranges from 0.5%/°C to 2%/°C, but usually from about 70°C.
2. *Tolerance:* Due to the good control over metal depositing and laser trimming, it is possible to provide metal film resistors with good tolerance. Typical available tolerances are ±5 down to ±0.1%.
3. *Temperature coefficient:* The temperature coefficient of metal film resistors depends on what type of metal is used to form the film. For this reason there is a great variation in this specification. In the worst case the TC will be on the order of 200 ppm/°C. In the best case the TC will be nominally zero but with an uncertainty of ±20 ppm/°C. This means that the resistance may increase or decrease with increasing temperature, but in either case the change will not exceed 20 ppm/°C. This is a fraction of ±0.00002 (0.002%) of the nominal resistance!
4. *Applications:* The metal film resistor may be used in any application in electronics for which very high resistances are not required and power dissipation is low to moderate (<10 W). If the spiral cut is not extensive, inductance will be small so that high-frequency effects are not a problem.

Since the metal resistivity is typically not high, it is difficult to fabricate metal film resistors with resistances above 1 MΩ.

Metal film resistors are available in standard sets of values that are much more extensive than the traditional set given in Table 4.3.

Chip Resistors   Chip resistors are metal film resistors that have been fabricated on small ceramic chips instead of cylinders. These are available in both thick (>10^{-6} in.) and thin (<10^{-6} in.) films. The resistance element, as shown in Figure 4.14, is deposited on one side of the chip and solder connections are placed at

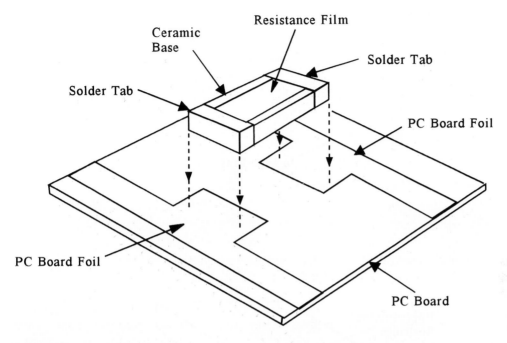

FIGURE 4.14
Chip resistors are metal film on ceramic for printed circuits.

each end. The chip resistor can be soldered directly onto a printed circuit board. Characteristics of this resistor are much the same as the metal film resistor.

Resistor Networks   Resistors are often fabricated using integrated circuit (IC) technology and thin-film technology and placed in standard 14- and 16-pin IC sockets as shown in Figure 4.15. These resistors find applications as pull-up resistors in digital circuits for example. Various internal wiring configurations are used including isolated resistors and an internal common tie-point for all the resistors. The characteristics are much the same as those of thin-film resistors.

4.4   VARIABLE RESISTORS

In many instances it is desirable to have a resistance in some circuit that can be mechanically changed to adjust for varying circumstances. This is provided by a variable resistor or *potentiometer*, often just called a *pot*.

4.4.1   Variable Resistor Specifications

Figure 4.16(a) shows the schematic symbol for a variable resistor. Note that in its most basic form it is a three-terminal device. The resistance between the endpoints,

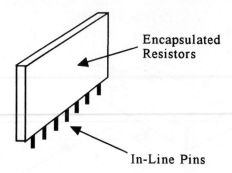

Encapsulated Resistors

In-Line Pins

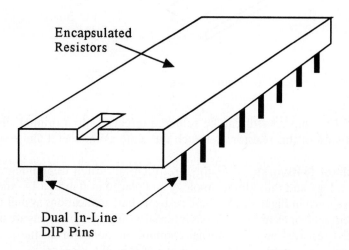

Encapsulated Resistors

Dual In-Line DIP Pins

FIGURE 4.15
In-line or IC resistor sets.

1 and 2, is fixed and equal to the total resistance of the pot. Terminal 3 is connected to the *wiper*, which can "tap" resistance smoothly between terminals 1 and 2. Thus the resistance between terminal 3 and either 1 or 2 varies as the wiper is moved from one end of the pot to the other.

In some cases we may connect the wiper and one end together as shown in Figure 4.16(b). This may be done externally by the user or internally by the manufacturer. In this case we have a two-terminal device whose resistance varies from zero to the maximum as the wiper is moved from one extreme to the other. The schematic symbol is sometimes modified for this case to that shown in Figure 4.16(c).

FIGURE 4.16
Symbols for variable resistors.

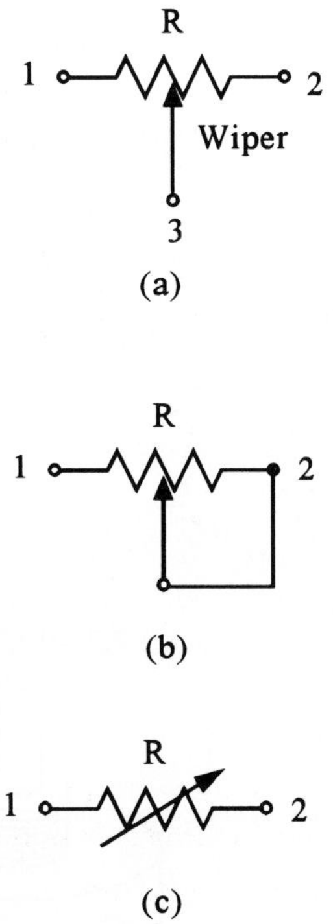

(a)

(b)

(c)

Resistance   The resistance of a pot is just the net fixed resistance of the device, such as that between terminals 1 and 2 of Figure 4.16(a).

Controls   Variable resistors used for controls, as with the volume of a television for example, are designed for repeated and frequent adjustment of the wiper. They are most often rotary, i.e., a shaft is rotated to move the wiper and vary the resistance. In some cases a sliding or lever type of design is used. Figure 4.17 shows typical control type pots.

For rotary pots the *number of turns* to move the wiper from one end to the other is part of the specifications. For the simplest form of pots, less than one full turn of the shaft is requried. Although called a *single-turn* pot, in general these pots sweep out the full resistance in only 230°.

Multiple-turn variable resistors are designed so that many turns of the shaft are required to sweep through the full range of resistance. A ten-turn pot is the most common. Obviously this allows a much finer adjustment of the resistance than the single-turn pot.

Trimmers   Another type of variable resistor is used in cases where very infrequent adjustment of the resistance is required. For example, it may be necessary to make slight adjustments of some resistance in an instrument circuit during calibration. In some other case, it may be necessary to set some resistance value one time,

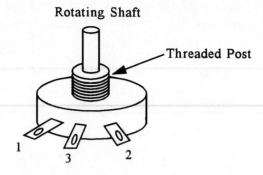

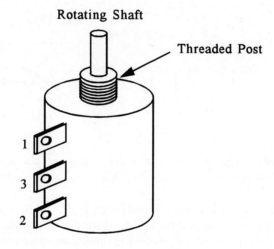

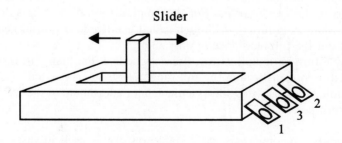

FIGURE 4.17
Appearance of typical variable resistors.

for example to achieve a desired gain, but further adjustments are never required so the shaft is glued, or *staked,* in that position. The trimmer variable resistor is used in these cases.

In general, these units are very small and a screwdriver slot is provided in the rotating shaft to make required variations in resistance. Some trimmers are single-turn (230°) rotary units and some are multiple-turn (usually ten) units. Figure 4.18 shows typical single- and multiple-turn trimmers.

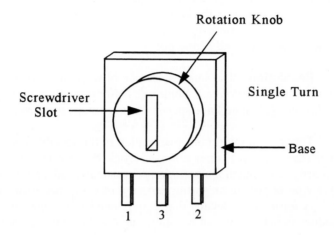

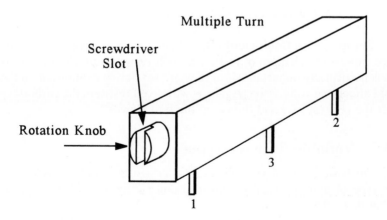

FIGURE 4.18
Physical appearance of miniature trimmer resistors.

Power Rating   Variable resistors have a power rating that specifies the amount of power which the unit can dissipate without damage from overheating. For small units used in electronic circuits, this rating is general from $\frac{1}{2}$ to 2 W.

Larger power variable resistors can have power ratings of hundreds of watts. In general, such units are typically wire-wound (see Section 4.4.2) and are often called *rheostats.*

As with fixed resistors, pots have power derating curves that reduce the amount of power which can be dissipated as the temperature of the unit increases. Furthermore, note that the power rating is over the entire resistance element. Care must be taken to insure that only a fraction of that power is dissipated between the wiper and one end of the pot.

Resolution   The smallest change of resistance that a variable resistance can offer is called the *resolution.* Some units offer virtually continuous resistance (infinite resolution) and some have specific discrete steps in resistance change.

Taper   An important specification of the variable resistance is how the resistance changes as the wiper is moved from one extreme to the other. In the case of a rotary pot we mean the variation of resistance with angle. In the case of a sliding, lever type of pot we mean the variation of resistance with slider position.

A *linear* taper means that equal changes of angle or slider position always produce equal changes of resistance across the range of motion. This is shown in Figure 4.19, curve A.

An *audio* taper is nonlinear so that equal changes of angle or slider position do *not* produce equal changes of resistance across the range of motion. Curve B of Figure 4.19 shows a logarithmic audio taper. Note that at the low end of the wiper motion only small changes of resistance occur for wiper motion, whereas at the high end very large changes of resistance occur. This is used in volume control to provide fine loudness control at low levels.

Temperature Coefficient (TC)   Just as with fixed resistors a variable resistance can exhibit changes of resistance with temperature. In the case of trimmers this is particularly important since they are set during calibration and any changes could affect that calibration. Again as with fixed resistors the carbon type resistors offer the poorest TCs and the metal the best (lowest).

4.4.2   Variable Resistor Types

The two basic types of variable resistor design are continuous and wire-wound. The type of material used for the resistance element and the type of wire vary within these classifications.

Continuous   Figure 4.20 illustrates the essential features of the continuous variable resistor. The resistance material is deposited on a ceramic or other insulating substrate and connections are made to each end. These comprise leads 1 and

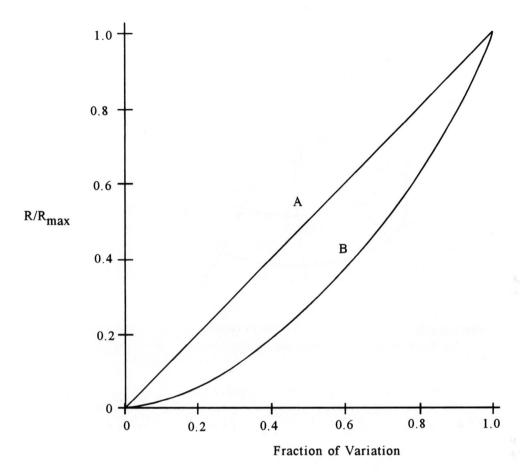

FIGURE 4.19
Variable resistance tapers.

2 of Figure 4.16. A wiper slides on the resistance element and is connected to lead 3 through a bearing type of connection. The resolution of such a pot is essentially infinite, i.e., an infinitesimal change in wiper position will produce an infinitesimal change in resistance between the wiper and either end connection.

The actual resistance material is typically carbon or cermet. Cermet is a combination of metal and ceramic (or glass) mixed, heated, and bonded to a ceramic substrate. The mixture ratio determines the resistance per unit length. The cermet types can dissipate more power than carbon and typically have low TCs, perhaps down to 20 ppm/°C.

Wire-Wound   Figure 4.21 shows a typical wire-wound variable resistor. Note that the wiper slides over turns of a coil of wire. Leads 1 and 2 of Figure 4.16 are

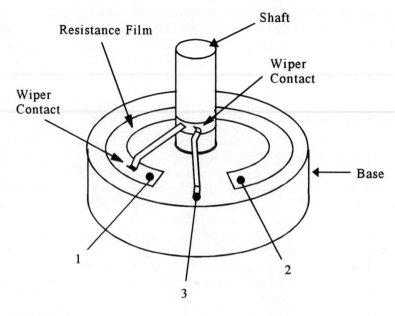

FIGURE 4.20
Physical structure of a rotary carbon film variable resistor.

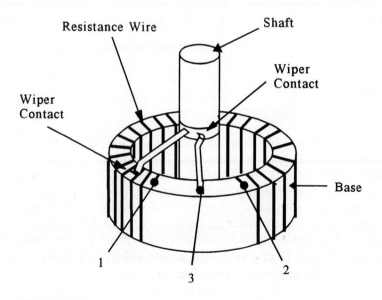

FIGURE 4.21
Physical structure of a wire-wound variable resistor.

connected to either end of the wire coil and comprise the fixed resistance of the wire. Lead 3 is connected to the wiper, which slides over the coil.

Clearly, the resolution in this case is the resistance of a single turn of the wire. That is, as the wiper is moved the resistance will change in discrete steps, ΔR, each of which is the resistance of a single turn of the coil.

Note that the wire-wound pot can have significant inductance since it is a coil and this limits its usefulness in high-frequency circuits.

4.5   SPECIAL RESISTANCE DEVICES

This section gives a very brief description of a number of other resistance-based devices in use in electricity and electronics. A resistance was defined as a device for which the ratio of voltage drop to current was a constant. Thus the resistance is constant with respect to voltage or current changes. Many of the devices discussed in this section have resistances that change due to some other parameter, such as temperature, light intensity, etc., but they are still resistances.

4.5.1   Thermistors and RTDs

Thermistors and RTDs are components whose resistance varies with temperature. They are used for the purpose of measuring temperature. In the case of resistors, variation of resistance with temperature was a nonideal characteristic. In these devices, that variation is employed for temperature measurement.

A thermistor is made from semiconductor material. Its resistance decreases nonlinearly with temperature. It can have a very large sensitivity, as much as $-10\%/°C$.

An RTD is made from metal wire. Its resistance increases very nearly linearly with temperature. It has a low sensitivity, on the order of 0.4%/°C.

Both of these devices have a maximum power dissipation rating and a dissipation constant, P_D. This constant tells the amount of self-heating for every watt of dissipated power. Thus, a dissipation constant of 20 mW/°C means that if you put current through the device and dissipate 20 mW, it will self-heat by 1°C.

4.5.2   Strain Gauges

A strain gauge has a resistance whose value is changed by physical deformation, such as by stretching to change its length. The fractional change in length is called the *strain*. The relation between resistance change and length change is given by the *gauge factor* (GF).

$$\Delta R/R = \text{GF}(\Delta L/L) \tag{4.9}$$

where $\Delta R/R$ is the fractional change in resistance and $\Delta L/L$ is the fractional change in length, i.e., the strain.

FIGURE 4.22
IV characteristic of a varistor.

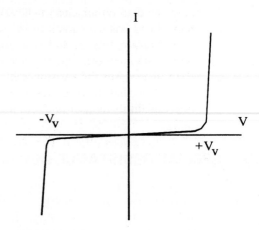

For strain gauges made from metal wire, the GF is approximately constant with a value of about 2. For semiconductor strain gauges, GF varies with strain and is negative. A typical maximum is GF ≈ -160.

4.5.3   Photoconductive Cells

A photoconductive cell component has a resistance that changes with light intensity. The photoconductive cell is made from semiconductor material. One of the most common materials is cadmium sulfide (CdS). The device exhibits a very large change in resistance. It may have a resistance of many mega-ohms in the dark but drop down to only a few thousand ohms at typical room light intensity.

The photoconductive cell resistance changes only for light in a certain range of wavelengths. CdS cells respond to light in the visible spectrum, in much the same manner as the human eye.

4.5.4   Varistors

A varistor is not really a resistor at all; it has the characteristic that its resistance is a function of applied voltage. The curve of Figure 4.22 shows a typical varistor

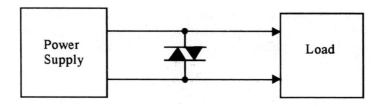

FIGURE 4.23
Use of a varistor for transient protection.

current versus voltage characteristic. Varistors are used to protect electronic equipment from voltage surges.

Consider Figure 4.23 where a varistor has been placed across a power supply input to some electronic circuit. For voltage less than $\pm V_V$ as shown in Figure 4.22 the resistance of the varistor is essentially infinite. It thus draws no current and has virtually no effect on a circuit. However, if the supply voltage should surge above $\pm V_V$, Figure 4.22 shows that the varistor resistance will essentially become zero and all current from the supply will be shunted through the varistor, protecting the electronic equipment from the surge.

5

CAPACITORS*

A capacitor is a component of electrical and electronic circuits that exhibits the property of capacitance. In this chapter, capacitance is defined and the characteristics of capacitors are given. This discussion is followed by a presentation of the major types of capacitors, their special characteristics, and application notes.

5.1 BASIC DEFINITIONS

In this section, the dynamic definition of capacitance is given for the ideal capacitor. This is followed by a physical definition of capacity and properties of capacitor combinations.

5.1.1 Concept of Capacitance

The basic definition of capacitance is found from the measured behavior of the component in an electric circuit. Consider two metal plates separated by an insulator and connected to a source of some voltage, V, as shown in Figure 5.1. Current will flow in the circuit to build up charges on the plates in response to the voltage.

We find that at every instant of time the ratio of charge on the plates, Q (in coulombs), to voltage across the plates, V (in volts), is a constant. Thus, if the voltage changes, the charge on the plates will also change such that the ratio remains constant. This ratio is called the *capacitance*. The capacitance or capacity of the component is usually denoted by the variable C. The unit of capacitance is the coulomb/volt, which is given the name *farad* and the unit symbol F. A component that has this characteristic is called a *capacitor*.

---

* This chapter was written by Dr. Curtis D. Johnson, College of Technology, University of Houston.

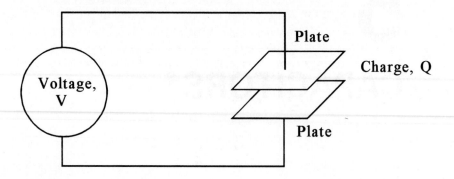

FIGURE 5.1
Basic structure of a capacitor.

The basic definition of capacity is expressed by the following equation:

$$C = \frac{Q}{V} \qquad (5.1)$$

where Q = charge on the plates in coulombs (C)
             V = voltage across the plates in volts (V)
             C = capacitance in farads (F)

In electrical and electronics work we do not often deal with charge. Instead, the current, I, which is the rate of flow of charge, is measured and used. Thus, the definition of Eq. (5.1) is not of much practical value.

It is possible to use calculus to express the definition in terms of voltage and current, but the equations are still of limited value to practical circuit work. For reference, these equations are, first, that the voltage across a capacitor is given by the integral of the current over time divided by the capacitance,

$$V = \frac{1}{C} \int_0^t I(t)\, dt + V(0) \qquad (5.2)$$

where $V(0)$ is the initial voltage on the capacitor at $t = 0$. Second, the current through a capacitor is given by the capacitance times the rate of change, or derivative, of the voltage:

$$I = C \frac{dV}{dt} \qquad (5.3)$$

EXAMPLE 5.1

The voltage across a 1.5-μF capacitor is caused to increase at a rate of 0.5 V/s. How much current will the voltage source have to provide?

Solution

Since the rate of change of voltage is given, we can use Eq. (5.3) to find the current:

$$I = (1.5\ \mu\text{F})(0.5\ \text{V/s}) = 0.75\ \mu\text{A}$$

Unit Prefixes As given earlier, the unit of capacity is the farad (F), which represents one coulomb of deposited charge per volt across the plates. This unit is very large in comparison with everyday electricity and electronics. Therefore, the metric (SI) prefixes are very commonly used with the farad to represent practical values. The common prefixes used for capacity are as follows:

milli-	(m)	10^{-3}
micro-	(μ)	10^{-6}
nano-	(n)	10^{-9}
pico-	(p)	10^{-12}

Before standardization of prefixes, capacitors were often given a designation of "micro-micro-farads," or $\mu\mu$F. This is really just a picofarad, since pico- is the proper prefix.

Charged When a voltage source is connected to a capacitor, a current will flow while charges build up on the plates of the unit. When the current has stopped flowing, then the amount of charge on the plates is the proper value, as given by Eq. (5.1), for the value of the voltage and the capacitance. When this condition has been reached, we say the capacitor is *charged*. A charged capacitor can be disconnected from the circuit and "placed on the shelf" but the charges still reside on the plates and a voltmeter will still indicate a voltage across the plates.

If a load, such as a resistor, is placed across the capacitor leads, the charges will flow out of the device. This is referred to as *discharging* the capacitor.

Energy Storage A very important characteristic of capacitors is the fact that energy is stored in a capacitor when it is charged. This can be seen from the fact that if the charges on the plates of a capacitor are allowed to flow out of the device then that will constitute a flow of charges, i.e., a current, which can do work. Thus, the stored charges represent stored energy that can be retrieved for later use.

The amount of stored energy, in the SI unit of energy, joules (J), is given by the relation

$$W = (1/2)CV^2 \tag{5.4}$$

where W = stored energy in joules (J)
C = capacitance in farads (F)
V = voltage across the capacitor in volts (V)

EXAMPLE 5.2 A 50-μF capacitor is charged to 250 V. What is the stored energy?

Solution
The stored energy can be found directly from Eq. (5.4):

$$W = (1/2)(50\ \mu\text{F})(250\ \text{V})^2$$
$$= 1.5625\ \text{J}$$

5.1.2 Physical Definition of Capacitance

Capacitance can also be defined in terms of the physical size, shape, and materials used to construct the component. In Figure 5.2 a basic capacitor is illustrated. The capacitor consists of two plates of area A separated by a distance d in which an insulator resides. To define the capacity, we must first define a characteristic of the insulator that separates the plates.

FIGURE 5.2
Physical definition of a capacitor.

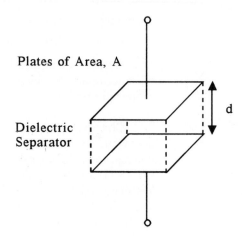

Plates of Area, A

Dielectric
Separator

d

Dielectric Constant Even though the material between the plates is an insulator, it will respond to the electric field from an applied voltage. This response is a measure of the deformation of atoms, which make up the material. The *permittivity* is a measure of this response. The permittivity of materials is most often given in comparison to that of a perfect vacuum. The ratio of the permittivity of a material to that of a vacuum is called the *dielectric constant.* In general then, the permittivity of a material is given by

$$\varepsilon = K\varepsilon_0 \qquad (5.5)$$

where $\varepsilon_0 = 8.85 \text{ pF/m} = $ permittivity of a vacuum
$K = $ material dielectric constant
$\varepsilon = $ permittivity of the material

You can see that the units of permittivity are F/m in keeping with the fact that this unit will be associated with the capacitance of a capacitor. Table 5.1 lists the dielectric constants of a number of materials commonly used for the insulator in capacitors.

Capacitance Now the capacitance of the capacitor given in Figure 5.2 is given by

$$C = K\varepsilon_0 \frac{A}{d} \qquad (5.6)$$

TABLE 5.1

Dielectric constants

Material	Dielectric Constant
Vacuum	1
Dry air	1.0006
Teflon	2
Polystyrene	2.5
Mylar	3
Polycarbonate	3
Waxed paper	2.2–4.9
Mica	5
Glass	6
Aluminum oxide	7
Tantalum oxide	25
Ceramic materials	10–10,000

where, specifically, A = area shared by the two plates (m^2)
d = separation of the plates (m)
K = dielectric constant of the insulator between the plates
ε_0 = 8.85 pF/m = vacuum permittivity

The capacitance is directly proportional to the area of the plates and the dielectric constant of the insulator. It is inversely proportional to the distance between the plates.

EXAMPLE 5.3 A capacitor is made from two plates of area 0.01 m^2. The plates are separated by 0.1 mm. Compare the capacity if the insulator is vacuum or mylar.

Solution

For a vacuum, $K = 1$ and so the capacity is given by

$$C_{vac} = (1)(8.85 \text{ pF/m})(0.01 \text{ m}^2)/(0.0001 \text{ m})$$
$$= 885 \text{ pF}$$

When mylar is used for the insulator, the air capacity is simply multiplied by the dielectric constant, $K = 3$. Therefore, the capacity is

$$C_{mylar} = KC_{vac} = (3)(885 \text{ pF}) = 2655 \text{ pF}$$
$$= 2.655 \text{ nF} = 0.002655 \ \mu\text{F}$$

Voltage Rating The existence of a difference of potential, i.e., a voltage, across two plates means that an electric field exists between the plates. Electric field E is measured in volts per meter (V/m) and is simply found by the voltage across the plates divided by the distance between the plates. As an equation this is

$$E = V/d \tag{5.7}$$

The electric field is important because it is the size of the electric field that determines when an electrical breakdown of an insulator will occur. For a capacitor, this means that there will be an upper limit on the voltage that can be placed across the plates. This is the voltage rating.

Table 5.2 gives the breakdown electric field of some common insulators. The field is given in kilovolts/millimeter (kV/mm) to indicate values appropriate to actual capacitor dimensions.

TABLE 5.2
Typical dielectric strength

Material	Strength (kV/mm)
Dry air	3
Ceramic materials	3–10
Waxed paper	16
Mylar	20
Teflon	60
Glass	120
Mica	200

EXAMPLE 5.4

What is the maximum voltage that can be placed across the capacitor of Example 5.3 when mylar is used as the insulator?

Solution
Table 5.2 gives the breakdown electric field of mylar as 20 kV/mm. From Eq. (5.7) we find the voltage as

$$V_{max} = Ed = (20 \, \text{kV/mm})(0.1 \, \text{mm})$$
$$= 2 \, \text{kV}$$

5.1.3 Capacitor Combinations

Combinations of capacitors in series or parallel will have an *equivalent capacitance*. This means that the combination acts like a single capacitor of the equivalent value. The importance of such combinations is the ability to obtain any effective value of capacitance by series and parallel combinations.

Series When capacitors are placed in series, as shown in Figure 5.3, the equivalent capacitance is found from the inverse of the sum of inverses. In equation form, for three capacitors, this means

$$\frac{1}{C_{eq}} = \frac{1}{C_1} + \frac{1}{C_2} + \frac{1}{C_3} \tag{5.8}$$

where C_{eq} = the equivalent capacitance (F).

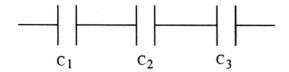

FIGURE 5.3
Capacitors in series.

Note that in the case of only *two* capacitors in series Eq. (5.8) can be simplified to the following form:

$$C_{eq} = \frac{C_1 C_2}{C_1 + C_2} \tag{5.9}$$

The equivalent capacitance of a series of capacitors will *always* be less than the least value of capacity.

EXAMPLE 5.5

The three capacitors in Figure 5.3 have values of 270, 47, and 1200 pF. What is the equivalent series capacity?

Solution
We simply use Eq. (5.8) directly:

$$1/C_{eq} = 1/(270\,\text{pF}) + 1/(47\,\text{pF}) + 1/(1200\,\text{pF})$$
$$= 3.7 \times 10^9 + 2.1 \times 10^{10} + 8.3 \times 10^8$$
$$= (0.37 + 2.1 + 0.083) \times 10^{10} = 2.55 \times 10^{10}$$

Thus,

$$C_{eq} = 1/(2.55 \times 10^{10}) = 3.9 \times 10^{-11} = 39\ \text{pF}$$

Note that this is less than the 47-pF capacitor.

EXAMPLE 5.6

What is the equivalent capacitance of a 0.01-μF capacitor in series with a 0.05-μF capacitor?

Solution
Since there are only two capacitors, we can use the simplified equation for series equivalent capacitance, Eq. (5.9). Then,

$$C_{eq} = (0.01)(0.05)/(0.1 + 0.5) = 0.0083\ \mu\text{F}$$

EXAMPLE 5.7

What capacitance needs to be placed in series with a 2-μF value to get a 0.5-μF value?

Solution
Again, since there are two capacitors, we can use Eq. (5.9), but in this case the answer will come more easily from Eq. (5.8) because we can write

$$(1/0.5 \ \mu F) = 1/(2 \ \mu F) + 1/C_2$$
$$2 \times 10^6 = 0.5 \times 10^6 + 1/C_2$$

or

$$1/C_2 = 1.5 \times 10^6$$

So,

$$C_2 = 1/(1.5 \times 10^6) = 0.67 \ \mu F$$

Parallel Capacitors Figure 5.4 shows three capacitors in parallel. In such a case the equivalent capacitance is simply the sum of all of the capacitances. To avoid confusion with the series case, we will call this the net capacitance.

$$C_{net} = C_1 + C_2 + C_3 \tag{5.10}$$

FIGURE 5.4
Capacitors in parallel.

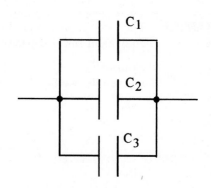

EXAMPLE 5.8 What would be the net capacitance if the three capacitors of Example 5.5 were placed in parallel?

Solution
Using Eq. (5.10), we find

$$C_{net} = 270 \ \text{pF} + 47 \ \text{pF} + 1200 \ \text{pF} = 1517 \ \text{pF}$$

5.2 PRACTICAL CAPACITOR CHARACTERISTICS

An ideal capacitor has infinite breakdown voltage, infinite resistance, and zero inductance. Real capacitors depart from these ideals as well as possess a few other nonideal characteristics that affect their use in circuits.

5.2.1 Capacitor Value Marking

Obviously, it is necessary to provide information that indicates the value of a capacitor, perhaps with other characteristics such as breakdown voltage and tolerance. This is done by the markings placed on the component. Some standardization exists in marking codes.

Actual Stamped Value In some cases, the actual value of capacitance is written or stamped on the unit. Often the maximum working voltage and tolerance are included. This is mainly true on physically large units, such as those found in power supply filters.

In many cases the prefix is left off. If it is physically a medium to large unit, the stamped value is probably in microfarads (μF). Figure 5.5 shows some examples. Thus, if it is a silver or aluminum "can" that says "470" this means 470 μF. If it is a small tan disc marked ".01" it is 0.01 μF.

If it is very small it is probably in picofarads (pF). So a unit about the size of a 0.5-W resistor marked "39" most likely means 39 pF. As you can tell there is some uncertainty about direct value marking unless the actual prefix is included.

   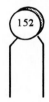

FIGURE 5.5
Value markings on capacitors.

FIGURE 5.6
Coded value markings on a capacitor.

Coded Stamped Value It is quite common to mark the capacitor with numbers that are in fact a "code" to convey the actual value. The code numbers have the same meaning as the standard colored bands markings used for resistors and other components. There will typically be three numbers. The first two are the digits or significant values and the third is a power of ten multiplier, i.e., the number of zeros. This combination gives the value of capacitance in picofarads (pF). Figure 5.6 shows a disc capacitor marked with 152. As shown, the significant figures are 1 and 5 and the multiplier is $10^2 = 100$. Thus the value is 15×100 pF = 1500 pF = 1.5 nF = 0.0015 μF.

EXAMPLE 5.9

Three capacitors are marked 104, 472, and 223. What are the values of capacitance?

Solution

Thinking of these as colored bands provides the following identification of value:

$$104 \rightarrow 10 \times 10^4 = 10 \times 10000 = 100000 \text{ pF} = 0.1 \ \mu\text{F}$$
$$472 \rightarrow 47 \times 10^2 = 47 \times \quad 100 = \quad 4700 \text{ pF} = 0.0047 \ \mu\text{F}$$
$$223 \rightarrow 22 \times 10^3 = 22 \times \quad 1000 = \quad 22000 \text{ pF} = 0.022 \ \mu\text{F}$$

Color-Coded Value   Just as with resistors, capacitors often have colored bands to indicate their value and tolerance. The relation between bands and numbers or multipliers is shown in Table 5.3. The band nearest one end of the capacitor gives the most significant value. That and the next band give the two digits or significant values of the capacity. The third band gives the multiplier as a power of ten. The value of capacity is in picofarads (pF). The fourth band, if present, indicates the *tolerance* of the capacitor (see the next section).

TABLE 5.3
Color code

Color	Value	Multiplier
Black	0	1
Brown	1	10
Red	2	100
Orange	3	1,000
Yellow	4	10,000
Green	5	100,000
Blue	6	1,000,000
Violet	7	10,000,000
Gray	8	
White	9	
No color	±20% Tolerance	
Silver	±10% Tolerance	
Gold	± 5% Tolerance	

EXAMPLE 5.10   What is the capacitance of the color-coded capacitor shown in Figure 5.7?

Solution

The color nearest one end of the unit is green and the next is brown. From Table 5.3 this gives significant digits of 51. The third band is red and Table 5.3 shows this

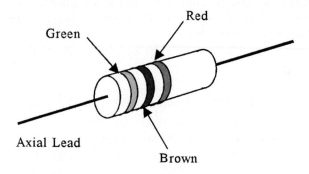

FIGURE 5.7
Color-coded capacitor for Example 5.10.

to be a multiplier of 100. Therefore the capacitance is given by

$$C = 51 \times 100 \text{ pF} = 5100 \text{ pF}$$

A small cylindrical object with a color code could be a resistor, capacitor, inductor, or diode. If the resistance is high (say, >10 MΩ) when measured in both directions with a source of at least 1 V, the unit is probably a capacitor.

5.2.2 Tolerance

A manufacturer produces many thousands of capacitors of an intended value in a typical production run. It is to be expected that there will be some variation of value among these thousands of capacitors. The manufacturer provides information about this variation through the tolerance placed on the unit value.

Tolerance is expressed as a plus-or-minus percent variation from the marked value. This means that any random unit drawn from a given batch will have an *actual* value that is within the plus-or-minus percent tolerance of the marked value.

For capacitors simply used for rough filtering or dc blocking, the actual value is often not very important. For this reason, capacitors are available with very large positive tolerance and smaller negative tolerance. Thus, a unit may be indicated to be 470 μF with a tolerance of $+80\%$ and -10%. This means the actual value is within a range of 470 μF $+ 0.8(470\ \mu$F$) = 846\ \mu$F down to 470 μF $- 0.1(470\ \mu$F$) = 423\ \mu$F. But if used for smoothing, for example, the only thing that is really important is that the capacity is large.

For more stringent applications, standard tolerances for capacitors are employed. These are indicated by direct marking, a letter code, or a color code as follows,

$\pm 20\%$ = M = no fourth colored band

$\pm 10\%$ = K = silver band

$\pm 5\%$ = J = gold band

Smaller tolerances are also available.

EXAMPLE 5.11 A capacitor is marked 682K. What range of values could this unit actually have?

Solution
The numbers indicate a value of 68×10^2 pF, which is $68 \times 100 = 6800$ pF. The K indicates a tolerance of $\pm 10\%$, so the actual value of the capacitor is in the range of

$$6800 \pm 0.1(6800) \text{ pF} = 6800 \pm 680 \text{ pF}$$

or anywhere from 6120 to 7480 pF.

5.2.3 Working Voltage

A capacitor has a maximum voltage at which it can be operated without chance of breakdown of the dielectric insulator. This value will be somewhat lower than the actual breakdown voltage to provide for a margin of safety.

In some cases the working voltage will be marked directly on the unit. For instance, a marking of 100 or 100V refers to a working value of 100 volts. Often, however, this marking is left off and manufacturer specification sheets must be consulted to determine the working voltage.

5.2.4 Temperature Coefficient

The capacitance of a capacitor is a function of temperature. This is due to temperature variations in the physical dimensions of the device and variations in the dielectric properties. For many capacitors, as employed in power supply filters, for example, this specification is not provided since the actual value of capacitance is not critical.

In general, the variation of capacitance with temperature is not linear and is indicated by a curve of variation, in percent, versus temperature. Figure 5.8 shows typical curves that might appear as part of a capacitor specification. You can see that large temperature variations are possible for some capacitors.

One way in which capacitor temperature coefficients are represented is by the nominal parts-per-million (ppm) variation per degree Celsius. Thus the designation N750 would mean that the capacity variation is negative (decreasing with increasing temperature) at 750 ppm/°C, or −0.075%/°C. This is usually valid over some specified temperature range, say, −30 to 70°C for example.

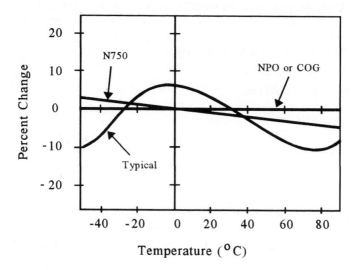

FIGURE 5.8
Typical curves of capacitance variation with temperature.

The designation NPO or COG refers to capacitors with variations of less than ± 30 ppm/°C over the range -55 to 85°C. These are very high quality, low temperature coefficient capacitors.

Many other temperature coefficient standards and classification schemes are employed by manufacturers.

5.2.5 Polarity

The ideal capacitor has no polarity, i.e., it makes no difference which end of the capacitor is connected to the positive terminal of a source. Many types of practical capacitors have a polarity requirement.

Outer Foil Marking   One type of polarity marking is used to indicate which terminal is connected to the "outer" foil of a rolled capacitor. Looking ahead, Figure 5.11 shows a drawing of a capacitor made by winding two metal foils separated by an insulator into a spiral cylinder. Notice that one of the foils ends up on the outer side of the finished component. In many circuit applications it is important to know which terminal connects to this outer foil. A black band is used to mark this terminal. In the case of color-coded components, the colored bands are on the end nearest this terminal.

Such a capacitor is still bipolar, i.e., either terminal can be connected to the positive terminal of a source.

Polarized Capacitors   Some capacitors, as discussed in the next section, actually have a preferred polarity and should be used with the more positive voltage connected to a particular terminal.

The required polarity connection may be shown by imprinting a "+" sign near the *positive* terminal. This would mean that when used in a circuit this terminal must be connected to the more positive voltage with respect to the other terminal. Generally, the value of capacitance indicated for such a unit is valid only when used with the correct polarity connection.

Other methods of indicating polarity are a black band near the *negative* terminal, the colored bands nearest the negative terminal, and a colored dot near the negative terminal.

5.2.6 Leakage Current

A practical capacitor will exhibit some finite current flow when connected to a circuit. This means that the insulator is not perfect. The leakage current possibility is often quoted as the *insulation resistance.*

High-quality capacitors may have an insulation resistance of more than 10,000 MΩ. Capacitors intended for filtering and power supply applications may have an insulation resistance of ≈ 1 MΩ.

You can see that in the first case operation at 10 V would mean a leakage current of 1 pA, whereas in the second case the leakage current would be 10 μA.

5.2.7 Dissipation Factor and Circuit Models

When capacitors are used in circuits at high frequencies, it is helpful to have an equivalent circuit model of the component. This model takes into account the effective dissipative resistance of the device as well as the capacity.

Even though the insulation resistance may be very high, the dissipative resistance at high frequency can be lower because of frequency-related losses in the dielectric.

Dissipation Factor A common representation of a capacitor is the effective series resistance (ESR) of the device, R_s, and the series capacitance, C_s, as shown in Figure 5.9. Ideally, $R_s = 0$. The dissipation factor (DF) is defined as the percentage of the effective series resistance to the capacitive reactance:

$$DF = 100(R_s/X_C) \qquad (5.11)$$

where DF = dissipation factor in percent
R_s = effective series resistance in ohms (Ω)
$X_C = (1/\omega C)$ = capacitive reactance in ohms (Ω)

Dissipation factors of capacitors, at 1 kHz, for example, can range from as low as 0.1% to 10%. Clearly, the ideal is zero.

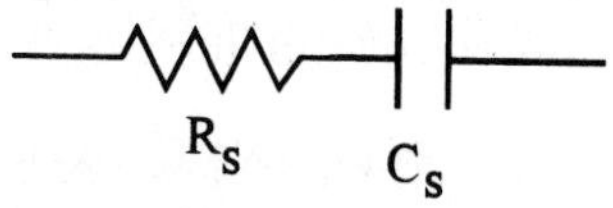

FIGURE 5.9
Dissipative circuit model of a capacitor.

The *quality, Q,* of a capacitor is the inverse of the dissipation factor, $Q = 1/DF$. In this case the ideal is for the quality to be very large.

Power Factor Capacitors are often used in applications of energy storage and retrieval. In this case, the losses associated with charging and discharging the capacitor are important. These losses are expressed by the power factor (PF). The power factor simply expresses the percentage of energy lost in charging a capacitor. A PF of 2% means that an attempt to store 100 J in a capacitor will result in a loss of 2 J to heating and other lossy mechanisms.

5.3 TYPES OF FIXED CAPACITORS

In this section many of the common types of fixed capacitors are presented along with their particular characteristics. Capacitors differ in two ways: construction and type of dielectric.

5.3.1 Stacked Capacitors

In the stacked configuration of capacitors, a series of stacked plates is employed. These plates are separated by the dielectric insulator. This capacitor, often called

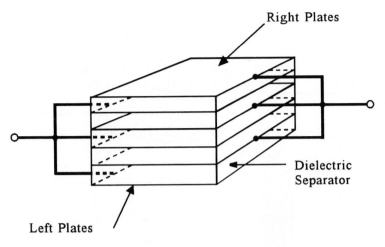

FIGURE 5.10
Multilayer, stacked capacitor.

multilayer, is illustrated in Figure 5.10. Each layer acts like a single capacitor and the entire stack is like a parallel arrangement. Since capacitors in parallel add, the net capacitance is the sum of the individual members. Of course, the assembly is encapsulated in glass, ceramic, or a high-quality plastic.

Mica A common stacked or multilayer capacitor uses mica as the dielectric and insulator. These capacitors are available in the range of 1 pF to 0.1 μF. The dissipation constant is generally very low and temperature variation is also small. They have found use for many years in communication and other high-frequency applications.

Ceramic A new type of multilayer capacitor is formed by alternate layers of deposited conductor and ceramic coating. This type of capacitor is an outgrowth of integrated circuit metal vaporization and deposition process technology. "Chip" capacitors used in hybrid thick-film and thin-film circuit development are ceramic multilayer capacitors. They are generally available in the range of 1 pF to 0.1 μF and have low DF and low temperature variation.

5.3.2 Tubular Capacitors

If the capacity must be large, we need a large plate area and small separation between the plates with a high dielectric constant material. A very common technique for making the area large, yet keeping the total size down, is to form the plates as two long strips separated by an insulator. This assembly is then rolled into a tubular form. Figure 5.11 illustrates the basic concept.

The two strips extend beyond the insulator, one on each side. This allows a secure connection to the plates with, typically, axial leads. Thus, such a capacitor

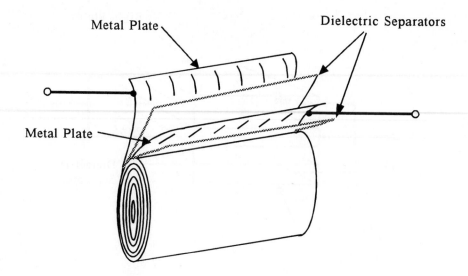

FIGURE 5.11
Tubular, rolled capacitor.

often looks like a resistor, although it is usually larger. The tubular form is then encapsulated in plastic or ceramic.

Paper   A common insulator and dielectric is paper that has been saturated or impregnated with a wax or resin material. The first tubular capacitors were made in this manner. In general the temperature range and stability, insulation resistance, and breakdown voltage characteristics are not as good as newer capacitors made with plastic insulation.

Plastic Film   Modern tubular capacitors are made using an insulator formed of plastic film material. Various materials are used, each with different characteristics with respect to dielectric constant, resistance, and thermal properties. Examples of materials employed are polystyrene, mylar (polyester), and polycarbonate.

In modern practice, the metal plates are actually deposited on each side of the plastic film. This assures a separation between the plates that is as uniform as the plastic material itself.

5.3.3   Disc Capacitors

A disc capacitor is formed by depositing metal film on each side of a ceramic dielectric insulator. Since the ceramic can be made very thin, it is possible to make the effective plate separation small to increase the capacitance. In some cases several such discs are stacked to increase the capacitance. The complete package is then encapsulated in a ceramic or plastic insulation.

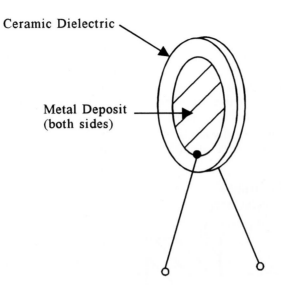

FIGURE 5.12
Disc ceramic capacitor.

Figure 5.12 shows the typical construction of such a capacitor. By variation of the composition of the ceramic, it is possible to achieve a wide variety of characteristics such as dielectric constant, temperature sensitivity, and resistance.

High dielectric disc capacitors (high-k) can provide capacitances as large as 2 to 5 μF in a small package. On the negative side, their temperature coefficients can be as large as 1%/°C (10,000 ppm/°C) and the tolerance on the marked value can be as large as -20 to 80%. They are primarily used for bypass applications and other cases where the exact capacitance value is not critical.

Low dielectric disc capacitors (low-k) are fabricated with special ceramics that have low dielectric constants and much improved temperature and tolerance characteristics. High-quality low-k disc capacitors can have a zero temperature coefficient (NPO) over the range of -55 to 85°C!

5.3.4 Electrolytic Capacitors

In general, all of the capacitor types defined thus far are limited to a maximum capacity of a few microfarads. It is possible to produce larger capacitance in such units but only if they are made very large. In many applications, such as power supply filtering, it is desirable to have capacitance values of hundreds or even thousands of microfarads. The electrolytic type of capacitor can provide these values in a modest size.

Polarized   The electrolytic capacitor must be connected to a circuit with one lead at a higher potential than the other, i.e., it is *not* bipolar with respect to its leads. In general, a "+" sign is printed by the lead or terminal that must be held at the higher potential. Sometimes a colored dot marks the positive terminal. If an electrolytic capacitor is reverse connected, at the least the capacitance will be less, perhaps much less, than the quoted value and at worst the device may rupture or even explode. Failure occurs because when it is reverse connected, it is possible for considerable leakage current to flow, which causes the device to heat up.

Figure 5.13(a) shows two metal plates with an electrolyte separator. The electrolyte is of course a conductor, so the device is not a capacitor—yet. When connected to a voltage source of the correct polarity, for the first time, a current

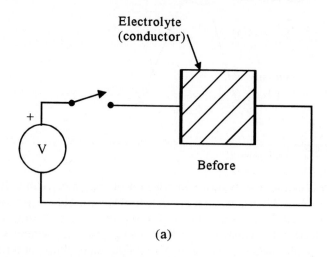

(a)

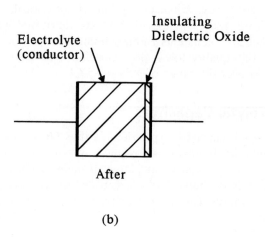

(b)

FIGURE 5.13
(a) Unpolarized electrolytic capacitor. (b) Polarized electrolytic capacitor.

will flow. This current will cause an electrolytic/metallic, electrochemical reaction, which results in the formation of a thin layer of the metal oxide on one plate. Since this is an insulator, current flow and reaction will cease when the layer covers all of the exposed surface of the metal plate.

Now the device, as shown in Figure 5.13(b), is a capacitor with a very thin insulator (the oxide layer). The remaining electrolyte serves now only as an extension of the other metal plate. Equation 5.6 now shows that with d very small the capacity can be very large. Making the area large also helps increase the capacity.

Reverse connection to the capacitor may cause depletion of the oxide layer and a subsequent current flow.

Computer Grade   Capacitor characteristics are very widely varied with respect to temperature variation, working voltage, shelf life, reliability, and a host of other considerations. Certain industrial standards have been adopted by the EIA and the military to classify capacitors. One such standard is called the computer grade. For a capacitor to be called *computer grade,* it must meet these standards. In general, this is a high-quality capacitor.

AC Electrolytic Capacitors   A simple electrolytic capacitor cannot be operated in a pure ac application because polarity reversals will occur in each cycle. For such applications, special manufacturing techniques are employed to deposit an oxide layer on *each* metal plate. Although the capacity is reduced by a factor of two, the device can now be used in ac applications. Such a capacitor, even though electrolytic, is bipolar.

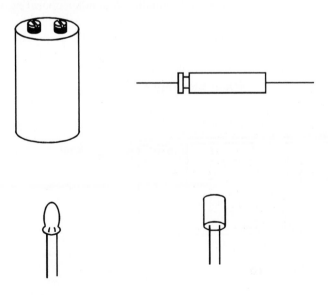

FIGURE 5.14
Typical appearance of electrolytic capacitors.

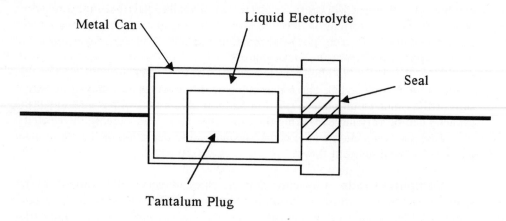

FIGURE 5.15
Wet slug tantalum electrolytic capacitor.

Aluminum   One of the most common metals for electrolytic capacitors is aluminum. The design is most often tubular, i.e., rolled aluminum plates with the electrolyte saturated in a separation material such as thick paper.

These capacitors come in a great variety of sizes and shapes as shown in Figure 5.14. There is typically a large tolerance on the capacity such as −20 to +80% of the marked value. The voltage rating may be as small as a few volts to several hundred volts. In general, the marked capacitance value is valid at or near the rate working voltage, i.e., it may be smaller at smaller operating voltages.

Tantalum   Another common metal for electrolytic capacitors is tantalum. In general, tantalum capacitors can be made much smaller than aluminum capacitors

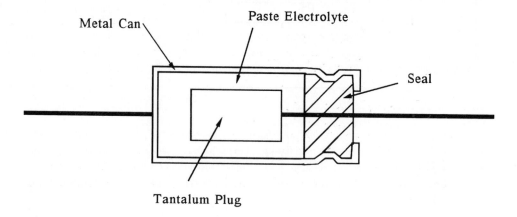

FIGURE 5.16
Dry slug tantalum electrolytic capacitor.

and yet have similar values of capacitance because the oxide has a larger dielectric constant.

There are three basic construction techniques for tantalum capacitors. One is the same as aluminum, i.e., rolled strips of metal with an electrolytic soaked paper separator. An oxide layer is formed on one of the strips.

The *wet slug* variety of tantalum capacitor, which uses a sponge plug of tantalum metal, is illustrated in Figure 5.15. The plug is quite porous and so has a very large surface area of exposed metal. An oxide layer is formed on this entire inner and outer surface. The plug is sealed in, but insulated from, a metal can, which is filled with liquid electrolyte. One connection is made to the plug itself (the "+") and another to the can.

The *solid plug* or *dry slug* type is made using the same kind of spongy plug of tantalum. In this case, however, after the oxide layer is formed the plug is filled with a paste electrolyte, which is then connected to the metal can. Figure 5.16 illustrates the basic structure of such a capacitor.

In general, the working voltage of tantalum capacitors is less than that of aluminum ones of comparable size.

5.4 VARIABLE CAPACITORS

There are many instances when the value of capacitance must be varied to alter circuit performance. We classify two modes of such variation. A *tuning* capacitor is one for which regular and repeated variation of capacity is required in some applications. An example is the tuning of different stations in an AM radio receiver. A *trimmer* variable capacitor, on the other hand, is one for which only a single time, or infrequent variation in capacitance, is necessary.

5.4.1 Basic Principles

Dependence of capacitance on physical configuration is illustrated by Eq. (5.6), which shows that variation in capacity can be effected by changing the plate separation d, changing the shared plate area A, or changing the dielectric constant of the insulator K. All three techniques are employed to provide variable capacitance.

5.4.2 Tuning Capacitors

The most common technique for providing a variable capacitor for tuning applications is to vary the area shared by the two plates. Figure 5.17 shows the basic structure of a variable tuning capacitor. Note that as the knob is rotated one set of plates is rotated with respect to the other. The effective area between the plates is thus also varied.

The dielectric between the plates is air. The stacked arrangement of plates is effectively like many capacitors in parallel so that the net capacity is the sum of each individual member. The range of capacity variation is determined by the size

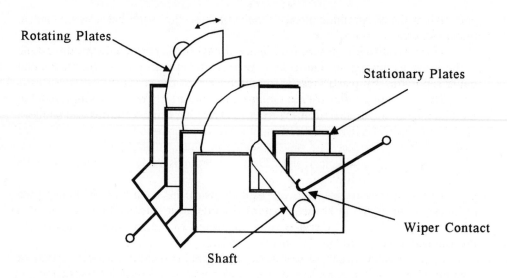

FIGURE 5.17
Rotary variable capacitor.

of the plates, the plate separation, and the number of plates. Rotation is usually over 180°.

Tuning capacitors with a maximum in the range of 1 to 50 pF will consist of small-area plates and only a few plates may be involved.

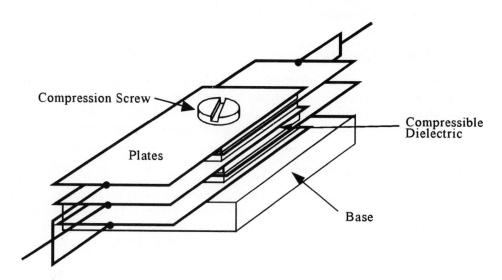

FIGURE 5.18
Compression or trimmer variable capacitor.

When the maximum capacity is in the range of 50 to 500 pF, many ganged plates of larger area are used.

Working voltage is determined by the breakdown voltage between the plates.

Miniature tuning capacitors, often used in small radios, employ very thin metal plates separated by plastic insulators. In some cases, one set of metal plates is actually deposited on the plastic. In this case, the principle is the same but the dielectric constant of the plastic will be greater than air so the capacity will be larger than that of a similar unit using air only.

5.4.3 Trimmer Capacitors

Trimmer capacitors are usually employed in applications where small adjustments of circuit capacitance are required after a system has been assembled. For this reason, the maximum value and range of adjustment are usually not large, probably not exceeding 150 pF.

Some trimmers use the same principle of tuning capacitors, with several plates and variable area. They are simply miniature versions of the tuning capacitor.

A very common type of tuning capacitor uses a set of stacked plates with a plastic or mica insulator, as shown in Figure 5.18. Variation is accomplished by a screw adjustment that squeezes the stack together, effectively changing the separation d between the plates and hence the capacity.

Another type of trimmer capacitor uses a screw to move a metal plug in a cylindrical ceramic form. The outer surface of the ceramic has a deposited metal film. As the metal slug moves up and down the cylinder, the effective area between the slug and the outer metal film is changed and thus so is the capacity.

6

INDUCTANCE*

A basic principle of physics teaches that charges in motion produce magnetic fields. Therefore, when an electric current passes through a circuit, magnetic fields are produced. Another principle of physics shows that the motion of charged particles is affected by magnetic fields. Consequently, the magnetic fields generated by the currents in a circuit affect the characteristics of current flow in the circuit. The impact of these phenomena in electric circuits is defined by the concept of *inductance*.

6.1 IDEAL CHARACTERISTICS

Under certain circumstances, a changing electric current in a circuit can induce a voltage to appear in the circuit. The relationship between the changing current and induced voltage is defined by the inductance that exists in the circuit. In this section, the basic principles of inductance are presented. In real implementations of inductance there are significant deviations from the ideal principles presented in this section. Other sections describe the practical issues of inductance construction and application.

6.1.1 Self-Inductance

It is possible for changing currents in an isolated electric circuit to induce voltages within that same circuit, i.e., within itself. This is called *self-inductance*. A circuit component designed specifically for the purpose of providing self-inductance is called an *inductor*.

* This chapter was written by Dr. William F. Schallert, Parks College, Saint Louis University.

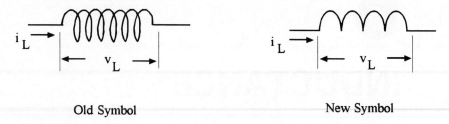

| Old Symbol | New Symbol |

FIGURE 6.1
Schematic symbols and voltage/current definitions for inductors.

Dynamic Definition   Alternative schematic symbols for the inductor are shown in Figure 6.1, along with current and voltage definitions associated with the component. The inductor symbol represents a coil that is assumed to be wound of ideal wire having no electrical resistance. Therefore, Ohm's law does not predict any voltage drop to appear across the coil, i.e., $V = IR = 0$. This turns out to be true, but only when the current through the coil is constant. The concept of self-inductance arises when we find that if the current is caused to vary in time, a voltage will be generated or *induced* across the coil terminals. The induced voltage is related to the rate of change of current by a constant. This constant is called the *inductance* and the unit is the henry (H), which is equal to one volt per (amp per second), i.e., 1 H = 1 V/(A/s). In equation form, this voltage is expressed using the derivative of the current as

$$v_L(t) = L \frac{di_L}{dt} \tag{6.1}$$

where $v_L(t)$ = voltage induced in volts (V)
L = inductance in henrys (H)
di_L/dt = rate of change of current through coil in amperes per second (A/s)

Equation (6.1) is the dynamic definition of inductance because it is defined in terms of voltages and currents in the circuit and not by the physical characteristics of the component itself.

Physical Definition   It is also possible to define the self-inductance of a coil in terms of its physical structure by making use of certain principles of physics. The coil in Figure 6.2 is wound of N turns of ideal wire to a length L_c on a core of cross-sectional area A. Physics tells us that if a current i is passed through the coil a magnetic field is generated in the coil, as shown by the dashed lines in Figure 6.2. The magnetic field intensity, H, is approximately given by the equation

$$H = \frac{Ni}{L_c} \tag{6.2}$$

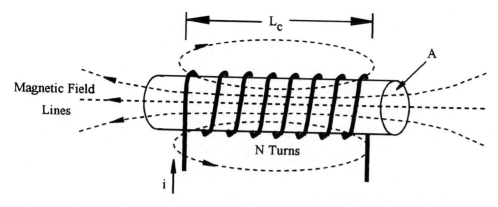

FIGURE 6.2
Physical structure of an inductor.

where H = magnetic field intensity in amperes/meter (A/m)
  i = current in amperes (A)
  L_c = coil length in meters (m)

The magnetic flux density, B, is found from the magnetic field intensity and the permeability, μ, of the core material through which the magnetic field permeates:

$$B = \mu H \tag{6.3}$$

where B = magnetic flux density in tesla (T), which is 1 weber/m² (Wb/m²), and μ = permeability of the core (T-m/A). The permeability of vacuum is given by a constant, $\mu_0 = 4\pi \times 10^{-7}$ T-m/A. The unit given for permeability is also equal to henrys/meter (H/m), showing the linkage with inductance. Permeability of materials other than a vacuum are often expressed by a relative permeability, μ_r, such that the material permeability is given by

$$\mu = \mu_r \mu_0 \tag{6.4}$$

The actual magnetic flux, Φ, passing through the coil is approximately given by the following equation:

$$\Phi = BA \tag{6.5}$$

where Φ = magnetic flux in webers (Wb)
  B = magnetic flux density in tesla (T = Wb/m²)
  A = cross-sectional area (m²)

By combining Eqs. (6.2), (6.3), and (6.5), the flux can be expressed in terms of the current through the coil by the equation

$$\Phi = \mu \frac{NA}{L_c} i \tag{6.6}$$

where all the terms have been previously defined. Finally, a basic law of physics, called Faraday's law, states that a voltage will be induced across a coil of N turns if the magnetic flux through the coil is changing in time. This is expressed as

$$v_L(t) = N\frac{d\Phi}{dt} = N\frac{d\Phi}{di}\frac{di}{dt} \tag{6.7}$$

Combining Eqs. (6.1), (6.6), and (6.7) provides the basic equation for the physical inductance of a coil:

$$L = N\frac{d\Phi}{di} = \frac{\mu N^2 A}{L_c} \tag{6.8}$$

EXAMPLE 6.1

Find the inductance of an air core coil of 100 turns, length 5 cm, and diameter 2 cm.

Solution

The permeability of air is virtually the same as that of a vacuum so that we can use μ_0 in Eq. (6.8). The area is $A = \pi(2 \times 10^{-2}\text{ m})^2/4 = 3.14 \times 10^{-4}\text{ m}^2$. Thus, the inductance is

$$L = \frac{(4\pi \times 10^{-7}\text{ H/m})(100)^2(3.14 \times 10^{-4}\text{ m}^2)}{(5 \times 10^{-2}\text{ m})}$$

$$= 7.9 \times 10^{-5}\text{ H} = 79\ \mu\text{H}$$

Energy   When current is flowing in an inductor, energy is stored in the magnetic field formed by the current. When action is taken to change the current, the magnetic field and, therefore, the stored energy will also be changed. If the current is reduced, energy will reenter the circuit from that stored in the magnetic field. If the current is increased, energy will be taken from the circuit and stored in the magnetic field. The amount of stored energy is given by the relation

$$W = \frac{1}{2}Li^2 \tag{6.9}$$

where W = stored energy in joules (J)
L = inductance in henrys (H)
i = current in amperes (A)

Large-scale inductors are sometimes used for temporary energy storage in electrical power applications. The problem with this is that real inductors have some dc resistance so that the steady current required to store the energy will be accompanied by an I^2R energy loss in the resistance.

EXAMPLE 6.2

A large inductor with a 14.5-H inductance and a resistance of 0.15 Ω carries a current of 5 A. How much energy is stored in the inductor and what is the rate of energy loss due to resistance?

Solution

The stored energy can be computed from Eq. (6.9) as

$$W = (14.5 \text{ H})(5 \text{ A})^2/2 = 181.25 \text{ J}$$

The energy loss is given by the power loss in a resistor, $P = (5 \text{ A})^2(0.15 \text{ }\Omega) = 3.75 \text{ W} = 3.75 \text{ J/s}$. Thus, it costs 3.75 J/s to store 181.25 J in the magnetic field of the inductor.

6.1.2 Mutual Inductance

Equation (6.7) predicts that if the magnetic flux through a coil changes, a voltage will be developed across the coil. Consider two coils arranged near each other so that the flux from either passes through the other. We say they are magnetically linked. If the current through one coil changes and therefore its magnetic field, a voltage will be induced in the second coil, and vice versa. This is described as a mutual inductance, M, between the coils.

Figure 6.3 shows two magnetically linked coils. If the mutual inductance between the coils is M, then the voltage induced in coil 1 from current changes in coil 2 is given by

$$v_{12} = M \frac{di_2}{dt} \tag{6.10}$$

where $\quad v_{12}$ = voltage induced in coil 1 from current changes in coil 2 (V)
$\quad\quad\quad M$ = mutual inductance in henrys (H)
$\quad\quad\quad i_2$ = current in coil 2 (A)

In a similar way, a voltage is induced in coil 2 from current changes in coil 1:

$$v_{21} = M \frac{di_1}{dt} \tag{6.11}$$

The value of the mutual inductance is related to the inductance of each coil and a geometrical factor. This factor depends on the degree of magnetic coupling between

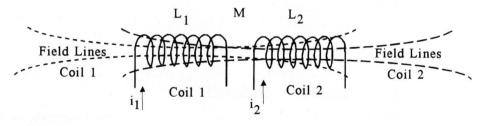

FIGURE 6.3
Flux linkage for mutual inductance.

FIGURE 6.4
Use of dots to denote aiding or impeding flux linkages.

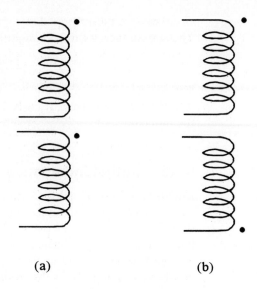

(a)   (b)

the coils. In general, the mutual inductance is given by

$$M = k\sqrt{(L_1 L_2)} \qquad (6.12)$$

where k = coupling coefficient. The coupling coefficient has a value between 0 and 1 so that the maximum value of the mutual inductance is $\sqrt{(L_1 L_2)}$.

The mutual inductance has a polarity with respect to the self-inductances of the two coils. That is, the mutual inductance can induce a voltage that is in opposite polarity to that induced from self-inductance. This is generally specified by showing a "dot" on the schematic to indicate the end of the coil which an entering current would induce a voltage to enhance the self-induced voltage of the second coil. Figure 6.4(a) shows coils for which current introduced into the top of each coil produces an induced voltage of the same sense as the self-induced voltage. In Figure 6.4(b) the opposite is true.

EXAMPLE 6.3

Find the mutual inductance between two coils with $L_1 = 20$ mH and $L_2 = 12$ mH if the coupling coefficient is $k = 0.72$.

Solution
The mutual inductance is found from Eq. (6.12):

$$M = 0.72[(20 \text{ mH})(12 \text{ mH})]^{1/2} = 11.2 \text{ mH}$$

6.1.3 Inductor Combinations

The net inductance of inductors in series and parallel depends on whether the inductors are magnetically isolated or coupled and whether they are connected to enhance or reduce the fields.

Isolated Inductors   Inductors are magnetically isolated when they are placed far apart so they do not share magnetic flux or when they are arranged at 90° or otherwise arranged so that no sharing of magnetic flux occurs. When inductors are magnetically isolated, the net inductance of a series connection is simply the sum of the inductances. Thus, if three inductors are connected in series and are not coupled magnetically, the net inductance is given by

$$L_{net} = L_1 + L_2 + L_3 \qquad \textbf{(6.13a)}$$

If inductors are connected in parallel and are magnetically not coupled, then the net inductance is the inverse of the sum of the inverses, as for resistors. Therefore, if three inductors are connected in parallel and are not magnetically coupled, the net inductance is found from

$$\frac{1}{L_{net}} = \frac{1}{L_1} + \frac{1}{L_2} + \frac{1}{L_3} \qquad \textbf{(6.13b)}$$

Coupled Inductors   When inductors are connected in series or parallel and they are coupled magnetically, then the effective mutual inductance must be taken into account. Furthermore, account must be taken for whether the field relations are such as to enhance or reduce the field due to self-inductance effects. Thus, if two inductors are connected in series and the magnetic coupling gives rise to a mutual inductance, M, then the net inductance will be

$$\text{Fields aiding:} \quad L_{net} = L_1 + L_2 + 2M \qquad \textbf{(6.14)}$$

$$\text{Fields opposing:} \, L_{net} = L_1 + L_2 - 2M \qquad \textbf{(6.15)}$$

If these same inductors were connected in parallel, the net inductance would be given by the expressions:

$$\text{Fields aiding:} \quad L_{net} = \cfrac{1}{\cfrac{1}{L_1 + M} + \cfrac{1}{L_2 + M}} \qquad \textbf{(6.16)}$$

$$\text{Fields opposing:} \, L_{net} = \cfrac{1}{\cfrac{1}{L_1 + M} - \cfrac{1}{L_2 + M}} \qquad \textbf{(6.17)}$$

EXAMPLE 6.4   Determine the total inductance of the series–parallel combination of inductances shown in Figure 6.5. Assume there is no magnetic coupling.

Solution
First we find the parallel combination net inductance:

$$L_p = \cfrac{1}{\cfrac{1}{20\,\text{mH}} + \cfrac{1}{80\,\text{mH}}} = 16\,\text{mH}$$

Now the series of this with L_1 gives the net:

$$L_{net} = 16\,\text{mH} + 16\,\text{mH} = 32\,\text{mH}$$

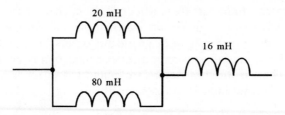

FIGURE 6.5
Inductor combinations for Example 6.4.

EXAMPLE 6.5 The two coils in Figure 6.6 have self-inductances of 35 and 122 mH. If the mutual inductance is found to be 12 mH, what is the net inductance of the combination?

FIGURE 6.6
Inductor combinations for Example 6.5.

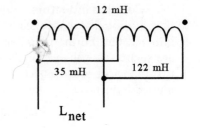

Solution
The schematic shows that the coils are wired in parallel and that they are wired in a fields opposing manner. Therefore, the net inductance is found from Eq. (6.17):

$$L_{net} = \cfrac{1}{\cfrac{1}{35 + 12} - \cfrac{1}{122 + 12}} = 72.4 \text{ mH}$$

Note that it is possible to determine the mutual inductance between two inductors by measuring the net inductance when they are connected series aiding and then series opposing. To do this, we connect the two inductors in series aiding and measure the net inductance, getting, say, L_A. We then reverse the connections for a series opposing arrangement and measure the net inductance, getting L_B. Then it is easy to show from Eqs. (6.14) and (6.15) that the mutual inductance is given by

$$M = \frac{L_A - L_B}{4} \qquad \textbf{(6.18)}$$

Note that it is not necessary to measure or even know the values of the individual inductances.

6.2 PRACTICAL ISSUES

The presentations of the previous section are correct in presenting the spirit of inductance and inductors but are deficient in issues of practical application. For example, Eq. (6.8) for finding the inductance gives only an approximate answer for a real coil and can fall far short for a coil with a magnetically active core material.

Some of the practical issues that must be considered in computing inductance are (1) real wire has some finite resistance, which will change the response of the inductor in a circuit; (2) real wire has finite dimensions, which will change the distribution of magnetic flux; (3) magnetically active core material, such as iron, has a permeability that varies depending on the strength of the magnetic field intensity, i.e., the permeability is not a constant; and (4) fringing of the field due to the finite size of wire and lack of tightness of windings means the flux density is not constant across the length or cross-sectional area of the coil.

Derived equations such as Eq. (6.8) and others can provide an estimate of the inductance. For more accurate representations we resort to semiempirical equations and measurement. Design equations are presented in later sections of the chapter for construction of various types of inductors.

6.2.1 Q of Inductors

An ideal inductor is constructed from wire with no electrical resistance. The effect of finite wire resistance was illustrated in Example 6.2 for a dc case, where the power loss in wire resistance is a nonideal effect. This finite resistance also affects the behavior of an inductor in ac problems. The result is often described in terms of the "quality" or Q of the coil.

The ac impedance of an ideal inductor would be given in terms of a pure inductive reactance,

$$Z_{\text{ideal}} = jX_L = j\omega L$$

The presence of wire resistance in any real inductor gives rise to a real part of the inductance. Furthermore ac effects in the core, called *eddy currents,* add more dissipative effects, which appear effectively as an increased resistive part of the impedance. Thus the actual impedance of an inductor is given by

$$Z_L = R_e + j\omega L$$

where Z_L = impedance of the inductor
R_e = effective resistance of the inductor at angular frequency ω
ω = angular frequency
$= 2\pi f$ with f as the frequency in hertz (Hz)

A measure of the effective resistance of a coil is given by the quality or Q of the coil, which is defined as the ratio of reactance to effective resistance:

$$Q = \frac{\omega L}{R_e} \tag{6.19}$$

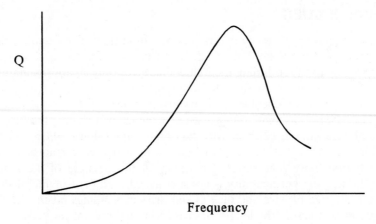

FIGURE 6.7
Q of an inductor versus frequency.

If R_e simply represented the ohmic resistance of the coil wire, this equation would predict that the Q should rise linearly with frequency. However, the effective resistance includes anything that causes a loss of energy from the inductor. Therefore, eddy current losses in the core, loses in dielectric materials forming the inductor assembly, and skin-effect increases in resistance are included. All of these effects increase at higher frequencies. For this reason, a coil Q more often has the appearance like that shown in Figure 6.7 with a maximum at some frequency.

6.2.2 Equivalent Circuits

The equivalent circuit of an inductor is one that expresses the real properties of an inductor in terms of resistance, inductance, and capacitance. Such models allow analysis of the circuit to be made using network laws. For lower frequencies, where

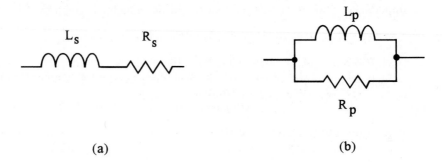

FIGURE 6.8
(a) Series and (b) parallel equivalent circuits of inductors.

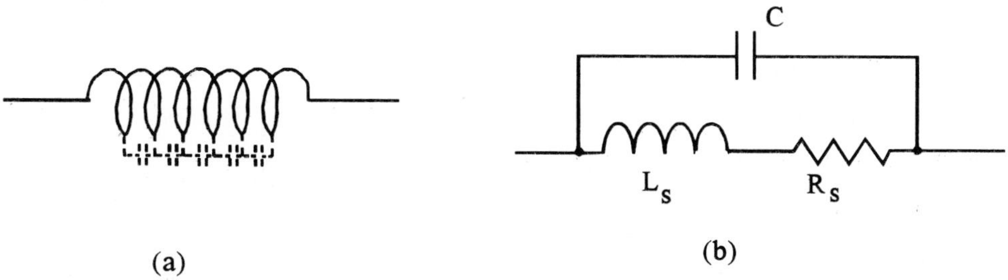

FIGURE 6.9
(a) Distributive capacity in inductors and (b) the equivalent circuit.

distributed capacity can be neglected, the equivalent circuit can be described as an ideal inductance in either series or parallel with an effective resistance as shown in Figure 6.8.

For the models shown in Figure 6.8 the values of the model components are given by equations involving Q. If the series model has an inductance L_s and resistance R_s, then the parallel model will have values given by

$$L_p = L_s \frac{1 + Q^2}{Q^2} \tag{6.20}$$

$$R_p = R_s(1 + Q^2) \tag{6.21}$$

The inductor Q will be the same whether the series or parallel equivalent circuit is used, but the equation for Q in terms of the parallel model is different:

$$Q = \frac{R_p}{2\pi f L_p} \tag{6.22}$$

For Q values greater than about ten, L_s and L_p are approximately equal. Note that because of Q, the models are valid for only a selected frequency.

At higher frequencies the distributed capacity between windings of the coil becomes important. Figure 6.9(a) shows how a capacity exists between every turn of the coil. Figure 6.9(b) shows an equivalent circuit, which is more appropriate for cases when the distributed capacity is important.

EXAMPLE 6.6 Calculate the Q of an inductor with a series resistance of 8.18 Ω and a series inductance of 8.72 mH at a frequency of 1 kHz.

Solution
From Eq. (6.19), we find

$$Q = \frac{2\pi f L_s}{R_s} = \frac{2\pi(1000)(0.00872)}{8.18} = 6.7$$

EXAMPLE 6.7

Determine the parallel representation of the inductor of Example 6.6.

Solution

From Eq. (6.20) and (6.21) and the value of Q determined in Example 6.5, we find

$$L_p = L_s \frac{1 + Q^2}{Q^2} = 8.72 \frac{1 + 6.7^2}{6.7^2} = 8.91 \text{ mH}$$

$$R_p = R_s(1 + Q^2) = 8.18(1 + 6.7^2) = 375 \text{ } \Omega$$

6.3 AIR CORE INDUCTORS

Inductors with an air core are used in applications requiring relatively small inductance and Q but greater stability. They are particularly useful in high-frequency tuned circuits and in the construction of inductance standards.

Most air core inductors are wound as single or multilayer coils on cylindrical forms. In other cases they are wound on square or polygonal forms. In addition they are sometimes wound on toroids and as flat spirals.

6.3.1 Skin Effect

Current carried in wires changes in distribution across the conductor as a function of frequency. This changes the inductance. At low frequency the current is distributed uniformly across the conductor but as the frequency is increased the current tends to flow in a thin layer at the wire surface. Since the current is carried in a smaller cross-sectional area of the conductor, the resistance presented to current flow is larger.

Skin effect depends on the frequency, resistivity of the conductor, geometry of the conductor, and permeability of the conductor. Estimates of the skin effect on round wire resistance can be made using Figure 6.10. This figure shows the variation of ac to dc resistance ($K_x = R_{ac}/R_{dc}$) with a quantity x defined by

$$x = \sqrt{\frac{8\pi f \mu_r}{R_x \times 10^9}} \tag{6.23}$$

where
f = frequency (Hz)
μ_r = relative permeability of the conductor
R_x = dc resistance per centimeter of the conductor (Ω/cm)

From Figure 6.10 you can see that when the frequency is low, and therefore x is small, K_x is equal to one so that ac and dc resistance are the same. As the frequency increases, x gets larger and K_x increases so that the ac resistance is larger than the dc resistance. For x larger than about 4 to 5, the ac resistance increases approximately linearly with x. The approximate value of K_x for $x > 4$ is given by

$$K_x = 0.3536x + 0.2286 \tag{6.24}$$

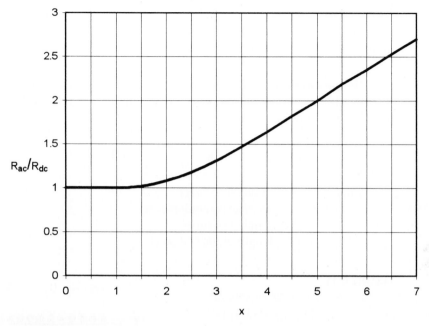

FIGURE 6.10
Skin effect described as ac to dc resistance ratio.

EXAMPLE 6.8

Contrast the ac to dc resistance ratio for AWG #20 copper wire at 1 kHz, 100 kHz, and 10 mHz. The relative permeability of copper is one, $\mu_r = 1$.

Solution

From Appendix 2, the dc resistance of AWG #20 copper wire is 10.1 Ω/1000 ft. This is converted to the required ohms per centimeter by

$$R_x = (10.1 \ \Omega/1000 \ \text{ft})(1 \ \text{ft}/12 \ \text{in.})(1 \ \text{in.}/2.54 \ \text{cm}) = 3.31 \times 10^{-4} \ \Omega/\text{cm}$$

Now the value of x can be expressed in terms of frequency using Eq. (6.23):

$$x = \left[\frac{8\pi}{3.31 \times 10^{-4} \times 10^9} \right]^{1/2} \sqrt{f} = 8.71 \times 10^{-3} \sqrt{f}$$

Now the ratio of ac to dc resistance, K_x, can be found from Figure 6.10 and/or Eq. (6.24):

$$
\begin{array}{llll}
f = 1 \ \text{kHz} & x = 0.27; & \text{from Figure 6.10,} & K_x = 1 \\
f = 100 \ \text{kHz} & x = 2.76; & \text{from Figure 6.10,} & K_x = 1.25 \\
f = 10 \ \text{MHz} & x = 27.5; & \text{from Eq. (6.24),} & K_x = 9.95
\end{array}
$$

Thus you can see that the ac resistance has increased to nearly ten times the dc value at 10 MHz.

6.3.2 Self-Inductance Calculations

Semiempirical equations have been developed that modify Eq. (6.8) to allow quite accurate calculations of air core coil inductance.

Single-Layer Air Coil   Figure 6.11 shows a single-layer coil. The wire diameter is considered very small compared to the coil radius. In this case, the inductance in μH can be found from the equation

$$L = \frac{0.394r^2N^2}{9r + 10L_c} \; \mu\text{H} \qquad (6.25)$$

where
L = inductance (μH)
r = average coil radius (cm)
N = number of turns
L_c = length of the coil (cm)

FIGURE 6.11
A single-layer cylindrical air core coil.

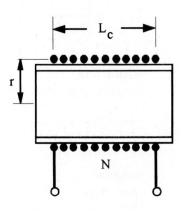

EXAMPLE 6.9

A single-layer coil is required with an inductance of 30 μH, a length of 5.0 cm, and a diameter of 1.0 cm. How many turns are required?

Solution
From Eq. (6.25) we need to solve for N since everything else is known. The radius will be 0.5 cm. Solving for N,

$$N = \left[\frac{L(9r + 10L_c)}{0.394r^2}\right]^{1/2} = \left[\frac{30[9(0.5) + 10(5)]}{0.394(0.5^2)}\right]^{1/2} \approx 13 \text{ turns}$$

Multilayer Coil   The inductance of a multilayer air core coil can be found by reference to the geometry shown in Figure 6.12. Here r is the average radius of the coil layer and d is the width of the coil as shown, both in centimeters. In this

FIGURE 6.12
A multilayer cylindrical air core coil.

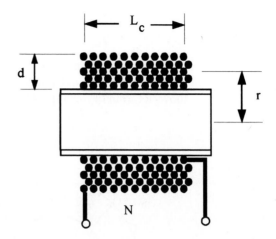

case, and using the same definitions as for Eq. (6.25), the inductance in μH is given by the following equation:

$$L = \frac{0.315r^2N^2}{6r + 9L_c + 10d}\mu\text{H} \tag{6.26}$$

Flat Coil A flat coil is one that is wound in a plane in a spiral. This configuration is shown in Figure 6.13. The inductance for such a coil can be found from the equation

$$L = \frac{0.394r^2N^2}{8r + 11d} \tag{6.27}$$

where Figure 6.13 shows how d describes the width of the coil.

In all of the preceding equations, you should note that the parameters are not independent since the actual diameter of the wire will affect the maximum number of turns. For example, if the wire diameter is D and the number of turns in the coil is N, then the minimum length of a single-layer coil, L_c, will be given

FIGURE 6.13
A single-layer wafer coil.

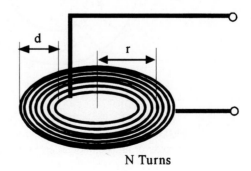

by $L_c = ND$. Thus N and L_c are not independent unless we want to keep the length greater than ND.

Toroidal Coils For toroidal coil inductors the equations given next assume that the coil is wound uniformly around the toroid in one layer. Figure 6.14(a) shows a toroid of rectangular cross section and Figure 6.14(b) shows one of circular

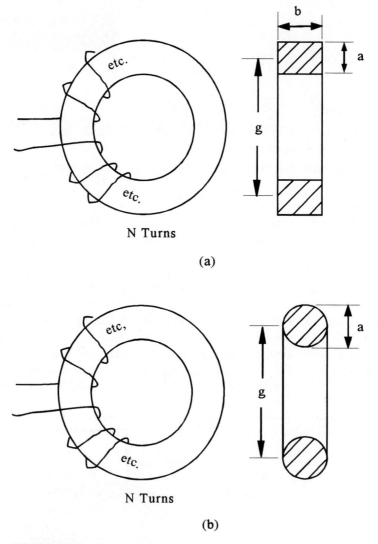

FIGURE 6.14
Toroids of (a) rectangular and (b) circular cross-section.

cross section. This figure also serves to define the quantities appearing in the following equations.

For the rectangular cross section,

$$L = 2N^2 b \log_e \left(\frac{g+a}{g-a} \right) \text{nH} \tag{6.28}$$

where you should note that the inductance is given in nano-henrys (10^{-9} H).

When the cross section of the toroid is circular, the inductance can be calculated from the equation

$$L = 0.4\pi N^2 [g - \sqrt{(g^2 - a^2)}] \, \mu\text{H} \tag{6.29}$$

where the inductance comes out in micro-henrys (10^{-6} H).

EXAMPLE 6.10

Determine the inductance of a toroid with a rectangular cross section with $g = 6.25$ cm, $a = 1.5$ cm, and $b = 2.05$ cm. A single layer of 86 turns is wound on the toroid.

Solution
From Eq. (6.28) the inductance is found as

$$L = 2(86^2)(2.05) \log_e \left(\frac{6.25 + 1.5}{6.25 - 1.5} \right) \text{nH}$$

$$= 14845 \text{ nH} = 14.845 \, \mu\text{H}$$

Coaxial Cable A coaxial cable is one having an inner conductor of radius r_i in the center of a hollow outer conductor of radius r_o as shown in Figure 6.15. The inductance depends on what kind of insulator exists between the conductors. For an air insulator the *inductance per meter* of such a coil is approximately given by

$$L_l = \frac{1}{20} \left(1 + 4 \log_e \frac{r_o}{r_i} \right) \mu\text{H/m} \tag{6.30}$$

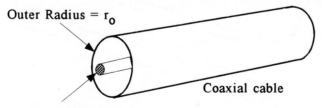

Outer Radius = r_o

Coaxial cable

Inner Radius = r_i

FIGURE 6.15
Coaxial cable has inductance per unit length.

EXAMPLE 6.11

Determine the inductance of 5000 m of a cable with $r_i = 0.055$ cm and $r_o = 0.225$ cm.

Solution

The inductance per unit length is given by Eq. (6.30):

$$L_l = \frac{1}{20}\left(1 + 4\log_e\frac{0.225}{0.055}\right)\mu H/m = 0.33\ \mu H/m$$

$$L = (5 \times 10^3\ m)(0.33\ \mu H/m) = 1.66\ mH$$

6.3.3 Mutual Inductance Calculations

Mutual inductance represents the electrical linking of two circuit components through magnetic fields generated by currents in the circuits. The value of the mutual inductance relates the magnitude of induced voltage in one circuit to the rate of change of current in another circuit. There are semiempirical equations which allow calculation of approximate values of mutual inductance.

Parallel Wires   Two parallel wires have a mutual inductance between them which is best expressed as the mutual inductance per unit length. The following equation allows one to calculate the approximate mutual inductance per meter:

$$L = \frac{1}{10}\left[1 + 4\log_e\left(\frac{d}{\sqrt{r_1 r_2}}\right)\right]\mu H/m \qquad (6.31)$$

where, as shown in Figure 6.16, d is the separation between wires, and r_1 and r_2 are the wire radii. This inductance represents a common source of noise in electrical circuits where two lines run parallel to each other for some distance.

Calculations of mutual inductance for other configurations, such as coils separated by a gap or concentric coils of different radii, are quite difficult. Figure 6.17 shows three common coil configurations, two for single-layer coils and one for

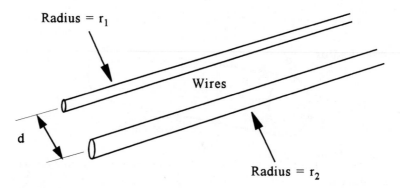

FIGURE 6.16
Parallel wires have a mutual inductance.

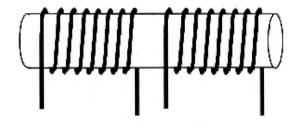

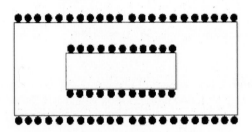

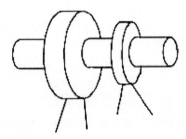

FIGURE 6.17
Three examples of mutual inductance coil arrangements.

multiple-layer coils. A good way of finding the mutual inductance is to measure the inductance of the coils when they are series aiding, L_A, and series opposing (reverse the leads), L_B. Now, from Eq. (6.18), the mutual inductance can be computed. Note that the inductance of the individual coils need not be measured.

EXAMPLE 6.12

Two small rf coils are wound on a common form as shown in Figure 6.18. When connected in series one way the net inductance is 67 μH, and when the connections are reversed the inductance turns out to be 135 μH. What is the mutual inductance?

FIGURE 6.18
Coil for Example 6.12.

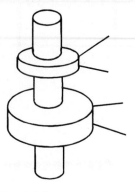

Solution
From Eq. (6.18) we find

$$M = (L_A - L_B)/4 = (135 - 67)/4 \, \mu\text{H}$$
$$= 17 \, \mu\text{H}$$

6.4 IRON CORE INDUCTORS

The use of iron and other magnetically active materials for coil cores can result in a significant increase in inductance over air core types. The basic equation for inductance given by Equation (6.8) remains the same but the value of permeability can be much greater than that of air. In this section the principles and applications details of iron-core inductors will be described.

6.4.1 Basic Concepts

A magnetically active material is one that can enhance the magnetic flux density which would be generated by some current-induced magnetic field intensity in the material. When used for a core material, this leads to an increase in the effective inductance of a coil.

Relative Permeability Equation (6.3) shows that the magnetic flux density in a material is given by the permeability of that material times the current-generated magnetic field intensity. For some materials the value of relative permeability, defined by Eq. (6.4), can be much greater than unity. Table 6.1 gives the average relative permeability for a number of materials. The consequence of this is that the magnetic flux density will be large and hence so will the magnetic flux given

TABLE 6.1
Relative permeability

Material	μ_r
Vacuum	1.0
Air	1.0
Transformer iron	3000–5000
Cobalt	60
Nickel	50
Cast iron	60–90
Pure iron	4000–8000
Silicon steel	18,000

by Eq. (6.5). This leads to an increase in the inductance by the same factor as shown by Eq. (6.8).

EXAMPLE 6.13   Suppose the core used in Example 6.1 is changed to iron with a relative permeability of $\mu_r = 4000$. What is the new inductance?

Solution
The inductance for a relative permeability of unity was found in Example 6.1 to be 79 μH. Therefore, with the iron core we have

$$L_{Fe} = 4000 L_{air} = 4000(79\ \mu H)$$
$$= 316\ mH$$

You can see that magnetically active materials can lead to a significant increase in inductance for the same coil configuration. This shows that use of iron for a coil core can provide a given inductor of a much smaller size and fewer turns that an equivalent air core type.

BH Curves   A difficulty with using iron and similar materials for inductor cores is that the permeability is not a constant with respect to current through the inductor. This variation is described by revising Eq. (6.3) to show that the permeability depends on the magnetic field intensity:

$$B = \mu(H)H \tag{6.32}$$

Now, since H depends on the current as shown in Eq. (6.2), μ also depends on the current. Since μ shows up in the inductance, the inductance also depends on the current. Thus, if this inductor is used in a circuit where the current varies, the inductance will vary also.

Figure 6.19 shows a typical curve of magnetic flux density B versus magnetic field intensity H for a magnetically active material. This shows that the permeability,

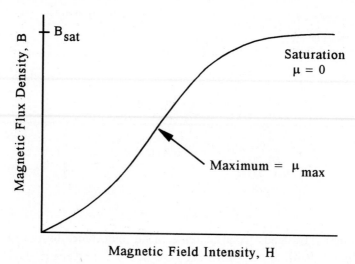

FIGURE 6.19

Typical *B–H* curve for a magnetically active material.

which is the slope of the curve, starts out with some modest value and then increases to a large value at the inflection point labeled μ_{max}. Further increases in H (or equivalently in the current i, which produces H) cause the permeability to decrease until it is virtually zero. Further increases in current and magnetic field intensity cause no further increases in flux density. We say the material is *saturated*. Equation (6.8) showed that the inductance depends on the rate at which flux changes with current. This is just the slope of the *B–H* curve such as that given in Figure 6.19.

From the preceding argument you can see that the inductance will vary with current in such a manner as to go through a maximum at the same current that produces the inflection point in Figure 6.19. Thus, Figure 6.20 shows how the

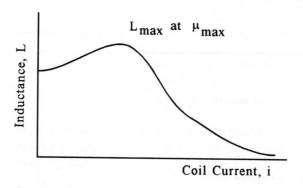

FIGURE 6.20

The inductance of an iron core inductor varies with current.

FIGURE 6.21
An air gap is built into some iron core inductors.

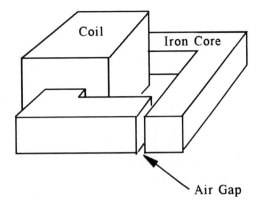

inductance of a typical iron core inductor will vary with current. You can see that it goes through a maximum when the permeability is a maximum and actually goes to zero as the magnetic flux density becomes saturated. That is, at a high enough current the inductance of the "inductor" will be zero!

Because of the variation of inductance with current, specifications of iron core inductors will give the inductance at a particular current. This is often the maximum value of inductance as shown in Figure 6.20. If operated at higher or lower currents, the inductance will be less.

Filter Choke   One common application of inductors is as a filter to block frequency components from passing into a circuit. For power supplies this reduces the 60-Hz ripple from a rectifier, whereas in rf applications they block high frequencies from passing into the power supply. An inductor in such an application is called a *choke*. You can see that if the dc current at which the choke operates is allowed to vary much, its inductance will change and thus reduce its effectiveness as a filter element.

Air Gaps   In some cases air gaps are intentionally built into the core of inductors. Figure 6.21 shows an iron core inductor with a small air gap. The consequence of this is a reduction of the effective permeability of the core and a resulting reduction of the inductance. The advantage of the gap is that it effectively increases the range of magnetic field intensity, i.e., current, over which the device can operate before saturation. This effect is illustrated in Figure 6.22. This figure shows the *B–H* curve with and without the gap. Since the saturation value of *B* in the iron is the same with or without the gap, the range of *H* over which the core is *not* saturated is increased when the gap is included. Of course, the slope of the *B–H* curve is also smaller so the permeability is smaller than with no gap.

Eddy Current Losses and Laminations   Figure 6.23(a) shows a cross section of an inductor iron core through which the magnetic flux generated by current in the coil passes. The iron is a conductor and therefore has free electrons. When

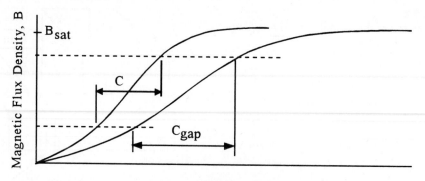

FIGURE 6.22
The air gap extends the range of nonsaturation of inductors.

the electrons move perpendicular to the magnetic field, they will be forced by the field to move in circles, as noted in the figure by dashed lines. This then represents a transverse current in the iron, called an *eddy current*. Heating of the core will occur from such current due to i^2R losses in the iron so the overall efficiency of the inductor is reduced.

To combat eddy losses, a number of approaches are taken but they all involve using insulating material in the core to block current flow. The most common is to laminate the core as shown in Figure 6.23(b). Each lamination is separated from the next by a layer of insulating lacquer so that no transverse current can flow.

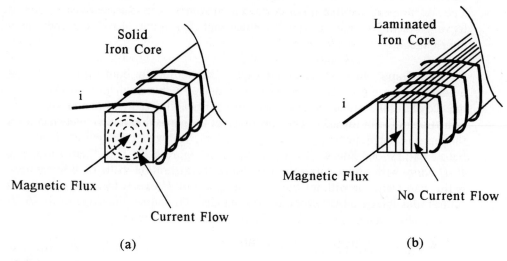

FIGURE 6.23
Eddy currents are reduced by lamination of the core.

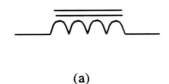

(a)

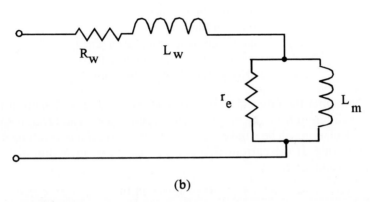

(b)

FIGURE 6.24
(a) The schematic symbol of an iron-core inductor and (b) its circuit model.

Another method binds iron powder with an insulator, which is formed into the core shape under heat and pressure.

Equivalent Circuit of Iron Core Inductor The schematic symbol for an inductor is modified slightly when an iron core is used. Figure 6.24(a) shows the symbol for an iron core inductor. Equivalent circuits for inductors were given earlier for series and parallel models. When iron cores are used with inductors a better model is the one shown in Figure 6.24(b). This includes losses in the core. In this equivalent circuit, R_w is the actual dc winding resistance whereas r_e represents the resistance losses due to eddy currents in the core. The reactance, X_w, is the actual winding reactance, and X_m represents the reactance contributed by the core.

Frequency Effects The permeability of magnetically active material such as iron diminishes with ac current frequency and consequently so does the inductance. Generally inductors are designed for operation in a certain range of frequencies and should not be used beyond that range.

6.4.2 Calculation of Inductance

It is very difficult to calculate the inductance of iron core inductors because of the variation of permeability with magnetic field or current. If a *B–H* curve for the material is known and the magnetic field intensity can be determined, then the slope of the B–H curve will give the permeability. From this the inductance can be calculated by the following equation.

The basic equation for determination of the inductance is given by

$$L = \frac{2\pi A_m N^2 \mu_r}{5 l_m \times 10^8}$$

(6.33)

where L = inductance in henrys (H)
N = number of turns
A_m = magnetic cross-sectional area (cm^2)
l_m = magnetic mean length of core (cm)
μ_r = relative permeability of the core

Note that the physical cross-sectional area and mean length of the core may be different from the magnetic cross-sectional area and length. Of course, this calculation will be valid only for a specific current and resulting magnetic field intensity. If the current changes, so will the permeability and consequently the inductance.

EXAMPLE 6.14

An inductance bridge is used to measure inductance of a coil at 1 kHz. Find the effective permeability if the inductance is found to be 7.92 mH with A_m = 4.58 cm^2, N = 200 turns, and l_m = 25 cm.

Solution
For this problem we solve Eq. (6.33) for the permeability and then substitute the appropriate values:

$$\mu_r = \frac{5 L l_m \times 10^8}{2\pi A_m N^2} = \frac{5(7.92 \times 10^{-3})(25) \times 10^8}{2\pi(4.58)(200)^2} = 86$$

6.4.3   Core Types

Iron core inductors are available in a wide range of inductance values and current carrying ability. They range in size from a few milligrams to many kilograms. Inductors are classified by the type of core (stacked laminations, tape wound, powdered), the frequency of application (60, 400, 800, 1000 Hz, audio frequencies, radio frequencies, and higher), by the application (choke, filter, reactor, impeder, power), and by the core material (grain oriented, ferrite, iron powder).

Iron Powder and Ferrite Core Inductors   Ferrite and iron powder cores are two types of powdered core inductors. Cores are made of processed particles and suitable binders and are molded into desired shapes, including the toroid. Each

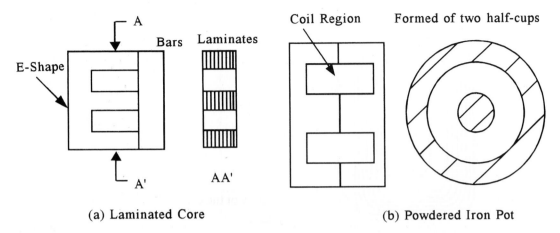

(a) Laminated Core   (b) Powdered Iron Pot

FIGURE 6.25
Two types of inductor iron cores.

will have different characteristics such as permeability, frequency range, saturation flux density, and temperature coefficient.

Iron powder cores are made from fine processed iron particles. Relative permeabilities that range from 5 to 90 are typical. They are used as inductors for high and very high frequencies. Iron powder core inductors have a moderate Q and good stability over a wide range of temperature, dc current, and flux.

Ferrite cores are made from a combination of manganese, magnesium, cobalt, nickel, iron, and zinc oxides. These cores have relative permeabilities ranging from below 100 to more than 10,000. The various core types are shown in Figure 6.25. Cores are available from core manufacturers in many forms. They provide catalog listings for many shapes, sizes, and types of magnetic materials.

Tape-Wound Cores   Tape-wound cores are simply cores in the form of a long strip of thin, lacquered iron. This is then wound, like tape, into the desired shape and baked to make the system rigid. The coil is then wound around the now laminated core. Figure 6.26 shows a tape-wound core.

Magnetic Cross-Sectional Area   The magnetic cross-sectional area will not be the same as the physical cross-sectional area of a core. This is because the lacquer between the laminates and the binder in the powdered forms take up some of the area. A stacking factor (SF) is given by the core manufacturer to specify the fraction of the physical area that is magnetically active:

$$A_m = (SF)A \tag{6.34}$$

where A is the physical cross-sectional area.

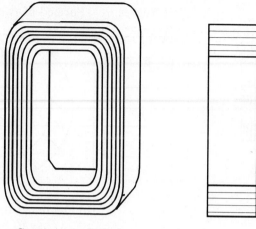

Continuous Spiral

FIGURE 6.26
The tape-wound iron core.

EXAMPLE 6.15

A toroid of rectangular cross section, as shown in Figure 6.14(a), has $g = 4$ cm and $a = b = 1$ cm. The relative permeability is found from a B–H curve to be 90 and the stacking factor is given to be SF = 0.92. If the toroid is wound with 256 turns, find the inductance.

Solution

The inductance can be found from Eq. (6.33). First the physical cross-sectional area is given by $A = ab = (1$ cm$)(1$ cm$) = 1$ cm^2 so the magnetic cross-sectional area is found to be

$$A_m = (SF)A = (0.92)\ 1\ \text{cm}^2 = 0.92\ \text{cm}^2$$

The mean magnetic length is the circumference represented by a diameter of g, so,

$$l_m = \pi g = \pi(4\ \text{cm}) = 12.6\ \text{cm}$$

Now the inductance can be found as

$$L = \frac{2\pi(0.92\ \text{cm}^2)(256)^2(90)}{5(12.6\ \text{cm})(10^8)} = 5.4\ \text{mH}$$

6.4.4 Variable Inductors

A variable inductor is one for which the value of inductance can be changed by external, usually mechanical, means. Several methods are used to accomplish this but the most common is to provide a core that can be screwed into and out of the central part of an inductor as illustrated in Figure 6.27. In many cases this is used

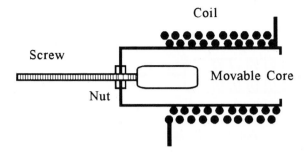

FIGURE 6.27
A variable inductor has a tuning slug of magnetic material.

to make only small changes in the value of inductance, i.e., to trim its value. Variation ranges from ±10 to ±80% of the nominal value.

Another type of variable inductor is one that varies the mutual inductance between two coils. This is done by physically moving one coil so that it shares a varying amount of magnetic flux with the other coil.

6.4.5 Inductor Specifications

Specifications for inductors depend on the application intended and the type of construction involved in their manufacture.

Air Core Typical specifications for air core inductors are simply the inductance in henrys and the electrical resistance. This allows the series model to be determined quickly. Care should be taken about the maximum amount of current that is placed through the inductor so that damage from overheating does not occur. In some cases the appropriate range of frequencies for which the indicated inductance is valid is also given.

Chokes Generally chokes are inductors that have an iron core. For low-frequency applications, these are usually iron laminates, tape-wound laminates, or powdered iron. For higher frequency applications the powdered iron and ferrite cores are used.

Since these inductors use a magnetically active material the specification will give the inductance at a particular current, as well as the dc resistance. Often the rf chokes will include the Q specified at a particular frequency. In that case the effective series resistance can be calculated.

Many miniature inductors have the appearance of a normal cylindrical resistor, even to the extent of having the inductance value indicated by colored bands. To distinguish the component from a resistor, a double-wide silver band is placed at one end of the device. This is followed by the usual four bands indicating first and second significant figures, the multiplier, and the tolerance. The value of inductance

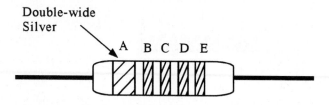

FIGURE 6.28
Inductors are often marked with a color code to indicate their values.

is specified in micro-henrys (μH). Figure 6.28 shows the physical appearance and Table 6.2 gives the color code scheme for inductors.

Note that if one of the significant figure bands is gold, it represents an inductance of less than 10 μH and a decimal point. The third band is then the second significant figure and there is no multiplier.

TABLE 6.2
Inductor color code

Color	Band B and C	Band D	Band E
	Significant Figures	Multiplier	Tolerance
Black	0	1	
Brown	1	10	±1%
Red	2	100	±2%
Orange	3	1000	±3%
Yellow	4	10,000	±4%
Green	5		
Blue	6		
Violet	7		
Gray	8		
White	9		
None			±20%
Silver			±10%
Gold	Decimal point		±5%

EXAMPLE 6.16   A component has five bands. A double-wide silver band is followed by a red band, a green band, an orange band, and a gold band. What is the inductance?

Solution

The double silver indicates that the device is an inductor. The next two bands give 25 and the multiplier is found from orange to be 1000. Therefore, the inductance is 25,000 μH or 25 mH. The tolerance is $\pm 5\%$ from the gold band at the end.

EXAMPLE 6.17   A small cylindrical object has a double-wide silver band, a gray band, a gold band, a red band, and another red band. What is the inductance?

Solution

Again the double silver indicates an inductor. The gray band is an eight and the next band being gold indicates a decimal point followed by a red, which will be the second significant figure of two. There is no multiplier so the last band indicates a tolerance of $\pm 2\%$. The value is 8.2 μH.

Of course, in both cases the resistance of the inductors could simply be measured with a dc ohmmeter.

7

TRANSFORMERS*

7.1 BASIC CONSIDERATIONS

7.1.1 Transformers

Transformer action occurs whenever two or more coils are magnetically coupled. Transformers are used extensively in both ac power applications and in electronics. There are four basic uses for these devices:

1. Voltage transformation
2. Current transformation
3. Impedance transformation
4. Electrical isolation.

Transformers can be classified as follows:

Frequency: power, audio-frequency, intermediate frequency, radio-frequency
Core: air core, iron core, or other ferromagnetic material
Phase: single phase, three phase
Other: pulse, flyback, auto, input, output, interstage, etc.

Figure 7.1 shows several schematic symbols used to represent transformers.

7.1.2 Self-Inductance and Mutual Inductance of Two Coils

Figure 7.2 shows two coils with magnetic coupling. In this situation, not all of the flux generated by one coil links the other. The degree of coupling is defined as

* This chapter was written by Dr. William F. Schallert, Parks College, Saint Louis University.

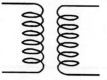
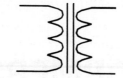
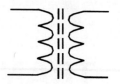

(a) Air Core, Old Symbol   (b) Iron Core, New Symbol   (c) Powdered Core

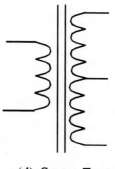

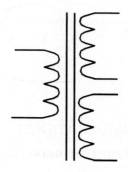

(d) Center Tapped

(e) Multiple Secondary

FIGURE 7.1
Schematic symbols for transformers.

$$k = \frac{M}{\sqrt{L_1 L_2}}$$ (7.1)

where M = mutual inductance of the two coils
         L_1 = self-inductance of N_1
         L_2 = self-inductance of N_2

FIGURE 7.2
Inductive coupling.

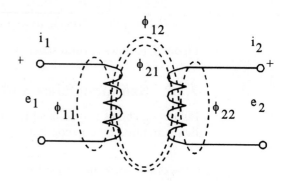

7.1.3 Ideal Transformer

A real transformer can be approximated by an ideal transformer. Further, ideal transformer relations are easier to use than mutual inductance relations. Figure 7.3 shows the schematic symbol and nomenclature for the ideal transformer. The assumptions for this model are that there is no leakage flux, the coils contain no resistance, and the core contains no losses. The winding connected to the source is called the *primary* and the winding connected to the load is called the *secondary*.

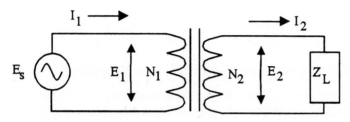

FIGURE 7.3
Definitions of voltage and current for the ideal transformer.

The ideal transformer relationships relating to the primary and secondary are

$$\frac{E_1}{E_2} = \frac{N_1}{N_2} \tag{7.2}$$

$$\frac{I_1}{I_2} = \frac{N_2}{N_1} \tag{7.3}$$

$$\frac{Z_1}{Z_2} = \left(\frac{N_1}{N_2}\right)^2 \tag{7.4}$$

where E_1 = rms value of primary voltage
E_2 = rms value of secondary voltage
I_1 = rms value of primary current
I_2 = rms value of secondary current
N_1 = number of turns in primary
N_2 = number of turns in secondary
Z_1 = impedance of the primary (E_1/I_1)
$Z_2 = Z_L$ = impedance of the secondary (E_2/I_2)

EXAMPLE 7.1   A transformer having a turns ratio (N_1/N_2) equal to 6 is connected to a 120-V, 60-Hz line. What voltage is supplied by the secondary?

Solution
Using Eq. (7.2),

$$E_2 = \frac{E_1}{N_1/N_2} = \frac{120}{6} = 20 \text{ V}$$

EXAMPLE 7.2

An impedance of $60\angle130°\,\Omega$ is connected to the secondary of a transformer having a turns ratio (N_1/N_2) of $20:1$. What will be the primary current when connected to 208 V?

Solution

Using Eq. (7.4),

$$Z_1 = \left(\frac{N_1}{N_2}\right)^2 Z_2 = (20)^2 \times 60\angle30° = 24\angle30°\,\text{k}\Omega$$

$$I_1 = \frac{E_1}{Z_1} = \frac{208\angle0°}{24 \times 10^3\angle30°} = 8.67\angle-30°\,\text{mA}$$

7.1.4 Effects of Core Materials and Windings

Chapter 6 provides the basic ideas of circuits containing ferromagnetic core materials. Combining this information with the ideal transformer covered in Section 7.1.3 provides a model that describes transformer action with core losses included.

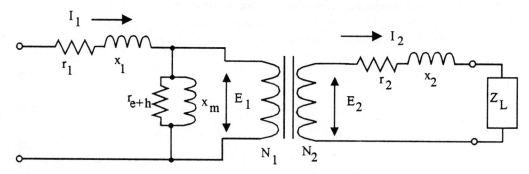

FIGURE 7.4
Equivalent circuit of a transformer showing iron and winding losses.

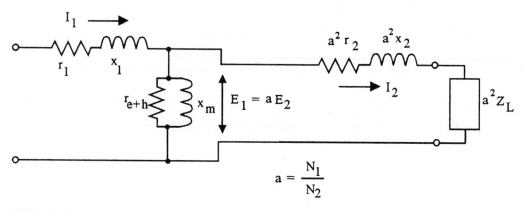

FIGURE 7.5
Equivalent circuit of a transformer with all values referred to the primary side.

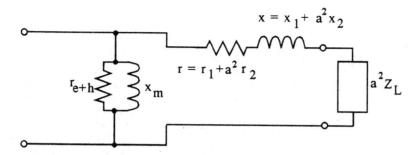

FIGURE 7.6
Simplified equivalent circuit of a transformer.

In Figure 7.4, r_1 and r_2 represent the primary and secondary winding resistances, x_1 and x_2 represent the primary and secondary reactances (self-inductance or leakage reactance), r_{e+h} represents the real power loss of the core (hysteresis and eddy current), and x_m represents the reactive component of the core.

The equivalent current can be simplified by referring all values to the primary (or the secondary) side of the ideal transformer as shown in Figure 7.5 where $a = N_1/N_2$. Figure 7.5 can be reduced further for simplified analysis (with some loss of accuracy) by moving the core loss representation to left (or right) as shown in Figure 7.6.

7.2 PRACTICAL ISSUES

7.2.1 Short-Circuit and Open-Circuit Tests

The parameters of the transformer equivalent circuit can be obtained by calculations derived from wire tables and loss curves for the core material. If the transformer or a prototype is available, the equivalent circuit can be derived from two specific tests: the *short-circuit test* and the *open-circuit test.*

The short-circuit test provides information for determining the winding resistance and reactance. Since the core loss impedance is large compared to the winding impedance, the equivalent circuit can be reduced to the representation shown in Figure 7.7 for rated current. The electronic wattmeter/voltmeter/ammeter (WVA) is generally connected in the low-voltage side of the transformer. Separate instruments could be used but this would require additional hookup time.

For the short-circuit test a variable ac supply voltage (ac rated frequency) is slowly increased until the ammeter reads the rated current of the winding in which the instrumentation is connected. Power, voltage, and current are then recorded as P_{sc}, V_{sc}, and I_{sc}. With this information

$$r_w = \frac{P_{sc}}{I_s^2} \quad [\Omega] \tag{7.5}$$

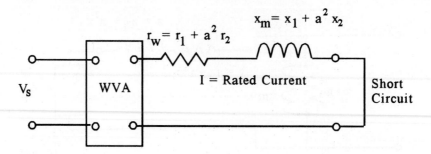

FIGURE 7.7
Short-circuit test.

$$S_{sc} = V_{sc}I_{sc} \quad [VA] \tag{7.6}$$

$$Q_{sc} = \sqrt{S_{sc}^2 - P_{sc}^2} \quad [VAR] \tag{7.7}$$

$$x = \frac{Q_{sc}}{I_{sc}^2} \quad [\Omega] \tag{7.8}$$

assuming that

$$r_1 = a^2 r_2 \tag{7.9}$$

$$x_1 = a^2 x_2 \tag{7.10}$$

Then for the winding resistances

$$r_1 = \frac{r_w}{2} \tag{7.11}$$

and referred to the secondary

$$r_2 = \frac{r_w}{2a^2} \tag{7.12}$$

and for the winding or leakage reactance

$$x_1 = \frac{x_w}{2} \tag{7.13}$$

and referred to the secondary

$$x_2 = \frac{x_w}{2a^2} \tag{7.14}$$

Safety Precaution High ac voltages may be present. This would be especially true if the supply were connected to the low-voltage side and the short circuit on the secondary were removed.

The open-circuit test provides information for determining the core loss resistance r_{e+h} and reactance r_m. For the open-circuit test, the winding resolution and reactance can be neglected, resulting in the parallel equivalent circuit of Figure 7.8.

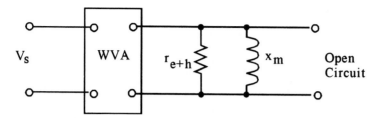

FIGURE 7.8
Open-circuit test, parallel equivalent.

When specifying this test it is important to indicate the desired representation, i.e., whether it is the parallel equivalent as shown in Figure 7.8 or the corresponding series representation. For power transformers the parallel representation is generally used.

For this test, the voltage applied is generally the rated voltage of the winding. Power, voltage, and current are then recorded as P_{oc}, V_{oc}, and I_{oc}. With this information,

$$r_{e+h} = \frac{V_{oc}^2}{P_{oc}} \quad [\Omega] \tag{7.15}$$

$$S_{oc} = V_{oc}I_{oc} \quad [\text{VA}] \tag{7.16}$$

$$Q_a = \sqrt{S_{oc}^2 - P_{oc}^2} \quad [\text{VAR}] \tag{7.17}$$

$$x_m = \frac{V_{oc}^2}{Q_{oc}} \quad [\Omega] \tag{7.18}$$

Also

$$I_e = \text{exciting current} = I_{oc} \tag{7.19}$$

$$I_m = \text{magnetizing current} = V_{oc}/X_m \tag{7.20}$$

EXAMPLE 7.3

Open-circuit and short-circuit tests are made on a 2-kVA, 120/36-V, 60-Hz power transformer. An electronic WVA meter connected on the high side provides the following readings:

Open-circuit: 50.0 W, 120 V, 545 mA

Short-circuit: 47.5 W, 3.79 V, 16.7 A

Determine the ac equivalent circuit with core loss (r_{e+h}, r_m, r_1, x_1, r_2, and x_2).

Solution

$$r_{e+h} = \frac{V_{oc}^2}{P_{oc}} = \frac{120^2}{50} = 288\ \Omega$$

$$S_{oc} = V_{oc}I_{oc} = (120)(0.545) = 65.4\ \text{VA}$$

$$Q_{oc} = \sqrt{S_{oc}^2 - P_{oc}^2} = \sqrt{65.4^2 - 50^2} = 40.7\ \text{VAR}$$

$$S_{sc} = V_{sc}I_{sc} = (13.79)(16.7) = 63.3 \text{ VA}$$

$$r_w = \frac{P_{sc}}{I_{sc}^2} = \frac{47.5}{16.7^2} = 170 \text{ m}\Omega$$

$$S_{sc} = V_{sc}I_{sc} = (3.79)(16.7) = 63.3 \text{ VA}$$

$$Q_{sc} = \sqrt{S_{sc}^2 - P_{sc}^2} = \sqrt{63.3^2 - 47.5^2} = 41.8 \text{ VAR}$$

$$X_w = \frac{Q_{sc}}{I_{sc}^2} = \frac{41.8}{16.7} = 2.51 \text{ }\Omega$$

$$r_1 = \frac{r_w}{2} = \frac{170}{2} = 85.0 \text{ m}\Omega$$

$$a = \frac{N_1}{N_2} = \frac{120}{36} = 3.33$$

$$r_2 = \frac{r_w}{2a^2} = \frac{170}{2(3.33)^2} = 7.67 \text{ m}\Omega$$

$$x_1 = \frac{x_w}{2} = \frac{2.51}{2} = 1.26 \text{ }\Omega$$

$$x_2 = \frac{x_w}{2a^2} = \frac{2.51}{2.(3.33)^2} = 113 \text{ m}\Omega$$

EXAMPLE 7.4   Determine the exciting current and the magnetizing current of the transformer in Example 7.3.

Solution

$$I_c = I_{oc} = 545 \text{ mA}$$

$$I_m = \frac{v_{oc}}{X_m} = \frac{120}{335} = 358 \text{ mA}$$

7.2.2 Efficiency and Regulation

Regulation   The output voltage of a transformer will change from the no-load condition to the full-load condition due to the winding impedance $r_w + jx_w$ even if the source voltage remains the same. These load changes are generally undesirable. Figure 7.9 shows the equivalent circuit and corresponding phasor diagram of the transformer used to calculate the regulation.

The regulation of a transformer describes the phenomenon and is calculated as

$$\text{Regulation} = \frac{V_{NL} - V_{FL}}{V_{FL}} \times 100\% \tag{7.21}$$

where V_{FL} is the full-load or rated voltage of the transformer. The no-load voltage is calculated as

$$V_{NL} = V_{FL} + (r_w + jx_w)(I_{FL}\angle - \theta°) \tag{7.22}$$

where I_{FL} is the rated current.

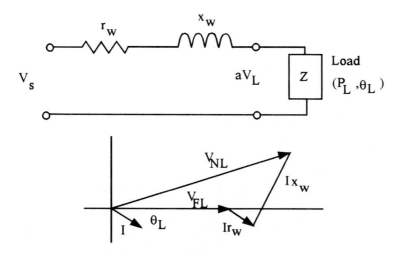

FIGURE 7.9
Voltage regulation.

EXAMPLE 7.5   Calculate the regulation of a 50-kVA, 7200/240-V transformer (r_w = 15.56 Ω and x_w = 41.48 Ω) at full load 0.8 PF lagging.

Solution

$$\theta = \cos^{-1}(\text{PF}) = \cos^{-1}(0.8) = 36.87°$$
$$V_{\text{NL}} = 7200\angle 0° + (15.56 + j41.48)(6.94\angle -36.87°) = 7463\angle 1.28° \text{ V}$$
$$\text{Regulation} = \frac{7463 - 7200}{7200} = 3.65\%$$

Efficiency   Most transformers have high efficiencies ranging from 70 to 98%. The efficiency of a transformer is given by

$$\eta = \frac{\text{Output}}{\text{Output} + \text{Losses}} \tag{7.23}$$

where the output and losses refer to real power quantities. This may be rewritten as

$$\eta = \frac{P_{\text{out}}}{P_{\text{out}} + P_{\text{Fe}} + P_{\text{Cu}}} \tag{7.24}$$

where
$$P_{\text{out}} = E_L I_L \cos\theta = S \cos\theta \, P = S(\text{PF})$$
$$S = \text{complex power } EI \text{ in VA}$$
$$P_{\text{Fe}} = \text{iron loss}$$
$$P_{\text{Cu}} = \text{copper loss} = I_L^2 r_w$$
$$\text{PF} = \text{power factor}$$

EXAMPLE 7.6

The results of a short-circuit and open circuit test are (with meters on the low side) as follows:

Open-circuit: 100 W, 240 V, 0.8 A

Short-circuit: 120 W, 140 V, 10 A

Determine the efficiency for a 0.9 PF-leading rated load.

Solution

The magnitude of the complex power for the rated load will be

$$|S| = E_{\text{rated}} I_{\text{rated}} = (240)(10) = 2.4\,\text{kVA}$$
$$P_{\text{Fe}} = P_{\text{oc}} = 100\,\text{W}$$
$$P_{\text{cu}} = P_{\text{sc}} = 120\,\text{W}$$
$$\eta = \frac{S\cos\theta}{S\cos\theta + P_{\text{Fe}} + P_{\text{Cu}}}$$
$$= \frac{(2400)(0.9)}{(2400)(0.9) + 100 + 120} = 90.75\%$$

7.2.3 Frequency Dependence

The general frequency response of a transformer is shown in Figure 7.10. The equivalent circuit to account for the decrease in $E_{\text{out}}/E_{\text{in}}$ at low and high frequencies is given in Figure 7.11, where r_1 and x_1 represent the primary impedance; r_2 and x_2 represent the secondary impedance; r_{e+h} and x_m represent the core; C_1 and C_2

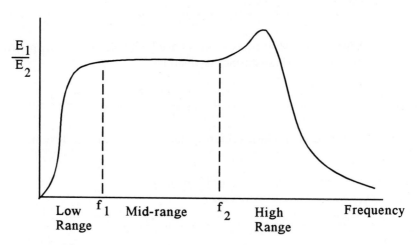

FIGURE 7.10
Transformer frequency response.

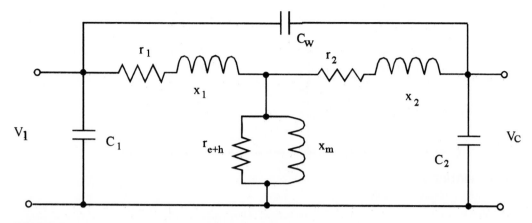

FIGURE 7.11
Equivalent circuit for low- and high-frequency effects.

represent the primary and secondary circuits; and C_w represents the capacitance between the windings.

For low frequencies, x_1 and x_2 can be ignored and x_m becomes dominant: $x_m = 2\pi f L_m$ becomes small with decreasing frequency and bypasses the input since it is in parallel with the input as shown in Figure 7.12. For mid-frequencies, the transformer can be represented as shown in Figure 7.13. For high frequencies C_w and C_2 are dominant. Figure 7.14 shows an approximate equivalent circuit when C includes the effects of C_w and C_2.

The peaking observed in the high-frequency region is due to the series resonant circuit of C, x_w, r_w, and r_{e+h}.

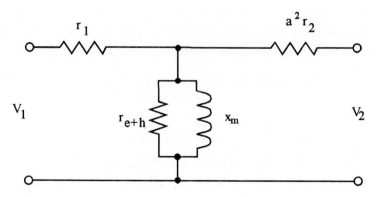

FIGURE 7.12
Low-frequency equivalent.

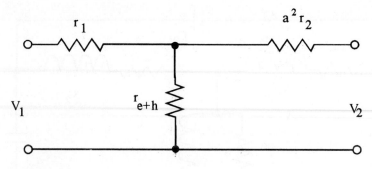

FIGURE 7.13
Mid-frequency equivalent.

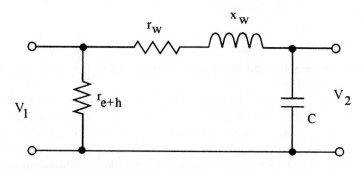

FIGURE 7.14
High-frequency equivalent.

7.2.4 Power and Current Rating

The power rating of a transformer is primarily determined by heat dissipation. The heat is the result of the iron (Fe) and winding (Cu) losses in the transformer. These are, in turn, determined by the voltages (V_1 and V_2), the operating current, and the frequency.

The rating specifies the apparent power S in volt-amperes, the input and output voltages V_1/V_2 in volts, and the frequency f in Hz.

EXAMPLE 7.7

What is the meaning of the following transformer rating?:

$$10 \text{ kVA}, 4160/208\text{V}, 60 \text{ Hz}$$

Solution
The transformer specified is a three-phase device having a capacity of 10 kVA (for all three phases). The rated line voltage on the primary side is 4160 V and the rated

Y-connected secondary line voltage is 208 V. The rated secondary line current is

$$I_{rated} = \frac{S}{\sqrt{3}E_1} = \frac{10000}{\sqrt{3}(208)} = 27.8 \text{ A}$$

7.2.5 Core Loss

The equivalent circuit showing the core loss of a transformer is the same as that for the inductor (Chapter 6). This circuit is shown in Figure 7.15, where r_{e+h} represents the real power loss and results in heating that must be dissipated. This loss is due to eddy currents and hysteresis in the iron core. The X_m term represents the reactive "loss" in the core. Since the magnetizing current I_m is 90° in phase from the voltage no real loss occurs. Energy shuttles back and forth between X_m and the line but does not have a heating effect on the core.

FIGURE 7.15
Core loss equivalent circuit.

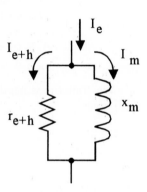

7.3 DESIGN AND CONSTRUCTION METHODS

7.3.1 Basic Design

The following voltage equation relates the voltage across the primary of the transformer to the flux density in the core. This equation is a consequence of Faraday's law and is written as

$$V_{rms} = 4.44A_e N f B_{max} \times 10^5 \tag{7.25}$$

where A_e = effective magnetic area in cm^2
f = frequency (Hz)
B_{max} = maximum value of the sinusoidal flux density (kG)
N = number of turns on the primary

Equation 7.25 can be rearranged to represent the number of volts per turn (V/N) as

$$V/N = 4.44A_e f B_{max} \times 10^5 \tag{7.26}$$

The equation is also used with ϕ rather than B_{max} ($\phi_{max} = B_{max}A_e$) as

$$V/N = 4.44 f \phi_{max} \times 10^8 \qquad (7.27)$$

where ϕ_{max} is given in maxwells.

For cross-sectional area in terms of in.^{2} for A_e, use

$$V/N = 25.8 A_e N f B_{max} \times 10^5 \qquad (7.28)$$

where A_e is in in.^{2}, f is in Hz, and B_{max} is in kG.

The apparent power handling capacity of the transformer is given by the equation

$$P_a = 4.55 f J A_w A_e B_{max} \times 10^{-8} \qquad (7.29)$$

where
P_a = apparent or complex power (VA)
f = frequency (Hz)
J = current density (A/in.^{2})
A_w = window area (in.^{2})
A_c = core area (in.^{2})
B_{max} = maximum flux density (G)

We sometimes need to calculate the $A_w A_c$ product. Rearranging Equation (7.29) gives

$$A_w A_c = \frac{P_a}{4.55 f J B_{max}} \times 10^8 \qquad (7.30)$$

7.3.2 Windings

The wire forming the various coils of the transformer must fit through the window area A_w. The required area is determined by the geometry of the core used. Figure 7.16 shows examples of simple tape-wound cut cores.

The area of wire required can be estimated using the area of the wire required to carry rated current in that winding and the number of turns required:

$$A_N = A_{Cu} N \qquad (7.31)$$

where
A_N = estimated area (m^2)
A_{Cu} = wire area (m^2)
N = number of turns

The window area is determined by multiplying by a winding factor k_w:

$$A_w = k_w A_{Cu} N = K_w A_N \qquad (7.32)$$

Table 7.1 lists the wire areas of a variety of stranded copper wire sizes. A value of $K_w = 1.33$ provides for a conservative estimate of A_w.

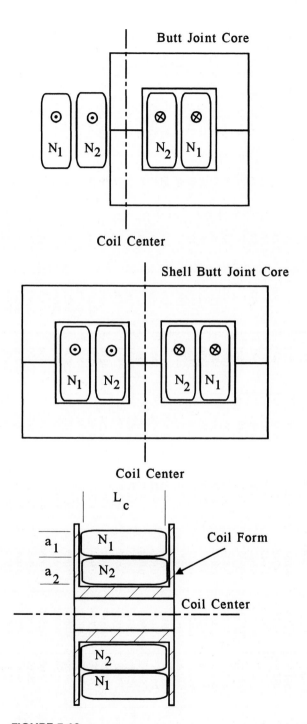

FIGURE 7.16
Windings on tape-wound cut cores.

TABLE 7.1
Wire table

Wire Size AWG	Circular (mils)	Diameter (in.)		Resistance (Ω/1000 ft)	Weight (lb/1000 ft)	Random Winding		Layer Winding				Wire Size AWG
		Single Insulation	Heavy Insulation			Single Turns/in.²	Heavy Turns/in.²	Single Turns/in.	Heavy Turns/in.	Layer Insulation (in.)	Edge Distance (in.)	
10	10384		0.106	0.999	31.7	86	75	9	8	0.010	0.250	10
11	8226		0.094	1.26	25.2	108	95	10	9	0.0100	0.250	11
12	6529		0.084	1.59	20.1	133	130	11	11	0.0100	0.250	12
13	5184		0.075	2.00	15.9	162	159	12	12	0.0100	0.250	13
14	4109	0.0658	0.067	2.52	12.6	212	193	14	13	0.0100	0.188	14
15	3260	0.0587	0.060	3.18	10.0	255	248	15	15	0.0100	0.188	15
16	2581	0.0524	0.054	4.02	7.95	324	316	17	17	0.0100	0.188	16
17	2052	0.0468	0.048	5.05	6.32	405	394	19	19	0.0070	0.188	17
18	1624	0.0418	0.043	6.39	5.02	525	487	22	21	0.0070	0.125	18
19	1289	0.0373	0.039	8.05	3.99	641	596	24	23	0.0070	0.125	19
20	1024	0.0334	0.035	10.13	3.16	850	792	27	26	0.0050	0.125	20
21	812.3	0.0298	0.031	12.77	2.51	1055	982	30	29	0.0050	0.125	21
22	640.1	0.0266	0.028	16.20	1.99	1340	1210	34	32	0.0050	0.125	22
23	510.8	0.0238	0.025	20.30	1.59	1370	1260	38	36	0.0050	0.125	23
24	404.0	0.0213	0.022	25.67	1.26	1730	1550	42	40	0.0020	0.125	24
25	320.4	0.0190	0.020	32.37	1.01	2150	1940	47	45	0.0020	0.125	25
26	252.8	0.0170	0.018	41.02	0.799	2990	2700	53	50	0.0020	0.125	26
27	201.6	0.0152	0.016	51.44	0.634	3700	3550	59	55	0.0020	0.125	27
28	158.8	0.0136	0.014	65.31	0.504	4680	4180	66	62	0.0015	0.125	28
29	127.7	0.0122	0.013	81.21	0.401	5900	5160	73	68	0.0015	0.125	29
30	100.0	0.0109	0.012	103.7	0.318	7500	6560	82	77	0.0015	0.093	30
31	79.21	0.0097	0.011	130.9	0.254	9270	8090	91	85	0.0015	0.093	31
32	64.00	0.0088	0.010	162.0	0.202	11400	10000	100	94	0.0013	0.093	32
33	50.41	0.0078	0.009	205.7	0.161	14500	12500	113	105	0.0013	0.093	33
34	39.69	0.0070	0.008	261.3	0.127	18800	16250	128	119	0.0010	0.093	34
35	31.36	0.0062	0.007	330.7	0.101	24000	20600	144	133	0.0010	0.093	35
36	25.00	0.0056	0.0060	414.8	0.0803	29650	25000	158	145	0.0010	0.093	36
37	20.25	0.0050	0.0055	512.1	0.0641	37400	30900	177	161	0.0010	0.093	37
38	16.00	0.0045	0.0049	648.2	0.0509	46700	39300	198	181	0.0010	0.062	38
39	12.25	0.0039	0.0043	846.6	0.0403	62700	51500	226	205	0.0007	0.062	39
40	9.61	0.0035	0.0038	1079	0.0319	89600	72000	262	226	0.0007	0.062	40
41	7.84	0.0031	0.0034	1323	0.0252	107800	89800	274	250	0.0007	0.062	41
42	6.25	0.0028	0.0030	1659	0.0199	133500	116500	304	283	0.0005	0.062	42
43	4.84	0.0025	0.0027	2143	0.0159	167000	143000	340	315	0.0005	0.062	43
44	4.00	0.0022	0.0025	2593	0.0127	217000	168500	386	340	0.0005	0.062	44

A more accurate method of determining the coil area is to use tabulated values of the number of turns/in.[2]. Equation (7.32) becomes

$$A_w = \frac{k_w N}{k_A} \tag{7.33}$$

where k_A = turns/in.[2] taken from Table 7.1 and N = number of turns. With A_w, the length of the coil, and the inside diameter of the coil known, the mean length of a coil turn and the total wire length can be estimated. The build of each coil, which is the coil width, can be determined by using

$$A_w = l_c a_c \tag{7.34}$$

where a_c = build of the coil and l_c = mean length of the coil or

$$a_c = \frac{A_w}{l_c} \tag{7.35}$$

The mean length of a turn (MLT) is calculated from the geometry of the winding as shown in Figure 7.17:

$$l_{m1} = \text{MLT}_1 = 2(r + l_f) + 2(s + l_f) + \pi \left(\frac{a_1}{2} + a_2 \right) \tag{7.36a}$$

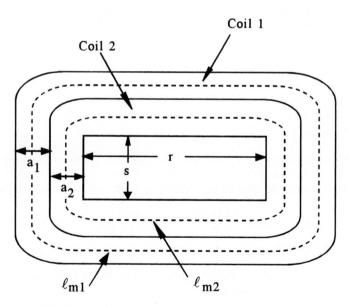

FIGURE 7.17
Cross section of windings to define mean coil wire length per turn.

where l_{mn} = mean length of coil n and l_f = thickness of the coil form. Also,

$$l_{m2} = \text{MLT}_2 = 2(r + l_f) + 2(s + l_f) + \frac{\pi a_2}{2} \qquad \textbf{(7.36b)}$$

and the length of wire required is calculated by

$$l_w = l_m N \qquad \textbf{(7.37)}$$

7.3.3   Core Configuration

Some of the various core types used for transformers are illustrated in Chapter 6 and include the following types:

Pot core

Stacked EI laminations (shell core)

Tape-wound toroidal

Tape-wound cut core

Tape-wound shell core

Both tape-wound cores made from high-mu grain-oriented silicon steel and stamped laminated nonoriented steel cores are common at the power and audio-frequencies.

7.3.4   Design

Most transformer applications can be satisfied using off-the-shelf transformers. If a stock transformer is not available, it may be desirable to design and construct a working unit. The following example provides a simplified approach to transformer design.

EXAMPLE 7.8   Design a power transformer to provide 18 V at 1.5 A from a 120-V, 60-Hz line. Use a tape-wound core made from 12-mil grain-oriented steel. Use B_{max} = 10 kG for this material.

Solution

Step 1: Determine the WA or $(A_w A_c)$ product using Eq. (7.30) where J has been replaced by $1/S$ and the unit of area changed from square inches to circular mils:

$$\text{WA} = A_w A_c = \frac{17.26 S P_a}{f B_{max}} = \frac{(17.26)(800)(27)}{(60)(10{,}000)} = 0.6214 \text{ in.}^4$$

$$\text{RPHC} = \text{Relative power handling capacity} = \text{WA} \times 100 = 62$$

where S = circular mils per ampere (use 800 as a conservative estimate)
$P_a = EI = (18)(1.5) = 27 \text{ VA}$
f = frequency = 60 Hz
B_{max} = flux density in gauss = 10,000

Step 2: Select a core having an RPHC $\geq$ 62 from a manufacturer's table. For this core, use

$D = 1$ in.
$E = 0.5$ in.
$F = 1.5$ in.
$G = 2.5$ in.
$Wt = 0.96$ lb
$A_c = 0.5$ in.^{2}
$l_m = 7.35$ in.
$A_w = F \times G = 1.25$ in.^{2}.

Step 3: Calculate the turns needed in the primary and secondary where

$$N_1 = \frac{3.5 \times 10^6}{fA_mB_{max}}$$

and

A_m = effective area of the core
$A_m = (SF) A_c$
SF = stacking factor = 0.95
$A_m = (0.95)(2.4) = 2.01$ in.^{2}

So,

$$N_1 = \frac{(3.5 \times 10^6)(120)}{(60)(0.475)(10,000)} = 1474 \text{ turns}$$

$$N_2 = \frac{N_1 V_2}{V_1 \sqrt{\eta}}$$

where η is approximately 0.75 from Table 7.2.

$$N_2 = \frac{(1474)(18)}{(120)\sqrt{0.75}} = 255.3 \text{ or } 255 \text{ turns}$$

TABLE 7.2
Estimates of efficiency of η

Power Out (W)	η (%)
20	65
30	75
40	78
50	80
90	85
100	86
150	90
200	92

and the no-load output voltage will be approximately

$$V_2 = \frac{N_2}{N_1} V_1 = \frac{255}{1474} \times 120 = 20.8 \text{ V}$$

Step 4: Select wire gauges for the windings for this example: $I_2 = 1.5$ A. So

$$I_1 = \frac{N_2}{N_1} I_2 = \frac{(255)(1.5)}{1474} = 259 \text{ mA}$$

for the primary $A_{Cu} = (0.259)(800) = 2072$ circular mils. Size 27 (Table 7.1) wire with an area of 201.6 is very close for the secondary. $A_{Cu} = (1.5)(800) = 1200$ circular mils. Size 19 wire with an area of 1289 is close.

Step 5: Calculate the area build and mean length per turn for each coil. From Table 7.1 using the column for Heavy Turns/in.²:

Winding	Wire Size	Turns/in.²	N Turns	G (in.)	A (in.²)
N_2	19	596	255	$3\frac{1}{8}$	0.428
N_1	27	3550	1474	$3\frac{1}{8}$	0.415

where A_1 and A_2 are calculated as the number of turns divided by the number of turns per inch:

$$A_2 = \frac{255}{596} = 0.428 \text{ in.}^2, \qquad A_1 = \frac{1474}{3550} = 0.415 \text{ in.}^2$$

The total area of copper required is

$$A_T = A_1 + A_2 = 0.428 + 0.415 = 0.842 \text{ in.}^2$$

Allowing for a space factor of 1.30, the window area required will be

$$A_w = 1.30 A_T = (1.30)(0.842) = 1.10 \text{ in.}^2$$

The selected core has a window area of 1.25 in.² providing ample room for the winding and winding form. Calculate the winding builds $a_1 + a_2$ using the method of Section 7.3.2:

$$a_1 = \frac{A_1}{G} = \frac{0.415}{2.5} = 0.166 \text{ in. (build for coil } N_1)$$

where G = width of core (kg).

$$a_2 = \frac{A_2}{G} = \frac{0.428}{2.5} = 0.171 \text{ in. (build for coil } N_2)$$

Neglecting the coil form thickness

$$l_{m1} = 2E + 2D + \pi \left(\frac{a_1}{2} + a_2 \right)$$

$$= 2(0.5) + 2(1) + \pi \left(\frac{0.166}{2} + 0.171 \right) = 3.80 \text{ in.}$$

$$l_{m2} = 2E + 2D + \pi \frac{a_2}{2}$$

$$= 2(0.5) + 2(1) + \pi \frac{0.171}{2} = 3.09 \text{ in.}$$

This completes the basic design of the transformer. Using this information, the length of conductor for each coil can be determined. The resistance of each coil can then be calculated along with the winding loss. Core losses can be calculated by using the apparent and real loss curves for the core material used. Using this information, the equivalent circuit can be determined along with the efficiency and regulation of the transformer.

7.4 APPLICATIONS

7.4.1 Transformer Specifications

Figure 7.18 shows a power transformer with multiple primary and secondary windings. The terminals are identified by the symbols H_n and X_n, where H indicates high-voltage leads and X indicates low-voltage leads. Either set (H_n or X_n) can be used as the primary. If H_n is used as the primary, the transformer is called a *step-down transformer*. The smaller number identifies the starting point for each winding

FIGURE 7.18
Terminal markings for power transformers.

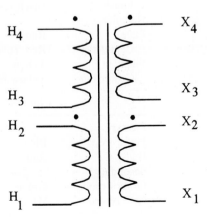

FIGURE 7.19
General terminal markings.

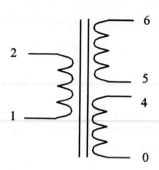

and the number sequence indicates the direction of a voltage rise. The dots placed in Figure 7.18 also indicate the phase relationships.

Terminals are sometimes specified by a sequence of numbers as shown in Figure 7.19. In most cases these are specified on the case of the transformer. Color codes are also used when only wire leads are provided. Table 7.3 shows an early code used for radio transformers. Power transformers usually carry significant data regarding the characteristics of the device. The following table shows the type of transformer nameplate data information that may be supplied along with an example

TABLE 7.3
Color codes for transformer leads

Radio Power Transformers			
Primary	Black	General Use	
If tapped:		Filament No. 1	Green
Common	Black	Center tap	Green-Yellow
Tap	Black-Yellow	Filament No. 2	Brown
Finish	Black-Red	Center tap	Brown-Yellow
Rectifier		Filament No. 3	Slate
Plate	Red	Center tap	Slate-Yellow
Center tap	Red-Yellow		
Filament	Yellow		
Center tap	Yellow-Blue		

Intermediate-Frequency Transformers			
Primary		For full-wave transformers:	
Plate	Blue	Second diode	Violet
B+	Red	Old standard is same as above, except:	
Secondary		Grid return	Black
Grid or diode	Green	Second diode	Green-Black
Grid return	White		

for a specific transformer:

Nameplate Data	Example
P_a or S	50 kVA
Temperature rise	110°C
Primary volts	4160
Secondary volts	120/240
Frequency	60 Hz
Phase	Single or 1Ø
Type	Dry
Model	EC 538
Manufacturer	ECCO
Windings	
Primary	4160 line on H_1 and H_2
Secondary	lines on X_1 and X_4
	120 V connect X_1–X_3 and X_2–X_4
	240 V connect X_2–X_3
Weight	625 lb

7.4.2 Autotransformers

The autotransformer is a special connection of a transformer such that energy transfers from the primary to the secondary through the dual process of electrical conduction and transformation. Use of this connection greatly reduces the size of a transformer for a given power handling capacity. It can only be used, however, when the normal isolation provided by the transformer is not required.

For the transformer of Figure 7.20:

$$\frac{E'_2}{E_1} = \frac{N_2}{N_1} \tag{7.38}$$

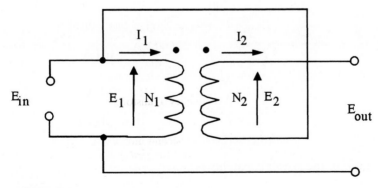

FIGURE 7.20
Wiring for an autotransformer.

or

$$E'_2 = \frac{N_2}{N_1} E_1 = \frac{E_1}{a} \qquad (7.39)$$

In Figure 7.20

$$E'_2 = E_{out} = E_1 + E_2 \qquad (7.40)$$

or

$$E_2 = E_1 + \frac{E_1}{a} = \left(1 + \frac{1}{a}\right) E_1 \qquad (7.41)$$

Also

$$I_{in} = I_2 - I_1$$
$$S_{out} = E_2 I_2$$
$$S_{in} = E_1 I_{in}$$

The rating of the transformer in the autotransformer configuration is

$$S_{auto} = \frac{E_{out}}{|E_{out} - E_{in}|} S_{trans} = K S_{trans} \qquad E_{out} \neq E_{in}$$

when $E_{out} = E_{in}$ power transfer occurs by conduction alone.

EXAMPLE 7.9

A 25-kVA, 2400/240-V transformer is connected in the autotransformer configuration of Figure 7.20. Determine the power handling capacity of this connection and compare it to the transformer capacity.

Solution

$$\frac{E_1}{E_2} = \frac{2400}{240} = 10$$

$$E_{load} = E_1 + E_2 = 2400 + 240 = 2640 \text{ V}$$

$$I_{load} = I_{2rated} = \frac{25,000 \text{ VA}}{240} = 104 \text{ A}$$

$$I_{1rated} = \frac{N_2}{N_1} I_{2rated} = \frac{1}{10} 104 = 10.4 \text{ A}$$

$$I_{in} = I_1 + I_2 = 104 + 10.4 = 114.4 \text{ A}$$

$$S_{out} = (2400)(114.4) = 275 \text{ kVA}$$

The ratio of the power capability is

$$\frac{S_{auto}}{S_{trans}}$$

This could have been calculated by the ratio

$$\frac{2640}{2640 - 2400} = 11$$

The variable autotransformer is also a useful variation of the standard transformer. In this case, a single winding is placed on a toroidal core. One side of the coil is ground to expose the bare copper winding. A brush is used to make the moving contact with the winding. Figure 7.21 shows a schematic diagram of this construction. The output voltage can be made to vary from 0 volts to a voltage above E_{in}.

Variable autotransformers are sold under a variety of trade names and are found in most electrical labs. The desired voltage range, maximum output current, and frequency must be known to specify the required autotransformer. Autotransformers can be operated at higher than rated frequencies, but not at lower frequencies unless the primary voltage is derated.

7.4.3 Rectifier Circuits

One of the most extensive applications of the transformer is in power supplies for which we need to transform an ac voltage to a specified dc voltage level. The ability of the transformer to modify the voltage level is used for this application. Another reason to use the transformer is the electrical isolation provided between the input and output of the rectifier.

7.4.4 Audio Coupling

Transformers are sometimes used to couple the audio-frequency signal from one circuit to another without altering the bias levels required for proper operation. Two examples of transformer use are shown in Figure 7.22. These transformers must pass a wide range in frequencies. In the circuit T_2 must carry a large dc

FIGURE 7.21
Variable autotransformer.

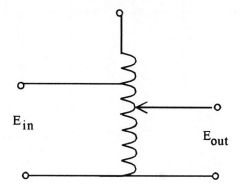

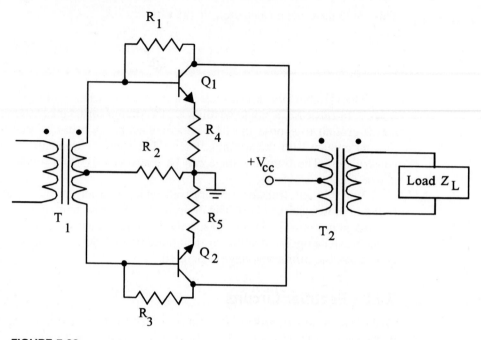

FIGURE 7.22
Transformer coupled circuits.

component in its primary and must have a turn ratio to match the impedance Z_2 of the load with those of the transistors Q_1 and Q_2. T_1 also provides for the phase shift to drive Q_1 and Q_2 in a push–pull operation.

Transformers are listed for their intended use, frequency range, and power capability in electronics catalogs.

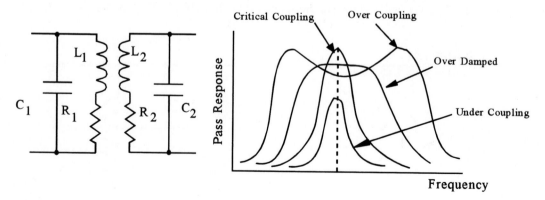

FIGURE 7.23
IF transformer circuit and frequency response.

FIGURE 7.24
RF transformer application.

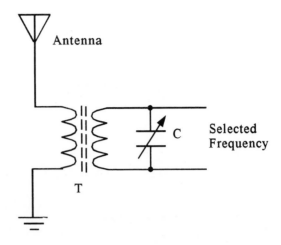

7.4.5 IF and RF Coupling

An IF (Intermediate frequency) results in radio receiver circuits when an adjustable local oscillator frequency is mixed with an incoming rf (radio-frequency) to provide a constant frequency output. The local oscillator is varied with the rf tuning circuit so that the difference in the two frequencies is always constant. A coupled tuned circuit using an air core transformer is generally used for this purpose. Figure 7.23 shows the coupled transformer with corresponding frequency response. The IF transformer is determined by the circuit configuration and IF frequency to be used (e.g., 455 kHz is the standard AM broadcast IF frequency). This transformer circuit must pass a small band of frequencies and reject all others.

The rf frequency circuit must be capable of passing a large range of rf frequencies but selecting a small band of these frequencies while rejecting all others. The basic tuned circuit of Figure 7.24 provides a selected center frequency depending on the value of C. These rf transformers are available from electronic parts manufacturers. The range of C for a specified frequency band will be indicated for each rf transformer. Many of the rf coils have a powdered iron core.

8

MAGNETIC DEVICES AND MOTORS*

8.1 BASIC MAGNETIC CONCEPTS

8.1.1 Flux and Magnetomotive Force

The presence of a magnetic field is often illustrated through the concept of lines of force. The total number of lines is a measure of the amount of flux, ϕ, or the strength of the field. However, because it is not possible to show all of the lines in a given magnetic configuration, we use the concept of the number of flux lines passing perpendicularly through a given cross-sectional area. This is called the *flux density, B*, and is a measure of the magnetic field strength. In SI units, flux is given in webers so flux density is in webers per square meter or teslas. In English units, flux is measured in lines or kilolines so flux density is given in lines or kilolines per square inch (1 Wb = 10^8 lines). The basic formula is then:

$$B = \phi/A \tag{8.1}$$

where B = flux density (T)
         ϕ = flux (Wb)
         A = cross-sectional area (m^2).

The lines of magnetic flux are shown for a permanent magnet in Figure 8.1. These lines follow some fundamental rules that help to explain the behavior of magnetic circuits and devices. These rules follow:

1. Lines of force form closed loops. They continue within the body of the magnet in order to be closed.
2. Lines of force always leave from a north pole and enter at a south pole. Internal to the magnet, the lines are from the south pole to the north pole.

---

* This chapter was written by Dr. Warren Hill, Weber State College.

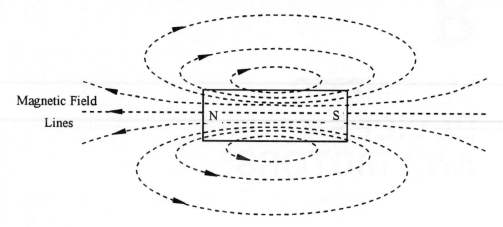

FIGURE 8.1
Lines of magnetic flux.

3. The lines do not intersect and no two lines can occupy the same space at the same time.
4. Lines of force repel each other.
5. Lines of force attempt to follow the shortest possible path. The combination of rules 4 and 5 means that there will be a balance of forces such that for any given configuration, an equilibrium condition will be reached.
6. Lines of force will try to follow the shortest path consistent with the rest of the preceding rules.

In terms of their magnetic properties, practically speaking there are only two basic types of materials, magnetic and nonmagnetic. Only a few various alloys of iron and nickel are magnetic; all other materials are nonmagnetic. The basic difference between the two is that magnetic materials provide an easy path for lines of flux to follow, whereas nonmagnetic materials do not.

8.1.2 Magnetomotive Force, Permeance, and Reluctance

To produce a magnetic field in most magnetic devices, a current is established in a coil of wire. The flow of current produces a magnetic field around the wire. Consider a coil of wire where the coil is formed into a toroid as shown in Figure 8.2. When current flows in the coil, a magnetic field is formed within the toroid. This current produces a magnetomotive force directly proportional to the amount of current that flows and the number of turns in the coil. The basic formula for magnetomotive force is given by:

$$F = NI \qquad (8.2)$$

where F = magnetomotive force (mmf) in amp turns
N = number of turns in the coil
I = coil current (A).

N Turns

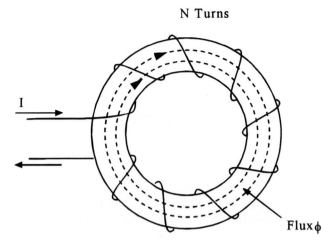

I

Flux φ

FIGURE 8.2
Magnetic toroid.

Dividing the magnetomotive force by the average length of the flux path gives the magnetic field intensity H or:

$$H = NI/L \qquad\qquad (8.3)$$

where H = magnetic field intensity in ampere turns per meter (A-t/m)
           N = number of turns in the coil (dimensionless)
           I = coil current (A)
           L = average length of the flux path (m).

The ratio of the flux density B to the magnetic field intensity H for a given magnetic field is defined as the permeability μ or:

$$\mu = B/H \qquad\qquad (8.4)$$

where μ = permeability in teslas per A-t/m
           B = flux density (T)
           H = magnetic field strength in A-t/m.

For nonmagnetic materials the permeability is a constant, i.e., the ratio of B to H is a constant. For magnetic materials the permeability varies nonlinearly as H is changed. The permeability is a measure of how easily a magnetic field can be established in a given material. To compare the permeability of different materials, the concept of relative permeability was developed where the relative permeability of a material is given by

$$\mu_r = \mu/\mu_0 \qquad\qquad (8.5)$$

where μ_r = relative permeability (dimensionless)
           μ = actual permeability of the material in teslas per amp-turn per meter

μ_0 = permeability of a vacuum equal to $4\pi \times 10^{-7}$ teslas per amp-turn per meter.

The relative permeability of all nonmagnetic materials is unity, whereas for magnetic materials the relative permeability runs from the low hundreds up into the thousands.

The resistance of a given material to the establishment of a magnetic field within the material is called the *reluctance*. The reluctance is given by

$$R = L/\mu A \tag{8.6}$$

where R = reluctance
L = average length of the flux path (m)
μ = permeability of the material in which the flux exists
A = cross-sectional area of the material (m^2).

Because of the definition of μ, the expression for R can also be written as

$$R = F/\phi \quad \text{or} \quad F = \phi R \tag{8.7}$$

This latter expression is frequently referred to as Ohm's law for magnetic circuits. The reciprocal of reluctance is called *permeance* and is designated by the symbol P.

8.1.3 *B–H* Curves

For magnetic materials, the relationship between the magnetic field intensity and the magnetic flux density is nonlinear. This relationship for three common magnetic materials is shown in Figure 8.3 along with the same data for nonmagnetic materials. The steeply rising portion of the curves is the nonsaturated region, whereas the almost horizontal region is called saturation. Because the ratio of the flux density at any point on the curve to the magnetic field intensity at that point is the permeability of the material, we can see that the permeability is not a constant for magnetic materials.

8.1.4 Magnetic Circuits, Air Gaps, and Fringing

A typical magnetic circuit, as shown in Figure 8.4, contains an air gap. Usually we need to know either what mmf is required to obtain a certain flux density in the air gap or, conversely, given a certain mmf, what flux density will be obtained. The first question can be solved by an orderly procedure, whereas the second requires a cut-and-try approach. For the circuit of Figure 8.4, the flux that is established in the circuit by the current in the coil is the same everywhere. This flux will produce an mmf drop across each portion of the circuit such that the sum of these drops will equal the total applied mmf. This is sometimes referred to as Kirchhoff's voltage law for magnetic circuits.

For the circuit in Figure 8.4, we can label the points shown such that each point is midway between the inside and outside edges. These points then define an

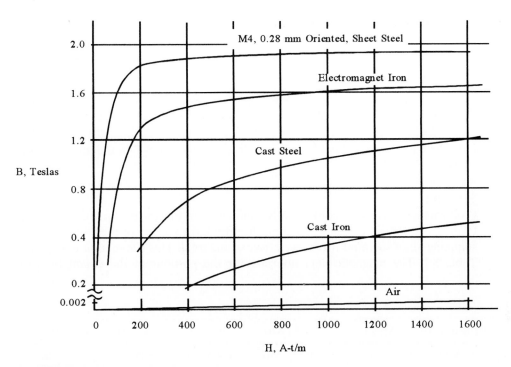

FIGURE 8.3
B–H curves.

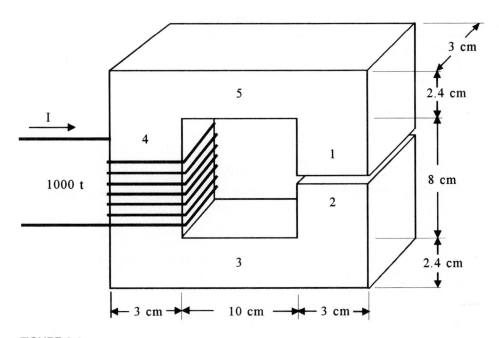

FIGURE 8.4
Series magnetic circuit.

average flux path such that a mean length for the flux path can be determined. The mmf relation for this circuit is then:

$$F_{tot} = F_1 + F_2 + F_3 + F_4 + F_5 + F_{ag} \qquad (8.8)$$

EXAMPLE 8.1

Find the current necessary to produce a flux density of 0.8 T in the air gap of the circuit in Fig. 8.4 if the material is cast steel.

Solution

Problems such as this are conveniently handled by putting the information in tabular form. Thus, we can calculate the mmf drop for each portion of the circuit, and by summing all of the mmf drops, we arrive at the total. This summation is shown in Table 8.1. The magnetic field intensity for the portions of the circuit that are

TABLE 8.1
Results for the circuit of Figure 8.4

Path	Length (cm)	Area (cm²)	ϕ (Wb)	B (T)	H (A-t/m)	$F = HL$ (A-t)
1	5.2	9	0.72×10^{-3}	0.8	500	26
2	5.2	9	0.72×10^{-3}	0.8	500	26
3	13	7.2	0.72×10^{-3}	1.0	800	104
4	10.4	9	0.72×10^{-3}	0.8	500	52
5	13	7.2	0.72×10^{-3}	1.0	800	104
Air gap	0.2	9	0.72×10^{-3}	0.8	63×10^4	1273

magnetic are found from the B–H curve for that material. The mmf for each section is then found by using the average length for that portion of the circuit. The field intensity for the air gap is found by using the permeability of air, which is a constant. The sample calculations are as follows:

$$H_{ag} = \frac{B_{ag}}{\mu_0} = \frac{0.8}{4\pi \times 10^{-7}} = 63 \times 10^4$$

so

$$F_{ag} = 63 \times 10^4 \times 2 \times 10^{-3} = 1273 \text{ A-t}$$

Applying Eq. (8.8) to all of the paths gives

$$F_{tot} = 26 + 26 + 104 + 52 + 104 + 1273 = 1585 \text{ A-t}$$

Thus, the current required is

$$I = \frac{F_{tot}}{N} = \frac{1585}{1000} = 1.59 \text{ A}$$

If Example 8.1 is turned around so that the total mmf is given, then it is apparent that because of the nonlinearity of the $B-H$ curve, the division of the mmf for each section of the circuit cannot be found. A reasonable approach for a problem of this type is to assume that most or all of the applied mmf appears across the air gap as a starting point and then to adjust the flux obtained from that assumption up or down as required.

EXAMPLE 8.2
Given the circuit of Figure 8.4, if the coil current is 1.8 A or the mmf is 1800 A-t, what is the air gap flux density?

Solution
Again putting the information in tabular form for each attempt at a solution, the result can be adjusted until the solution is within the desired range. Starting with the assumption that 85% of the total mmf is across the air gap, the results are shown in Table 8.2. As can be seen for this simple circuit, only four attempts are necessary to achieve results within 5%.

TABLE 8.2
Results for Example 8.2

Try	F_{ag}	H_{ag}	B_{ag}	ϕ	$B_{1,2,4}$	$B_{3,5}$	$H_{1,2,4}$	$F_{1,2}$	F_4	$F_{3,5}$	F_{tot}
1.	1530	76.5×10^4	0.96	0.87×10^{-3}	0.96	1.2	730	1580	38	76	2092
Total too high—assume 80% mmf drop across the air gap.											
2.	1440	72.0×10^4	0.90	0.81×10^{-3}	0.90	1.13	610	1380	32	64	1926
Total still too high—assume 75% mmf drop across air gap.											
3.	1350	67.5×10^4	0.85	0.77×10^{-3}	0.85	1.06	540	960	24	48	1696
Total too low—assume 78% mmf drop across air gap.											
4.	1400	70.0×10^4	0.88	0.79×10^{-3}	0.88	1.10	580	1070	30	60	1798

EXAMPLE 8.3
Given the circuit of Figure 8.5, what mmf is necessary to create an air gap flux density of 0.6 T if the material is electromagnetic iron?

Solution
For magnetic circuits with parallel branches, the mmf has to be the same across each branch. Again putting the solution in tabular form, we get the results shown in Table 8.3. Here, once the mmf across the center leg is known, the mmf across the right-hand leg has to be the same. These mmfs then determine the flux for each of these paths. The flux in the left-hand leg then has to be the sum of the fluxes

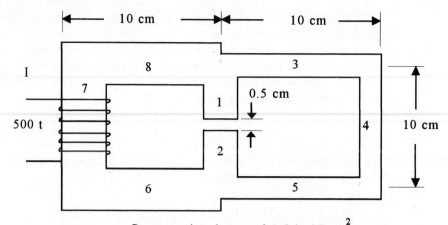

FIGURE 8.5
Magnetic circuit for Example 8.3.

TABLE 8.3
Results for Example 8.3

Path	Area	Length	B	ϕ	H	$F = NI$
Air gap	25	0.5	0.6	0.0015	48×10^4	240
1	25	5	0.6	0.0015	80	4
2	25	5	0.6	0.0015	80	4
3 + 4 + 5	25	30	1.59	0.0040	827	248
6 + 7 + 8	48	30	1.14	0.0055	150	45

in the center and right-hand legs. These sums are shown in Table 8.3. Note that in this example, the right-hand leg is saturated.

8.1.5 Fringing and Leakage Flux

Whenever an air gap is present in a magnetic circuit, the lines of flux will spread out in the vicinity of the gap to create an effect known as *fringing* as shown in Figure 8.6. The effect of fringing is to increase the effective area of the air gap. Thus, for a required gap flux density, there must be an increase in flux, which in turn requires an increase in magnetomotive force. The amount of fringing also increases with the length of the air gap. Because fringing flux requires greater mmf, attempts are made to minimize fringing by keeping air gaps as short as possible. Except in very simple cases, fringing is almost impossible to account for analytically.

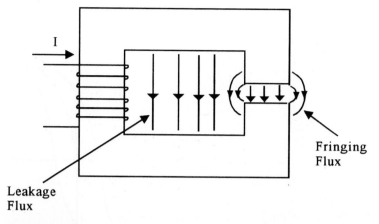

FIGURE 8.6
Fringing and leakage fluxes.

For very short air gaps, a useful approximation that accounts for fringing is to add the gap distance to both dimensions of the gap area. If fringing is a concern, several approximate methods have been developed to account for it (see entry 9 in Section 8.4, Bibliography).

Figure 8.6 also shows that portion of the total flux produced by the coil which does not appear in the magnetic circuit. This flux is the leakage flux, ϕ_l, and as in the case of the fringing flux, is not normally useful flux. If necessary, leakage fluxes can be approximated or an empirical factor can be used to account for that portion of the flux which is leakage.

8.1.6 Hysteresis and Eddy Currents

If the current used to generate the B–H curve of Figure 8.7 is increased to point *a* as shown and then reduced back to zero, the flux density will not retrace the original path but instead will return to a nonzero value. This point is called the *residual flux density, B_r*. If the current is then increased in the opposite direction, the flux density will drop to zero. The magnetic field intensity required to reduce the flux density to zero is called the *coercive force, H_c*. If the current is further increased in the opposite direction, the flux density will build up in the other direction until saturation again occurs. Reducing the current from this point will again cause the flux density to decrease to a negative residual value. Increasing the current again in the positive direction will cause the flux density to increase to its original value. This process traces a closed path called a *hysteresis loop,* shown in Figure 8.7.

Hysteresis is a result of realignment of the magnetic domains in the material when the magnetic field intensity changes. Because this realignment requires energy, hysteresis causes energy loss in magnetic devices, particularly in devices operated with ac because a complete loop is generated for every cycle of the ac waveform.

FIGURE 8.7
Hysteresis loop.

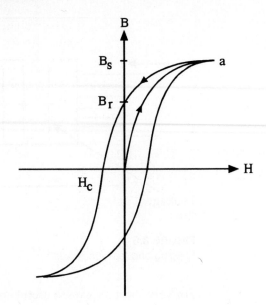

The expression for hysteresis loss is an empirical one and is given by the Steinmetz equation:

$$P_h = 10^{-7} \, \eta \, \text{Vol} \, f B^n \tag{8.9}$$

where P_h = power loss (W)
   η = a constant that depends on the type of material
   Vol = volume of the material (cm^3)
   f = frequency (Hz)
   B = flux density (G)
   n = an empirical constant ranging from 1.5 to 2.5.

The area of the hysteresis loop is a measure of the hysteresis loss. Materials that have a narrow loop (i.e., low coerciveness) are called *low-retentivity materials* and those with a wide loop are called *high-retentivity materials*. The former are useful in machines and transformers in order to reduce hysteresis loss; the latter are useful for permanent magnets (see Section 8.1.8).

In an ac circuit where the flux density changes every half cycle, voltages are induced in the magnetic material. The currents that flow as a result of these voltages are called *eddy currents* and give rise to another power loss in magnetic circuits. This loss is given by

$$P_e = 10^{-7} \, \text{Vol} \, (\pi f t B_{max})^2/(6\rho) \tag{8.10}$$

where P_e = eddy current loss (W)
   Vol = volume of the material (cm^3)
   f = frequency (Hz)
   t = thickness of the material (cm)

B_{max} = maximum flux density (G)

ρ = lamination resistivity ($\mu\Omega$-cm).

In Eq. (8.10), the flux is assumed to pass through the material perpendicular to the thickness dimension. Hence, if the material is made thinner, eddy current losses can be reduced. Note also that eddy current losses are proportional to the square of both the flux density and the frequency.

EXAMPLE 8.4

Given low-carbon sheet steel laminations 0.036 cm thick, find the hysteresis and eddy current losses per cubic centimeter of material for a flux density of 0.6 T at a frequency of 60 Hz.

Solution

For low-carbon sheet steel, the Steinmetz hysteresis coefficient is 0.003 and, unless otherwise known, the exponent is taken to be 1.6. Since 1 T = 10,000 G, 0.6 T = 6000 G and the hysteresis loss per cubic centimeter at 60 Hz becomes

$$P_h = 10^{-7} \times 0.003 \times 60 \times (6000)^{1.6}$$

or

$$P_h = 1.109 \times 0.003 \times 60 = 0.2 \text{ W/cm}^3$$

For low-carbon sheet steel, the resistivity is 13 $\mu\Omega$-cm and the eddy current loss per cubic centimeter at 60 Hz is given by

$$P_e = 10^{-7} \frac{(\pi f t B_{max})^2}{6\rho}$$

so

$$P_e = 10^{-7} \frac{(3.14 \times 60 \times 0.036 \times 6000)^2}{6 \times 13}$$

Thus,

$$P_e = 2.12 \text{ W/cm}^3$$

Note that for this example, the eddy current losses are ten times the hysteresis losses.

8.1.7 Related Fundamental Laws

Many magnetic devices depend on certain related fundamental principles for their operation. The first of these is Faraday's law, which states that the voltage induced in a loop of wire is proportional to the rate of change of flux through the loop. For a loop consisting of N turns, the expression for the voltage is given by

$$e = -N \, d\phi/dt \tag{8.11}$$

where e = the induced voltage (V)
          N = the number of turns through which the flux passes
$d\phi/dt$ = rate of change of flux (Wb/s).

The minus sign results because of conservation of energy. If the loop is closed so that current can flow due to the induced voltage, and if the sign were positive, this current would produce a magnetic field that would add to the original field. This, in turn, would increase the voltage, which would cause more current to flow and the voltage would increase indefinitely.

There are several ways to consider the effects of Faraday's law. For example, when a single conductor is moving through a magnetic field, the voltage equation is given by

$$e = BLv \tag{8.12}$$

where e = induced voltage (V)
          B = flux density (T)
          L = length of the conductor lying within the field (m)
          v = velocity of that portion of the conductor moving
               perpendicular to the field (m/s).

If this single conductor is formed into a loop and the flux remains constant while the loop is moved, the loop still sees a rate of change of flux and so a voltage is induced. Mechanically, the simplest solution is to rotate the loop or coil so that generator action is achieved. When the coil is rotated, the induced voltage is

$$e = N\phi\omega \cos\omega t \tag{8.13}$$

where e = generated voltage (V)
          N = number of turns in the coil
          ϕ = flux (Wb)
          ω = rotational speed of the coil (rad/s).

The conservation of energy mentioned earlier is described by Lenz's law, which states that the induced voltage will cause a current to flow in a direction such that the magnetic field produced by that current will be in a direction that opposes the field that caused the current to flow in the first place. It is a result of Lenz's law that torque is required in a generator or produced in a motor.

Lenz's law is quantified by the Biot–Savart force relationship, which states that the force on a current-carrying conductor in a magnetic field is given by

$$F = BLI \tag{8.14}$$

where F = force on the conductor (N)
          B = flux density (T)
          L = length of the conductor lying within the magnetic field (m)
          I = current in the conductor (A).

EXAMPLE 8.5

Find the rms value of the voltage induced in a 200-turn square coil 10 cm on a side rotating 300 rpm in a uniform magnetic field of 0.5 T.

Solution
Since $\phi = BA$, the flux is given by

$$\phi = 0.5 \times (10 \times 10^{-2})^2 = 5 \text{ mWb}$$

and

$$E_{rms} = 0.707 E_m = 0.707 \times N \times \phi \times \omega$$

Converting the speed to radians per second gives

$$300 \frac{\text{rev}}{\text{min}} \times 2\pi \frac{\text{rad}}{\text{rev}} \times \frac{1 \text{ min}}{60 \text{ s}} = 31.4 \text{ rad/s}$$

Thus,

$$E_{rms} = 0.0707 \times 200 \times 5 \times 10^{-3} \times 31.4 = 22.2 \text{ V}$$

EXAMPLE 8.6

If the coil in Example 8.5 is connected to an external circuit such that the maximum current is 0.5 A, find the maximum torque on the coil.

Solution
From Eq. (8.14), the force on each conductor is given by $F = BLI$. The torque on the loop is the force on each conductor times the radius times the number of conductors:

$$T = 2NBLI_r = NIBLd = NIBA = NI\phi$$

Thus, the maximum loop torque is

$$T = 200 \times 0.5 \times 5 \times 10^{-3} = 0.5 \text{ N-m}$$

8.1.8 Permanent Magnets

Permanent magnets are magnetic materials whose domains have been partially aligned so that a magnetic field exists permanently without the need for an external current source. When used in a device such as a loudspeaker, motor, or meter movement, they operate in the second quadrant of the hysteresis curve. A typical demagnetization curve for a common permanent magnet material, Alnico V, is shown in Figure 8.8. If used in a device with an air gap as is normally the case, the intersection of the air gap line and the demagnetization curve determines the operating flux density. This is because the mmf provided by the magnet must equal

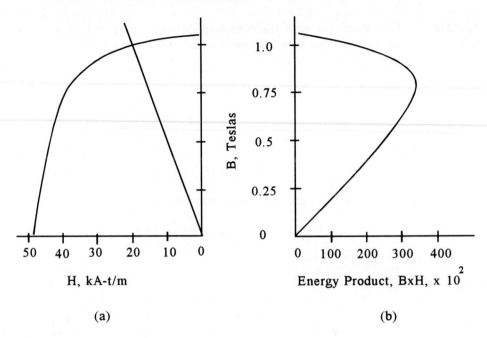

FIGURE 8.8
Permanent magnet curves: (a) Demagnetization curve and (b) energy product curve.

the mmf drop across the air gap. Equating these two gives the basic equation for a permanent magnet circuit, which is

$$H = -\frac{B_m A_m g}{\mu_0 A_g L_m} \qquad (8.15)$$

where
$\quad H$ = magnetic field intensity (A-t/m)
$\quad B_m$ = flux density in the magnet (T)
$\quad A_m$ = magnet area (m^2)
$\quad g$ = length of the air gap (m)
$\quad \mu_0 = 4\pi \times 10^{-7}$, permeability of air
$\quad A_g$ = area of the air gap (m^2)
$\quad L_m$ = mean length of the magnet (m).

To make maximum use of the magnet material, it is desirable to have the air gap line intersect the demagnetization curve near the peak of the energy product curve. The actual volume of magnet material required is increased by both the leakage flux and the fringing flux. In permanent magnet circuits, leakage flux is usually quite high, requiring from 2 to 20 times the calculated magnet volume, whereas fringing may require from 1.1 to 1.5 times the calculated magnet volume. Leakage flux can be minimized by placing the magnet material as close as possible to the air gap and completing the circuit with a magnetic material rather than having the air gap connected to the magnet by the magnetic material.

**EXAMPLE
8.7**

Find the flux density in the air gaps of the magneto of Figure 8.9 for the armature in the position shown. Neglect the reluctance of the armature and the fringing and leakage fluxes. The gap length is 0.065 in.

Solution

The flux density can be found by plotting the air gap line as found from Eq. (8.15) onto the demagnetization curve as shown in Fig. 8.8(a). This example is complicated by the fact that the air gap cross-sectional area is a section of a cylinder. The length of the arc of a circle is given by $s = r\theta$ where θ is the included angle and is given by $L = 2r \sin\theta/2$ where L is the length of the cord. The length of the arc times the depth then gives the air gap area.

Substituting in these equations the values from Figure 8.9 then gives

$$0.75 = 2(0.5) \sin\frac{\theta}{2} \quad \text{or} \quad \theta = 98° = 1.71 \text{ rad}$$

The length of the arc is then

$$s = r\theta = 0.5 \times 1.71 = 0.86 \text{ in.} = 2.17 \text{ cm}$$

The gap area is

$$A_g = 2.17 \times 1.0 \times 2.54 = 5.52 \text{ cm}^2$$

The gap length is

$$g = 0.065 \times 2.54 = 0.165 \text{ cm}$$

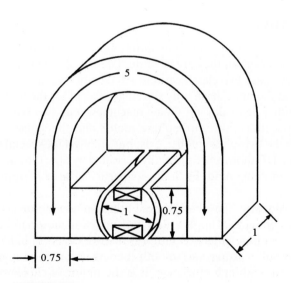

FIGURE 8.9
System for Example 8.7 (not to scale).

The magnet area is

$$A_m = 0.75 \times 1 \times (2.54)^2 = 4.84 \text{ cm}^2$$

Substituting the above into Eq. (8.15) and neglecting the mmf drop across the armature and pole pieces gives

$$H = -B \frac{4.84 \times 2 \times 0.165}{4\pi \times 10^{-7} \times 5.52 \times 5 \times 2.54} = -18,000B$$

From where the air gap line crosses the demagnetization curve, we can see that the air gap flux density is about 1.0 T.

8.2 MOTORS AND GENERATORS

8.2.1 Application Considerations

Figure 8.10 shows most of the important considerations for the proper choice of a motor. Of particular importance are the mechanical considerations of speed, starting torque, running torque, and maximum torque. Electrical considerations include three-phase or single-phase motors, allowable voltage, and maximum current. Except for those applications requiring variable speed control or battery operation, ac motors are generally preferred over dc machines because of the high cost premium of the dc motor. With the decreasing cost of variable speed controls for ac motors, a very careful cost analysis is necessary to determine the most economical choice between an ac or dc motor for variable speed applications.

8.2.2 Motor Types

Although motors can be classified in many different ways, perhaps the easiest is to classify them by whether or not they are ac or dc. There is also a major difference in construction between these two classifications: All dc motors require commutators, whereas ac motors, at the most, have only slip rings (except for the universal motor, which can run on either ac or dc). Of the ac machines, only two types are used in the majority of applications, the three-phase motor and the single-phase motor. Other types, such as the two-phase motor, exist but are used for special applications such as servo motors. DC motors are normally classified by how they are connected, either shunt, series, or compound. Each has its particular area of application.

Three-Phase Motors   Two basic types of three-phase motors are commonly used in industry and commercial applications, the induction motor and the synchronous motor. The three-phase induction motor is the workhorse of industry because it is rugged, low cost, self-starting, and smooth running. It has good starting torque, good speed regulation, and high efficiency. It is the motor of choice over a range of horsepower from fractional to 5000 or more, except where constant speed is required. For constant speed applications, the synchronous motor is the motor of

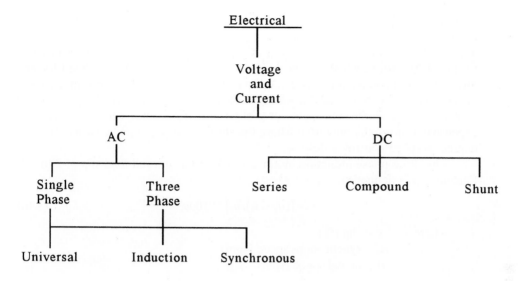

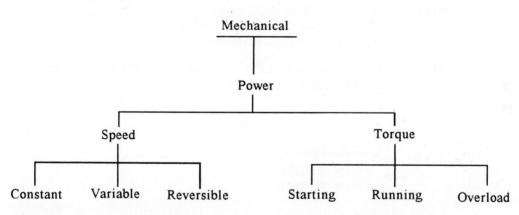

FIGURE 8.10
Motor application considerations.

choice and has the further advantage of being able to do power factor correction.
It has the disadvantage of being considerably more expensive than the induction
motor.

The operation of a three-phase induction motor depends on a rotating flux,
which is automatically established by the application of three phase voltages to a
set of windings equally displaced in space. The rotating flux produced cuts the rotor

conductors, inducing a voltage in these conductors. If the rotor conductors are short circuited so that current can flow, this current will produce a flux that will tend to align itself with the rotating flux produced by the stator. This in turn produces a torque on the rotor, which can be used to drive a load. There is no need for any direct electrical connection between the stator and the rotor. The rotor may consist of bars shorted at both ends in what is called a *squirrel cage winding* or may be wound with the windings brought out to slip rings. The wound rotor, which is more expensive, is normally only used where the starting torque must be varied or some degree of speed control is desired.

The three-phase induction motor normally runs at a full load slip of 2 to 5%. Percent slip is defined by

$$s = [(n_s - n)/n_s] \times 100\% \tag{8.16}$$

where s = slip (%)
            n_s = synchronous speed (rpm)
            n = actual speed (rpm).

The synchronous speed is a function of the frequency of the applied voltage and the number of poles in the machine and is given by

$$n_s = 120 f/P \tag{8.17}$$

where n_s = synchronous speed (rpm)
            f = frequency of the applied voltage (Hz)
            P = number of poles in the machine.

From this equation, we can see that the maximum speed of a two-pole motor is 3600 rpm at 60 Hz. For higher speeds, a gear box can be used with a resultant reduction in torque for the same power. Recall that power is given by

$$hp = nT/5252 \tag{8.18}$$

where   hp = horsepower
            n = speed (rpm)
            T = torque (lb-ft).

Figure 8.11 shows two common equivalent circuits used to represent the three-phase induction motor on a per phase basis using the machine constants of the motor. The circuit of Figure 8.11(a) is based on the representation of the motor as a transformer neglecting the leakage resistance. The right-hand resistor in this schematic represents the mechanical load on the motor. This circuit is useful for finding the line current, power factor, and input power.

The circuit in Figure 8.11(b) is found from the circuit of Figure 8.11(a) by taking the Thévenin equivalent circuit of everything to the left of resistor R_r. Combining the resistors on the right-hand side of Figure 8.11(a) gives the resistor R_r/s in Figure 8.11(b). This circuit is useful for finding input current, input power, and power factor as a function of slip.

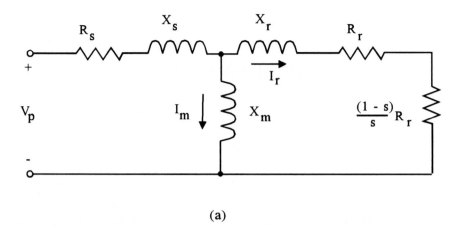

(a)

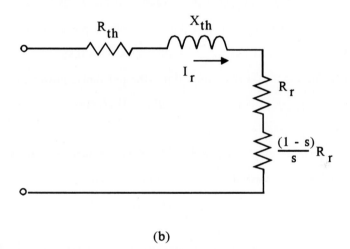

(b)

FIGURE 8.11
Equivalent circuits for the induction motor: (a) Transformer equivalent and (b) Thévenin equivalent.

EXAMPLE 8.8

Consider a 15-hp, 440-V, three-phase, 60-Hz, eight-pole induction motor that is Y connected. The motor losses are 340 W and are assumed to be constant over load. At a slip of 4% for the motor constants given, find (1) input current, (2) power factor, (3) input power, (4) rotor current, (5) output horsepower, (6) efficiency, and (7) torque. The motor constants are $R_s = 0.5\ \Omega$, $R_r = 0.6\ \Omega$, $X_s = 1.2\ \Omega$, $X_r = 1.2\ \Omega$, and $X_m = 40\ \Omega$.

Solution
(1) Since the motor is Y connected, the per phase voltage is the line voltage divided by 3. As was pointed out before, the sum of the two rotor resistances is just R_r/s.

The total rotor impedance is then just

$$R_r/s = 0.6/0.04 = 15 \ \Omega$$

The rotor impedance is in parallel with the leakage reactance such that

$$X_m \| (R_r/s + jX_r) = j40 \| (15 + j1.2) = 12.48 + j5.71$$

The total impedance on a per phase basis is then

$$Z_{tot} = R_s + jX_s + [X_m \| (R_r/s + jX_r)]$$

or

$$Z_{tot} = 0.5 + j1.2 + 12.48 + j5.71 = 14.7 \angle 28° \ \Omega/\text{phase}$$

The input current per phase is the phase voltage divided by the impedance per phase or

$$I_s = (V_p/\sqrt{3})/Z_{tot} = (440/\sqrt{3})/(14.7 \angle 28°) = 17.3 \angle -28° \ \text{A}$$

(2) The power factor is just the cosine of the phase angle or

$$PF = \cos 28° = 0.88$$

(3) The input power is three times the per phase power or

$$P_{in} = 3V_p I_s \cos\theta$$

so

$$P_{in} = 3 \times 254 \times 17.3 \times 0.88 = 11.6 \ \text{kW}$$

(4) The rotor current can be found by current division as

$$I_r = I_s [jX_m/(jX_m + R_r/s + jX_r)]$$

so

$$I_r = 17.3 \angle -28°[j40/(15 + j41.2)] = 15.8 \angle -8° \ \text{A}$$

(5) The output horse power is the per phase power delivered to the mechanical load times three or

$$P_o = 3I_r^2 \frac{(1-s)}{s} R_r - P_l$$

so

$$P_o = 3(15.8)^2 \frac{(1-0.04)}{0.04} (0.6) = 340 = 10.49 \ \text{kW} = 14 \ \text{hp}$$

(6) The efficiency is the ratio of the output power to the input power or

$$\eta = \frac{10.44}{11.6} \times 100\% = 90\%$$

(7) The torque is found from Eq. (8.18) once the motor speed is known. The motor speed is found from Eq. (8.16) by first finding the synchronous speed from Eq. (8.17). Solving these three equations gives

$$n_s = 120 \times 60/8 = 900 \text{ rpm}$$

so

$$n = (1 - s)n_s = (1 - 0.04)900 = 864 \text{ rpm}$$

and

$$T_o = \frac{5252}{n} P_o = \frac{5252}{864} \times 14 = 85.1 \text{ ft-lb}$$

The Thévenin equivalent circuit can be used to determine such quantities as the slip at maximum torque, slip at maximum power, and the resulting maximum torque and power. It can also be used to determine the current at starting since at starting the rotor is not moving and thus the slip is unity.

EXAMPLE 8.9

For the motor of Example 8.8, find the input current at starting.

Solution

Taking the Thévenin equivalent of the circuit to the left of resistor R_r, we get a Thévenin equivalent voltage of

$$E_{Th} = V_p(jX_m)/(jX_m + R_s + jX_s)$$

or

$$E_{Th} = 254(j40)/(j40 + 0.5 + j1.2) = 247\angle 0° \text{ V}$$

The Thévenin equivalent impedance is given by

$$Z_{Th} = jX_m \| (R_s + jX_s) + jX_r$$

or

$$Z_{Th} = j40 \| (0.5 + j1.2) = 1.26\angle 68.1°$$

The total impedance is then given by

$$Z_{tot} = R_{Th} + jX_{Th} + R_r/s$$

or

$$Z_{tot} = 0.47 + j1.17 + 0.6/1.0 = 1.59\angle 47.6°$$

The current at starting is thus

$$I_s = V_{Th}/Z_{tot} = (247\angle 0°)/(1.59\angle 47.6°) = 155.3\angle -47.6°$$

The National Electrical Manufacturers Association (NEMA) has classified squirrel cage induction motors according to starting current, starting torque, and torque at rated load. The characteristics of the four most common classes are shown and defined in Figure 8.12.

The synchronous motor is similar to the induction motor in that it depends on the rotating flux set up by the application of three-phase voltages to the stator winding. However, during running, a dc voltage is applied to the rotor through slip rings. This voltage produces a flux, which is fixed relative to the rotor and causes the rotor to align itself with the rotating flux produced by the stator. Thus, the synchronous motor runs at a constant speed but is not self-starting.

To start a synchronous motor, a squirrel cage winding is built into the rotor such that the motor starts as an induction motor. When the motor reaches a slip of 5 to 8%, the dc field is applied to the rotor and the rotor locks into synchronism.

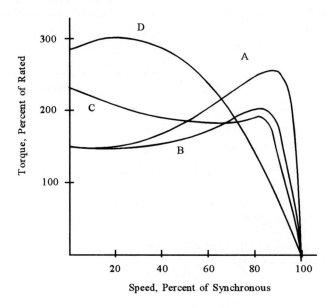

Class A: Normal starting torque
        Normal starting current

Class B: Normal starting torque
        Low starting current

Class C: High starting torque
        Low starting current

Class D: High starting torque
        High slip

FIGURE 8.12
NEMA induction motor classes.

If the load torque exceeds the maximum torque that the motor can deliver, the motor will fall out of synchronism but still run as an induction motor.

The per phase equivalent circuit and the phasor diagram for a lagging power factor are shown in Figure 8.13. Associated with the synchronous motor is a torque angle, δ, which is the angle between the applied voltage per phase and the back electromotive force (emf) produced by the rotor field cutting the stator windings. The motor equations associated with this equivalent circuit are

$$P_p = \frac{E_p V_p}{X_s} \sin\delta \qquad (8.19)$$

and

$$T_p = \frac{E_p V_p}{\omega X_s} \sin\delta \qquad (8.20)$$

where
P_p = power per phase (W)
E_p = back emf per phase (V)
V_p = applied voltage per phase (V)
X_s = synchronous reactance per phase (Ω)
δ = torque angle (deg)
T_p = developed torque per phase (N-m)
ω = motor speed (rad/s) (synchronous speed) as found from Eq. (8.17).

The back emf is found by subtracting the drop across the synchronous reactance from the applied voltage. The actual output power differs from that found from Eq. (8.19) by the amount of the losses.

One of the advantages of a synchronous motor is that its power factor can be varied by varying the field current. The variation of stator current as a function of field current generates a set of curves known as V curves as shown in Figure 8.14. The dashed lines, called *compounding curves,* show the variation in power factor of the motor with different field excitation levels at different loads.

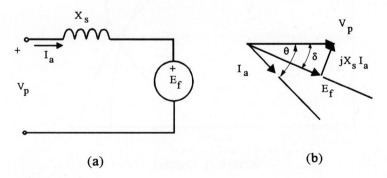

(a) (b)

FIGURE 8.13
Synchronous motor equivalent circuit. (a) Equivalent circuit and (b) phasor diagram.

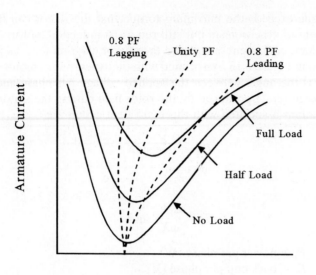

FIGURE 8.14
Synchronous motor *V* curves.

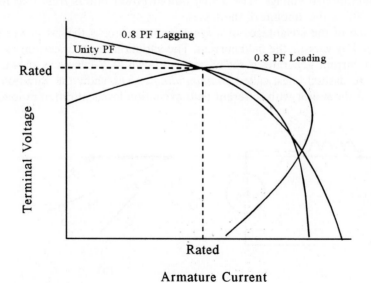

FIGURE 8.15
Synchronous generator curves.

The synchronous generator is a synchronous motor in which power flows from a mechanical source into the electrical system. This type of generator is used to produce essentially all commercial electrical power in addition to being the primary power source on all modern automobiles. The equivalent circuit for a synchronous generator is exactly the same as the one shown in Figure 8.13 for the synchronous motor. In the case of the generator, the generated voltage exceeds the line voltage by the drop across the synchronous reactance and power is delivered to the line. Figure 8.15 shows the variation in terminal voltage as a function of load power factor and load current. The terminal voltage is adjustable by varying the dc field current.

Single-Phase Motors   The single-phase induction motor is used for applications requiring relatively high horsepower (0.5 to 2) at speeds at or below 3600

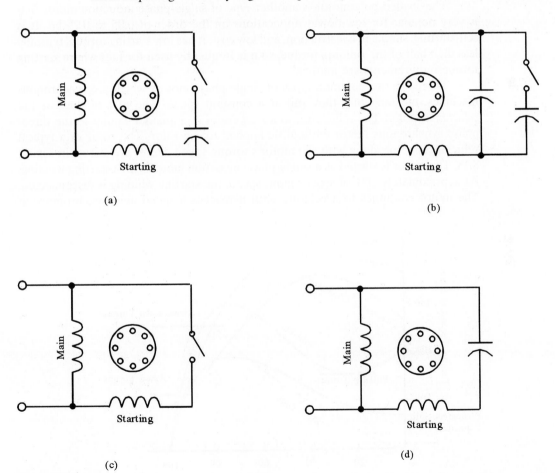

FIGURE 8.16
Starting methods for single-phase induction motors: (a) Capacitor start, (b) two value capacitor, (c) split phase, and (d) permanent split capacitor.

rpm. The single-phase induction motor by itself is not self-starting so an auxiliary starting winding must be added to the basic machine to provide starting torque. Once started, the motor will run as an induction motor on the main winding only or the starting winding may be left connected. The torque speed curves vary with the type of starting method used. Figure 8.16 shows some of the various starting methods. Most of these methods use a centrifugal switch to disconnect the starting winding near 80% of synchronous speed. Similar to the three-phase induction motor, full load speed is at a slip of 2 to 5%.

Figure 8.17 shows the variations in starting torque obtained from the different starting methods. Also shown in Figure 8.17 is the speed torque relationship for the main winding only, which is the running winding after the starting winding is switched out of the circuit by the centrifugal switch.

The shaded pole motor is another type of single-phase induction motor that is very popular for low-power applications on the order of 0.05 to 0.25 hp. It is self-starting, simple in construction, and low cost. It has low starting torque, typically less than half of the running torque, so it is frequently used for fans where starting torque requirements are minimal.

There are two common types of single-phase motors which are synchronous machines, meaning that they run at a constant speed. The first of these is the *reluctance motor.* The reluctance motor operates in a manner similar to the three-phase synchronous motor without dc applied to the rotor. The rotor of a typical reluctance motor along with the motor's torque speed curves are shown in Figure 8.18. The motor is started as a single-phase induction motor with a starting winding. At approximately 70% of synchronous speed, the starting winding is disconnected. The motor continues to accelerate until it reaches a speed near synchronous, at

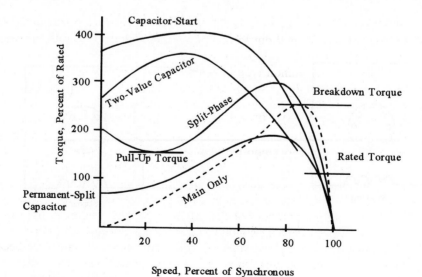

FIGURE 8.17
Speed torque characteristics for various types of single-phase induction motors.

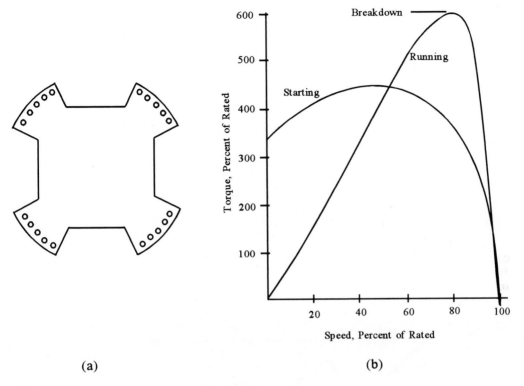

(a)   (b)

FIGURE 8.18
Reluctance motor: (a) Rotor construction and (b) torque speed curve.

which point it will snap into synchronism and run at synchronous speed. If it is loaded beyond the pull-out torque, it will drop out of synchronism but still run as an induction motor.

The other common type of single-phase synchronous motor is the *hysteresis motor* or *clock motor*. The rotor in a hysteresis motor is a permanent magnet and the stator is arranged to provide a rotating flux, normally by a permanent capacitor or through shaded poles. The rotor will attempt to align itself with the rotating flux and thus rotate at synchronous speed. Smaller sizes are used for clock motors while larger sizes make very smooth and quiet drives for record players and tape drives.

A third common type of single-phase motor is the *universal motor,* which is really a series dc motor designed to operate on ac. To reduce eddy current and hysteresis losses, the field poles and the armature are laminated. Because torque in a universal motor does not depend on a rotating flux field, but rather on the force on a current-carrying conductor, it is capable of very high speeds. The universal motor can provide high horsepower in a relatively small frame size because of its high-speed operation. Such motors find wide application in such devices as electric drills, vacuum cleaners, blenders, mixers, etc. Ease of speed control and ability to reverse direction are further advantages of the universal motor. Figure 8.19 shows the equivalent circuit of the universal motor when operated on ac along with the

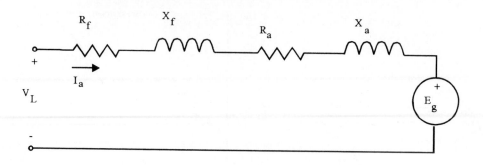

(a)

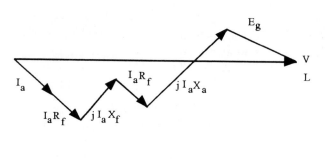

(b)

FIGURE 8.19
Universal motor: (a) Equivalent circuit and (b) phasor diagram.

associated phasor diagram. In this figure, the subscripts f and a refer to the field and armature, respectively.

DC Motors and Generators   Although they have been largely replaced by ac motors for many applications, the dc machine is unsurpassed in its wide variety of operating characteristics, particularly for speed control. In a dc motor, torque is produced due to the magnetic action between a fixed stator field and a rotating armature field. The basic motor equation is:

$$T_a = K_a \phi I_a \qquad\qquad (8.21)$$

where T_a = torque (N-m)
K_a = torque constant
ϕ = field flux per pole (Wb)
I_a = armature current (A).

The torque constant K_a is given by

$$K_a = (pZ)/(2\pi a) \tag{8.22}$$

where K_a = torque constant
         p = number of poles in the motor
         Z = number of conductors in the armature
         a = number of parallel paths of the armature conductors.

The dc motor generates a back emf because of the rotation of the armature in a magnetic field. This voltage opposes the applied voltage so as to limit the armature current. The magnitude of this voltage is given by

$$E_a = K_b \phi \omega \tag{8.23}$$

where E_a = back emf (V)
         K_b = back emf constant = K_a
         ϕ = field flux per pole (Wb)
         ω = rotational velocity (rad/s).

EXAMPLE 8.10

Consider a six-pole motor with 240 conductors and eight parallel paths, a flux per pole of 0.02 Wb and an armature current of 7.3 A. If this motor rotates at a speed of 1710 rpm, find the torque, back emf, and developed power.

Solution
Substituting directly into Eq. (8.22) gives the motor constant

$$K_a = \frac{pZ}{2\pi a} = \frac{6 \times 240}{2\pi 8} = 28.6$$

The speed must be converted to radians per second by multiplying by $2\pi/60$ or

$$\omega = (1710 \times 2\pi)/60 = 179 \text{ rps}$$

The back emf is then found from Eq. (8.23):

$$E_a = 28.6 \times 0.02 \times 179 = 102.5 \text{ V}$$

and the torque from Eq. (8.21) as

$$T = 28.6 \times 0.02 \times 7.3 = 4.18 \text{ N-m}$$

The power in SI units is then the torque times the speed or

$$P = \omega T = 179 \times 4.18 = 748 \text{ W}$$

In English units, since 1 hp = 746 W, the developed horsepower is given by

$$P = 748/746 = 1 \text{ hp}$$

These equations are equally valid for a dc generator except that the generated voltage, E_a, exceeds the terminal voltage so that current flows out of the machine.

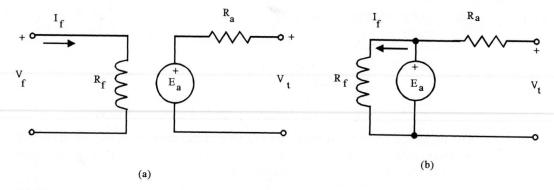

(a)

(b)

FIGURE 8.20
Basic dc machine: (a) Separately excited and (b) self-excited.

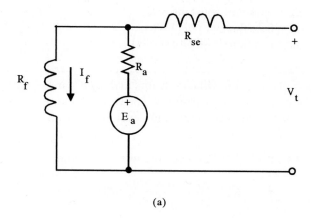

(a)

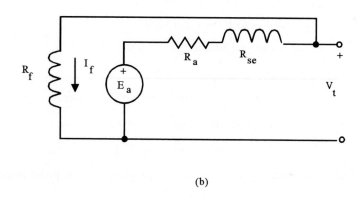

(b)

FIGURE 8.21
Compound dc machine connections: (a) Short shunt connection and (b) long shunt connection.

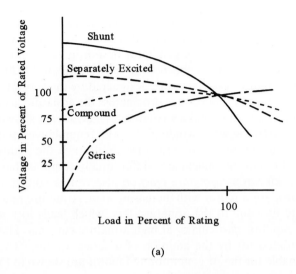

(a)

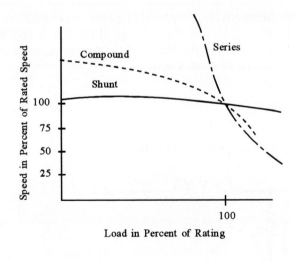

(b)

FIGURE 8.22
DC machine characteristics: (a) Generator characteristics and (b) motor characteristics.

In either a generator or a motor, the armature current causes a voltage drop across the armature resistance, which increases with increasing load. This voltage drop causes a drop in terminal voltage in the generator or a drop in speed in the motor if not compensated for. The basic dc machine is shown in Figure 8.20 where the field resistances and armature resistances are designated by the subscripts f and a, respectively. To compensate for the voltage drop across the armature, a compensating winding is frequently added to make a compound machine, as shown in Figure 8.21. The mmf generated by the series winding, R_{se}, adds (cumulative compounding) to the field mmf to compensate for the armature IR drop. If the additional mmf subtracts from the field mmf, the machine is differentially compounded.

A phenomenon known as *armature reaction* also causes voltage drop for a generator or speed drop for a motor with increasing load. Armature reaction is the distortion of the field flux due to the armature flux, which leads to a net overall reduction of the air gap flux, particularly at high armature currents. This decrease may also be compensated for by the addition of a series winding. The different characteristics obtainable for the dc generator and motor are shown in Figure 8.22. The series connection is for a machine that has only a series winding without a shunt field winding. In a series machine, no appreciable flux is available until significant armature current flows.

EXAMPLE 8.11

Consider a short shunt cumulatively compounded motor rated at 240 V, 10 A, 1200 rpm with $R_f = 180\ \Omega$, $R_a = 2\ \Omega$, and $R_{se} = 1.2\ \Omega$. If the loss is 160 W (friction, windage, and core), find at full load the back emf, power rating, efficiency, torque rating, developed torque, and no-load speed.

Solution

Pictorially, this motor is shown in Figure 8.23. The input current produces a voltage drop across the series resistance given by

$$V_{se} = I_l R_{se} = 10 \times 1.2 = 12\ \text{V}$$

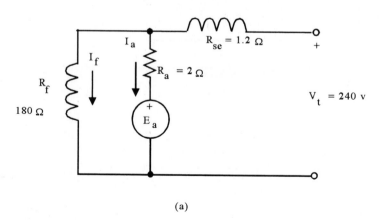

(a)

FIGURE 8.23
Schematic for Example 8.11.

The voltage across the field and the armature is then the difference between the line voltage and the series field drop or

$$V_f = V_t - V_{se} = 240 - 12 = 228 \text{ V}$$

The field current subtracts from the line current to give the armature current:

$$I_a = I_l - I_f = 10 - (228/180) = 8.73 \text{ A}$$

The back emf is then the field voltage minus the drop across the armature resistance or

$$E_a = V_f - I_a R_a = 228 - 8.73 \times 2 = 210.5 \text{ V}$$

The power rating is the input power minus the losses. The losses include the I^2R losses in the series winding, the field winding, and the armature resistance along with the friction, windage, and core losses. These losses are, respectively,

$$P_l = I_l^2 R_{se} + I^2 R_f + I_a^2 R_a + P_{f\omega c}$$

or

$$P_l = (10)^2(1.2) + (1.27)^2(180) + (8.73)^2(2) + 160 = 722 \text{ W}$$

The efficiency is the ratio of the output power to the input power or

$$\eta = \frac{P_o}{P_{in}} = \frac{(P_{in} - P_l)}{P_{in}} = \frac{(240 \times 10 - 722)}{240 \times 10} \times 100\% = \frac{1678}{2400} \times 100\% = 70\%$$

The torque rating is found from the power rating as

$$T = \frac{5252P}{n}$$

Thus,

$$T = \frac{5252}{1200} \times \frac{1678}{746} = 9.84 \text{ ft-lb}$$

The developed torque can be found from the developed power where the developed power is the output power plus the friction, windage, and core losses. The developed power is also the back emf times the armature current. Thus,

$$P_d = E_a I_a = 210.5 \times 8.73 = 1838 \text{ W}$$

so

$$T_d = \frac{5252}{1200} \times \frac{1838}{746} = 10.78 \text{ ft-lb}$$

The no-load speed can be approximated by neglecting the armature reaction so that the no-load air gap flux is assumed to be the same as the full-load air gap flux. At no load, the only power required is that necessary to overcome the friction, windage, and core losses. Thus, the input current will be small, and to a first approximation the current through the field winding can be neglected. The developed power is equated to the stray losses to give

$$P_d = E_a I_a = P_{f\omega c} = 160 \text{ W}$$

The voltage drop across the armature resistance and the series field is then given by (neglecting the field current)

$$V_l - E_a = I_a(R_a + R_{se}) = (P_d/E_a)(R_a + R_{se})$$
$$= (160/E_a)(2 + 1.2) = 512/E_a$$

Solving this expression for E_a gives

$$E_a^2 - 240E_a + 512 = 0$$

so

$$E_a = \frac{240 \pm [(240)^2 - 4 \times 512]^{1/2}}{2} = 237.8 \text{ V}$$

Then from Eq. (8.23) we have

$$E_{ab} = K_a \phi \omega_o \text{ at no load}$$
$$E_{af} = K_a \phi \omega_f \text{ at full load}$$

Taking the ratio of these two equations gives

$$\frac{E_{ao}}{E_{af}} = \frac{K_a \phi \omega_o}{K_a \phi \omega_f}$$
$$= \frac{\omega_o}{\omega_f} = \frac{n_o}{n_f}$$

so

$$n_o = \frac{E_{ao}}{E_{af}} n_f = \frac{237.8}{210.5} \times 1200 = 1356 \text{ rpm}$$

Specialty Motors Two basic special types of motors to be discussed are the stepping motor and the linear motion motor. The stepping motor is basically a synchronous motor that is excited by pulses such that the rotor rotates in incremental steps of fixed size. There are two basic types of stepping motors, the variable reluctance type shown in Figure 8.24 and the permanent magnet type shown in Figure 8.25. The stator normally consists of two, three, or four phases. The number of poles on the rotor depends on the step size, and when one of the stator phases is energized, the rotor will align itself with the stator flux.

The variable reluctance type of motor in Figure 8.24 uses a toothed rotor, made from some type of magnetic material, and stator windings, which are equally skewed in space with respect to each other. When one of the stator windings is energized, the rotor will seek the minimum reluctance position and align itself with

Stator Windings

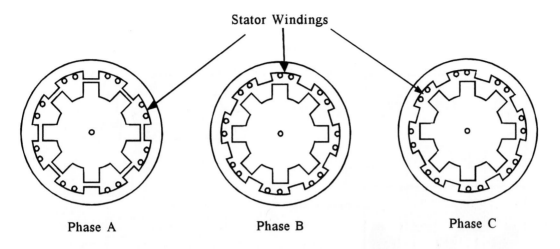

Phase A   Phase B   Phase C

Stators

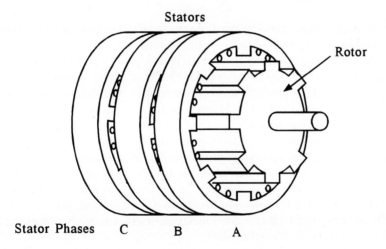

Rotor

Stator Phases   C   B   A

FIGURE 8.24
Variable reluctance stepping motor showing phase A energized.

the energized stator poles. For this type of stepper to have any holding torque, the stator winding has to remain energized.

The permanent magnet type of stepper shown in Figure 8.25 uses a two-phase winding. In this type of motor, the rotor contains permanent magnets that will align themselves with the energized stator winding. Because of the permanent magnets on the rotor, the motor will have holding torque without having to have the stator

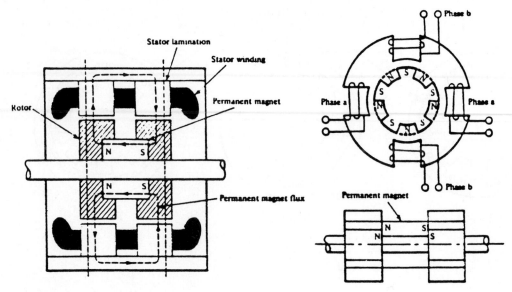

FIGURE 8.25
Permanent magnet stepping motor.

energized. Reversal of the order of energization of the coils in either type of motor will produce a change in direction of rotation.

Application considerations for stepping motors include many of the same considerations as discussed for ordinary motors. Other important considerations include step angle, stepping rate, holding torque, and detent torque. Typical step angles range from 0.9° to 15°. Stepping rates can go as high as 15,000 steps per second for small step angles.

Several versions of the linear motion motor exist, depending on the application. One of the simplest is the moving coil motor analogous to the loudspeaker where force is provided by the interaction of the field produced by the current in a moving coil and the field provided by a permanent magnet. The advantage of this type of motor is that since normally no iron is present, there are no nonlinear effects and force is directly proportional to input current. Such motors are frequently used in pen or strip chart recorders.

Another version of the linear motion motor is the linear induction motor, which can provide very high forces. The linear induction motor can be viewed as a rotary induction motor that has been sectioned and laid out flat as shown in Figure 8.26. As shown in Figure 8.26, either the "stator" or the "rotor" can be the moving part. Such motors have applications in high-speed transport, materials handling, and aircraft catapults.

A third version of the linear motion motor is the basic dc machine which has again been sectioned and laid out flat. Another version is shown in Figure 8.27, which is a linear actuator. The magnetic core that acts as the armature is wound with a single-layer winding, which reverses direction at the center. The armature

Rotor

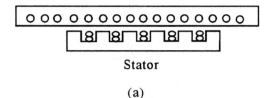

Stator

(a)

Rotor

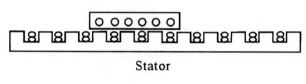

Stator

(b)

FIGURE 8.26
Linear induction motor: (a) Short stator machine and (b) short rotor machine.

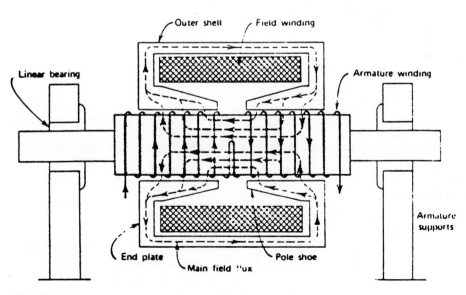

FIGURE 8.27
Direct current linear actuator.

is surrounded by a field winding, which produces a radial air gap flux. The air gap flux reacts with the flux produced by the current in the armature winding to produce a linear force. The dc type of linear machine has found application in closed loop table positioning applications where relatively large loads have to be positioned with high accuracy.

8.3 SOLENOIDS AND ELECTROMAGNETS

8.3.1 Application Considerations

A wide variety of solenoids and electromagnets are available for many different applications. Figure 8.28 shows some of the different considerations and variations in applying and using solenoids and electromagnets. Many of these are predetermined for a given application, but others such as speed are often neglected in the initial design.

Most solenoids fall into two basic categories, linear motion and rotary motion with the linear type being much more common. Most rotary solenoids are linear devices with mechanical conversion to rotary motion. The four most popular types of linear motion solenoids are the flat faced plunger, the conical faced plunger, the

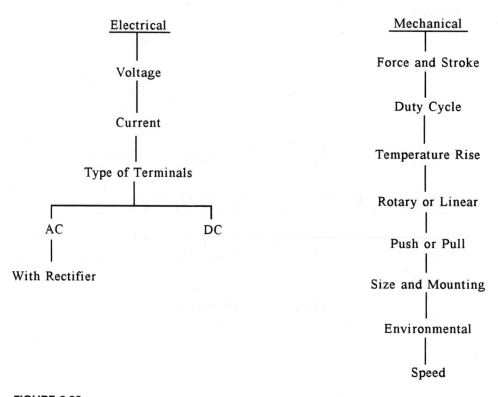

FIGURE 8.28
Solenoid selection parameters.

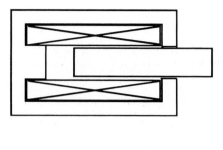

(a)

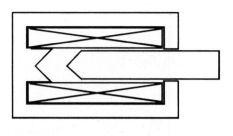

(b)

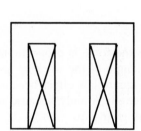

(c)

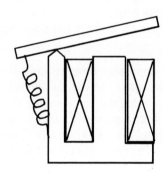

(d)

FIGURE 8.29
Different types of solenoids: (a) Flat faced plunger, (b) conical faced plunger, (c) lifting solenoid, and (d) relay type.

lifting, and the relay types. These different types are shown in cross section in Figure 8.29. Each of these exhibits slightly different force-stroke characteristics, which are shown in Figure 8.30.

The basic force equation for a solenoid is given by

$$f = A_g B_g^2 / (2\mu_0) \qquad (8.24)$$

where f = force (N)
A_g = area of the air gap (m^2)
B_g = flux density in the air gap (T)
μ_0 = permeability of air ($4\pi \times 10^{-7}$).

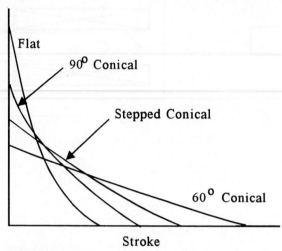

FIGURE 8.30
Solenoid force-stroke characteristics.

The important thing to note about this equation is that the force is proportional to the square of the flux density. If nothing except the length of the working gap is changed, the force stroke characteristic will follow a square law effect because the flux density will change in proportion to the gap length. Mainly because of fringing, Eq. (8.24) is only approximate, especially for long air gaps.

The force equation for the conical faced plunger (including fringing) is

$$f = B_g^2 \pi r_1^2 \sin^2\alpha / (2\mu_0) \qquad (8.25)$$

where f = force (N)
B_g = flux density in the air gap (T)
r_1 = radius of the plunger (m)
α = angle of plunger face from the horizontal in degrees
μ_0 = permeability of air ($4\pi \times 10^{-7}$).

EXAMPLE 8.12

Given a 30-deg conical faced plunger type of solenoid with a plunger radius of 0.25 in. and an air gap flux density of 74.2 kMx/in.^{2}, find the force generated by the plunger.

Solution
To use Eq. (8.25), the flux density must be converted to teslas or webers per square meter. Since there are 64.5 kMx per square inch in a tesla, the flux density is 74.2/64.5 or 1.15 T. The generated force is then

$$f = \frac{(1.15)^2 \times \pi \times (0.25/39.37)^2 (\sin 30°)^2}{2 \times 4\pi \times 10^{-7}} = 16.65 \text{ N} = 3.75 \text{ lb}$$

For comparison purposes, the force generated by a flat faced solenoid of the same plunger diameter is

$$f = \frac{(1.15)^2 \times \pi \times (0.25/39.37)^2}{2 \times 4\pi \times 10^{-7}} = 66.6 \, \text{N} = 15 \, \text{lb}$$

The difference is due to the change in effective gap area between the two types of solenoids.

For both the flat and conical faced plunger types of solenoids, there are two air gaps, a working gap whose length changes with the position of the plunger, and a nonworking gap whose length usually remains constant. The nonworking gap is necessary to allow plunger movement. Good solenoid design requires that the nonworking gap be as short as possible in order to limit the mmf drop. Lifting and relay types of solenoids only have a working gap but that gap may consist of more than one flux path.

There are many varied applications for solenoids. A very common application is in solenoid-operated valves, which control the flow of either fluids or gases. Another very common application is to provide some type of mechanical motion in response to an electrical input such as automatic door locks in an automobile. Another common automotive application is in the control valve for exhaust gas recirculation.

Mechanical Application Considerations   The primary mechanical application consideration is the force-stroke characteristic. To assure adequate built-in force, the solenoid must be sized for both worst case voltage and highest temperature operation. Increasing temperature decreases the pulling force, primarily due to the increase in coil resistance, which decreases the coil current at any given voltage. Decreasing coil current decreases the flux, reducing the force as per Eq. (8.24).

Other mechanical considerations include duty cycle, which relates primarily to self-heating, maximum on time, push or pull operation, speed, mounting, life expectancy, and the environment in which the solenoid is to operate. The duty cycle is defined as the ratio of the on time to the sum of the on time plus the off time as given in Eq. (8.26):

$$\text{Percent duty cycle} = \frac{\text{On time}}{\text{On time} + \text{Off time}} \times 100\% \qquad \textbf{(8.26)}$$

Solenoids are basically pull-type devices and converting a given solenoid to push-type operation will generally reduce the force by 20% due to the loss of working gap area. This area is lost due to the inclusion of a push rod, which must pass through the working gap, thus reducing its area.

Speed can be increased by venting the stop end to prevent compressing the air in the gap during closure. This will also result in a slight loss of gap area. Another way to vent the air gap area is to put grooves in the plunger to allow the air to pass by during closure. Other ways to increase speed include lubricating the plunger with some type of dry lubricant.

Electrical Application Considerations Primary electrical considerations include ac or dc operation, operating voltage, speed, and type of electrical connections. The speed of operation for an ac solenoid will vary depending at which point in the ac wave the solenoid is energized. A dc solenoid should be used if consistent

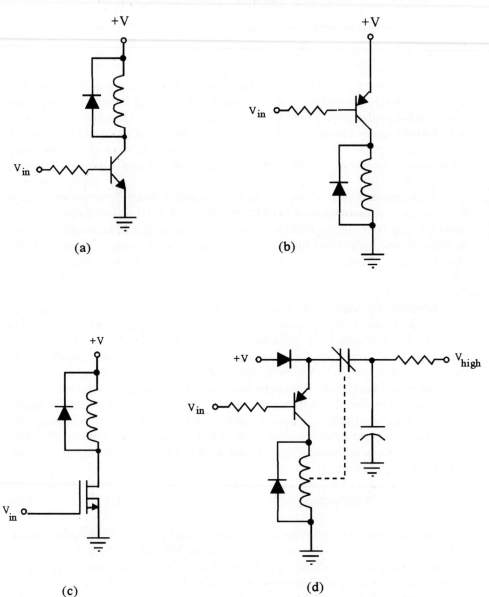

FIGURE 8.31
Driver circuits for solenoids: (a) NPN driver, (b) PNP driver, (c) FET driver, and (d) high-voltage source.

speed of operation is desired. The speed of a solenoid may also be increased by energizing at a higher voltage and reducing the voltage at closure to minimize heating (see Section 8.3.2). A dc solenoid can be driven from an ac source by including a rectifier and some capacitive filtering.

8.3.2 Solenoid Drive Circuits

Solenoids are frequently driven directly from the line voltage through a contact closure. A typical example would be the solenoid valves in a washing machine or a dishwasher where the contacts controlling the valves are part of the timer. However, solenoids can also be controlled through transistors, SCRs, or TRIACs. In particular, transistors can be used to drive dc solenoids, and SCRs or TRIACs can be used to drive ac devices. Because a solenoid is an inductive device, care must be taken to protect the controlling device against inductive kickback on turnoff. It is not unusual for a 12-V solenoid to produce a 600-V pulse when turned off.

Figure 8.31 shows a variety of drive circuits for controlling solenoids. Of particular interest is the circuit of Figure 8.31(d), which provides high-voltage energization for increased speed while providing a means of switching to a lower voltage after closure to prevent overheating. Many solenoid manufacturers will provide a set of contacts that actuates on solenoid closure for just this purpose.

8.4   BIBLIOGRAPHY

The following list of references represents an incomplete listing of the wide variety of books that cover the topics discussed in this chapter.

1. Fitzgerald, A. E., Arivn Grabel, and David E. Higginbotham, *Basic Electrical Engineering*, McGraw-Hill, 1981.
2. Friedman, M. H., and J. Rosenblatt, *Direct and Alternating Current Machinery*, Charles E. Merrill, 1984.
3. Chaston, A. N., *Electric Machinery*, Prentice Hall, 1986.
4. Kosow, I. L., *Electric Machinery and Transformers*, Prentice Hall, 1972.
5. Gingrich, H. W., *Electrical Machinery, Transformers, and Controls*, Prentice Hall, 1979.
6. Nasar, S. A., *Electric Machines and Transformers*, Macmillan, 1984.
7. Ryff, P. F., D. Platnick, and J. A. Karnas, *Electrical Machines and Transformers*, Prentice Hall, 1987.
8. Matsch, L. W., and J. D. Morgan, *Electromagnetic and Electromechanical Machines*, Harper & Row, 1986.
9. Roters, H. C., *Electromagnetic Devices*, John Wiley, 1941.
10. Del Toro, V., *Electromechanical Devices for Energy Conversion and Control Systems*, Prentice Hall, 1968.
11. Bozorth, R. M., *Ferromagnetism*, Van Nostrand, 1951.
12. Greenwood, D. C., *Manual of Electromechanical Devices*, McGraw-Hill, 1965.

13. Hadfield, D., *Permanent Magnets and Magnetism,* John Wiley, 1962.
14. Pericles, E., *Motors, Generators, Transformers, and Energy,* Prentice Hall, 1985.
15. Richardson, D. V., and A. J. Caisse, Jr., *Rotating Electric Machinery and Transformer Technology,* Prentice Hall, 1987.
16. Kuo, B. C., *Theory and Applications of Step Motors,* West Publishing, 1974.

9

DIODES*

This chapter presents the basic characteristics of diodes used in electricity and electronics. Application notes and selection guidelines will be presented for many of the diode types.

9.1 INTRODUCTION

In the simplest view, an *ideal diode* is simply a two-terminal device that acts as an electrical unidirectional switch for the passage of electrical current. In other words, it allows current to pass in only one direction.

Ideal Diode   In Figure 9.1 the schematic symbol of a simple diode is given, showing the anode connection and the cathode connection. Also shown in Figure 9.1 is the graph of the current versus voltage of an ideal diode. Note that when the applied voltage is positive on the anode, called *forward biased,* the diode is like a closed switch and passes (conventional) current without resistance. When the voltage polarity is reversed so that the anode is negative, or *reverse biased,* the diode is like an open switch (infinite resistance) and no current is passed.

Real Diodes   Of course, the real diodes used in circuits do not match the current versus voltage response of Figure 9.1. It is important to take into account the nonideal properties. The rest of this chapter is devoted to a presentation of real diodes in terms of how they differ from the ideal and how this affects their applications.

---

* This chapter was written by Dr. Curtis D. Johnson, College of Technology, University of Houston.

FIGURE 9.1
The symbol and I-V characteristics of an ideal diode.

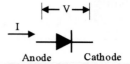

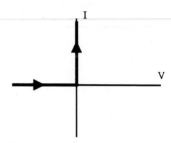

9.1.1 Diode Standards

Thousands of types of diodes are in use in the electrical and electronics industry. There are a number of loosely applied standards to aid in classification and identification of diodes.

Schematic Symbol   One obvious standard is the schematic representation of diodes. A basic diode is indicated by the symbol given in Figure 9.1. There are many variations of this symbol to denote special diode types. The standard schematic symbols for diodes are given in Table 9.1.

Registered Numbers   A company that manufactures diodes assigns unique, internal "part numbers" to each type of diode manufactured. If a particular diode type meets certain defined national standards, it may be registered as an EIA (Electronics Industries Association) or JEDEC (Joint Electron Devices Engineering Council) type. In this case it is assigned a number that starts with "1N" and is followed by digits. Thus, for example, 1N4004 is a registered diode type, which, regardless of the manufacturer, satisfies certain minimum specifications.

Packaging   The many types and applications of diodes result in many methods of packaging diodes. Figure 9.2 shows some of the standard packaging forms.

The smallest diode forms shown in Figure 9.2 are often encapsulated in glass. These are usually *signal diodes,* which are used for very low power, signal processing applications.

Small, low- to medium-power, *rectifier* and *Zener diodes* are often in the form of small black plastic cylinders.

Higher current rectifiers and Zeners are packaged in an assembly with a lug connection on one end and a threaded bolt at the other, as shown in Figure 9.2. These are designed to be bolted to a heat sink to prevent overheating.

Very high current diodes are larger versions of this last design, often with a

TABLE 9.1
Diode schematic symbols

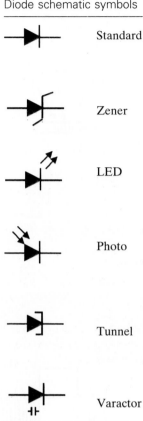

Standard

Zener

LED

Photo

Tunnel

Varactor

heavy, flexible copper braided assembly instead of a rigid lug. Again heat sinking is important.

Marking We must know the type number of the diode and which connection is the anode and which is the cathode. There are no universal standards for providing this information. Figure 9.2 shows some of the standard marking techniques:

1. Print the symbol of a diode on the body of the package. This of course clearly identifies which end is the cathode and which end is the anode.
2. Place a black band or ring or a series of colored bands nearest the *cathode* end of the diode.

The diode must also be marked with its manufacturer's part number or EIA/JEDEC registered type number. This is done by several different methods:

1. Print the actual number on the diode. In some cases the 1N is left off so a 1N756 might be labeled 756.

Cathode

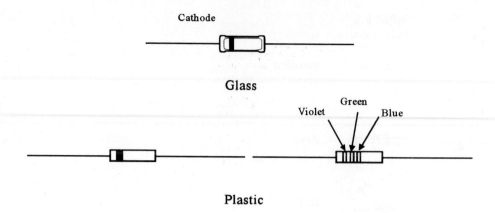

Glass

Violet Green Blue

Plastic

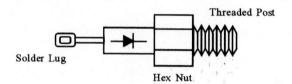

Solder Lug

Threaded Post

Hex Nut

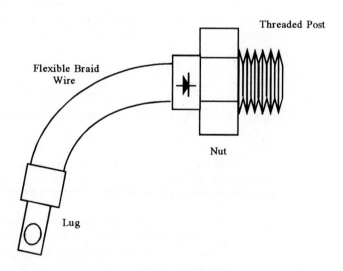

Flexible Braid
Wire

Threaded Post

Nut

Lug

FIGURE 9.2
Some of the many sizes and shapes of diodes.

TABLE 9.2
Color code

Color	Number
Black	0
Brown	1
Red	2
Orange	3
Yellow	4
Green	5
Blue	6
Violet	7
Gray	8
White	9

2. Use a series of colored rings placed nearest the cathode and read from cathode to anode. The numbers corresponding to the colors are the same as for resistor and capacitor color coding, but all rings are numbers, i.e., there are no multipliers. Table 9.2 explains the color code. Suppose the diode in Figure 9.2 has rings with colors violet (7), green (5), and blue (6). Therefore, this is a 1N756 and the cathode is that end which is nearest the colored bands. Obviously, confusion can arise over a small cylindrical component with colored bands. Is it a resistor or a capacitor or a diode? Section 9.2.4 on testing diodes shows how resistance measurements can be used for identification.

9.1.2 Vacuum Tube Diodes

Prior to the development of solid-state semiconductor technology, diode action was provided by a two-element vacuum tube. This diode uses a heated filament as the cathode and as a source of conduction electrons. If the anode or plate, as shown

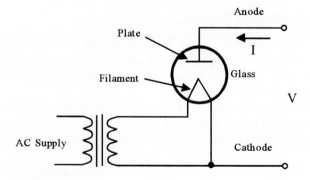

FIGURE 9.3
Representation of a vacuum tube diode.

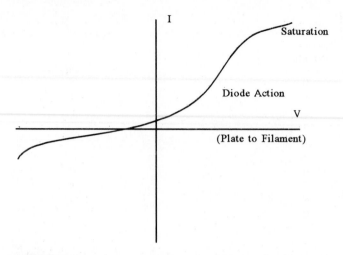

FIGURE 9.4
The I-V characteristic of a vacuum tube diode.

by the schematic symbol of Figure 9.3, is at a positive voltage with respect to the cathode, electrons are attracted to the plate, current flows, and the diode is then forward biased. If the plate is negative, electrons are repelled from the plate and current can not flow so that the diode is reverse biased.

Figure 9.4 shows a typical curve of current versus voltage for a vacuum tube diode. It is clearly quite different from the ideal diode. Of special note is the saturation that occurs with increasing plate voltage, i.e., the current saturates and diode action stops at some critical voltage. Note also that a small reverse bias must be applied before the *forward* current is driven to zero.

The vacuum tube diodes are made in many shapes and sizes to handle the great variety of diode applications from power rectifiers to microwave detectors.

9.1.3 Semiconductor Diodes

Modern diodes are constructed by the junction of two semiconductor materials, one *p*-doped and the other *n*-doped. The physical diagram and schematic symbol of Figure 9.5(a) show that the *p*-doped material is the anode and the *n*-doped material is the cathode.

The first semiconductor diodes were made using "point-contact" technology. Figure 9.5(b) shows that a fine wire makes contact with a small island of *p*-doped material on an *n*-doped substrate. These types of diode are little used today.

Materials Early semiconductor diodes were constructed of stacks of plates made with selenium or copper oxide, both of which are semiconductor materials.

FIGURE 9.5
(a) Semiconductor junction and (b) point-contact diodes.

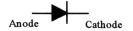

Symbol

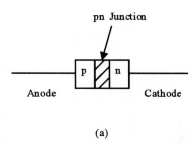

(a)

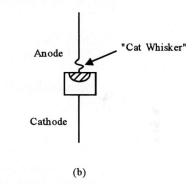

(b)

Now diodes are most often constructed using crystals of doped germanium or silicon. Certain other special semiconductor materials are used for special-purpose diodes.

Properties The following paragraphs summarize the general properties of semiconductor diodes. The I-V curve of Figure 9.6 is typical of a semiconductor diode. It is also clearly not ideal but in many respects is closer to ideal than the vacuum tube version. Since germanium and silicon are the most common materials, properties will be given specifically for these materials.

Forward-Bias Voltage (V_{FB}) Figure 9.6 shows that the real diode does not begin to conduct in the forward-biased configuration until a certain minimum voltage

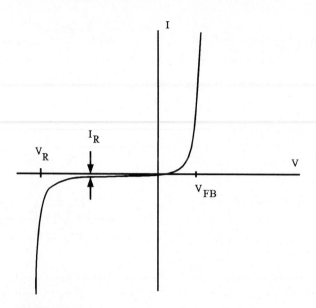

FIGURE 9.6
Typical I-V characteristics of a semiconductor diode.

is reached. This voltage is *temperature dependent* and *decreases* at approximately 2 mV/°C. For silicon at 25°C, $V_{FB} \approx 0.7$ V, and for germanium at 25°C, $V_{FB} \approx 0.3$ V.

Forward Resistance (r_F)   Once the diode has sufficient forward-bias voltage to conduct, the current rises with a steep but finite slope as voltage is increased. This means that the diode, even though forward biased and conducting, presents a small resistance to current flow.

The forward resistance will either be given in the diode specifications directly or can be computed from information given. Often the specification provides the *net* forward voltage drop at some current. The resistance can then be found from

$$r_F = \frac{(V_F - V_{FB})}{I_F} \tag{9.1}$$

where V_F = total forward voltage drop at I_F
V_{FB} = forward-bias voltage, defined earlier
I_F = forward current at V_F.

The *net forward voltage drop* is the sum of the forward-bias voltage and, from Ohm's law, the voltage dropped across the forward resistance.

EXAMPLE 9.1   A 1N916 diode is to be used in a circuit with a current of 25 mA. What is the forward resistance and the net forward voltage drop? If the temperature increases to 60°C, what will the forward voltage drop become?

Solution
A 1N916 specification sheet gives $V_F = 1$ V at $I_F = 10$ mA and that it is a silicon diode. We know that at room temperature ($\approx 25°C$) $V_{BF} \approx 0.7$ V, so, from Eq. (9.1),

$$r_F = (1 - 0.7)/10 \text{ mA} = 30 \text{ }\Omega$$

The net forward voltage drop at 25 mA is found from

$$V_F = 0.7 + (0.025)(30) = 0.7 + 0.75 = 1.45 \text{ V}$$

If the temperature increases to 60°C, then the forward-bias voltage will decrease:

$$V_{FB} = 0.7 - (0.002 \text{ V/°C})(60 - 25°C)$$
$$= 0.63 \text{ V}$$

so, $V = 0.63 + 0.75 = 1.38$ V. We assume that the forward resistance does not change, i.e., the slope of the conduction part of the curve is the same.

Reverse Current (I_R)   When the real diode is reverse biased, the current flow is not exactly zero as in the case of the ideal diode. Figure 9.6 shows that a relatively small but finite current flows in the reverse-biased configuration. For a high-current (250-A) rectifier such as the 1N3740 this may be as high as 12 mA, whereas a low-power signal diode such as the 1N916 will have a reverse current of only 5 μA.

Reverse Breakdown Voltage (V_R)   When the reverse voltage reaches some critical value, the diode "breaks down" and begins to conduct current in the reverse direction. This is shown in Figure 9.6 by the fact that once the reverse-bias voltage reaches and exceeds V_R current begins to flow.

The reverse breakdown voltage varies over a great range for the many types of diodes available. It may range from only 10 to 50 V for a signal diode to more than 1000 V for rectifiers. Having the reverse voltage reach or exceed V_R does not *destroy* the diode. The term *breakdown* simply means that normal diode blockage of reverse current terminates. As long as the current does not exceed limits causing overheating of the diode, which could also occur in the forward direction, the diode is not damaged.

Reverse Resistance (r_R)   For the ideal diode there is no current when reverse biased. A real diode has some small reverse current prior to reverse breakdown, as defined earlier. This means that the diode has some effective finite reverse resistance. The diode specifications will often give the value of this resistance. The specifications give the reverse current, I_R, at some specified reverse voltage, V_R. Then the reverse resistance is

$$r_R = V_R/I_R \tag{9.2}$$

As an example for the 1N916 signal diode we have $V_R = 75$ V and $I_R = 5$ μA, so

$$r_R = 75 \text{ V}/5 \text{ }\mu\text{A} = 15 \text{ M}\Omega$$

The 1N4001 rectifier has $V_R = 50$ V and $I_R = 0.03$ mA, so

$$r_R = 50 \text{ V}/0.03 \text{ mA} \approx 1.7 \text{ M}\Omega$$

Average Forward Current (I_0)   This specification is mostly appropriate for diodes used as rectifiers but is a general consideration of all diodes. A forward-conducting diode will have some net forward voltage drop. Therefore, power will be dissipated by the device. This power dissipation is given by

$$P_D = I_0 V_F \tag{9.3}$$

where $P_D =$ the power dissipated (W)
$I_0 =$ the average forward current
$V_F =$ the net forward voltage drop.

The diode is designed to dissipate a certain maximum amount of heat. If this is exceeded, the diode will overheat and the device will fail. This is usually specified by the maximum average forward current that the diode can carry.

Average current is used because the most common condition under which this specification is important is in rectification of ac voltage. The value of I_0 varies from a few milliamperes for some types of rectifiers to hundreds of amperes for large, high-power devices.

Peak Surge Current (I_S)   Most rectifiers have a specification for the peak, one-time current surge that can be conducted by the diode without failure. This is typically 10 to 30 times the average forward current. This is most appropriate for diodes in rectifiers also and is usually the maximum single cycle current that can occur without damage.

Reverse Recovery Time (t_r)   If a diode is forward biased and conducting, time is needed for it to convert to nonconduction if it is suddenly reverse biased. The reverse recovery time indicates how long it will take the diode to respond to this reversal. This is important in switching applications. The reverse recovery time is usually specified for signal diodes and varies from hundreds of microseconds to a few nanoseconds over the many diode types available.

9.2   DIODE CLASSIFICATION

Even though virtually all diodes are simply *pn*-junction devices, considerable differences are seen in performance specifications tailored to particular applications. We summarize here the most common classifications and how they are applied.

9.2.1   Rectifier Diodes

One class of diode is intended primarily for the rectification of ac voltage to dc voltage. These diodes are used in power supplies. Figures 9.7(a) and (b) show a simple half-wave diode rectifier and a diode bridge, full-wave rectifier, respectively. Of course, a large capacitor or other smoothing filter is usually used on the output, as shown, to smooth the dc voltage across the load. The important criteria for selection and application of these diodes follow:

1. *Reverse voltage:* In the simple half-wave application of Figure 9.7(a), the rectifier diode is exposed to a reverse voltage of twice the peak of the ac source every cycle. The diode selected must have a V_R equal to or greater than this peak voltage. Note that in the bridge each diode sees a reverse voltage of just the peak of the source.
2. *Average forward current:* The selected diode must have an average forward current rating equal to or greater than the maximum ever expected to be drawn by the load.
3. *Surge current:* The diode will experience surge currents, particularly when turned on. The selected diode must have a surge current rating equal to or greater than the maximum single-cycle current to be drawn.

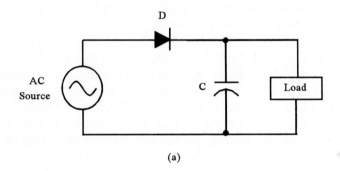

(a)

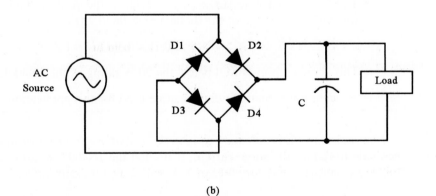

(b)

FIGURE 9.7

Application of diodes in rectifier circuits: (a) Half-wave diode rectifier and (b) diode bridge, full-wave rectifier.

EXAMPLE 9.2

See whether a 1N4001 diode could be used in a half-wave rectifier of a 6.3-Vac (rms) source and a 15-Ω load with a 100-μF filter capacitor.

Solution

From specification sheets we find for the 1N4001 the following specifications: $V_R = 50$ V, $V_F = 1.6$ V, $I_0 = 1$ A, $I_{surge} = 30$ A. Let's just check all the criteria:

1. A 6.3-Vac transformer has a peak voltage of

$$V_P = \sqrt{2}\,V_{rms} = (1.414)(6.3 \text{ V}) = 8.9 \text{ V}_{peak}$$

When the input is -8.9 and the capacitor is charged (at a maximum) to $+8.9$ V, then the reverse voltage across the diode will be

$$V_R = 2V_P = 2(8.9) = 17.8 \text{ V}$$

Since the 1N4001 has a peak reverse voltage rating of 50 V there is no problem here.

2. If we assume for worst case that the full 8.9 V appears across the load, the current to the load would be

$$I_0 = 8.9 \text{ V}/15 \ \Omega = 0.59 \text{ A}$$

The voltage across the load will certainly be somewhat smaller than the peak voltage. Thus, the average forward current will be 0.59 A or less, which is well below the rating of 1 A for the 1N4001.

3. The surge current will occur upon turn-on when the 100-μF capacitor is uncharged. During the first cycle the effective load of the diode will be 0 Ω! The current surge will be limited by the forward resistance of the diode. This can be estimated by the 1.6-V forward voltage, using Eq. (9.1):

$$r_F \approx (1.6 - 0.7) \text{ V}/1 \text{ A} = 0.9 \ \Omega$$

Thus, the worst case surge current will be the peak voltage minus the forward bias voltage of the diode across the 0.9-Ω diode resistance,

$$I_{surge} \approx (8.9 - 0.7) \text{ V}/0.9 \ \Omega = 9.1 \text{ A}$$

This surge current is well below the 30-A rating of the 1N4001.

The conclusion is that this diode can be used for this application.

In some cases it is necessary to insert a *current-limiting resistor* in series with the diode to reduce the surge current. Note that the 1N400X series of diodes are standard circuit rectifiers for cases of 1 A or less average current.

EXAMPLE 9.3

A spot welder will be constructed as shown in Figure 9.8. What diode specifications are appropriate? What diode could be used?

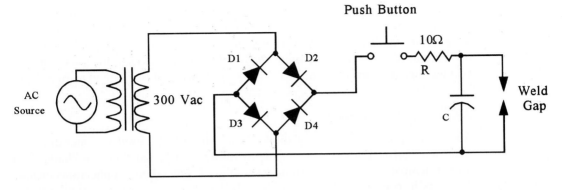

FIGURE 9.8
Spot welder circuit for Example 9.3.

Solution
By means of the criteria given earlier, we find the following:

1. The 300-Vac rms source means that the peak voltage will be

$$V_{\text{peak}} = \sqrt{2}(300) = (1.414)(300) = 424.2 \text{ V}$$

Thus, each diode has this value across its terminals so the required reverse voltage is $V_R > 424.2$ V.

2. The average forward current is not as important as the surge current in this application since the charging action will occur with significant delays, compared to the 60-Hz period, between each welding cycle.

3. The surge current will be determined by the initial placement of voltage across the uncharged capacitor. Ignoring the couple of volts dropped across the two diodes in the forward-biased state, the surge current will be

$$I_{\text{surge}} \approx 424.2 \text{ V}/10 \text{ }\Omega = 42.4 \text{ A}$$

The 1N4005 has a V_R of 600 V but cannot be used since its surge current is only 30 A. A 1N4438 has $V_R = 600$ V and $I_{\text{surge}} = 100$ A, so it could be used. There are many, many others and price or availability would probably force the decision as to which to choose.

Heat Sinks When rectifier diodes are used in high-current applications, a significant amount of heat can be dissipated by the diode itself. For example, a 1N3170 can handle a 240-A average forward current and has an average forward voltage drop of 1.3 V. This means that the diode is dissipating a huge amount of power:

$$P = \text{IV} = (240 \text{ A})(1.3 \text{ V}) = 312 \text{ W } !!!$$

Such a diode *must* be bolted to a good heat sink to carry away this heat.

Power Derating   The rated maximum current or power dissipation of a diode is specified up to a certain temperature. If the diode is operated at a higher temperature, the maximum power must be *derated* by a factor for each additional rise in temperature. For example, the 1N400X series diodes are good for 1-A average forward current up to 75°C but beyond this must be derated at 10 mA/°C.

9.2.2   Signal Diodes

Another major classification of diodes is for those to be used for operations on electronic signals. For these diodes the forward current is small (just milliamperes) so that dissipation is not important. Examples of signal diode applications include detectors in communication circuits, peak detectors in data acquisition systems, signal clamps and clippers, and a host of other communications and signal conditioning requirements.

The following parameters are most important in the selection and application of diodes in this classification.

1. *Reverse voltage or peak reverse voltage:* Of course, the maximum reverse voltage will be important because diode action is terminated and reverse current flows if the peak reverse voltage rating of the diode is exceeded. Many signal circuits involve inductive elements, which can create very high reverse voltages across diodes.
2. *Forward voltage drop at forward current:* In signal applications the forward voltage drop and forward resistance of the diode are important. The specifications give the forward voltage drop at a specific forward current so that the forward resistance can be determined.
3. *Reverse current:* In many applications of signal diodes, it is essential to keep reverse current at an absolute minimum. Therefore, the reverse or *leakage* current is of importance.
4. *Reverse recovery time:* Signal diodes are often used in applications where high frequencies are involved, as opposed to the relatively low frequency of rectifiers ($f \approx 10$ to 400 Hz). Signal diodes are used from audio frequencies ($f \approx$ kHz) through the higher microwave frequencies ($f \approx$ GHz). This means that the diodes must often have a very fast reverse recovery.

EXAMPLE 9.4   Figure 9.9 shows a signal diode that will be used in a positive peak detector circuit. At the end of a sample period of 10 ms (determined by other circuitry) the capacitor will be charged to the peak value of the input. The input range is given as ±10 V at 100-kHz maximum. What are the required characteristics of the diode? What diode could be used?

Solution

Let's look at each of the criteria given earlier. First the peak reverse voltage is 20 V since the capacitor could be charged to +10 V and the signal setting at −10 V.

FIGURE 9.9
Signal diode used in the peak detector circuit of Example 9.4.

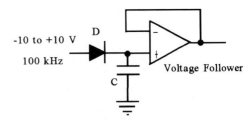

The forward resistance will be important since it will determine the maximum sampling rate. The charging time constant will be,

$$\tau_c = r_F C$$

i.e., the RC of the forward conducting diode and the capacitor. To "track" a signal at a 100-kHz or 10-μ_s period, the value of τ_c must be ten times faster or 1 μs. Thus,

$$r_F = \tau_c/C = 1 \ \mu s/0.01 \ \mu F = 100 \ \Omega \ \text{(or less)}$$

Most diode specifications give the forward current at a net forward voltage of 1 V. The forward current will be (assuming silicon)

$$I_F = (1 - 0.7) \ \text{V}/100 \ \Omega = 3 \ \text{mA (or greater)}$$

The reverse resistance must be sufficiently large so that the capacitor does not discharge significantly within one sample period, 10 ms. The discharge time constant is given by $\tau_d = r_R C$. If we make τ_d ten times the sampling period, then $\tau_d = 100$ ms. Thus, the reverse resistance must be at least

$$r_R = \tau_d/C = 100 \ \text{ms}/0.01 \ \mu F = 10 \ \text{M}\Omega$$

At the maximum voltage this corresponds to a maximum leakage current of

$$I_R = 10 \ \text{V}/10 \ \text{M}\Omega = 1 \ \mu A \ \text{(maximum)}$$

The reverse recovery time should be about ten times faster than the maximum signal period, 10 μs, or about 1 μs.

So, we are looking for a diode with $V_R > 20$ V, $I_F > 3$ mA at $V_F = 1$ V, $I_R < 1 \ \mu$A, and $t_r < 1 \ \mu$s. Many diodes could be used. For example, the 1N920 has $V_R = 36$ V, $I_F = 500$ mA at $V_F = 1$ V, $I_R = 0.25 \ \mu$A and $t_r = 0.3 \ \mu$s.

9.2.3 Zener and Reference Diodes

A third very important application of diodes is in the regulation of voltage sources. The specially fabricated diodes that perform this function are called *Zener diodes*. A similar function is provided by a *reference diode,* which is also discussed in this section.

Zener Diodes Zener diodes are used in a reverse-bias condition to provide some regulation of sources of voltage. By specialized doping techniques it is possible to fabricate diodes for which the reverse-bias breakdown has two important characteristics:

1. The voltage at which reverse breakdown occurs can be selected over a large range from a few volts to hundreds of volts.
2. The slope of reverse current versus voltage, after breakdown, can be made very large. This is the same as a very small reverse resistance after breakdown.

Figure 9.10 shows a typical Zener diode curve of current versus voltage. In the forward-bias condition it acts more or less like a standard diode, although it is not intended to be used as such. In the reverse-bias condition note the very sharp knee at the *Zener voltage, V_{Z0}*. Apart from the very small effect of finite reverse resistance, the voltage across the Zener will stay at V_Z even though the current changes.

Figure 9.11 shows a Zener connected to a source voltage, V_s, which is larger than V_{Z0}, through a resistor, R. A load is connected across the Zener. Ideally the voltage across the Zener will remain constant at V_{Z0}. Thus, the current through R is

$$I_R = (V_s - V_{Z0})/R \tag{9.4}$$

Also this current divides between the load and the Zener

$$I_R = I_Z + I_L \tag{9.5}$$

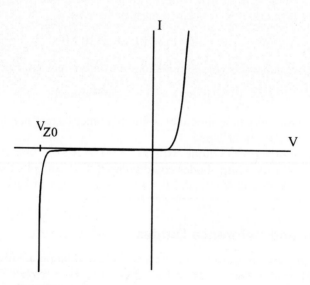

FIGURE 9.10
Typical I-V characteristic of a Zener diode.

FIGURE 9.11
Circuit showing the biasing of a Zener diode.

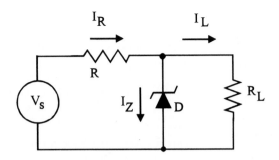

Variations in supply voltage and variations in load will not change the voltage across the load, i.e., the load voltage is regulated. If V_s changes then I_R will vary but, as Figure 9.10 shows, V_{Z0} and I_L will remain constant, so that I_Z changes to compensate. Alternately, if the load changes so that I_L changes, again I_Z will change but V_{Z0} will remain constant.

Practical Considerations   Use of the Zener diode as a regulator requires consideration of the following practical applications notes and selection criteria.

Tolerance.   It is very important to realize that Zener diodes, like resistors, have a tolerance on the indicated Zener voltage. Typical tolerances are ± 20, ± 10, and $\pm 5\%$. Some units are also available at $\pm 2\%$ and even 1%. Often a suffix (the letters A or B, for example) is added to the 1N number to indicate the tolerance. Unfortunately, there are no standard meanings for these suffixes yet.

Consider, for example, that if you buy a 1N3678 you expect a 9.1-V Zener. If this is a $\pm 20\%$ tolerance unit, the actual Zener voltage could be anywhere in the range from 7.28 to 10.92 V!

Temperature Coefficient.   Zener voltage will also vary with temperature. The change is indicated by the percent per degree Celsius of temperature. For standard Zeners this will typically range from as small as 0.04%/°C to as large as 0.1%/°C.

Power Dissipation.   There must be some current passing through the Zener for regulation to occur. This means that the Zener is dissipating power, given by

$$P_D = I_Z V_{Z0} \tag{9.6}$$

where I_Z is the Zener current. Zener diode specifications give a maximum power rating so that the maximum operating current can be determined.

EXAMPLE 9.5   A 1N1770 Zener diode has a Zener voltage of 9.1 V and a 1-W power rating. It is used as in Figure 9.11 with $V_s = 15$ V. What is the maximum Zener current? What resistance and power rating should R have?

Solution

The specifications indicate that the 1N1770 can dissipate 1 W at 9.1 V. Therefore, the maximum current is found from Eq. (9.6) to be

$$I_{Zmax} = 1 \text{ W}/9.1 \text{ V} \approx 110 \text{ mA}$$

At this current the value of R will be found from Eq. (9.4):

$$R = (15 - 9.1) \text{ V}/110 \text{ mA} = 53.6 \text{ } \Omega$$

This resistor must have a power rating of

$$P = I^2 R = (0.11 \text{ A})^2 (53.6 \text{ } \Omega) = 0.65 \text{ W}$$

so that a 1-W resistor would be used. Note that any load connected across the Zener will draw some current away from the Zener as shown by Eq. (9.5), so this is a worst case (no-load) design.

Typically, a Zener design uses the worst case (no-load) condition to determine R so that if something happens to disconnect the load the Zener will not fail from overheating.

Power Derating. High-power Zeners typically operate at an elevated temperature, in which case the power handling capability of the Zener is reduced. Power derating is expressed as a reduction in the power rating per degree Celsius temperature. For example, a 1N3020 (a 1-W Zener) is specified to be derated at 6.67 mW/°C above 25°C, whereas a 1N2804 (a 50-W Zener) is derated at 500 mW/°C above 75°C.

Finite Reverse Resistance. The fact that the Zener has a finite resistance will introduce some variation in voltage across the Zener, i.e., reduce the quality of regulation. In fact, the voltage across the Zener will be the Zener voltage plus the voltage dropped across its reverse resistance:

$$V_Z = V_{Z0} + I_Z r_Z \qquad (9.7)$$

where V_Z = actual voltage across the Zener
V_{Z0} = Zener voltage at the knee
I_Z = Zener current
r_Z = Zener reverse conduction resistance.

Specifications will give either the value of r_Z or the value of V_{Z0} (the minimum) and the value of V_Z at maximum Zener current (the maximum). Maximum Zener current is given or found from the power dissipation.

EXAMPLE 9.6

How can a Zener be used to provide a regulated 10 V from a 15-V unregulated source? The source is specified at 15 ± 1 V ($\approx$ ±7%). The load will nominally draw 20 mA. What is the regulation in percent?

Solution

If the load is 20 mA ($R_L = 500\ \Omega$), then we let the Zener draw the same under nominal conditions, which means that the Zener should be able to handle at least 40 mA (in case the load gets disconnected). This is a power of 0.4 W. So we should use a 0.5- to 1-W Zener. A 1N3020 is a 10-V, 1-W Zener. Specification sheets show that $r_Z = 7\ \Omega$. To find the value of R (see Figure 9.11), we use nominal conditions:

$$R = (15 - 10)\ \text{V}/0.04\ \text{A} = 125\ \Omega$$

The power rating should be $P = (0.04)^2(125) = 0.2$ W so we can use a 0.25-W or 0.5-W resistor. Now the maximum, nominal, and minimum Zener voltages can be determined from Eq. (9.7).

For $V_s = 16$ V, $I_R = (16 - 10)$ V/125 Ω = 48 mA so the Zener current will rise to 28 mA. Then the Zener voltage will be

$$V_{Z\text{max}} = 10 + (0.028)(7\ \Omega) = 10.196\ \text{V}$$

For $V_s = 15$ V, $I_R = (15 - 10)/125 = 40$ mA and $I_Z = 20$ mA so that the voltage is

$$V_Z = 10 + (0.02)(7) = 10.14\ \text{V}$$

Finally, for $V_s = 14$ V, $I_R = (14 - 10)/125 = 32$ mA so that $I_Z = 12$ mA. This gives

$$V_{Z\text{min}} = 10 + (0.012)(7) = 10.084\ \text{V}$$

Thus, we see that the nominal Zener voltage is 10.14 and the regulation is

$$10.14 \pm 0.056\ \text{V or} \pm 0.6\%$$

Note, however, that the 1N3020 is a $\pm20\%$ tolerance unit so that any selected sample may have a V_{Z0} of 8 to 12 V! A 1N3020A is $\pm10\%$ and a 1N3020B is $\pm5\%$.

Reference Diode   This diode, sometimes called a *temperature-compensated diode*, is a special Zener with very little variation in Zener voltage with temperature over some range of temperature. Whereas a standard Zener may have a temperature coefficient of 0.04 to 0.1%/°C, the reference diodes have coefficients that vary from a maximum of around 0.01%/°C to as small as 0.0005%/°C. Circuit design using reference diodes is the same as regular Zeners although operating currents are generally less than 10 mA.

9.2.4   Testing Diodes

The testing of diodes is complicated by the fact that they have such differing purposes and characteristics. Nevertheless, certain common testing approaches can be used to determine whether the device is a diode and its basic characteristics.

Basic Resistance Test   If you pick up a small cylindrical object with colored bands, it may be a resistor or diode. Also if you have a small black cylindrical object with a white band near one end, it is most likely a diode, but is it any good? A simple resistance measurement can often provide basic information.

First you must be sure the ohmmeter provides a sufficient voltage to overcome the forward breakdown voltage of the diode. This value must be at least 0.7 V to cover both germanium ($\approx$0.3 V) and silicon. Most modern digital voltmeters have one or more resistance measuring scales that are intended for diode tests by providing sufficient voltage. They have a small diode symbol by their scale settings.

Resistance is now measured in both directions through the device. If it is a resistor, it will show the same resistance both ways. If it is a diode, it will show a low resistance in one orientation and a very high resistance connected the other way. Of course, in the low-resistance position the diode lead connected to the positive source of the ohmmeter is the anode of the device.

Reverse Breakdown Voltage   The reverse breakdown voltage can be determined by connecting a variable voltage source, with a current-limiting resistor, to the diode. As the reverse-bias voltage is increased, a sudden increase in reverse current will be noted at the reverse breakdown voltage. Note that this could be several hundred volts. *Reverse leakage current* can be measured by this same test by measuring the current flow before the reverse breakdown voltage is reached.

9.3   SPECIAL-PURPOSE DIODES

This section presents brief discussions of a number of special-purpose diodes used in the electrical and electronics industry. In each case the basic characteristics are given followed by application notes and selection criteria.

9.3.1   Light-Emitting Diodes

A diode that is forward biased dissipates energy given by the product of the forward current and forward voltage drop. In the normal diode types, this power dissipation appears as a thermal effect, i.e., the diode heats up. A light-emitting diode (LED) is a special type of diode, constructed of special semiconductor materials, for which some of this power is dissipated in the form of electromagnetic radiation, in particular, visible and infrared light.

FIGURE 9.12
The LED is used in the forward-conducting mode.

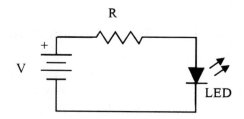

TABLE 9.3

Light sources

Source*	Intensity (cd)
Full sun	10,000
Carbon arc	4,500
150-W flood	7.4
60-W incandescent	0.55
20-W Fluorescent	0.041
Full moon	0.033
LED range	0.001–0.07

* 0.1-in.-diameter section.

Figure 9.12 shows that the LED is connected to a voltage source with a current-limiting resistor. The diode is forward biased. Characteristics of the LED include the following:

1. *Maximum forward current:* This determines the maximum light intensity the LED can provide. Current-limiting resistors are used to control the maximum forward current.
2. *Forward voltage drop:* Just as with any diode, conduction does not begin until a certain forward-bias voltage is applied. This voltage is typically from one to several volts and therefore a little larger than for silicon diodes.
3. *Light wavelength (color):* LEDs are available in many different colors, defined by the wavelength of the peak emission of light.

Infrared (invisible) >800 nm (nanometers)

Red ≈650 nm

Orange ≈630 nm

Green ≈560 nm

Yellow ≈590 nm

4. *Light intensity:* The light intensity of LED output at some current is specified by the number of candelas (cd), or more typically millicandelas (mcd) emitted. The candela is a measure of luminous intensity. It compares the emission of some source to that of a standard source, including both spectrum and intensity. For reference purposes, Table 9.3 compares the intensity of a 0.1-in.- (≈2.5-mm)-diameter section of various sources and typical LEDs.

EXAMPLE 9.7

An indicator light is needed to indicate when a 10-V power supply has been turned on. How can an LED be used for this light?

Solution

A parts catalog shows a red, panel-mount LED has specifications of 1.7 V_F at 20 mA, with 10 mcd. To operate at 20 mA means that we need a series limiting resistor of

$$R = (10 - 1.7) \text{ V}/20 \text{ mA} = 415 \ \Omega$$

9.3.2 Photodiodes

When a diode is reverse biased a very small amount of reverse current flows. This current is due to minority carriers, i.e., holes in the n-doped material and electrons in the p-doped material. Electromagnetic radiation of the correct wavelength, if allowed to illuminate the pn junction, can increase the number of minority carriers. This means that the reverse current will increase also. Thus, a photodiode is one that has been fabricated of special materials and with an illuminating window to the pn junction so that electromagnetic radiation in the infrared, visible, or ultraviolet bands will increase the reverse current.

A photodiode is connected to a source reverse biased, for example, as shown in Figure 9.13. The value of R will be very large, perhaps equal to the dark (no illumination) reverse resistance of the diode. In that case, the voltage across the diode will be half that of the source. If a pulse of light falls on the diode, the voltage across the diode will decrease due to the increased reverse current. Since the reverse resistance may be very large (approximately mega-ohms) a very high impedance measurement system must be used.

9.3.3 Varactors

A varactor, a diode-based device is a variable capacitor, with the capacity determined by the value of an applied voltage. The varactor is also called a voltage variable capacitor. A reverse-biased diode has a very small reverse current that is equivalent to a very large reverse resistance. Such a diode is in many respects like a capacitor. The high-resistance, reverse-biased pn junction acts like the insulator between the "plates" of p-doped and n-doped semiconductor material.

When the reverse-bias voltage across the diode is changed in value the effective width of the pn junction changes, i.e., the depletion region becomes smaller or larger. Since capacity is inversely proportional to the gap between plates, such a diode is effectively a variable capacitor, with voltage controlling the capacity value. This is the principle behind the varactor.

Varactor capacity is determined by the reverse-bias voltage from 0 V to the reverse breakdown voltage, V_R. Therefore, this voltage determines the maximum range of capacity variation.

FIGURE 9.13

A photodiode is used in the reverse-biased mode.

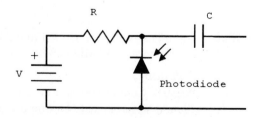

Specifications The varactor specifications provide information about the range of capacity and the reverse voltages that produce this range. The varactor is identified by a nominal value of capacity determined at some value of reverse voltage (often the minimum voltage). It is important to realize that there is a tolerance to this value determined by a suffix added to the EIA number. Typical values of tolerance are $\pm 1, 2, 5, 10$, and 20% of the nominal capacity.

Variation of capacity with reverse voltage is nonlinear. Specifications often give the capacity at a specified voltage, the ratio of maximum to minimum capacity, and the voltage range which produces that variation.

The 1N4787 has a $C_J = 8.2$ pF ($\pm 20\%$), and $C_{max}/C_{min} = 2.56$ over a reverse-bias voltage range of 0 to 4 V. This means that the capacity varies from 8.2 to 21 pF as the voltage is changed from 0 to 4 V. Of course, there is a $\pm 20\%$ tolerance for any given unit. The 1N4787C has a $\pm 2\%$ tolerance.

EXAMPLE 9.8 A 1N4787 varactor diode is used to vary the resonant frequency of a tank circuit as shown in Figure 9.14. What is the range of resonant frequency?

Solution
The resonant frequency of a parallel LC tank circuit is given by

$$f = 1/(2\pi\sqrt{LC})$$

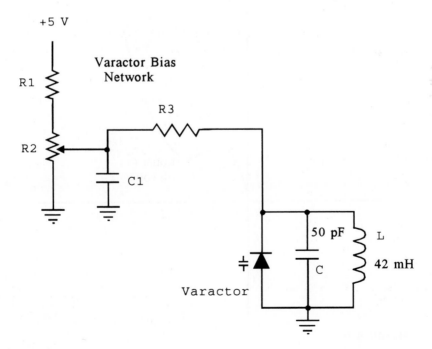

FIGURE 9.14
A varactor diode changes in capacity with reverse bias.

In this case $L = 42$ mH and the value of the parallel of C and the varactor is given by

$$C_p = C + C_v$$

The minimum of the varactor is 8.2 pF and the maximum is 21 pF, so when added to the parallel 50 pF we get a net capacity of

$$C_{min} = 8.2 \text{ pF} + 50 \text{ pF} = 58.2 \text{ pF}$$

and

$$C_{max} = 21 \text{ pF} + 50 \text{ pF} = 71 \text{ pF}$$

Thus, the range of resonant frequency is

$$f_{max} \approx 101.8 \text{ kHz} \quad \text{and} \quad f_{min} \approx 92.2 \text{ kHz}$$

9.3.4 Tunnel Diodes

A tunnel diode, also called an Esaki diode, is fabricated such that the depletion region is very small. This gives the device a very peculiar I-V characteristic, as shown in Figure 9.15, with a region of negative resistance (slope). These diodes are principally used in high-frequency communication circuits and high-speed computer

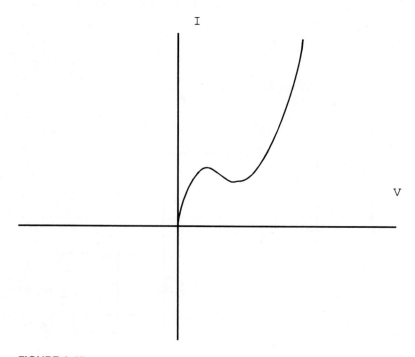

FIGURE 9.15
The tunnel diode has a region of negative resistance.

circuits. The region of negative resistance allows oscillator circuits to be implemented using the tunnel diode.

9.3.5 Schottky Barrier Diodes

This device, also called a hot-carrier diode, uses a junction formed between an *n*-doped semiconductor and a metal. A depletion region forms at the junction of the metal and semiconductor. The metal is the anode and the semiconductor the cathode. This diode has a much smaller forward-bias voltage than the germanium or silicon *pn* diode. Unfortunately, it also has a larger reverse current and a smaller reverse breakdown voltage.

The Schottky barrier diodes are used in high-speed communication and data processing equipment. Recent advances in the technology have allowed their application in power supplies and ac-to-dc converters.

10

TRANSISTORS AND OTHER SEMICONDUCTOR DEVICES*

This chapter summarizes bipolar and field-effect transistors (FETs) and other discrete, multijunction semiconductor devices in common industrial use.

10.1 COMMON CHARACTERISTICS

A consistent basis for description and specification of semiconductor devices is provided by the following organizations: Electronics Industry Association (EIA), Joint Electron Devices Engineering Council (JEDEC), and the International Electrotechnical Commission (IEC) through the American National Standards Institute (ANSI). More information on semiconductor standards may be obtained from:

EIA/JEDEC standards:	Electronic Industries Association 2001 Eye Street NW Washington, DC 20006 (202)659-2200
IEC standards:	American National Standards Institute, Inc. 1430 Broadway New York, NY 10018 (212)868-1220
Military standards:	Commanding Officer U.S. Naval Publications and Forms Center 5801 Tabor Avenue Philadelphia, PA 19120

* This chapter was written by Prof. Hal Broberg, Purdue University.

10.1.1 Major Manufacturers

Many producers of discrete devices also manufacture a wide variety of integrated circuits, both analog and digital. Some of the largest producers of discrete devices are given in the following list. Specifications from many of these corporations will be used throughout this chapter:

- Motorola
- Texas Instruments
- General Electric/RCA/Intersil
- Fairchild
- International Rectifier
- TECCOR
- Westinghouse
- ECG
- Siliconix
- Tandy Corporation
- International Devices, Inc.
- Clairex
- General Instrument
- TRW
- Lansdale
- National Electronics
- Sprague

10.1.2 Packaging

Many common device cases have been standardized through JEDEC. These cases have been assigned a number of the form TO-3, TO-92, or TO-220AB. Some of the most common cases used in discrete semiconductor devices are shown in Figure 10.1. Manufacturers also produce customized packages for use by major customers. There are more than 400 transistor outlines [1] and many more case types are used for thyristors, diodes, and other semiconductor devices. Pinouts are not standardized and data sheets must be consulted for each device type.

10.1.3 Temperature Effects

Use of Thermal Specifications One of the most important activities of semiconductor design and practice is ensuring that use of a device does not exceed its rated thermal capabilities/power dissipation. This section defines and discusses the thermal specifications usually provided on manufacturers' data sheets and provides examples of how to use them. Some of the most common thermal specifications are shown in Table 10.1.

The average power dissipated (P_D) will generate heat at the applicable junction, which will be removed at a rate that is directly proportional (with constant of proportionality, $R_{\theta JA}$) to the difference between the junction temperature, T_J,

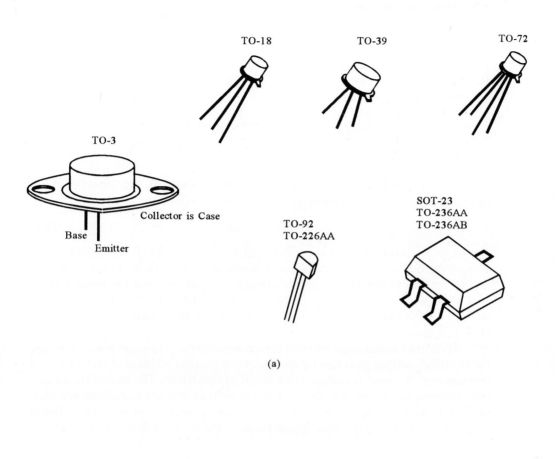

(a)

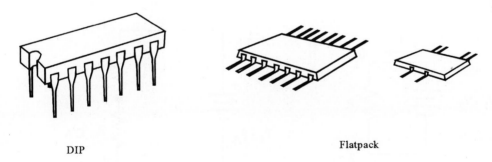

DIP

Flatpack

(b)

FIGURE 10.1

Some common semiconductor packages: (a) Single device packages and (b) multiple device packages.

TABLE 10.1

Thermal resistance, junction-to-case $(R_{\theta JC})$ in °C/W

Thermal resistance, junction-to-ambient $(R_{\theta JA})$ or junction-to-free-air in °C/W, or

  A linear derating factor, or

  A device dissipation derating curve

A maximum device dissipation at a specified ambient temperature, usually 25°C

A maximum device dissipation at a specified case temperature, usually 25°C

---

and the ambient temperature, T_A. Equation (10.1) follows directly from this statement and is applicable to almost all basic thermal calculations:

$$T_J - T_A = P_D R_{\theta JA} \tag{10.1}$$

Average power dissipated is calculated using $P = VI$ and is dependent on the area of the device that must dissipate the majority of the heat. For a transistor, this would be the power dissipated between the collector and the emitter $(P = V_{ce}I_c)$; for an FET, the power dissipated in the channel $(P = V_{ds}I_d)$; and for an SCR or TRIAC, the power dissipated between the anode and the cathode $(P = V_{ac}I_{on})$.

To further understand thermal conduction characteristics, it is useful to use the electrical analogy shown in Figure 10.2. Note that the addition of series thermal resistances is the same as adding series electrical resistances. The thermal resistance from junction to case and the device dissipation from junction to case are not used in the following example. These will be used in the section on heat sinks. Based on this brief review of thermal specifications, consider the following example.

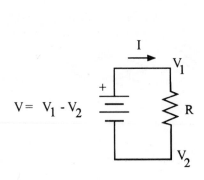

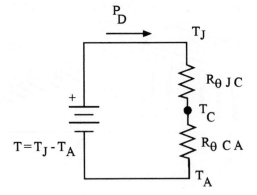

Electrical Term

V = voltage difference
I = current
R = resistance

Thermal Term

T = temperature difference
P = power
R_θ = thermal resistance

FIGURE 10.2
Equivalence of electrical and thermal terminology.

EXAMPLE 10.1

You have a design for a class A transistor amplifier to be used in equipment that has specifications requiring it to operate at up to 140°F. The transistor in the design is biased at I_c = 20 mA and V_{ce} = 6 V. A 2N3904 transistor is used in the design. Will this general-purpose transistor function under the specified operating conditions with no heat sink? What is the maximum ambient temperature at which the amplifier can operate with no heat sink?

Solution

1. You look up the specifications for the 2N3904 [2] and find that the maximum dissipation at an ambient temperature of 25°C is 625 mW and the device must be derated above 25°C at 5 mW/°C. The value of $R_{\theta JA}$ is given as 200°C/W. (Note that the value given for $R_{\theta JA}$ is the reciprocal of the derating factor, which represents the slope of the equivalent derating curve.) As shown in Table 10.1, this curve may be given instead of a linear derating factor or the thermal resistance. The key here is that the three factors— linear derating factor, thermal resistance, and device dissipation operating curve—are all expressions of the same effect and only one of the three is needed.
2. You calculate that 140°F is 60°C.
3. You derate the transistor by $(60 - 25)5 = 175$ mW so the transistor can dissipate only $625 - 175 = 450$ mW at 140°F (ambient).
4. For a class A amplifier, maximum power is dissipated with no signal, so

$$P = V_{ce}I_c = (6)(20) = 120 \text{ mW} \qquad (10.2)$$

Thus, the amplifier will operate with no problem at 140°F.
5. To find the maximum ambient temperature, we substitute directly from the specifications into Eq. (10.1) or use the thermal circuit diagram as shown in Figure 10.3.

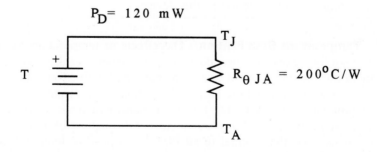

$$T_{J(max)} - T_{A(max)} = (120 \text{ mW})(0.2°C/mW) = 24°C$$

FIGURE 10.3
Equivalent thermal circuit.

Since we are looking for the maximum ambient temperature at which the device will dissipate 120 mW, we must know the maximum junction temperature. We can find this by using the derating specification since at the maximum junction temperature no power can be dissipated:

$$625 - (T_{J(max)} - 25)5 = 0 \qquad \textbf{(10.3)}$$

From this we find that

$$T_{J(max)} = 150°C \qquad \textbf{(10.4)}$$

Now substituting this and the 120-mW dissipation into Eq. (10.1), we find

$$120 - (150 - T_{A(max)})5 = 0 \quad \text{and} \qquad \textbf{(10.5)}$$

$$T_{A(max)} = 126°C \,(258.8°F) \qquad \textbf{(10.6)}$$

which is quite warm.

EXAMPLE 10.2

From Example 10.1, the final stage of the amplifier is a push–pull or complementary symmetry power amplifier that must deliver 2.25 W to an 8-Ω load. The design uses 2N4854 complementary dual amplifier transistors (note that this is two matched NPN/PNP transistors in a single package and not two separate transistors). Will these transistors support the requirement without a heat sink?

Solution

1. The specs [2] say that, at $T_A = 25°C$, the total device dissipation using both die (both transistors) can be 600 mW with a derating factor of 4 mW/°C above 25°C.
2. Derating the device we find that it can handle $600 - (60 - 25)4 = 460$ mW at 140°F. We know that this type of amplifier has a maximum efficiency of 78.5% so 2.25 W to the load will be 78.5% of the total power dissipated (the remainder is dissipated in the transistors). So $P_T = 2.866$ W and the two transistors dissipate $2.866 - 2.25 = 0.616$ W. Thus, a heat sink is required if we are to use this device.

Temperature Stabilization The effects of temperature are different for FETs than for bipolar junction devices. A *pn* junction's resistance decreases as its temperature increases and it is said to have a negative temperature coefficient. Thus, the current through the base-emitter junction of a transistor increases as the temperature increases and since this current controls a much larger current, thermal runaway is possible. The drain-source current through an FET is controlled by the gate voltage, and the channel in an FET has a positive temperature coefficient. This means that as the drain-source current increases, the channel resistance increases and no thermal runaway occurs.

Temperature stabilization is used to ensure that a transistor remains very close to the same operating point at different temperatures. This can be accomplished effectively by using negative feedback in the network (such as an emitter resistor)

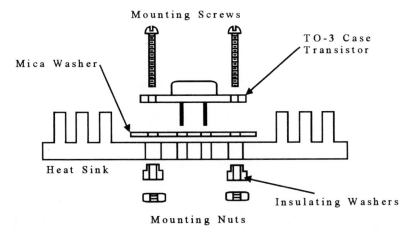

FIGURE 10.4
Mounting a heat sink.

since this will stabilize all of the circuit characteristics. Additionally, positive and negative temperature coefficient thermistors are often used to compensate for temperature effects in bipolar and field-effect transistors.

Use of Heat Sinks   The use of a heat sink is necessary to take full advantage of the power handling capability of any semiconductor device. Most devices classified for high-power use (e.g., power MOSFETs, SCRs, power transistors) give the thermal resistance from junction to case $R_{\theta JC}$, the maximum power dissipation at a case temperature of 25°C, and a derating factor at a case temperature of 25°C. A heat sink is electrically insulated from the case of the device using silicon grease and special washers that provide thermal contact but no electrical conduction. Figure 10.4 shows an example of how a heat sink could be mounted.

When using a heat sink, the thermal resistance from the device case to the sink must be considered. Typical values of $R_{\theta CS}$ (which is determined by the washers and silicone grease used) are 0.02 to 5°C/W [3]. Also, the thermal resistance of the sink itself must be considered. Typical values of $R_{\theta SA}$ are 0.55 to 9.15°C/W [4, 5]. These data can be obtained from the heat sink and the silicon grease/washer specs.

EXAMPLE 10.3

While building a prototype of a control circuit you decide to use an *n*-channel enhancement-type power MOSFET to control a small motor. A 50-W power output is required. The designer of the circuit specified that the RFM3N45 power MOSFET should be used. You look up the specs for this device [6] and see that it has a 75-W power dissipation rating at a case temperature of 25°C, and $R_{\theta JC}$ of 1.67°C/W, and that it must be derated by 0.6 W/°C above 25°C.

Solution
First use Eq. (10.1) to find the allowable dissipation without a heat sink. The maximum junction temperature for all RCA power MOSFETs is 150°C [6]. We

FIGURE 10.5
Thermal diagram.

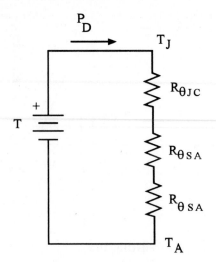

must also know the $R_{\theta CA}$ (thermal resistance from the case to air). Assuming a heat transfer of 125°C/W-in.^{2} from the case with an area of about 0.5 in.^{2}, $R_{\theta CA}$ is about 62.5°C/W. Now, using Eq. (10.1),

$$T_{J(\text{max})} - T_A = (P_D)(R_{\theta JA}) \tag{10.7}$$

or, substituting,

$$150 - 25 = (P_D)(62.5 + 1.67) \tag{10.8}$$

and $P_D = 1.95$ W. So the power MOSFET can dissipate only about 2.5 W without a sink!

Next you find that the heat sinks you have for this device have an $R_{\theta SA}$ (thermal resistance from sink to air) of 2°C/W and the silicone grease/washer has an $R_{\theta CS}$ of 0.08°C/W. Now, you can use Eq. (10.1) and the thermal diagram shown in Figure 10.5. Thus, $P_D = 33.3$ W and you must use a bigger/more efficient heat sink or a different power MOSFET. After consultation with the designer about the practicality of the recommendation you choose another device with a higher power rating.

The key point here is that use of the device at the rated power requires an ideal heat sink, which is not possible, so it is difficult to approach the maximum power ratings given in the manufacturer's specifications. Note that use of fans to circulate the air will also increase the cooling and allow the device power handling capability to approach the specifications.

10.2 BIPOLAR TRANSISTORS

10.2.1 Basic Concepts

The principal material used in semiconductors is silicon; however, germanium (lower *pn*-junction voltages of approximately 0.2 V versus about 0.6 V for silicon), gallium

FIGURE 10.6

The *npn* and *pnp* transistor symbols.

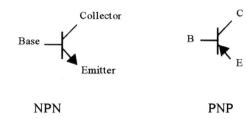

NPN   PNP

arsenide (used primarily at frequencies in the microwave region), and other materials are also used. Bipolar transistors have two *pn* junctions and come in two configurations as shown in Figure 10.6. Note that the arrow shows the direction of conventional current.

Manufacturers generally place transistors into several major groupings depending on whether they are designed for use as small-signal (relatively low-power) amplifiers, for high-power amplification, or for high-speed switching. Within these three groups there are a number of specific uses into which the transistors are further subdivided. Some of these uses follow:

1. *General purpose:* Usually the general-purpose transistors are low cost and usable for small-signal amplification or relatively low-speed switching.
2. *Frequency specific:* Frequency-specific uses include these ranges: rf, which usually indicates that the transistor will provide amplification throughout the normal AM and FM range (from 500 kHz to 110 MHz); HF (high frequency), which can mean above 100 MHz or the 3- to 30-MHz range of HF communications. VHF (very high frequency), 30 to 300 MHz; UHF (ultra high frequency), 300 MHz to 3 GHz; and microwave, which overlaps the UHF band of frequencies and extends into the tens of gigahertz.
3. *Switching speed:* For switching transistors, this can range from microseconds for high-power switching transistors to tens of nanoseconds for low-power (small-signal) switching devices.
4. *Analog versus digital switching:* Analog switches are generally designed to handle more power and higher voltages and have slower switching speeds.
5. *Low noise:* Low-noise transistors are designed to be used as the first stage (or a very early stage) in a communications receiver or other receiver that must amplify very low power input signals with small signal-to-noise ratios.
6. *High voltage:* High-voltage use means that a transistor has been internally designed to handle high-voltage signals without breakdown.
7. *Darlington:* Two transistors in a single package with the base of the second connected internally to the emitter of the first. The Darlington pair was one of the earliest integrated circuits and is discussed further in Example 10.6.
8. *Complementary pairs:* Again, we have two transistors in one package, one NPN and one PNP. The two have virtually the same characteristic curves because they are fabricated on the same chip. Their use is illustrated in Example 10.5.

9. *Phototransistors:* These types of transistors are activated by incoming light. This light can be in the visible, infrared, or other frequency range. These transistors can have two (no base lead) or three leads (see Example 10.7).
10. *Multiple transistor packages:* Dual and quad transistors are available in a single package. The principal advantage is size, and the main disadvantage is power handling ability.

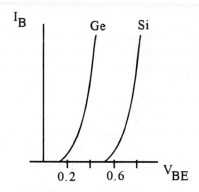

Typical Base Characteristics

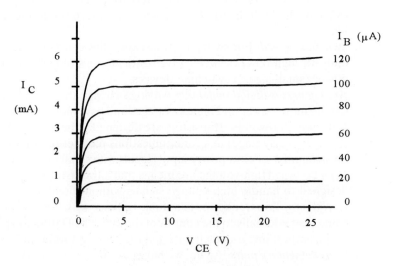

Typical Collector Characteristics

FIGURE 10.7
The *npn* I-V characteristic curves.

I-V Characteristics The collector current versus the collector-to-emitter voltage curves along with emitter current versus the base-to-emitter voltage are known as the I-V characteristics of a transistor. An example of this is shown in Figure 10.7. Note that the base-emitter characteristic is the same as the diode characteristic in Chapter 9.

In biasing a transistor for linear operation, the concept of dc and ac load lines can be useful. Figure 10.8 shows a common emitter amplifier, its biasing, and the

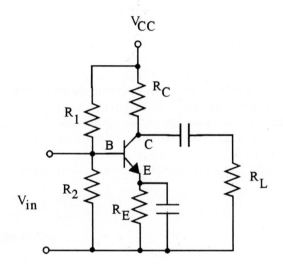

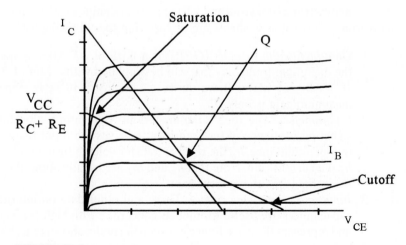

FIGURE 10.8
Transistor circuit load lines.

associated dc and ac load lines. Let's review these concepts:

1. The load lines shown in Figure 10.8 can be used with either NPN or PNP transistors with appropriate sign changes.
2. Both load lines represent the current and voltage range of a signal above and below the dc operating point. The operating point is the intersection of the load lines. Thus, dc changes will move up and down the dc load line (as long as none of the factors involved in drawing the original load lines is changed) and an ac signal will vary between the operating point (Q) and the upper and lower limits of the signal.
3. The dc values vary between the power supply voltage (V_{CC}) and the maximum current through the transistor with no voltage from collector to emitter [$V_{CC}/(R_C + R_E)$ in Figure 10.8].
4. The ac values vary from the operating point (which is determined by the base biasing resistors) to the maximum ac current, which occurs when the collector-emitter junction has no voltage and the total current is the supply voltage divided by the parallel combination of the collector and the load resistances ($R_C\|R_L$).
5. Bias point equations are shown next and correspond to Figure 10.8:

$$V_B = \frac{R_2}{R_1 + R_2}(V_{CC}) \tag{10.9}$$

$$V_B = V_E + 0.6 \text{ (for silicon)} \tag{10.10}$$

$$VE = \frac{I_C}{R_E} \tag{10.11}$$

$$\text{Slope of ac load line} = -\frac{1}{R_C\|R_L} \tag{10.12}$$

Specifications/Parameters   The practical significance of some of the major items from transistor spec sheets are discussed in the following list:

1. *Gain-bandwidth product (GBP):* If a GBP of 200 MHz is specified, then the maximum frequency that can be amplified with a gain of 200 is (200 × 10^6)/200 = 1 MHz. If we desire to amplify a 10-MHz signal, then the maximum gain is only 20.
2. *h parameters:* The four h parameters provide information on the small-signal performance of the transistor: h_{ie} gives the input resistance, h_{oe} the output admittance, h_{re} the reverse transfer voltage (usually very small), and h_{fe} the current gain. Note that h_{fe} and h_{FE} are usually about the same where h_{FE} = beta (β), the current gain.
3. *Input and output capacitance:* These and other data on internal transistor capacitances are useful for models concerned with high-frequency analysis and represent the capacitance between the nodes shown (e.g., base-emitter).
4. *Turn-on and turn-off time:* These parameters are used in switching applications and enable the user to determine the actual transistor switching speed.

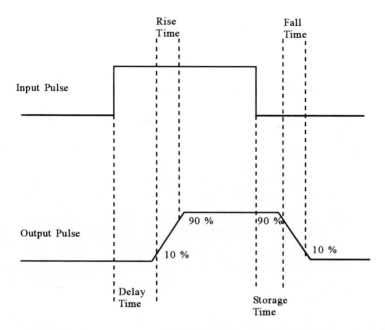

FIGURE 10.9
Transfer effects on an input pulse.

5. *Delay time, rise time, storage time, and fall time:* Figure 10.9 shows the meanings of these terms. If you put a square wave with rise and fall times much faster than those shown for the transistor into the test circuit shown on the spec sheet (and ensure that the oscilloscope input and lead capacitance are small enough that they do not affect the output), the output will have the characteristics shown.

6. *s parameters:* The *s* parameters are used with microwave devices and define the input and output impedances and reflection coefficients.

7. *Noise figure:* All devices add noise to a signal. This spec provides a comparison of the amount of noise power added to the signal as it is amplified by the transistor. Thus, a 2N3904 adds 5 dB (at the conditions specified in the specs):

$$5 \text{ dB} = 10 \times \log_{10}(P_{\text{out}}/P_{\text{in}}) \qquad (10.13)$$

which means that the input noise power is increased by a factor of 3.16. For comparison, a low-noise transistor may have a noise factor of 1 to 3 dB, which corresponds to increasing the input noise power by a factor of 1.26 to 2.

Testing   A few basic techniques for determining whether a bipolar transistor will still operate properly are given here:

1. *In-circuit testing:* With the circuit on, measure the base-emitter junction voltage. This should be about 0.2 V for Ge and 0.6 V for Si. The polarity

depends on whether the transistor is PNP (negative from base to emitter) or NPN (positive from base to emitter). This ensures that the base-emitter junction is not shorted or open. Next, short the base-emitter junction to remove the bias voltage and measure the voltage at the collector. This voltage should be virtually the same as the power supply voltage. If not, there is an internal short.

2. *Out-of-circuit testing* (without a transistor checker) [7]: For an NPN transistor (for a PNP reverse the polarities), *ensure that you know the polarity of the ohmmeter's leads and do not use a low (R × 1 or R × 10) scale.*

   Using an ohmmeter on the $R \times 100$ scale, connect the positive lead to the base and the negative lead to the emitter. The reading should be 35 to 50 Ω for power Ge, 200 to 500 Ω for small-signal Ge, 200 to 1000 Ω for power silicon, and 1000 to 3000 Ω for small-signal silicon.

   Using an ohmmeter on the $R \times 1000$ scale, connect the positive lead to the collector and the negative lead to the emitter. The reading should be several hundred ohms for power Ge, 10 to 100 kΩ for small-signal Ge, greater than 1 MΩ for power silicon, and very high (many mega-ohms) for small-signal silicon.

10.2.2 Applications and Design

Amplifiers   Five of the most common classes of transistor amplifier [8] are discussed in the following paragraphs.

Class A.   This is a small-signal, linear amplifier. It is biased near the center of the ac load line and is designed to remain in the linear region of its I-V characteristics for all input signals (which means that the input signal does not push the transistor into saturation or cutoff). The three basic class A amplifier configurations, common emitter (CE), common collector (CC), and common base (CB), are shown in Figure 10.10. The reason for choosing one of these over the others is the respective

TABLE 10.2

Gains and impedances of class A
transistor amplifiers

	CE	CC	CB
A_v	$-\dfrac{R_L}{r_{be}}$ (high)	$+1$	$-\dfrac{R_L}{r_{be}}$ (high)
A_i	β (high)	β (high)	$+1$
R_i	$br_{be}{}^a$	βR_E (high)^b	r_{be}
R_o	$1/h_{oe}$ (high)	$(R/\beta) + r_{be}$ (low)^c	$1/h_{oe}$ (high)

^{a} $r_{be} = 0.26/I_E$ is the internal resistance of the base-emitter junction.
^{b} R_E is the emitter resistor.
^{c} R is the resistance looking back from base to input source.

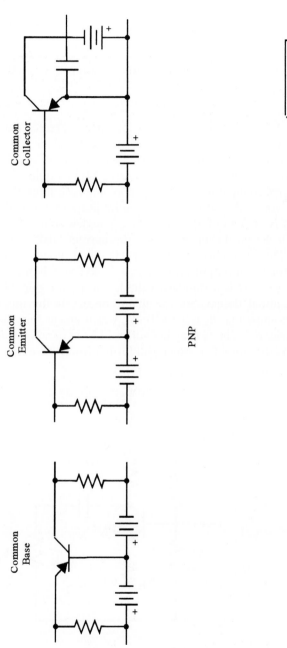

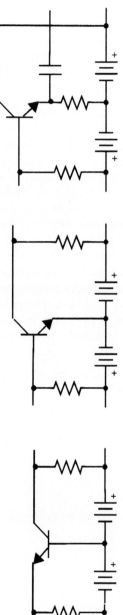

Common Base

Common Emitter

Common Collector

PNP

NPN

FIGURE 10.10

Bipolar transistor amplifier configurations.

317

input and output impedances and the current and voltage gains of the amplifiers. A chart showing the relative characteristics of each configuration is provided in Table 10.2. Note that only the CE configuration inverts the signal.

EXAMPLE 10.4

You are required to build a low-cost linear amplifier for a 1-MHz signal. The required power supply is 12 V.

Solution

The amplifier must feed a 4-kΩ load and have an input impedance greater than 20 kΩ. The input signal is very small and a power gain of greater than 100 is required. The choice of semiconductors in this case would be between bipolar transistors, FETs, and operational amplifiers. The inexpensive FETs would have less gain in a single common source configuration. A 1-MHz op-amp costs more than a single bipolar transistor. Thus, you settle on the 2N3904 general-purpose transistor.

A typical amplifier design is shown in Figure 10.11. Let's review some of the factors that went into this final design. From the specs, the current-GBP is 300 MHz and the h_{fe} is 100 to 400. Thus, amplification of a 1-MHz signal should be no problem. Matching the transistor to an output impedance of 4 kΩ merely means using a 4-kΩ collector resistor (3.9 kΩ is a standard value). The power gain will have to be checked after the initial design. So, the main concern is the input impedance of 20 kΩ. The specs show that h_{ie} is 1 to 10 kΩ, which means that the worst case of 1 kΩ must be considered. The biasing network will be in parallel with the input impedance of the transistor amplifier, which will further lower the overall

FIGURE 10.11
A common emitter amplifier.

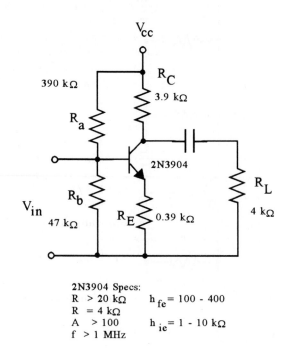

2N3904 Specs:
R > 20 kΩ h_{fe} = 100 - 400
R = 4 kΩ
A > 100 h_{ie} = 1 - 10 kΩ
f > 1 MHz

input impedance. You must therefore use an unbypassed emitter resistor to increase the input impedance. As shown in Table 10.2, the input impedance of a CC amplifier is approximately beta (β) times R_E. This same principle applies for the input resistance of a CE amplifier with an unbypassed emitter resistor. The input resistance is therefore 100 times the value of the emitter resistor. The biasing resistors are in parallel with each other and their combined resistance is in parallel with the input resistance of the transistor itself. So, the biasing resistors must have a combined (in parallel) resistance of greater than 40 kΩ for the overall input impedance of the amp to be greater than 20 kΩ. Load lines are probably not required if the amplifier is biased in the center of the dc load line, which would be for a V_{CE} of 6 V.

Since I_C is approximately equal to I_E, I_C = 6/4.29 = 1.4 mA. Therefore, the emitter voltage V_E is (1.4)(0.39) = 0.55 V and the base voltage V_B is 0.55 + 0.6 = 1.15 V. Selection of the bias point and use of the ac load line are critical only if we are required to get the maximum possible range of amplification from the transistor. Otherwise, as is done here, selection of the bias point generally in the center of the dc load line (or in the general center of the transistor's active region) will suffice for linear amplification. The values of R_a and R_b are found using Eqs. (10.14) and (10.15):

$$R_a = (R_P/V_B)V_{CC} \tag{10.14}$$
$$R_b = V_B R_a/(V_{CC} - V_B) \tag{10.15}$$

where V_{CC} is the supply voltage (+12) and the R_P is the parallel combination of the two bias resistors (here determined to be 40 kΩ). Once these have been found, the gain can be found. The ac voltage gain (base to load) is approximately the ratio of the ac load resistance to the emitter resistance, which is about 5. The ac current divides about equally between the bias resistors and the base of the transistor since both have the same resistance. It is then multiplied by h_{fe} = 100 (minimum) and divides again equally between the collector resistor and the load. So the current gain is about 25 (minimum). Thus, the power gain is the current gain times the voltage gain or about 125 (minimum).

Class B. The class B amplifier is biased at cutoff (no current) on the load line so that the transistor is turned on for one-half of the input waveform cycle and off for the other half. This produces a higher efficiency power amplification (78.5% ideal efficiency) but the output waveform is a distorted version of the upper half of the input waveform. The distortion is due to the nonlinear amplification of the input while the transistor is near cutoff. This distortion can be eliminated using class AB operation and the other half of the waveform can be added to the output using a push–pull type of amplifier as shown in Example 10.5.

EXAMPLE 10.5
You need to build an inexpensive audio amplifier that will provide at least 1 W of power to an 8-Ω speaker. Plus and minus 12 V are available for a power supply and maximum efficiency is desired.

FIGURE 10.12

A complementary symmetry amplifier.

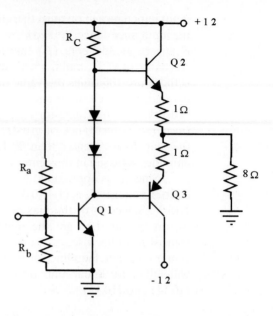

Solution

The class AB complementary (or push–pull) amplifier shown in Figure 10.12 [9] can be used. This design will produce little distortion because class AB means that the transistors are biased very close to, but not at, cutoff so that very little current is drawn unless the transistor is turned on by the input signal. Note that the two diodes provide the bias that keeps the two base-emitter (BE) junctions barely turned on. The stockroom has quite a few MHQ 6001 quad complementary transistors available should you decide to use these. This will allow use of the BE junctions of the transistors as matched diodes for biasing. The MHQ6001 [2] is a 14-pin DIP that can dissipate 1.9 W at an ambient temperature of 25°C, so that even with a 50% efficiency (and class AB usually provides 60 to 75%) the total device dissipation would be 1 W with 1 W to the speaker so the device should easily handle the load. The driver transistor ($Q1$) is an inexpensive 2N2222 general-purpose transistor and is biased as a class A linear amplifier. R_C is the collector resistor for the driver transistor. The two diodes are biased on and do not rectify. The emitter resistors of the two transistors shown in Figure 10.12 are frequently omitted because they dissipate power and lower the amplifier efficiency (they do, however, provide some stability). For an input signal of less than 0 Vac, the collector of $Q1$ becomes more positive, which increases the voltage at the base of $Q2$, turning it on and $Q3$ off (since the base voltage of $Q3$ increases, which tends to reverse bias this BE junction). With $Q2$ on and $Q3$ off, the input signal is amplified by $Q2$ and feeds the speaker. The opposite effect takes place for an input signal that is greater than 0 Vac. The two complementary transistors each amplify one-half of the signal and the resulting output to the speaker is a replica of the input with little distortion.

Class C.   This amplifier is biased below cutoff so that only a small portion of the input signal is amplified. High efficiency is obtained but it is very nonlinear. One application of this class is in amplification of modulated signals (see Chapter 18) where the output is an LC (usually a parallel LC tank circuit) tuned to the modulated frequency.

Class D.   This class of amplifier uses transistors as switches to produce high-efficiency output. A push–pull arrangement like that of Figure 10.12 can be used. The transistors can be biased at cutoff so that no dc current flows and the input signal alternately turns one then the other on. The difference is that the transistors are not turned on into their linear amplification region but are driven very rapidly into saturation. The output is a square wave at the fundamental frequency of the input. This output is filtered, as with a class C amplifier, using a tuned circuit so that the signal to the load is a sinusoid at the same frequency as the input. This class of amplifier is often used in radio transmitters.

Class S.   This forms the basis of switching regulators (see Chapter 3). Figure 10.13 shows a class S amplifier. The input is a series of pulses into the base that turn the transistor on and off. The output, taken at the emitter in this case, is filtered using the LC filter so that the signal to the load resistor is a dc voltage proportional to the duty cycle of the input waveform. The diode provides a current path for the inductive current when the transistor is switched off.

Cascading of amplifiers refers to the normal practice of using multiple amplifier stages. Normally cascaded amplifiers are class A with the final output stage a higher power, higher efficiency type such as a class B push–pull.

Coupling of amplifier stages to each other, to the input, and to the load can be accomplished via dc (also known as direct) coupling, capacitor coupling (frequently known as ac coupling), and transformer coupling (which, of course, is also ac coupling).

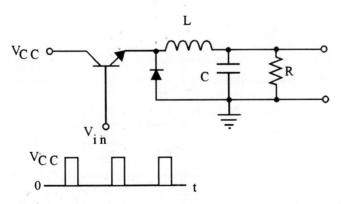

FIGURE 10.13
A class S amplifier.

Buffering of the input or of the load is a very important principle. This refers to the isolation of the input/load from the effect of the circuit to which it is attached. Many methods of buffering exist and Example 10.6 illustrates the use of a Darlington pair as a buffer amplifier.

EXAMPLE 10.6

You need to drive an 8-Ω load with a source that has a resistance of 5 kΩ; the input and output circuits need to be well isolated from each other; and a 12-V power supply is to be used. The circuit must be inexpensive, requires a high current gain but the input voltage level is satisfactory at the output (no voltage gain required), and must not use a transformer.

Solution
Basically, a buffer amplifier is required. Many buffers are available, and a unity-gain operational amplifier is relatively inexpensive; however, the Darlington circuit will provide very high current gain, and one can use two discrete transistors or use one of the many three-pin monolithic Darlingtons that are available. You decide to use the MPSA12 [2] Darlington, which has a TO-92 three-pin plastic case. Figure 10.14 shows a Darlington circuit that can be used to meet these requirements. The MPSA12 [2] has a beta (β) of greater than 20,000. For this calculation, let's assume that each transistor has a beta of 150 (150 $\times$ 150 = 22,500). You need to determine the values of the three biasing resistors. Since we want no loading of the input, the two base biasing resistors are chosen as 220 kΩ each (a choice that is somewhat arbitrary). For emitter biasing, a 0.81-kΩ resistor is chosen after several repetitions of the calculations shown below.

Biasing: Since the biasing resistors are equal, the base of $Q1$ is at 6 V. The emitter of $Q2$ is then at 6 $-$ 2(0.6) = 4.8 V, and the emitter current of $Q2$ is

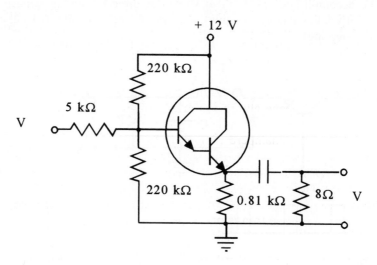

FIGURE 10.14
A Darlington transistor circuit.

4.8/0.81 = 5.9 mA. This is critical for the output resistance calculation, which is why several repetitions were required to determine initially the value of the emitter biasing resistor.

Input impedance: This is the parallel combination of the base biasing resistors, 110 kΩ in parallel with the input impedance of the transistors. This input impedance is like that of two emitter-follower (common collector) amplifiers with the emitter of the first connected to the base of the second. Since the 8-Ω load is far less than the 0.81-kΩ emitter resistor, the load resistance seen by the emitter is about 8 Ω. This 8 Ω multiplied by the minimum beta of the pair, 20,000, provides a minimum input impedance of 160 kΩ. This, in parallel with the base biasing resistors, provides an input resistance of 65 kΩ, which means very little loading of the source.

Output impedance: The ac emitter resistance of a diode is 25 mV/I_e (see Chapter 9). So the BE junction resistance of $Q2$ is 25 mV/5.9 mA = 4.2 Ω. The emitter current of $Q1$ is 5.9 mA/150 = 0.039 mA. The BE junction resistance of $Q1$ is 25 mV/0.039 = 641 Ω. The resistance seen looking back from the base of $Q1$ toward the source is 110 kΩ in parallel with the source resistance of 5 or 4.8 kΩ. Now calculating the output resistance of $Q1$ using the common collector formula from Table 10.2, 641 Ω + 4.8 kΩ/150 = 673 Ω. Using this same formula again to calculate the impedance seen by the load, 4.2 + 673/150 = 8.7 Ω. Thus, the source sees a high-impedance load and the load is fed by the Darlington circuit that looks like an 8.7-Ω source.

Phototransistors   These devices come in two- and three-lead packages. The two principal leads are the collector and the emitter. Leaving the base open provides maximum sensitivity but in the three-lead versions, a resistor can be connected between the base and ground to vary the sensitivity. Light striking the exposed collector-base junction increases the current through the junction in much the same way as heating a *pn* junction increases the reverse bias current through it. The difference between the photodiode (Chapter 9) and the phototransistor is sensitivity. The collector reverse-bias current in a phototransistor is a factor of beta times the reverse-bias current of a single *pn* junction (as in a diode). The phototransistor is, therefore, much more sensitive. The disadvantage of the phototransistor is that its switching time is generally in the microsecond range, whereas the photodiode switching time is in the nanosecond range. The phototransistor output current is usually in the milliampere range, whereas the photodiode's current is in the microampere range. One of the most comon uses of the phototransistor is as an optoisolator (or optical buffer). Phototransistors are available separately and are also packaged with an internal emitter diode and sold as phototransistor (and photo-Darlington) optoisolators or optocouplers. The following example shows one possible use of a phototransistor.

EXAMPLE 10.7   You have a conveyor belt that brings large items into the shop infrequently. Your normal workplace is in some adjoining rooms but when an item comes in it must be processed immediately. Some method of notification is necessary.

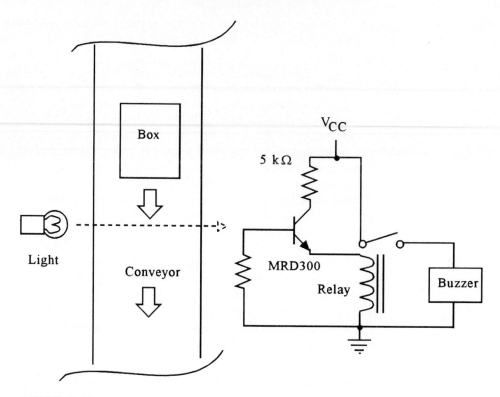

FIGURE 10.15
Use of a phototransistor.

Solution
Use of a light source (normal light bulb) on one side of the belt and a phototransistor on the other side of the belt as shown in Figure 10.15 is one easy way to do this if the room is fairly dark. The phototransistor used is an MRD300 [10], which produces sufficient current to drive a small relay (in the normally off position) that will activate a buzzer when the light to the phototransistor is interrupted. The phototransistor is on and current flows in this simple design as long as light reaches the phototransistor. The variable resistor is used to adjust the sensitivity of the phototransistor, if required.

Oscillators The many types of oscillator circuits all work on the principle of feeding back a portion of the output signal in phase with the input (there is normally no external input signal) so that regenerative action takes place. Normal transistor amplification takes place, but the positive (regenerative) feedback causes the output signal to build up at a frequency determined by a tuned (LC) circuit. Some of the most common oscillator circuits are shown in Figure 10.16. Feedback occurs as follows:

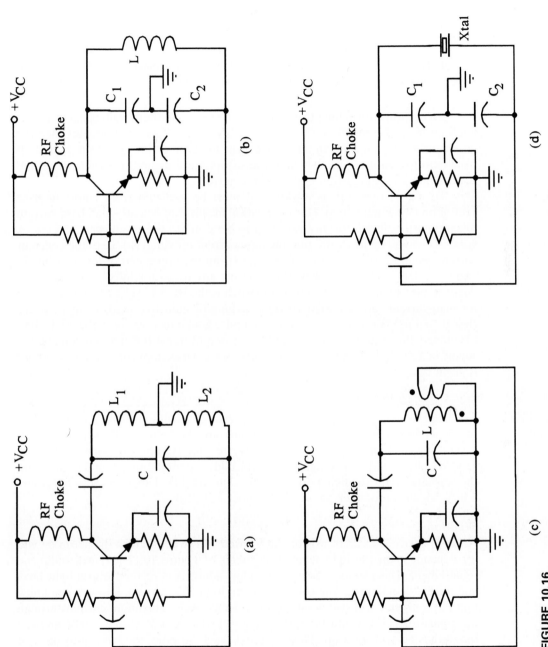

FIGURE 10.16

Some common oscillator circuits: (a) Hartley, (b) Colpitts, (c) Armstrong, and (d) crystal (Colpitts).

1. In the Hartley oscillator [Figure 10.16(a)], feedback is via a tapped inductance.
2. In the Colpitts [Figure 10.16(b)], feedback is via a capacitor divider circuit.
3. In the Armstrong [Figure 10.16(c)], feedback is via transformer coupling.
4. In the crystal oscillator [Figure 10.16(d)], feedback is via the capacitor divider circuit (thus, this is a Colpitts circuit).

Switching   Switching transistors can be used in either small-signal or power applications. All transistors can be used to switch currents much greater than the base current by providing an input that turns the base-emitter junction on and off from cutoff to saturation. Switching transistors are designed to switch faster and their region of amplification may be quite narrow.

To use a transistor as a switch, it must be switched from cutoff to hard saturation (see Figure 10.8). Hard saturation means a sufficiently high base current to saturate the device for any variation in beta due to the production process. A good rule would be to ensure that the base current provided is 1/10 of the collector current required to saturate the transistor (from the upper end of the load line in Figure 10.8, $I_C = V_{CC}/R_C$). This should drive any transistor into hard saturation. An example of a high-power transistor switch is the BUX47 [11], which is designed for high-current (greater than 10 A) switching of inductive loads at high speeds (less than 4 μs). Since switching power supplies generally operate in the 20- to 100-kHz range, this transistor can be used as the output transistor of a switching power supply in a configuration like that of Figure 10.13. An example of a switching circuit is given next.

EXAMPLE 10.8   You need to build an inexpensive battery-powered circuit to turn on an alarm when the lights go out in a certain area. You measure the light and dark resistance of several items of an available light-sensitive resistor (Chapter 4) such as the 276-116 CdS photocell [7]. The light resistance under room lighting is under 1 kΩ and the resistance with the lights off is greater than 100 kΩ. The alarm selected draws 5 mA, which is too much to use the photocell directly.

Solution
An inexpensive 2N2222 transistor can be used as a switch to turn the alarm on and off dependent on the light level. The circuit of Figure 10.17 [12] will fulfill the requirement. The variable resistor will allow adjustment of the turn-on light level and the 670-Ω resistor in the collector circuit limits the current. How do you know that the transistor will switch on and off? Think of the maximum current through the transistor. This would be 9 V/670 Ω = 13.4 mA if V_{CE}(sat) = 0 V and the buzzer has no voltage drop. Thus, to fulfill the 1:10 ratio, the base must draw at least 1.34 mA when the lights are off. If the variable resistor is 5 kΩ and the photocell is 10 kΩ, an initial calculation of the voltage at the base is 6 V, so there would be 3 V across the variable resistor, and 0.6 mA is drawn from the battery

FIGURE 10.17
A transistor used as a switch.

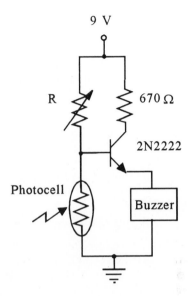

and is available to drive the transistor. The main point is that the base has sufficient current available to drive the transistor into hard saturation.

10.3 FIELD-EFFECT TRANSISTORS

10.3.1 Basic Concepts

The materials used in FETs are the same as those used for bipolar transistors including GaAs for higher frequency applications. These devices depend on the strength of an electric field across a *pn* junction to increase the number of current carriers (electrons or holes) and allow current to flow. They are therefore known as voltage-controlled current devices. There are two basic types in common use: junction field-effect transistors (JFETs) and metal-oxide semiconductor field-effect transistors (MOSFETs). Each of these comes in two configurations, *n*-channel or *p*-channel, and each of the MOSFETs can, in turn, be either depletion or enhancement mode. Figure 10.18 shows the schematic symbols for the six types of FETs.

Manufacturers group FETs according to the same functional areas as bipolar transistors, discussed in the previous section. These groupings include general-purpose amplifiers, switches and choppers, rf amplifiers, and mixers. The advantages of FETs are as follows:

1. Higher input impedance (due to the reverse-biased gate junction for JFETs and the insulated gate junction for MOSFETs)
2. Faster switching (due to not having to forward bias a junction to turn the device on)

FIGURE 10.18

Types of FETs.: (a) *n*-channel JFET, (b) *p*-channel JFET, (c) *n*-channel depletion MOSFET, (d) *p*-channel depletion MOSFET, (e) *n*-channel enhancement MOSFET, and (f) *p*-channel enhancement MOSFET.

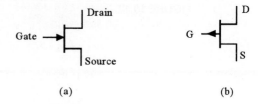

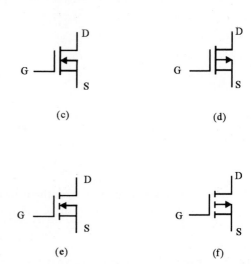

3. Lower noise (since the noise is primarily due to the resistance of the channel and the FET does not have added noise caused by the bipolar *pn* junctions)
4. Enhanced thermal stability and easier paralleling.

I-V Characteristics   Curves of the drain current, I_D, versus the voltage from the drain to the source, V_{DS}, and drain current versus the voltage from gate to source, V_{GS} (called transconductance), are known as the I-V characteristics of an FET. The *n*-channel FET characteristics are shown in Figure 10.19 along with typical load lines for each type and a common bias circuit that would produce the load line shown. I-V characteristics for *p*-channel devices are similar with appropriate polarity changes.

Figure 10.19(a) is for an *n*-channel JFET. Note that the maximum current is for zero voltage from gate to source. This is called a normally on device because, with no bias, the channel will conduct. The I_{DSS} means the current from drain to source with shorted gate and is the maximum current. If you force a positive gate-to-source voltage on a JFET, it becomes nonlinear and for higher voltages will break down. The most common method of biasing a JFET is known as self-bias and uses a source resistor R_S [as shown in Figure 10.19(a)] to ensure that the gate-source junction remains negatively biased at all times. The drain resistor R_D can be considered the load and the gate resistor R_G must be included to ensure that

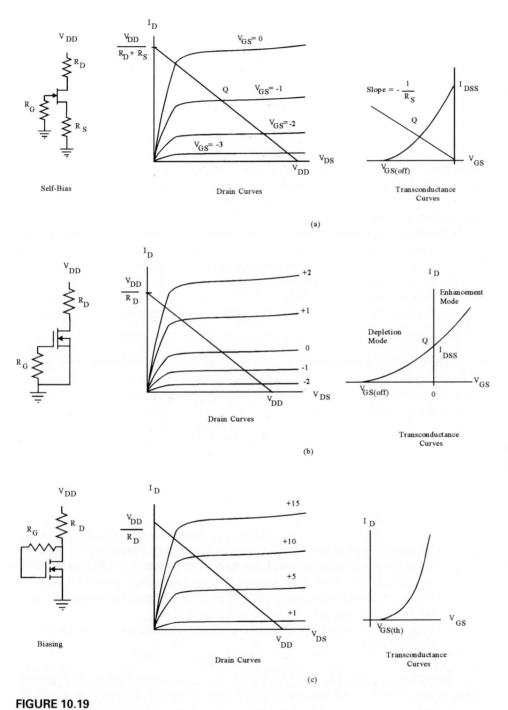

FIGURE 10.19

FET characteristic curves and biasing: (a) *n*-channel JFET, (b) *n*-channel depletion MOSFET, and (c) *n*-channel enhancement MOSFET.

there is a current path (even for the nanoamperes involved) so that a charge does not build up on these high input resistance devices. The load line is shown on the transconductance curve in some cases to illustrate the process. The transconductance curve in Figure 10.19(a) might be different for another JFET of the same type because of variations among the characteristics. It is easy to see that the bias point is at the same relative point on both curves so that this simple type of bias is quite effective.

Figure 10.19(b) shows typical n-channel depletion-mode MOSFET characteristics, circuit, and load line. This device can be biased, very conveniently, at 0 V from drain to source, which means it can operate for voltages above and below 0 Vdc. The bias circuit shown needs no resistance between the gate and source. The depletion mode MOSFET is, like the JFET, a normally on device and has characteristics similar to those of the JFET.

Figure 10.19(c) shows typical n-channel enhancement mode MOSFET characteristics, circuit, and load line. This device is known as a normally off type because in the absence of any gate-to-source voltage, the device does not conduct current. A positive-bias voltage is required from gate to source to turn the device on. As you will see in the discussion of the CMOS inverter circuit later and in Chapter 13, enhancement mode MOSFETs have become vital in modern, low-power, digital circuitry.

Specifications/Parameters   The practical significance of some of the major items from FET spec sheets are discussed in the following list (note that many spec sheet items such as noise figure and rise and fall times are the same as previously discussed for bipolar transistors):

1. Forward transfer admittance, g_{fs}, which is also known as forward transconductance. This is the ratio of the change in drain current to the change in gate-to-source voltage, or I_d/V_{gs}. This is used to find the gain in FET circuits much as beta was used in bipolar circuits. The variance in g_{fs} for FETs ranges from about 1 mmho in small-signal amplifier JFETs to several mhos in high-power MOSFETs.
2. Static drain to source on resistance or drain source on resistance (r_{ds}). Primarily found in switching FET specs, this is the resistance of the channel when the device is turned on. It can vary from hundreds of ohms in low-cost small-signal switching devices to tenths of an ohm in high-power MOSFETs.
3. Reverse transfer capacitance (C_{rss}). The capacitance between the gate and the drain with the gate to source short circuited.
4. Output admittance (y_{os}). A measure of the output resistance of the device ($z_{out} = 1/y_{os}$).

Handling   JFETs, like bipolar transistors, require no special handling precautions. Many MOSFETs, of which the CMOS that is so prevalent in computer circuits is a particular type, must be handled with caution. This is because of the extremely high input impedance of the device caused by the layer of metal oxide, an insulating material, between the gate lead and the semiconductor material. This layer of metal oxide allows the gate to build up charge like a capacitor and because of the small area involved (fractions of a millimeter in CMOS integrated circuits) can cause breakdown and burnout of the junction due to static electricity. Some MOSFETs have diodes built into the semiconductor material to provide overvoltage protection and decrease the sensitivity to static electricity and transient voltages. However, when handling these devices it is best to take appropriate precautions:

1. Keep the leads of the device in contact with a conductive material except during testing or operation.
2. Ground tools, soldering iron tips, and handling facilities.
3. Ensure that the power to the circuit is off and that all capacitors and inductors are completely discharged prior to insertion of the device because transients may cause damage.
4. Ensure that the power supply is on to provide bias prior to applying signals.

Testing   Prior to testing, you must determine whether the device is a JFET or a MOSFET, whether it is an *n*-channel or a *p*-channel device and, if it is a MOSFET, whether it is an enhancement or depletion type. The tests discussed next are out-of-circuit tests:

1. *JFETs:* The forward resistance of an *n*-channel device can be checked by connecting the positive lead of a low-voltage (on the $R \times 100$ scale) ohm-meter to the gate and the negative lead to the source (or the drain since many small-signal JFETs have an interchangeable source and drain). This just tests a forward-biased *pn* junction as with a diode or base-emitter of a transistor. To test the reverse resistance (which should be almost infinite unless the junction has broken down or is leaking) of an *n*-channel device, reverse the leads. For a *p*-channel device, reverse the leads in each case.
2. *MOSFETs:* The forward and reverse resistances can be checked using a low-voltage ohmmeter on the highest resistance scale. Both forward and reverse resistances should be almost infinite. Any lower reading indicates a breakdown in the insulation and a damaged MOSFET.

10.3.2   Applications and Design

Some of the principal applications of FETs are shown in Table 10.3 [8]. Some of these are discussed or examples provided in this section.

Amplifiers   FETs can be used in the same classes of amplifiers as bipolar transistors. The two basic configurations used with FETs are common source (CS) and common drain (CD), as shown in Figure 10.20. The CS circuit is similar to the bipolar common emitter and the CD is similar to the bipolar common collector.

The common gate circuit is not used in amplifiers but can be used to provide a voltage variable resistance.

The reason for choosing one configuration over another is based on impedance levels and gain, as with bipolar transistors. A chart showing the relative characteristics of each configuration is provided in Table 10.4. When comparing FET performance to bipolar (from Table 10.2), the FET has a lower voltage gain, a higher input impedance, and a similar output impedance for the common source but a generally higher output impedance for the common drain (also known as the source follower much as the common collector is known as an emitter follower).

FIGURE 10.20
JFET amplifier circuits: (a) Common source amplifier and (b) common drain amplifier.

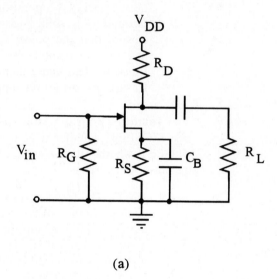

(a)

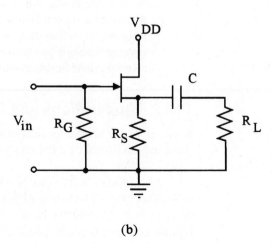

(b)

TABLE 10.3
FET applications

Application	Advantage
Buffer	High Z_{in}, low Z_o
RF amp	Low noise
Mixer	Low distortion
AGC amp	Easy gain control
Cascode amp	Low input capacitance
Chopper	No drift
Variable R	Voltage controlled
Low-frequency amp	Small coupling capacitor
Oscillator	Small frequency drift
CMOS	Small size, low power

TABLE 10.4
FET circuit comparisons

	Common Source	Common Drain
A_v	$-g_{fs}R_L$	$+g_{fs}R_L/(1 + g_{fd}2R_L)\ (\approx 1)$
A_i	NA	NA
R_i	R_G	R_G
R_o	R_D	$1/g_{fd}$

EXAMPLE 10.9

You need to build an amplifier with a very high input impedance so that an LC tank circuit is not loaded for use at about 10 MHz.

Solution

An FET is required for the high impedance so you select an MPF 108 [2] JFET, VHF unit, which has a maximum noise figure of 3.0 dB at 100 MHz. By comparison, the MPS6568A [2] is a VHF bipolar transistor with a noise figure of about 5.0 dB at the same frequency, thus illustrating the generally lower noise characteristics of FETs. One can find very low noise models of both bipolar transistors and FETs, but in similarly priced models the FETs should inherently provide lower noise characteristics. Actual front-end amplifier circuits may be quite complex due to automatic gain control and other considerations, but the basic common source amplifier can provide a reasonable circuit at 10 MHz. The input admittance of this device is 800 μmho at 100 MHz and should be quite high at 10 MHz although the 6.5-pF capacitance may have to be considered at this frequency. The g_{fs} (y_{fs} in the specs) is between 2000 and 7500 μmhos (a very wide range of tolerances!). Then, to obtain a voltage gain, A_v, of 10 from Table 10.4, you need a total ac resistance at the drain of between 1.33 and 5 kΩ. So if you select a 5.1-kΩ drain resistance and the input impedance to the next stage is 5 kΩ, the 2.5-kΩ resistance will provide a voltage gain of 3.3 to 18.75. The biasing of a JFET depends on its I-V

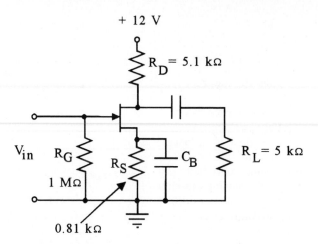

FIGURE 10.21
A JFET common source amplifier.

characteristics; however, if you bias the amplifier using a source resistor as in Figure 10.19(a), the bias point will remain reasonably constant for the wide variations in the characteristics. So, biasing at $V_{GS} = -1$ V requires $+1$ V at the source. With a 12-V power supply, using Kirchoff's voltage law,

$$12 = (5.1 \text{ k}\Omega)(I_D) + R_S I_D + V_{DS} \tag{10.16}$$

Solving this, with $R_S I_D = 1$ and $V_{DS} = 6$ V, since this biases the JFET in the center of its load line, $R_s = 1$ kΩ and $I_D = 1$ mA. The completed basic amplifier is shown in Figure 10.21.

Buffer Amplifier In the bipolar section, an example of a two-transistor Darlington circuit was used to illustrate a buffer amplifier. The CS or CD circuit can be used to buffer an input from the output because of the extremely high input impedance of the FETs. The common drain circuit has an input impedance in the mega-ohm range for JFETs or decades higher for MOSFETs and has a low output impedance (usually several hundred ohms) so that it can drive a low-impedance load.

EXAMPLE 10.10

You need to control an analog device with a computer (TTL) output. A buffer is needed so that the computer is not loaded and yet can easily drive an alarm directly.

Solution
The MOSFET called a FETlington™ from Siliconix (registered as the 2N7000) replaces a Darlington pair and provides a minimum g_{fs} of 100 mmho and (more important in this case where it is being used to switch a device on or off), a turn-on time of 10 ns maximum with a drain source ON resistance of 5 Ω. Figure 10.22 shows the very simple buffer/switch required to drive an alarm or lamp or relay

FIGURE 10.22
Use of a FETlington™.

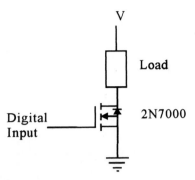

using a TTL input from a digital device. This device, as with any similar MOSFET, could be used to drive any high-current (in this case up to 200 mA) load that is almost completely buffered from the input.

Class B and C amplifiers are not as common using FETs primarily because they are high-power amplifiers and high gain is required. FETs have been developed to provide the high power, but the linearity and gain are still not generally as good as the bipolar.

The dual-gate MOSFET provides advantages over devices used previously. Figure 10.23(a) shows such a MOSFET used as a cascode amplifier and Figure 10.23(b) shows a dual-gate MOSFET used as a frequency mixer. The cascode amplifier is important because its capacitive effects are far less than those of other types of amplifiers. In Figure 10.23(a), a single dual-gate MOSFET is performing the function of two transistors or two JFETs. A common application of this type of amplifier is in rf circuits. If the g_{fs} of the dual-gate MOSFET is 10 mmho, the voltage gain is 18, the input impedance is approximately 1 MΩ (at relatively low frequencies where the capacitance effects are negligible), and the output impedance is 1.8 kΩ or about the value of the drain resistance. The frequency mixer of Figure 10.23(b) provides excellent output with very low distortion.

Oscillator circuits using FETs are very similar to those shown in Figure 10.16. Additional care must be taken with FET oscillators to ensure that the FET's gain is high enough (or the feedback factor is high enough) to provide a loop gain greater than 1. This is because of the generally lower gain factors of FET vice bipolar devices.

The various types of FETs make excellent voltage-controlled switches. An application of this is given in the next example.

EXAMPLE 10.11 You have +5- and +20-V available from the power supply operating a digital device and need a +12-Vdc voltage to operate an amplifier in the circuit. Your options are changing the power supply to provide another output or converting from the 5 V or from the 20 V to 12 V within the circuit.

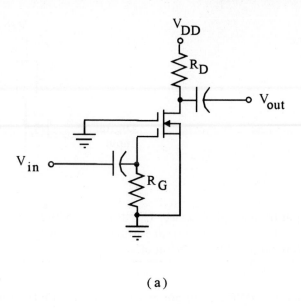

(a)

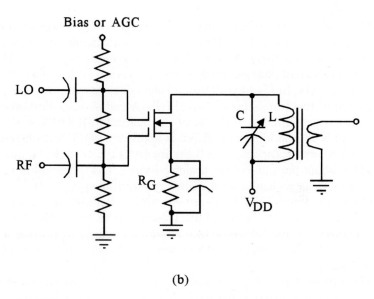

(b)

FIGURE 10.23
Dual-gate MOSFET circuits: (a) Cascode amplifier and (b) mixer.

Solution
Since the power supply is already a part of the unit, you decide on dc-to-dc conversion. Three methods appear feasible: use of a dc amplifier, a voltage divider, or a chopper. The 12-V output must be precise for the application and have virtually

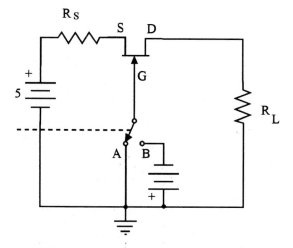

FIGURE 10.24
A JFET chopper circuit.

no drift so the dc amplifier is out. The voltage divider will waste too much power so a JFET chopper circuit is selected to convert the 5 Vdc into a square wave using a convenient digital output at 10 kHz to turn the JFET on and off. The JFET in Figure 10.24 provides the necessary chopping action, which breaks the dc input voltage up into a square-wave output. Note that the source resistance (which in this case would be a current-limiting resistor since high power is not required in this stage) and the load resistor (which is dependent on the input impedance of the next stage) are not specified because they depend on the particular application. This square-wave output can then be amplified by a standard transistor (or FET or op-amp) ac amplifier to about 15 V_{peak} and the output of the amplifier filtered and regulated (see Chapter 3) and a very stable 12-V output produced.

Complementary MOS (CMOS) circuitry has found its way into every aspect of modern life. The basis for the very low power consumption of this circuitry is the nature of the enhancement MOSFETs that make up each CMOS circuit. Just as complementary bipolar transistors can be used in a class D circuit (see the bipolar section of this chapter) to form a bipolar inverter like the one shown in Figure 10.25(a), enhancement MOSFETs can be used to form an inverter. Both Figures 10.25(a) and (b) operate similarly. If the input signal is a high enough positive voltage, the bottom device is turned on (saturation), the upper device is turned off (cutoff), and the output is essentially at ground potential (assuming a very small drain-source or collector-emitter resistance when each device is switched into saturation). When the input signal is a high enough negative voltage, the upper device is turned on and the lower device is turned off and the output is essentially at the supply voltage V_{CC}. Thus, the output is high for a low input and low for a high input, which is an inverter. Note that only one device is on at a time. The

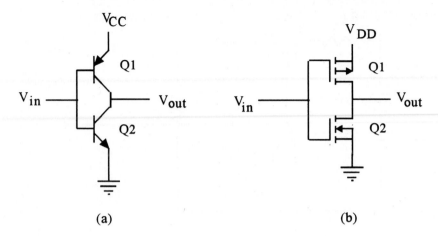

FIGURE 10.25
Inverter circuits: (a) Bipolar inverter and (b) CMOS inverter.

bipolar inverter has one device on at a time so that one base current is drawn at all times. The MOSFET (or CMOS) inverter has one device on at a time but the insulated gate draws virtually no current. Thus, in integrated circuits, if each bipolar draws 50 μA and 1000 circuits are on a chip and 100 chips are in a computer a considerable amount of power is required. The CMOS inverter draws only a few nanoamperes and, thus, battery-powered watches, computers, and other devices became feasible.

10.4   THYRISTORS

10.4.1   Basic Concepts

A thyristor is a *p-n-p-n* three-junction device that depends on regenerative (positive) feedback as defined by the International Electrotechnical Commission in 1964 [13]. Thyristors can also be thought of as multijunction devices that have internal positive feedback that can rapidly switch the device on and that can be used in a variety of trigger and oscillator circuits. Since a thyristor is either on or off, it is easy to see the derivation of the word *thyristor,* which is of Greek origin and means "door" [8]. The characteristics of all of the thyristors discussed here can be explained by referring to the basic *p-n-p-n* structure in Figure 10.26(a). First it can be considered a PNP/NPN transistor analog as in Figure 10.26(b) and then equivalent two-transistor latch (or regenerative feedback pair) as in Figure 10.26(c). A brief look at the figure shows that if the base current into $Q2$ is increased, this increases the $Q1$ base current, which increases the $Q1$ emitter current, which in turn increases the $Q2$ base current and the device is turned on (high current from anode to cathode) very rapidly. Thus, a very small increase in the $Q2$ base current rapidly turns on

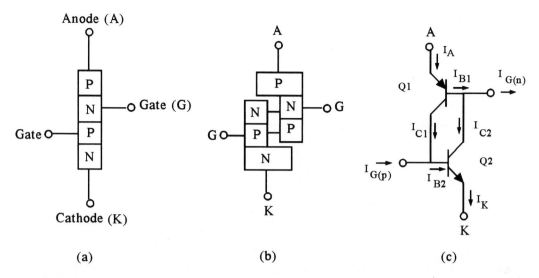

FIGURE 10.26
(a) A PNPN thyristor and (b) a two-transistor representation.

the switch and can turn on the device so that large currents flow from the anode to the cathode.

10.4.2 Unidirectional Thyristors

These devices conduct current in only one direction. Five members of this group are discussed in this section.

SCR Figure 10.26 showed the basic principle behind all thyristors and is the basis for the most commonly used and most well known of these devices, the silicon-controlled rectifier or SCR. From Figure 10.26 we can see that current can flow only from the anode to the cathode so that the device rectifies any voltage between the anode and the cathode. The SCR is a three-terminal device with a gate (for triggering the device and corresponding to the base of $Q2$ in Figure 10.26), an anode, and a cathode. Figure 10.27 shows the symbol and the I-V characteristics between the anode and the cathode.

Because the SCR is basically a switch that is turned on by a positive gate current and is only turned off when the anode-to-cathode voltage goes negative, the principal specs that are of interest are the gate trigger current and voltage, the maximum current that can be turned on through the device, and the switching speed. Another specification is the maximum rate of change of the gate voltage with respect to time (dv/dt in the specs). Any rate of change of the gate voltage greater than this dv/dt will cause the SCR to trigger on even without the required gate voltage and current. So one must be careful of transients or other waveforms

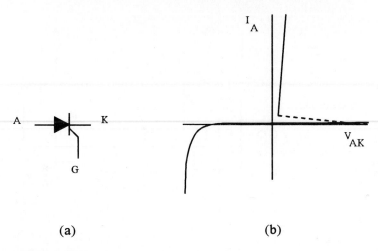

FIGURE 10.27
(a) The SCR symbol and (b) I-V characteristic curve.

that have very fast rates of increase with all of the thyristor devices. One of the main uses of SCRs has been control of the power to a device with an ac source as illustrated in Example 10.12.

EXAMPLE 10.12   You need to provide a simple circuit as a light dimmer for a 110-V, 100-W bulb that operates directly from a 220-Vac line.

Solution
Because SCRs provide rectification and most SCRs are high-power devices, this should not be a problem. The S2060 series [14] can handle 4-A_{rms} and the D, E, and M versions have peak reverse voltages in excess of 400 V. Since the 220-V_{rms} line has a peak voltage of $(220)(1.414) = 311$ V, any of these will provide the required capability. The S2060 has a maximum gate trigger current of 200 μA and a maximum gate trigger voltage of 0.8 V (typical voltage is 0.5 V) and is called a sensitive gate SCR because of these low triggering values. The circuit of Figure 10.28(a) provides the necessary light dimmer and illustrates the principle of phase control of ac voltages using SCRs, which has proven to be so vital for control of motors and other high-power devices.

Consider the effect of the positive half of the input ac waveform on the circuit: For $R = 0$, the capacitor charges to the trigger voltage (0.5 V for the gate trigger voltage plus 0.6 V to forward bias the diode or about 1.1 V) rapidly, the SCR is turned on, the device conducts, and the light is at its brightest. In this case, Figure 10.28(b) shows that one-half of the available power from the 220-V line is available to light the bulb and that this is the maximum available.

As R increases, the charging time constant of the RC circuit increases. This causes the capacitor to charge more slowly toward the input voltage value. Thus, there is a delay between the input voltage and the time it takes the capacitor to

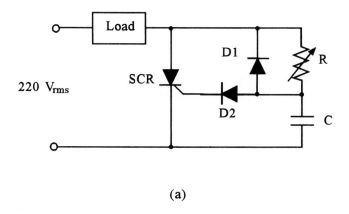

(a)

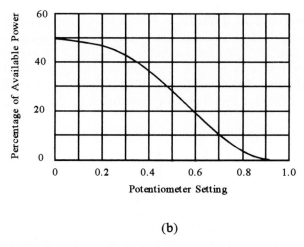

(b)

FIGURE 10.28
Half-wave phase control with an SCR.

charge to the trigger voltage of 1.1 V. If R is large enough that the long delay does not allow the capacitor to charge to the necessary 1.1 V to trigger the SCR, the light will not turn on at all [the zero power point in Figure 10.28(b)]. At an intermediate point [any other point on the graph of Figure 10.28(b)], the capacitor will charge to 1.1 V and turn on the light, but this may take an appreciable portion of the input period to do this, so that only a part of the input waveform (and hence the power) is available to turn on the light and it will be dim. Note that the capacitor does not charge beyond +1.1 V since this turns on the SCR and allows the main current to flow through it. This is called *half-wave phase control* because only the positive half of the input ac phase is used to power the load (in this case the bulb). *Full-wave phase control* can be achieved using two SCRs, with one for conduction in each direction. This is done, in fact, for very high power loads, but TRIACs (discussed later) are used for many full-wave control applications (as shown in Example 10.13).

Next consider the effect of the negative half of the input. The anode becomes negative with respect to the cathode, which turns off the SCR. The capacitor charges very rapidly to the negative peak voltage through diode $D1$, which resets the capacitor for the start of the next positive cycle where the capacitor will charge toward the gate turn-on voltage of 1.1 V.

Diode $D2$ provides reverse voltage protection for the SCR gate since the SCR is only rated for a 6-V reverse voltage.

Finally, consider the ratings required for the diodes and the capacitor. All must be high voltage and rated at a peak voltage greater than 311 V.

The symbols for the following members of the SCR family of unidirectional thyristors are shown in Figure 10.29.

Light-Activated SCR (LASCR) or Photo SCR   This is an SCR that uses light to trigger the gate. It is usually a three-terminal device—anode, cathode, and gate—with the gate left open for maximum sensitivity or connected to ground through a variable resistance for adjustable light sensitivity. A two-terminal device is known as a light-activated switch (LAS). With this device, an ac or dc load such as a bulb, alarm, or even a motor could be controlled by the ambient light.

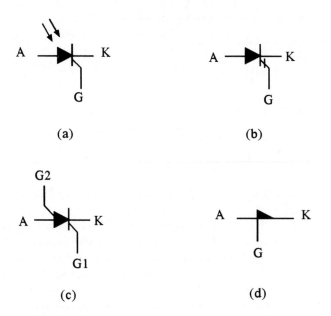

FIGURE 10.29
Other unidirectional thyristor symbols: (a) LASCR, (b) GCS, (c) SCS, and (d) SUS.

Gate-Controlled Switch (GCS) or Gate Turn-off SCR   This device can be turned on by a positive pulse at the gate (normal for SCRs) and turned off by a negative pulse at the gate (unique to this device). A series of positive and negative input pulses to the gate can provide a square (or rectangular) series of output pulses at the load.

Silicon-Controlled Switch (SCS)   This is a relatively low power device compared with the SCR. Two gates are provided, one called the anode gate [connected to the base of $Q1$ in the representation of Figure 10.26(c)] and the other called the cathode gate [connected to the base of $Q2$ in Figure 10.26(c)]. A forward bias or positive trigger at either gate can be used to begin the regenerative action and turn on the SCR. Light-activated SCSs have also been produced.

Silicon Unilateral Switch (SUS)   This is essentially a miniature SCR with an anode gate and a built-in Zener diode between the gate and the cathode. This device is primarily used in the basic relaxation oscillator circuit discussed in Section 10.4.4.

10.4.3   Bidirectional Thyristors

These three-junction devices differ from the SCR family because they conduct current in both directions. The symbols for the five members of this group, discussed next, are shown in Figure 10.30.

TRIAC   This device is like two SCRs in parallel and its construction is similar because it has two *p-n-p-n* configurations within the chip. Because of the more complex structure of the TRIAC, it does not have as high a power as the largest SCRs but is still considered a power control device. It operates in much the same manner with a gate that turns the device on at a specified, small gate current/ voltage. The TRIAC, when turned on, can conduct in either direction and is turned off only by the main terminal current dropping below a specified level for the particular device being used. It can, therefore, be used to provide full-wave phase control of loads with the half-wave phase control illustrated in Example 10.12. An example of a common TRIAC-DIAC full-wave phase control circuit is given in Example 10.13 following a brief discussion of the DIAC.

DIAC   This device is classified with thyristors (and usually with TRIACs) because of its bidirectional switching property and its common use, which is to switch TRIACs on. The device is composed of just three layers with two terminals. Think of an NPN transistor with only the collector and the emitter terminals available. The device turns on in both directions when a high enough voltage is reached to break down the appropriate *pn* junction. This voltage is known as the *breakover voltage* of the DIAC and this along with its maximum current and voltage capacity (off and on) and its leakage current (when off) are the main specs of interest for this device. The symbol is shown in Figure 10.30. Full-wave control of a load is shown in Example 10.13.

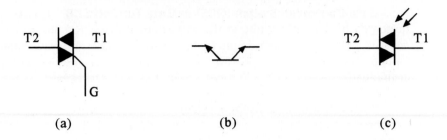

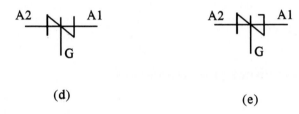

FIGURE 10.30

Bidirectional thyristor symbols: (a) TRIAC, (b) DIAC, (c) photo-TRIAC, (d) SBS, and (e) ASBS.

EXAMPLE 10.13

You need a light dimmer for a 700-W set of lights that operates on 110-Vac power. Full-wave control is necessary so that the maximum brightness can be achieved.

Solution

This can be accomplished using two SCRs in parallel, but a TRIAC can handle the job with a fairly simple circuit such as the one shown in Figure 10.31(a). Note the similarity of this circuit to Figure 10.28, which is half-wave phase control using an SCR. Since the TRIAC can conduct in both directions, a triggering device that will trigger for both positive and negative voltages is required. The ST2 DIAC [13] has a breakdown voltage of between 28 and 32 V and a peak pulse voltage (on breakdown) of at least 3 V, and is made to trigger all of the General Electric TRIACs. The SC241B TRIAC [13] will handle the 6 V_{rms} required of the 720-A_{rms} load with 120-V_{rms} as power. The variable resistor provides control of the brightness of the lamps by changing the time constant of the RC network. This changes the point in the input phase at which the capacitor is charged to the necessary 28 to 32 V required to switch the DIAC on. The DIAC switches on for both the positive and the negative halves of the input cycle when the capacitor charges to +28 to +32 V or to −28 to −32 V and full-wave control is achieved. The only problem with this circuit is that when the resistance is increased sufficiently to not trigger the TRIAC-

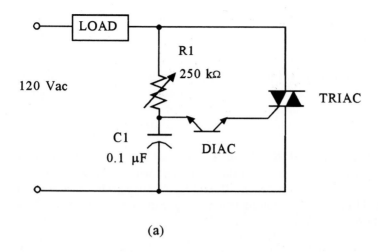

(a)

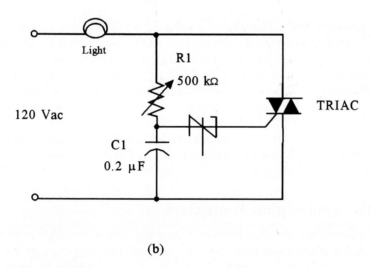

(b)

FIGURE 10.31
Full-wave phase control for (a) TRIAC-DIAC and (b) ASBS-TRIAC.

DIAC combination at all and then is again lowered a phenomenon called *snap-on* occurs. Snap-on means that the lights come on to full brightness as soon as the resistance is lowered to a value that will trigger the devices. After coming to this maximum brightness, the light dimmer will again operate over the full range of brightness. If this is annoying or not allowable, the circuit of Figure 10.31(b) can be used. It will be explained after a brief introduction to the silicon bilateral switch and the asymmetrical SBS.

Silicon Bilateral Switch (SBS)   This is essentially two identical SUSs (see SUS earlier in this section) arranged in inverse parallel (see Figure 10.30 for the symbol) so that it operates as a switch for either polarity of input. It can also be used as a triggering device for TRIACs or other bidirectional devices.

Asymmetrical SBS (ASBS)   This device (see Figure 10.30 for the symbol) acts like an SBS with a Zener diode in series with an SBS. The Zener diode causes the breakdown voltage to be different for opposite polarities of inputs. In the case of the ST4 [13], the switching voltage in one direction is 14 to 18 V and in the other direction is 7 to 9 V.

EXAMPLE 10.14   Use of an ASBS, as in the circuit of Figure 10.31(b), eliminates the snap-on of the lights after they are dimmed off and then adjusted back on. This circuit will provide continuous control of the light dimmer and operation is essentially the same as the DIAC dimmer.

Photo-TRIAC   The symbol for this device is shown in Figure 10.30. As with other light-activated devices, the photo-TRIAC can be used for all TRIAC device applications within its power limits. The MRD3010 [10] is a small, low-power device that can use ambient light or an LED as a trigger. As with transistors, many manufacturers produce optocouplers (also called optoisolators) that have a TRIAC as the output device. These devices are triggered by an input current that turns on an LED built into (not visible) the package. This LED then turns on the photo-TRIAC (also built into the package), which can turn on a high-voltage ac source. Using this type of device, direct logic control of 120/240 Vac sources can be achieved with complete isolation from the logic source.

10.4.4   Unijunction Transistors

The unijunction transistor, the complementary unijunction transistor, and the programmable unijunction transistor are grouped with thyristors in many cases but two of the three are considerably different from thyristors in construction as is explained later. Figure 10.32 shows the circuit symbols used to represent the three devices discussed.

Unijunction Transistor (UJT)   This is basically a single block of lightly doped n-type material (usually silicon) with a small heavily doped p-type area embedded into it as shown in Figure 10.33(a). It is readily apparent that the n-type material will act as a simple resistor and conduct if a voltage is placed across it.

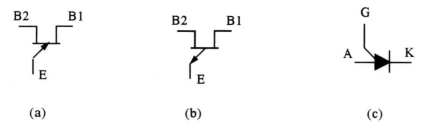

(a)   (b)   (c)

FIGURE 10.32
Unijunction transistors and circuit symbols: (a) UJT, (b) CUJT, and (c) PUT.

The resistance of the *n*-type material is known as the *interbase resistance* of the device and is on the order of 4 to 12 kΩ. The *p*-type area forms a diode *pn* junction, which splits the resistance into two resistances *R*1 and *R*2 as shown in Figure 10.33(b). If the emitter to base 1 *pn* junction is reverse biased, virtually no current flows through the *pn* junction and the only current is due to voltage across the interbase resistance. If the emitter to base 1 *pn* junction is forward biased, a current flows between the emitter and base 1 of the UJT. Because of the heavy doping of the *p*-type material and the lightly doped *n*-type material, many holes are injected into the *R*1 region. These holes provide very good conduction and make the resistance of *R*1 very small (a few ohms). Thus, a pulse of current flows as soon as the emitter to base 1 is forward biased. The ratio of resistances, $R1/(R1 + R2)$ is known as the *intrinsic standoff ratio* and is usually represented by the Greek letter eta, η. The intrinsic standoff ratio, which is just a voltage divider, is usually 0.5 to 0.8,

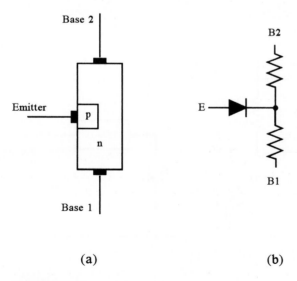

(a)   (b)

FIGURE 10.33
Unijunction transistor representation.

which shows that the *p*-type material is usually closer to base 1. Example 10.15 shows how the most common UJT circuit could be used.

EXAMPLE 10.15

You have a digital output available and this output must turn a high-voltage ac source on. One method of accomplishing this is by using an isolation circuit (or buffer) such as an FET between the digital source and a relay that is set to open when the digital output goes high. This relay can then be used to operate a UJT relaxation oscillator as shown in Figure 10.34. This oscillator produces periodic output pulses, each of which triggers the TRIAC to turn it on. If the pulse rate is much higher than the 60-Hz ac power, the UJT oscillator will continually trigger the TRIAC on to provide full power to the load.

Solution

An explanation of the operation of the relaxation oscillator shown in Figure 10.34 is given:

1. Assume that the relay is closed and then is opened by changing the digital output to a high at $t = 0$.
2. The capacitor starts to charge at $t = 0$ and continues to charge with a time constant of RC until it reaches the power supply value times the intrinsic standoff ratio. For the 2N2646 UJT, η is about 0.6. So at $(0.6)(12) = 7.2$ V, the diode *pn* junction between the emitter and base 1 becomes forward biased and at 7.8 V (0.6 V to turn on the silicon *pn* junction) the unijunction turns on.

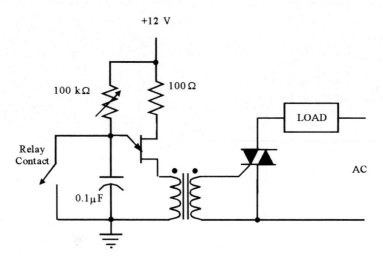

FIGURE 10.34
A unijunction relaxation oscillator circuit.

3. This produces a rapid discharge of the capacitor through base 1 and through the pulse transformer shown. This in turn triggers the TRIAC so it turns on.
4. An example of a time constant calculation follows: If the variable resistor R is adjusted to 10 kΩ, the time constant (RC) is 1 ms. Since the capacitor charges toward 12 V, the equation for the voltage across the capacitor is

$$V = 12[1 - \exp(-t/RC)] \qquad (10.17)$$

Substituting $V = 7.8$ and solving for the time at which the oscillator reaches the 7.8 V, $t = 1.38$ ms. The discharge is very fast so the period of the oscillation is about 1.4 ms. Since the period of a 60-Hz signal is 16.66 ms, the pulse rate is about 10 times the period of the signal and full power will be maintained.
5. Now, if the variable resistor is adjusted to 100 kΩ, the time constant becomes 10 ms, the pulse rate is about 14 ms and the TRIAC will be turned on only once each cycle. This will mean that the device will not get full power and provides a method to adjust the power to the load.

Complementary Unijunction Transistor (CUJT)   A CUJT is a UJT that uses a block of lightly doped p-type semiconductor material with a small area of heavily doped n-type material for the emitter. It operates in exactly the same way as the UJT with opposite polarity voltages.

Programmable Unijunction Transistor (PUT)   A PUT device is a thyristor with a p-n-p-n structure, but it operates much like the UJT after which it is named. The SCR (Section 10.4.2) has a cathode gate that is used to turn it on. The SCS (Section 10.4.2) has a cathode and an anode gate and either one can turn it on. The PUT is the third member of this family (despite its name) because it is structured like the SCR and SCS except that it has an anode gate to turn it on through regenerative action. If the gate of the PUT is maintained at a constant voltage, the device will remain in its off state until the anode voltage exceeds the gate voltage by the diode forward voltage drop. At this point the device turns on. The most important specs for the PUT are the valley point current I_v, which is the anode to cathode current below which the device turns off, and the peak point current, which is the minimum gate current required to turn the device on.

The PUT is used in relaxation oscillator circuits much like the UJT except that the voltage at which the device will turn on is adjustable by controlling the voltage at the gate usually using a voltage divider circuit as shown in Example 10.16.

EXAMPLE 10.16

You need a simple circuit to trigger a 12-V flasher at an adjustable rate from 5 to 50 times per second (so you need a free running oscillator). The circuit of Figure 10.35 is the basic PUT relaxation oscillator and will directly trigger a flasher requiring a trigger current of about 60 mA.

FIGURE 10.35
A PUT relaxation oscillator.

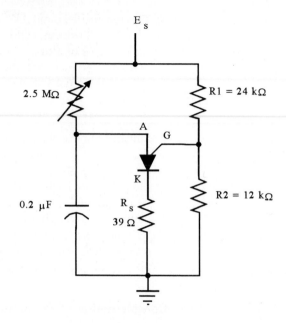

Solution
The operation of the circuit and the component values are discussed here:

1. The $R1$–$R2$ combination provides a voltage divider that maintains the gate at a constant voltage of

$$R2/(R1 + R2) = 0.33(12) = 4 \text{ V} \qquad \textbf{(10.18)}$$

2. The 2N6027 [13] has an $I_p = 5 \ \mu A$ and an $I_v = 70 \ \mu A$. With the variable resistor set at 2.5 MΩ, the time it takes the capacitor to charge to 4 V is determined by the following equation:

$$4 = 12[1 - \exp(-t/RC)] \qquad \textbf{(10.19)}$$

where R is the variable resistor. Solving for the time yields $t = 0.202$ s, which provides a flasher at about 5 times per second. Similarly a 50 times per second flash rate can be produced by adjusting the variable resistor to about 250 kΩ.

3. The R_s is a convenient value (39 Ω here) that will allow the capacitor to discharge rapidly and provide a large enough voltage pulse to trigger the flasher. The discharge time constant in this case is only 7.8 μs, which is negligible.

4. The 24-kΩ resistor can easily supply the 5 μA required to turn the device on.

5. The PUT, because it is like an SCR, will turn off only when the anode current is less than a certain value, in this case the valley current of 70 μA. The maximum current through the anode after the capacitor has charged to a high enough value to trigger the PUT must be less than this or the

PUT will trigger on and not turn off. The current through the anode for the 5- to 50-Hz operation is between 12/2.5 MΩ and 12/250 kΩ or between 4.8 and 48 μA, which is well within the required range (less than 70 μA).

10.5 BIBLIOGRAPHY

1. *D.A.T.A. Book—Transistors,* D.A.T.A., Inc., A Cordura Company, Orange, NJ.
2. *Motorola Small-Signal Transistor Data,* Motorola Inc., 1983.
3. Driscoll, F. F., and R. F. Coughlin, *Solid State Devices and Applications,* Prentice Hall, 1975.
4. Buchsbaum, W. H., *Complete Handbook of Practical Electronic Reference Data,* Prentice Hall, 1978.
5. *Newark Electronics Catalog,* Newark Electronics, Chicago, IL, 1985.
6. *RCA Power MOSFET's,* RCA Corporation, 1984.
7. *1986 Semiconductor Reference Guide,* Radio Shack, A Division of Tandy Corporation, Fort Worth, TX, 1985.
8. Malvino, A. P., *Electronic Principles,* McGraw-Hill, 1984.
9. Young, Paul H., *Electronic Communication Techniques,* Charles E. Merrill, 1985.
10. *Optoelectronics Device Data,* Motorola Inc., 1983.
11. *Power Products Data Book,* Texas Instruments, 1985.
12. Graf, R. F., *The Encyclopedia of Electronic Circuits,* Tab Books, 1985.
13. *SCR Manual,* 6th ed., General Electric.
14. *Power Devices,* RCA Solid State, 1981.

11

OPERATIONAL
AMPLIFIERS*

The operational amplifier or *op amp* has become the most common device for implementation of analog electronic functions. An op amp consists of an electronic circuit constructed of transistors, diodes, and resistors, most often in the form of an integrated circuit (IC). The purpose of this chapter is not to describe how the op amp is designed internally but rather to present practical details of circuit design using op amps. In this regard we consider the op amp to be a *circuit device* with certain given specifications, and we need be concerned with its internal configuration only to the extent that applications are affected.

11.1 INTRODUCTION

The op amp is a very high voltage-gain differential amplifier with high input impedance and low output impedance. The device is not used alone but in combination with other components in a *feedback system* so that many circuit functions can be realized.

11.1.1 Background

The concept of the op amp was developed in the late 1940s and 1950s. The first plug-in op amps were made using vacuum tubes and were used in analog computers. These units required bipolar 300-V power supplies. Smaller units developed using transistors led to more applications of op amps in the 1960s. Finally, development of the IC op amp in 8- or 14-pin dips brought about the widespread applications of its modern day use.

---

* This chapter was written by Dr. Curtis D. Johnson, College of Technology, University of Houston.

FIGURE 11.1

Schematic symbol and terminal identifications for an op amp.

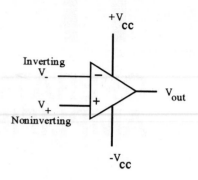

11.1.2 Basic Characteristics

The schematic symbol for the op amp is shown in Figure 11.1. There are two inputs, one labeled *inverting* and marked with a negative sign $(-)$, another labeled *noninverting* and marked with a positive sign $(+)$, and one output. All voltages are with respect to ground. Typically bipolar power supplies $(+V_{cc}$ and $-V_{cc})$ are required as shown but these are often not shown on a schematic.

Figure 11.2 shows the relationship between voltage applied to the inputs and the output voltage. Note that the output depends on the difference or *differential* voltage between the inputs. If the voltage on the inverting input is more positive than that on the noninverting terminal, the output voltage is negative. That is why it is called the inverting input.

The *open-loop gain*, A_V, of the op amp is the slope of the curve in Figure 11.2 as the output ranges between saturation values $(+V_{sat}$ and $-V_{sat})$. This gain is very high, ranging from 30,000 V/V to more than 1,000,000 V/V among various

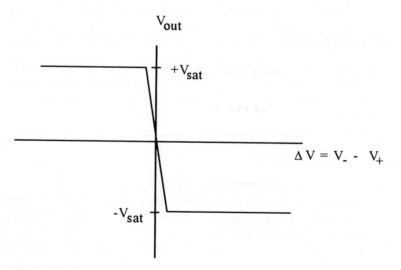

FIGURE 11.2

Op-amp output voltage versus differential input voltage.

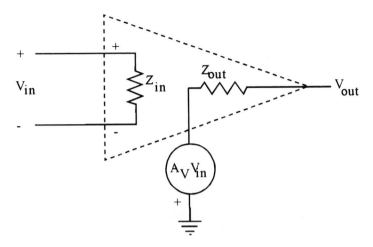

FIGURE 11.3
Simple model of an op amp for circuit analysis.

types of op amps. The *input impedance* of the op amp, between input terminals, ranges from 100,000 Ω to more than 10^{12} Ω among types of op amps, and the *output impedance* is typically in the vicinity of 100 Ω. Input impedance to ground is also very high.

Figure 11.3 shows a model of the op amp, which can be used for detailed circuit analysis. Often we can assume $Z_{in} \rightarrow \infty$ and $Z_{out} \rightarrow 0$ without causing significant errors in the analysis.

Most IC op amps have an output saturation voltage, as shown in Figure 11.2, of about ± 12 V. If we assume a typical voltage gain of, say, $A_V = 200{,}000$, then the differential input voltage between saturation limits varies in the range of ± 60 μV! So, by itself, the op amp is simple: Any input magnitude greater than about ± 100 μV will cause the output to saturate. Again, it is external feedback components that make the op amp useful.

11.1.3 Analysis of Op-Amp Circuits

An analysis of one of the most common op-amp circuits is presented here to provide an illustration of how the analysis is done and to show a useful feedback arrangement. In addition, this analysis provides an example for illustration of other specifications associated with op amps.

The example will be analyzed twice, once using the circuit model of Figure 11.3 and again using a simplified model suitable for most circuits. This analysis is for dc and low-frequency signals.

Figure 11.4 shows an op amp with two resistors. R_1 connects a single-ended, ground-based, input voltage to the inverting terminal. R_2 is a *feedback* resistor connecting the output back to the inverting input. The common connection is called the *summing point*. The noninverting terminal has been grounded. Following

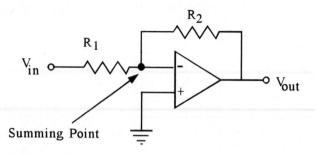

FIGURE 11.4
Basic inverting amplifier configuration using an op amp.

common convention, power supply connections have not been shown since they play no essential role in the analysis.

Op-Amp Model Analysis   The modeled circuit of Figure 11.5 can now be used to determine the output voltage in terms of the input voltage. The summing point has been assigned a voltage, V_s, and currents with arbitrary directions have been assigned. From Kirchhoff's current law (KCL) we can write the following equations:

$$I_1 + I_2 - I_3 = 0$$
$$I_2 + I_4 = 0$$

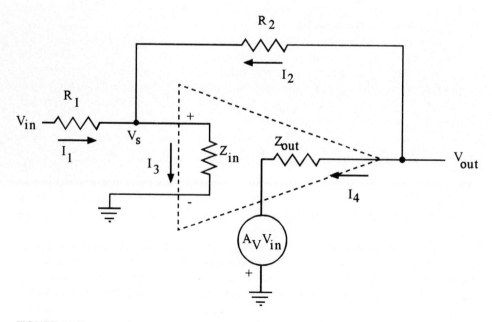

FIGURE 11.5
Using the op-amp model to analyze the inverting amplifier.

From Ohm's law we have the following relations between currents and voltages:

$$I_1 = (V_{in} - V_s)/R_1$$
$$I_2 = (V_{out} - V_s)/R_2$$
$$I_3 = V_s/Z_{in}$$
$$I_4 = (V_{out} + A_V V_s)/Z_{out}$$

This set of equations can be solved for V_{out} in terms of V_{in} and the resistors. After some algebra this relation can be written as

$$V_{out} = -\frac{R_2}{R_1}\frac{1}{1-\gamma}V_{in} \tag{11.1}$$

where the term γ is given by the expression

$$\gamma = \frac{(1 + Z_{out})(1 + R_2/R_1 + R_2/Z_{in})}{(A - Z_{out}/R_2)} \tag{11.2}$$

The circuit of Figure 11.4 is thus an *inverting amplifier* (gain < 0) with the gain given by the coefficient of V_{in} in Eq. (11.1). The amplifier can have gain greater than one, $R_2 > R_1$, or gain less than one (attenuation), $R_2 < R_1$. The gain would simply be $-(R_2/R_1)$ except for the term involving γ. Let us take a typical case of an amplifier with a resistor ratio of 10. Thus, we pick $R_2 = 10$ kΩ and $R_1 = 1$ kΩ. For a typical IC op amp (LM308) the other parameters in Eq. (11.2) are $A_V = 300{,}000$, $Z_{out} = 100$ Ω, and $Z_{in} = 40$ MΩ. Thus, $\gamma = 0.000056$! This means the correction in Eq. (11.1) is $[1/(1 - \gamma)] = 1.000056$ and, for all practical purposes, can be ignored.

This important result is typical of nearly all op-amp circuits: *the op-amp characteristics can be ignored* for circuit analysis, which considerably simplifies analysis.

Simplified Op-Amp Analysis   Analysis of nearly all op-amp circuits can be facilitated by assuming the output impedance is zero and imposition of the following two rules:

Rule 1: The differential voltage across the op-amp inputs is driven to zero, $\Delta V = V_- - V_+ = 0$.

Rule 2: The current through the op-amp inputs (summing point) is zero, $I_s = 0$.

The circuit of Figure 11.4 is now analyzed using the notations of Figure 11.6 and the preceding rules. Currents of arbitrary direction have been assigned to the resistors. Using rule 1 we can write

$$I_1 + I_2 = 0 \tag{11.3}$$

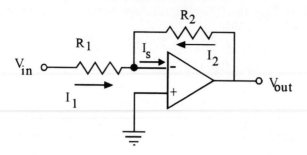

FIGURE 11.6
Current definitions for simplified op-amp circuit analysis.

since $I_s = 0$. From rule 2, since the noninverting terminal is grounded, $V_+ = 0$ and we have

$$V_s = 0$$

Now, from Ohm's law we can write $I_1 = V_{in}/R_1$ and $I_2 = V_{out}/R_2$. From Eq. (11.3),

$$V_{in}/R_1 + V_{out}/R_2 = 0$$

or

$$V_{out} = -(R_2/R_1)V_{in} \qquad\qquad \textbf{(11.4)}$$

which is the same as Eq. (11.1) if we ignore the op-amp characteristics.

Virtual Ground   The concept of virtual ground comes about from rule 2, which in this case stipulated that $V_s = 0$, i.e., a ground. Of course, it is not really a ground but, because of the effect of the feedback, it behaves like a ground.

11.1.4   Static Op-Amp Specifications

The following specifications of op amps relate to analysis and design of op-amp circuits. The inverting amplifier presented in the previous section is used to illustrate how the specification affects op-amp circuit performance. These specifications relate to static, i.e., non-time-dependent, effects.

Power Supply   Most op amps require a bipolar power supply. Typically, IC op amps require a nominal supply of ±15 V, but the devices will typically operate properly over a range, for example, of ±9 to ±18 V. Obviously, the saturation voltage depends on the supply voltage. Some op amps have internal circuitry that allows them to operate from a single power supply.

Open-Loop Gain (A_{OL})   This is the slope of the curve given in Figure 11.2 between saturation. Typical op amps will have an open-loop gain of 100,000 to 1,000,000 V/V. The value is often expressed in terms of V/mV, i.e., 200,000 V/V or 200 V/mV.

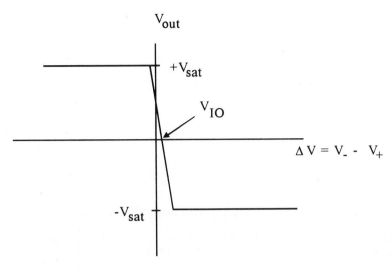

FIGURE 11.7
Nonideal input offset voltage for an op amp.

Input Offset Voltage (V_{IO})   The open-loop response of Figure 11.2 is ideal. For real op amps the curve does not pass exactly through zero. This is shown in Figure 11.7, where you can see that the output is not zero when the differential input voltage is zero. The input offset voltage is that differential input voltage for which the output will be zero. This voltage may be positive or negative and may vary with temperature.

The effect of this offset on the output of an amplifier can be computed by grounding the actual input and treating V_{IO} as an input.

EXAMPLE 11.1

The input offset voltage of an op amp is specified to be in the range of $V_{IO} = \pm 5$ mV. What effect will this have on the output of the inverting amplifier of Figure 11.6 if $R_1 = 10$ kΩ and $R_2 = 120$ kΩ?

Solution
The amplifier gain is given by Eq. (11.4) to be $-(R_2/R_1) = -12$. Thus, if the input is grounded the output should be zero, but because of the input offset voltage it will be in the range of

$$\delta V_{out} = -12(\pm 5 \text{ mV}) = \pm 60 \text{ mV}$$

The output for some input is thus

$$V_{out} = -12V_{in} \pm 60 \text{ mV}$$

Of course, a particular op amp will have a particular V_{IO} so this result tells us only that in general it will lie in this range.

Some op amps provide methods to compensate for input offset, but for others compensation circuitry must be provided. A later section shows input offset voltage compensation techniques.

The values of input offset voltage for the various IC op amps range from ±1 to ±5 mV.

Input Bias Current (I_{IB})   Variation from the assumption of no current into the op-amp input terminals is indicated by the input bias current specification. Figure 11.8 shows an inverting amplifier with the input bias currents, I_{IB-} and I_{IB+}, indicated on the schematic. If the input is grounded, as shown, and the noninverting terminal is grounded, then the only source of I_{IB-} will be through feedback resistor R_2. Assuming the summing point is at 0 V, we then have an output of

$$\delta V_{IB} = I_{IB-} R_2$$

By superposition this will add to any output derived when an actual input is connected:

$$V_{\text{out}} = -(R_2/R_1)V_{\text{in}} + I_{IB-} R_2$$

A similar analysis can be performed for other op-amp circuits to deduce the effect of input bias current. The next section presents a method that partially compensates for this offset.

Typical input bias currents for the various types of IC op amps range from 2 to 500 nA.

Input Offset Current (I_{IO})   The input offset current is the difference between the two input bias currents,

$$I_{IO} = I_{IB-} - I_{IB+} \tag{11.5}$$

Even with compensation techniques for the input bias current, the effect of the input offset current will show up on the output. For the inverting amplifier the effect is given by

$$\delta V_{IO} = \pm I_{IO} R_2$$

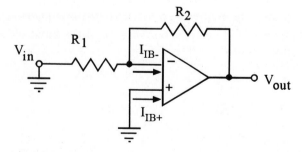

FIGURE 11.8
Input bias currents for an op amp.

Note that the input offset current can be in either direction so the specification is actually for a bipolar, plus-or-minus quantity.

For various IC op-amp types, typical values range from ±200 pA to ±200 nA.

Input Resistance (R_{in})　This is the open-loop resistance between the two inputs of the op amp. For modern IC op amps this value is very high—from 1 MΩ to more than 1000 MΩ. The input resistance of the *circuit* using the op amp can be very different from this value. For example, the effective input resistance of the inverting amplifier of Figure 11.6 is the input resistance, R_1, which may not be very large (about a few kilo-ohms).

Output Resistance　The output resistance of the op amp is typically less than 100 Ω. This is not usually an important specification for op amps because the effective output resistance of the *circuit* using an op amp is more dependent on the circuit than the op amp. Usually the circuit output resistance is smaller than the op amp by a factor of the circuit gain.

Output Swing　The plus-or-minus saturation voltage of the op amp depends on the magnitude of the bipolar power supply used and the load resistance. For the nominal IC op amp supply voltage of ±15 V, the output swing is typically ±13 to ±14 V.

Common-Mode Rejection Ratio (CMRR)　Figures 11.2 and 11.7 show that the output of the op amp is supposed to depend only on the differential input voltage, $\delta V = V_- - V_+$. In reality, the output will depend to a small degree on the nominal or common voltage present at the two input terminals. The common-mode voltage is often taken to be the average, $V_{CM} = (V_- + V_+)/2$. The alternative is to connect the two inputs and impose a common voltage on both.

This dependence is indicated by a specification of the ratio of the open-loop differential gain (A_{OL}) to the common-mode gain (A_{CM}):

$$CMRR = A_{OL}/A_{CM} \tag{11.6}$$

In the ideal situation $A_{OL} \to \infty$ and $V_{CM} \to 0$, so that CMRR $\to \infty$. Thus, a large CMRR is an indication of high quality, i.e., good common-mode rejection.

The common-mode rejection ratio is usually specified by expressing the CMRR in decibels rather than the simple ratio of Eq. (11.6). This is often called the common-mode rejection (CMR) although CMRR is also used. In any event the expression is given by

$$CMR = 20 \log(CMRR) \tag{11.7}$$

Of course, the *circuit* using an op amp will have its own CMRR, which will depend on the op-amp CMRR and other circuit components.

Typical values for the CMRR (or CMR) range from 40 to 110 dB.

Power Supply Rejection Ratio (PSRR)   The output of the op amp should be independent of the power supply magnitude (except for the saturation voltage). This is indicated in the specifications as the ratio of the change of output voltage caused by a change of supply voltage:

$$\text{PSRR} = 20 \log(\delta V_{\text{out}}/\delta V_{\text{supply}}) \qquad (11.8)$$

This quantity is typically in the range of 70 to 110 dB. If the supply is well regulated, the vast majority of op-amp circuits will not be significantly affected by power supply variation.

Temperature Drifts   Variation of operating temperature will affect the characteristics of the op amp. In most cases the concern is with variations of input offset voltage and current with temperature.

Variations of input offset voltage with temperature are expressed through the *input offset voltage temperature coefficient,* $\alpha(V_{IO})$, giving the volts per degree Celsius (V/°C). The value is typically in the range of 1 to 100 μV/°C. This quantity is very important since input offset voltage compensation is provided for only one specific temperature.

EXAMPLE 11.2

An LM308 has $\alpha(V_{IO}) = 6$ μV/°C. The op amp is compensated so that the input offset voltage is zero at 20°C and used in an inverting amplifier with a gain of -100. It is now operated at 40°C. What is the output due to this effect?

Solution
The input offset voltage changes by

$$\delta V_{IO} = (6 \ \mu\text{V/°C})(40°\text{C} - 20°\text{C}) = 120 \ \mu\text{V}$$

So, since the amplifier gain is -100 the output offset voltage is

$$\delta V_{\text{out}} = (-100)(120 \ \mu\text{V}) = 12 \ \text{mV}$$

Variation of input offset current with temperature is specified by the *input offset current temperature coefficient,* $\alpha(I_{IO})$. Since input offset current is not compensated, variations with temperature have a direct, but small, impact on the output. Typical values range from 0.5 to 10 pA/°C.

11.1.5   Dynamic Op-Amp Specifications

The previous specifications relate to conditions where variations in time do not occur. The characteristics of op-amp response to time variations in the input signals are very important.

Slew Rate   The slew rate specifies the maximum rate at which the output voltage of the op amp can change. If the input to the op amp is a very fast ($\rightarrow$

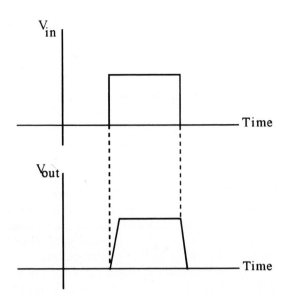

FIGURE 11.9
Effect of finite slew rate on op-amp output.

instantaneous) rising or falling signal, then the output will not be able to respond any faster than the slew rate. Figure 11.9 shows the input to an op amp and the resulting output, as limited by the slew rate.

Slew rate is specified as the number of volts per second, or, more commonly, volts per microsecond (V/μs), by which the output can change when the input has a step change. Typical values range from 0.5 V/μs to as fast as 100 V/μs.

The frequency and output amplitude of a signal are related by the slew rate in the following equation:

$$S = 2\pi f V_{out} \tag{11.9}$$

where S = slew rate (V/s)
f = frequency (Hz)
V_{out} = output amplitude (V).

EXAMPLE 11.3   An op amp with a 1.5 V/μs slew rate is to be used to amplify a signal at 80 kHz. What is the maximum output voltage allowed by the slew rate?

Solution
From Eq. (11.9) we have $V_{out} = S/2\pi f$, or,

$$V_{out} = (1.5 \text{ V}/10^{-6}\text{ s})/(2\pi \times 8 \times 10^4\text{ s}^{-1})$$
$$= 2.98 \approx 3 \text{ V}$$

This means that even though the saturation is ±13 V and even though 80 kHz may be well within the op-amp frequency response (see the next section), the output cannot rise above about ±3 V because of slew rate limitations.

Frequency Response   The frequency response specification of the op amp relates to performance when the input voltage varies periodically in time. The open-loop gain, A_{OL}, of the op amp will depend on the frequency of the input signal. A graph of the open-loop gain (expressed in decibels) versus frequency, on a log scale, is used to present this specification. This is called a *Bode plot.* In general, the phase shift versus frequency is also considered part of the Bode plot.

An op amp has two types of frequency response characteristics, compensated and uncompensated. It is possible for an uncompensated op amp to be unstable at some circuit gains, i.e., to break into spontaneous oscillations. External circuitry is then added to make the circuit stable. The early op amps were all uncompensated and many are still manufactured without built-in compensation.

Figure 11.10 shows a typical frequency response curve for a modern, compensated IC op amp. Note that at low frequency the gain is constant (flat). At some critical frequency (f_c) the gain begins to decrease (breaks) and then *rolls off* at about 20 dB for every tenfold increase in frequency (decade). This is the same as 6 dB for every doubling in frequency (octave). The curve crosses zero dB (unity gain, $A_{OL} = 1$) at another frequency, f_1, called the *unity-gain frequency.*

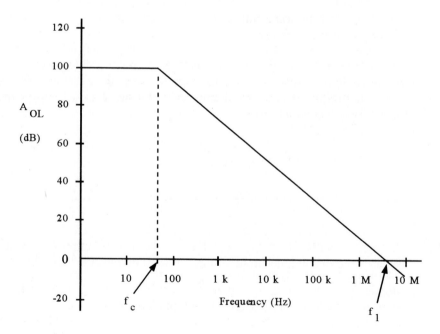

FIGURE 11.10
Typical open-loop frequency response of a compensated op amp.

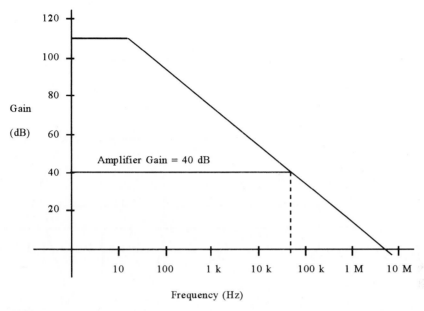

FIGURE 11.11
Bandwidth of an op-amp amplifier from the open-loop response.

Note that the Bode plot, such as the one in Figure 11.10, is really an approximation, and the break at f_c is really a smooth transition from a flat response to -20 dB/decade response.

The frequency response of a circuit using an op amp is limited by the frequency response of the op amp itself as well as the circuit components employed. For a simple amplifier such as the inverting amplifier of Figure 11.4, the frequency response is found by noting where the amplifier gain intersects the op-amp open-loop gain curve. This is shown in Figure 11.11 for an inverting amplifier with a gain of 100 (40 dB). Note that the amplifier gain breaks at about 45 kHz and then falls off at -20 dB/decade. Thus, the amplifier gain is flat (constant) to 45 kHz.

Uncompensated Frequency Response   Figure 11.12 shows the Bode plot for an uncompensated op amp. Note that the roll-off of gain versus frequency has two other breaks, -40 dB/decade starting at f_{c1} and -60 dB/decade starting at f_{c2}. If this op amp is used to make an amplifier with gain such that the gain curve intersects the op-amp curve after f_{c2}, the circuit will be unstable and oscillate. This is labeled as A_1 in Figure 11.12.

If the op amp is used to make an amplifier with a gain that will intersect the op-amp curve between f_{c1} and f_{c2} it will be marginally unstable, i.e., it may or may not break into oscillation, depending on other circumstances. This is shown as A_2 in Figure 11.12. If used for an amplifier with a gain to intersect before f_{c1}, the circuit will be stable, as A_3.

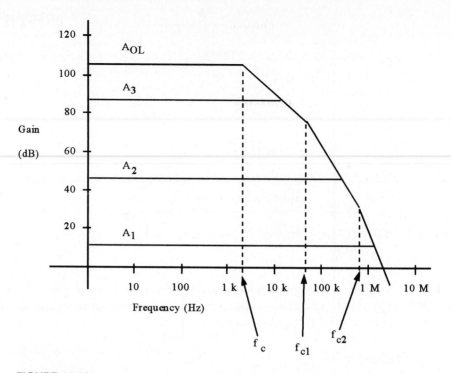

FIGURE 11.12
Open-loop frequency response of an uncompensated op amp.

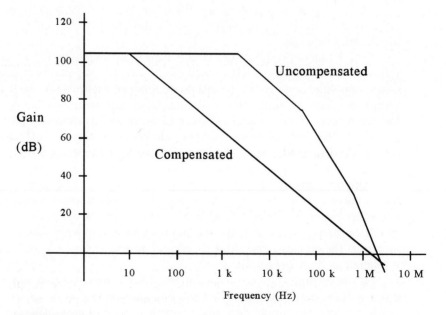

FIGURE 11.13
Compensation effectively reduces the bandwidth.

For the unstable cases it will be necessary to provide external compensation, in a way specified by the manufacturer. This will have the effect of reducing the bandwidth, as shown in Figure 11.13. This curve shows how compensation reduces the roll-off to −20 dB/decade.

11.2 PRACTICAL CONSIDERATIONS OF OP-AMP CIRCUITS

The following issues are presented as practical considerations involved in the design and operation of op-amp circuits.

11.2.1 Selection of Components

It is important to realize that an IC op amp can swing an output of only about ±14 V and can source or sink a current at an output of less than 20 mA. This means that the feedback current and load current cannot exceed ±20 mA, and for some op amps much less than that. As a consequence when using op amps we always *think milliamps and think kilo-ohms*.

EXAMPLE 11.4

Design an inverting amplifier to amplify a ±0.1-V signal to ±8 V.

Solution
This is very easy—we use the circuit of Figure 11.4 with $R_2/R_1 = 8/0.1 = 80$. So, suppose we pick $R_1 = 1\ \Omega$, then $R_2 = 80\ \Omega$. This is correct in principle. But wait! This will mean that when the input is, say, −0.1 V, the output will be 8 V and the feedback current will be $I_2 = 8\ \text{V}/80\ \Omega = 100$ mA! The op amp cannot do this and so the circuit response will not be as predicted. Thinking in milliamps and kilo-ohms, we select $R_1 = 1\ \text{k}\Omega$ and then $R_2 = 80\ \text{k}\Omega$, or any combination giving a ratio of 80 and in the kilo-ohm range. Now the maximum current will be 8 V/80 kΩ = 0.1 mA.

If more load drive current is required, circuit boosters that use transistors in conjunction with op amps can be employed.

11.2.2 Input Offset Compensation

Many op amps, such as the LM741, provide special pins and instructions for providing input offset compensation. As an example, Figure 11.14 shows an inverting amplifier using an LM741, with the input offset compensation provided as a variable resistance between specified op-amp pins and the −15-V power supply. The input of the *circuit* (not the op amp), V_{in}, is grounded and the pot is adjusted until the output is zero. The pot position can be glued or *staked* in that fixed setting now, and it will be valid unless a *different* LM741 is plugged into the circuit. The setting will remain valid for that particular op amp in that particular circuit except for some small variation due to temperature.

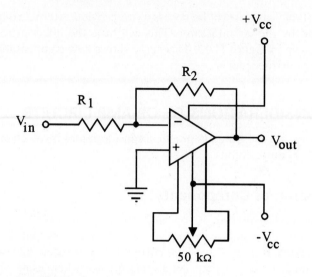

FIGURE 11.14
Input offset compensation for the 741 op amp.

This compensates for whatever effect, voltage or current, is causing the offset from zero with zero input. Now a signal can be connected to the input of the amplifier and the output will have a zero or minimized offset error.

If the op amp does not have provision for input offset compensation it can be provided by appropriate external circuitry. Figure 11.15 shows a typical system

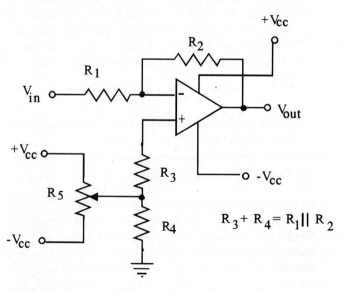

FIGURE 11.15
Input offset compensation using external circuitry.

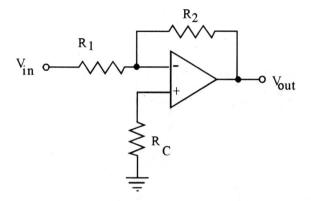

FIGURE 11.16
Input bias current compensation with an external resistor.

for external input offset compensation. In essence, these circuits are dividers that provide a small input voltage to cancel the effect of the input offset voltage and/ or current of the op amp.

11.2.3 Input Bias and Offset Current Compensation

The compensation facilities presented in the last section will zero out the effect of input bias and offset current also. The compensation system simply adjusts the output to zero with the amplifier circuit input grounded, regardless of the source of that offset.

  When a compensation adjustment circuit is not used, the effects of input bias current can be reduced by placing a resistor in series with the noninverting terminal that has the same value as that seen by the inverting terminal. In the case of the inverting amplifier of Figure 11.4, this means that a resistor of value

$$R_C = R_1 R_2 / (R_1 + R_2)$$

is placed from the noninverting terminal to ground. This is shown in Figure 11.16. This has the advantage of limiting the effects of temperature variation on the input offset current.

11.3 PRACTICAL OP-AMP CIRCUITS

Many manuals and books provide a great deal of information about a large number of op-amp circuits. The following is only a partial collection of types of op-amp circuits with examples to demonstrate the basic analog circuit functions provided by the op amp. It should be noted that power supply connections and offset compensation are not shown.

The input and output resistance of the op amp itself are given by Z_{in} and Z_{out} and the open-loop gain by A_{OL}.

11.3.1 Amplifier Circuits

The following circuits will provide either amplification (gain > 1) or attenuation (gain < 1) of an input signal. The gain, G, is simply the ratio of the output voltage to the input voltage:

$$G = V_{out}/V_{in} \qquad\qquad (11.10)$$

For each circuit the following parameters are given:

1. Gain
2. Circuit input resistance, R_{in}
3. Circuit output resistance, R_{out}
4. Special notes about the circuit.

Inverting Amplifier   This very common circuit is shown in Figure 11.4 and has been used as the sample circuit in the previous sections. In summary it has the following characteristics:

Gain: amplify or attenuate, $G = -(R_2/R_1)$

Input resistance: $R_{in} = R_1$ (so, not necessarily high—watch out for loading)

Output resistance: $R_{out} \approx (1 + R_2/R_1)Z_{out}/A_{OL}$ (typically much lower than Z_{out} itself).

EXAMPLE 11.5   An attenuator is required to reduce a signal by +0.4. How can this be provided by op amps?

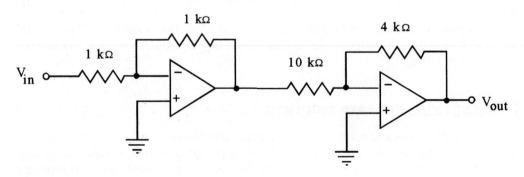

FIGURE 11.17
Amplifier system with an attenuation of 0.4.

Solution

We can use two inverting amplifiers, one with a gain of -1 and the other with a gain of -0.4, so the net effect is the required $+0.4$ attenuation. Figure 11.17 shows the schematic. Note the selection of resistors in the kilo-ohms.

This next example illustrates the important effect of *loading* in the use of an amplifier.

EXAMPLE 11.6   Suppose a 500-Hz source with a 5-V amplitude and a 1 kΩ output resistance is connected to the amplifier of Figure 11.17. What is the output voltage?

Solution

At first view one would simply say that since the amplifier has a gain of $+0.4$, the output should be

$$V_{out} = 0.4V_{in} = 0.4(5\ \text{V}) = 2\ \text{V}$$

But this is *not* the correct answer! Figure 11.18 shows the entire circuit with the modeled source and amplifier. Since the input resistance of the amplifier is only 1000 Ω the *actual* input voltage to the amplifier is loaded (divided) and is only

$$V_{in} = (5)(1000)/(1000 + 1000) = 2.5\ \text{V}$$

Therefore, the output voltage amplitude is only

$$V_{out} = 0.4(2.5\ \text{V}) = 1\ \text{V!}$$

Another way to look at this is to see, from Figure 11.18, that the *effective* inverting amplifier input resistance is now 2000 Ω so that the gain of that stage is -0.5 instead of -1. Now, when combined with the second-stage gain of -0.4, the net gain is 0.2.

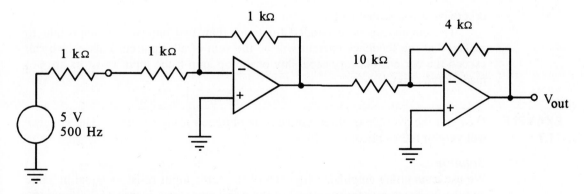

FIGURE 11.18
Connection of a source with a 1-kΩ output resistance to the amplifier.

Summing Amplifier A summing amplifier is simply an inverting amplifier with more than one input. The schematic is shown in Figure 11.19 for a case of two inputs. The characteristics are

Gain: $G = -(R_3/R_1)V_1 - (R_3/R_2)V_2$

Input resistance: $R_{in1} \approx R_1$, $R_{in2} \approx R_2$

Output resistance: $R_{out} \approx (1 + R_3/R_p)Z_{out}/A_{OL}$ where $R_p = R_1$ and R_2 in parallel.

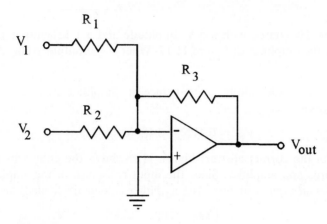

FIGURE 11.19
The inverting summing amplifier.

Note that the input resistance is not necessarily high, so loading may occur if the source resistance is not very low.

If all the resistors are equal we have a simple inverting adder:

$$V_{out} = -(V_1 + V_2)$$

otherwise it is a scaled adder.

The circuit can be expanded for more than two inputs. It is important to realize that the feedback current will be the sum of all input currents, which can exceed the current delivery capability of the op amp if the input resistors are not made large enough.

EXAMPLE 11.7

How can the average of three voltages be obtained using op amps? The voltages will vary in the range of 0 to 10 V.

Solution

We use a summing amplifier with each of the three input resistors equal in value and the feedback resistor one-third of the value employed. Figure 11.20 shows the solution. If all inputs were at 10 V, then our design has kept the resulting (maximum) feedback current at (conservatively) 10 mA. This means that the input resistors must be >3 kΩ.

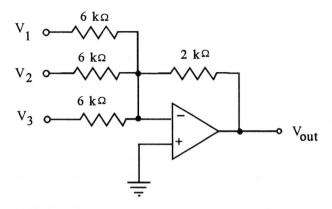

FIGURE 11.20
A circuit that takes the average of three voltages.

Noninverting Amplifier The circuit of Figure 11.21 does not change the sign of the input signal, but it *cannot* act as an attenuator, i.e., have a gain of less than one. It has a very high input resistance. The characteristics are

Gain: $G = 1 + R_2/R_1$

Input resistance: $R_{in} \approx Z_{in}$ (very high)

Output resistance: $R_{out} \approx GZ_{out}/A_{OL}$ (very low).

The very high input resistance of this amplifier means that it will not load the source, as long as its resistance is not also very large.

Voltage Follower This is a special case of the noninverting amplifier. Figure 11.22 shows that the feedback resistance is zero. This circuit is used for isolation of a source.

Gain: $G \approx 1$ (typically slightly less than one, ≈ 0.99)

Input resistance: $R_{in} \approx Z_{in}$

Output resistance: $R_{out} \approx Z_{out}/A_{OL}$

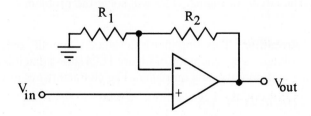

FIGURE 11.21
The noninverting amplifier.

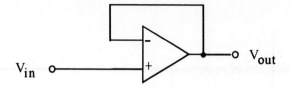

FIGURE 11.22
A voltage follower or unity-gain amplifier.

This circuit is also called an *impedance transformer* or an *isolation amplifier* since it makes a source voltage available at a very small output resistance.

EXAMPLE 11.8

A transformation is needed so that an input variation of 0.5 to 4.4 V becomes a variation of 0 to 10 V. The input source resistance is >100 kΩ.

Solution
This can be done using op amps by first finding an equation for the output voltage in terms of the input voltage and then putting op-amp circuits together to implement that equation. In this case, since the transformation is linear, we have

$$V_{out} = mV_{in} + V_0$$

where m is the slope and V_0 is the intercept. These constants are found by using the two known conditions,

$$10 = 4.4m + V_0$$
$$0 = 0.5m + V_0$$

Subtracting gives $10 = (4.4 - 0.5)m$, or $m = 2.564$. Then, from the second equation, $V_0 = -0.5m = -1.282$. Thus, we need circuits that give

$$V_{out} = 2.564V_{in} - 1.282$$

There are many ways this equation could be implemented. One way is to use an inverting amplifier with a gain of -2.564 followed by a unity-gain summing amplifier with the second input of 1.282 V. To prevent loading of the source, a voltage follower is used. The circuit of Figure 11.23 will solve the problem.

Differential Amplifier   This amplifier is designed to amplify only the difference between two voltages. The schematic of Figure 11.24 shows that it is a combination of an inverting and noninverting amplifier. The characteristics are

Gain: $V_{out} = (R_2/R_1)(V_2 - V_1)$
Input resistance: $R_{1in} \approx R_1 + R_2/2$, $R_{2in} \approx R_1 + R_2$
Output resistance: $R_{out} \approx (1 + R_2/R_1)Z_{out}/A_{OL}$

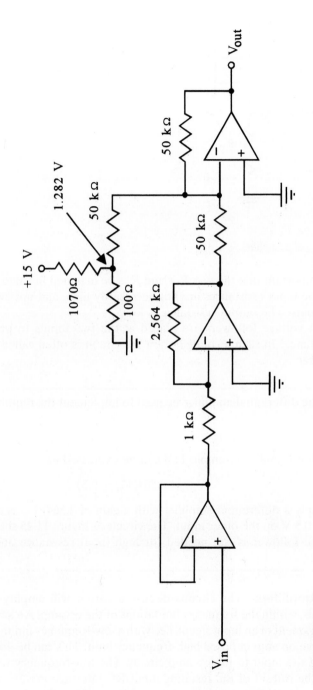

FIGURE 11.23
Solution to Example 11.8.

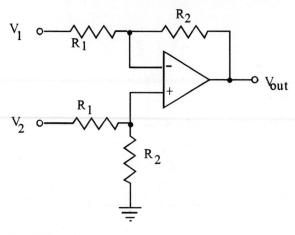

FIGURE 11.24
The differential amplifier.

It is very important that the two R_1's and R_2's be matched in value to provide good CMRR. The input resistances are not necessarily high, and not even the same for the two inputs. The output resistance is low.

Often voltage followers are placed at the two inputs to provide very high input resistance. In such a configuration the circuit is often called an *instrumentation amplifier*.

EXAMPLE 11.9
How can the differential amplifier be used to implement the requirement of Example 11.8?

Solution
The equation found in Example 11.8 can be expressed as

$$V_{out} = 2.564(V_{in} - 0.5)$$

This is simply a differential amplifier with a gain of 2.564, V_{in} as one input, and a constant +0.5 V on the other input. The circuit of Figure 11.25 shows the solution. The voltage follower is still needed for high input resistance and no loading of the source.

AC Amplifiers The circuits discussed earlier will amplify both ac and dc input signals, within the frequency limitations of the op amp. An ac amplifier blocks the dc component of an input signal, i.e., with a low-frequency limitation on response as well as the op amp imposed high-frequency limit. This can be done by capacitive coupling of the input to the op-amp circuit. The low-frequency response is determined by the roll-off of the resulting input RC filter action.

Figure 11.26 shows a noninverting AC amplifier. The low-frequency cutoff (down 3 dB) is given by $f_{c\,low} = 1/(2\pi R_3 C)$. Although the gain is still $(1 + R_2/R_1)$,

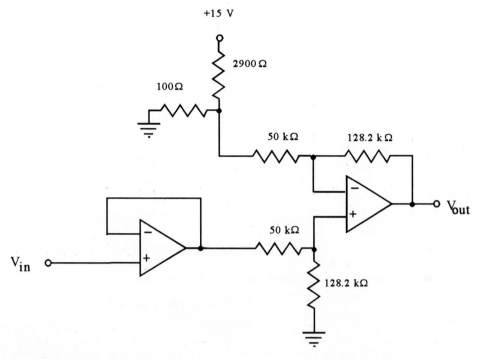

FIGURE 11.25
Another solution to Example 11.8.

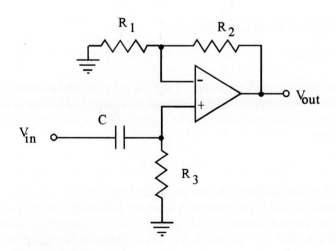

FIGURE 11.26
A simple ac amplifier.

the voltage amplified is that across R_3 so it may be reduced by the dividing action of C and R_3 for frequencies near the cutoff. The input impedance is no longer the input resistance of the op amp, Z_{in}, but instead that of C and R_3 in series.

EXAMPLE 11.10

An ac amplifier is needed with a gain of 100 and a flat response from 100 to 5000 Hz. How can an op amp provide this, and what is the high-frequency cutoff (the op-amp frequency response is given by Figure 11.10)?

Solution
Using Figure 11.26 for a guide, we first select the gain resistors:

$$1 + R_2/R_1 = 100$$

so we have $R_1 = 10 \text{ k}\Omega$ and then $R_2 = 99 \text{ k}\Omega$. Suppose we simply pick $C = 10 \text{ }\mu\text{F}$; then

$$R_3 = (2\pi \times 100 \times 10 \text{ }\mu\text{F})^{-1} = 1591 \text{ }\Omega$$

Note that the ac input resistance varies from

$$X_{in} = [(1591^2 + (2\pi \times 100 \times 10 \text{ }\mu\text{F})^{-2}]^{1/2}$$
$$= 2250 \text{ }\Omega$$

at 100 Hz to 1591 Ω at 5000 Hz. It is thus not high and so a voltage follower may be necessary on the input to prevent loading of the source. The high-frequency cutoff is found by drawing the gain 100 (40-dB) line on Figure 11.10 and noting that it intersects the op-amp open-loop frequency response at $\approx$30 kHz. Of course, many different capacitor/resistor combinations would work also.

11.3.2 Current Related Circuits

The following circuits are related to the current response of op amps.

Current-to-Voltage Converter Linearly proportional conversion of current to voltage can be accomplished by the circuit of Figure 11.27. This is an inverting, bipolar current-to-voltage converter. The relation between output voltage and current is given by

$$V_{out} = -IR$$

The circuit is bipolar, so that if the current is directed as shown (sinking) the output voltage is negative, and if the current is reversed (sourcing) the output is positive.

Maximum current is determined by output saturation and the value of R. The op-amp feedback current must match the input current so the input cannot exceed $\approx$20 mA. The resistor on the noninverting terminal is provided for input bias current compensation.

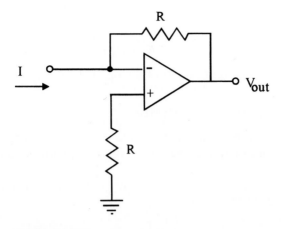

FIGURE 11.27
A current-to-voltage converter.

Voltage-to-Current Converter　A voltage-to-current converter can be made in many ways, depending on if the current is to be sourced, sinked, ground based, or floating. Figure 11.28 shows one circuit for a ground-based output current. The characteristics are

Output: $I = -(R_2/R_1R_3)V_{in}$ *provided* $R_2R_4 = R_1(R_3 + R_5)$

Minimum load resistance: $0\ \Omega$

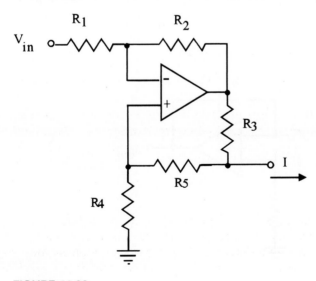

FIGURE 11.28
A voltage-to-current converter.

Maximum load resistance:

$$R_{Lmax} \approx \frac{(R_4 + R_5)(V_{sat}/I_{max} - R_3)}{R_3 + R_4 + R_5}$$

where I_{max} is the maximum current output of the op amp.

11.3.3 Integrator and Differentiator

It is possible to make op-amp circuits that provide the integral or derivative of an input voltage. In each case there are restrictions on their performance.

Integrator   The circuit of Figure 11.29 is an electronic inverting integrator. The output voltage is given by

$$V_{out} = -\frac{1}{RC}\int V_{in}\, dt$$

The input resistance is R so that loading may occur if this is not a large value. This basic circuit provides no way to reset by discharging the feedback capacitor. Even if used only for ac signals the output may gradually saturate due to dc offset voltage on the output.

A resistor is sometimes placed across the feedback capacitor to prevent offset voltages from saturating the amplifier output. This also limits the lowest frequency that can be integrated.

Differentiator   Figure 11.30 shows a differentiator circuit. Within the limited frequency range of differentiation, the output of this circuit is given by

$$V_{out} = -RC\frac{dV_{in}}{dt}$$

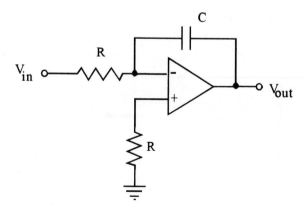

FIGURE 11.29
An op-amp integrator.

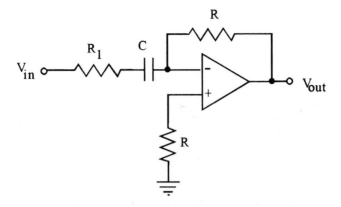

FIGURE 11.30
A compensated op-amp differentiator.

In principle, the input resistor, R_1, would be zero, i.e., it would not be present. In this case the circuit may be unstable and break into oscillation. Including R_1 makes the circuit stable by halting differentiation at some maximum frequency given by

$$f_{max} = (2\pi R_1 C)^{-1}$$

Active Filters   Many types of active filters can be constructed using op amps. Please see the chapter on filters in this handbook for examples.

12

SPECIAL FUNCTION INTEGRATED CIRCUITS*

12.1 INTRODUCTION

This chapter discusses various linear or analog integrated circuits (ICs) with specialized functions but still of sufficient importance they are found in a variety of circuits and systems. These ICs include comparators, timers, phase-locked loops, V/F and F/V converters, oscillators, and function generators. We first describe these systems using block diagram descriptions in terms of their basic operation. Several typical applications are then discussed and some of the various types available from different manufacturers are listed. The applications given and the available types listed are presented as a representative sample. Several specialized ICs are also identified.

12.2 COMPARATORS

12.2.1 Basic Operation

A comparator is basically a modified operational amplifier designed to operate open loop or with positive feedback in contrast to the normal operational amplifier, which is designed to operate with negative feedback. A block diagram of a typical comparator is shown in Figure 12.1. Another difference between the comparator and the operational amplifier is that the output stage in most comparators is an open collector device, whereas in most operational amplifiers it is a push–pull type of circuit. This allows the user to select the range of output voltage swing to fit the application. The schematic symbol for a comparator and its associated transfer function are shown in Figure 12.2. Because the comparator is a very high gain

---

* This chapter was written by Dr. Warren Hill, Weber State College.

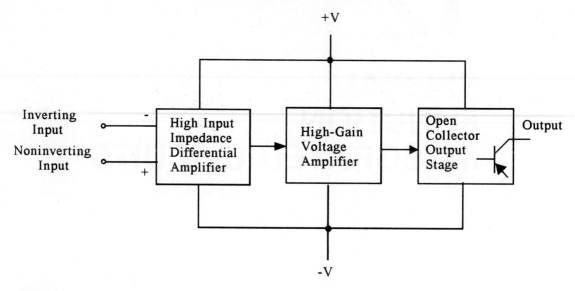

FIGURE 12.1
Block diagram of the comparator.

device, the slope of the portion of the curve passing through the origin is very steep, typically 200,000 V/V. By definition, a comparator compares the voltage at one input to the voltage at the other input, and if the noninverting input is at a higher potential than the inverting input, the output is high. If the noninverting input is at a lower potential than the inverting input, the output is low. The high and low output voltage levels can normally be set by the user.

Because the comparator typically has an operational amplifier type of front end, many of the same parameters used to specify operational amplifiers are used for comparators. These include such factors as input bias currents, input offset current, input offset voltage, common-mode rejection ratio, common-mode input voltage range, and maximum input voltages. These terms are defined in Chapter 11. A parameter of particular interest in a comparator is its slew rate or response time. The slew rate is the maximum rate of change of the output voltage with respect to time. Alternatively, the response time is the time required for the output to go from 10 to 90% of its final value in response to a step input. Response time is usually measured with a 100-mV step input plus a certain amount of overdrive, typically 5 mV. In other words, the actual input step is 105 mV. Under these conditions, the typical response time for a 311 comparator is 200 ns. Alternatively, the rise time for the 311 under the same conditions is 0.08 μs.

12.2.2   Typical Applications

When a comparator input is a fast changing signal from a low impedance source, the output of most comparators will change quickly and remain stable. However,

FIGURE 12.2

(a) Comparator symbol and (b) transfer function.

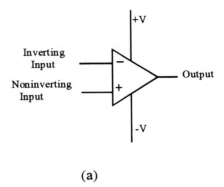

(a)

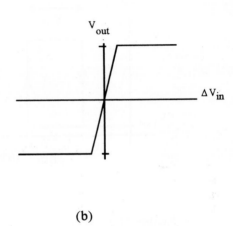

(b)

if the input changes slowly or comes from a high impedance source, there is a good possibility of oscillation about the switching point because of the comparator's wide bandwidth and high gain. The switching point is the input voltage point at which the output changes state. Thus, circuit layout and component placement are critical. In many cases, positive feedback is used to provide hysteresis and prevent oscillation. Figures 12.3 and 12.4 show the use of positive feedback for this purpose. For Figure 12.3(a), which uses positive level sensing,

$$V_{\text{UTP}} = V_{\text{REF}}\left(\frac{R_i + R_f}{R_f}\right) - \frac{R_i}{R_f}(V_{\text{Lsat}}) \tag{12.1}$$

$$V_{\text{LTP}} = V_{\text{REF}}\left(\frac{R_i + R_f}{R_f}\right) - \frac{R_i}{R_f}(V_{\text{Hsat}}) \tag{12.2}$$

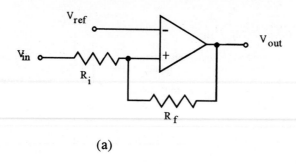

(a)

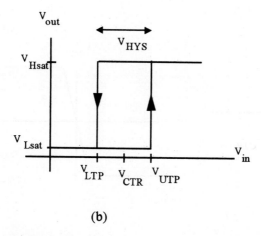

(b)

FIGURE 12.3
Positive level sensing comparator with hysteresis: (a) Schematic diagram and (b) output waveform.

For Figure 12.4(b), which uses negative level sensing,

$$V_{\text{UTP}} = \left(\frac{1}{R_i + R_f}\right)(R_f V_{\text{REF}} + R_i V_{\text{Hsat}}) \tag{12.3}$$

$$V_{\text{LTP}} = \left(\frac{1}{R_i + R_f}\right)(R_f V_{\text{REF}} + R_i V_{\text{Lsat}}) \tag{12.4}$$

where V_{UTP} = upper trip point (V)
V_{LTP} = lower trip point (V)
R_f = feedback resistor (Ω)
R_i = input resistor (Ω)
V_{REF} = reference voltage (V)

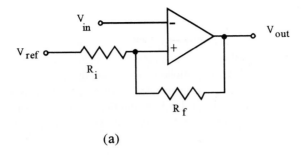

(a)

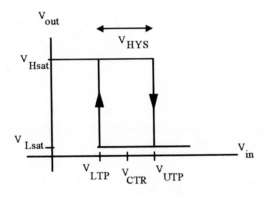

(b)

FIGURE 12.4
Negative level sensing comparator with hysteresis: (a) Schematic diagram and (b) output waveform.

$$V_{Lsat} = \text{low level output voltage (V)}$$
$$V_{Hsat} = \text{high level output voltage (V)}.$$

If the input voltage comes from a source with a large impedance compared to the input resistor in Figure 12.3(a), the source resistance can be included with the input resistor. If, in Figure 12.4(b), the reference voltage comes from a voltage divider, which will often be the case, the divider can be Thévenized and the Thévenin resistance included with the input resistance. Any pull-up resistor on the output can also be included with the feedback resistor if necessary.

EXAMPLE 12.1   Given a 311 comparator that has been connected to provide an output swing from 0.4 to 5 V, design a circuit that will cause the output to switch low when the input voltage exceeds 8.1 V and switch high when the input voltage drops below 7.9 V.

Solution

The problem description fits the circuit of Figure 12.4(b), negative level sensing, i.e., the output goes high for a low going input. Since we have three unknowns, the reference voltage value and the two resistance values, and two equations, one of the resistor values can be arbitrarily chosen. Note that in this and other examples where one of the element values can be arbitrary, that arbitrariness is still subject to current, size, power dissipation, and other practical considerations. Combining Eqs. (12.3) and (12.4) eliminates the reference voltage and allows us to solve for the ratio of R_i to R_f as

$$V_{\text{UTP}} - V_{\text{LTP}} = 8.1 - 7.9 = 0.2 \text{ V}$$

or

$$0.2 = \frac{R_i}{R_i + R_f}(V_{\text{Hsat}} - V_{\text{Lsat}}) = \frac{R_i}{R_i + R_f}(5 - 0.4)$$

Thus,

$$\frac{R_f}{R_i} = 22$$

Selecting R_i to be 10 kΩ, we can then substitute back into either Eq. (12.3) or (12.4) to solve for V_{REF}. Thus, $R_f = 220$ kΩ and

$$V_{\text{UTP}} = 8.1 = \frac{1}{R_i + R_f}(R_f V_{\text{REF}} + R_i V_{\text{Hsat}})$$

so

$$8.1 = \frac{220}{220 + 10}V_{\text{REF}} + \frac{10}{220 + 10} \quad (5)$$

Thus,

$$V_{\text{REF}} = \frac{23}{22}\left(8.1 - \frac{5}{23}\right) = 8.24 \text{ V}$$

Let us define the center voltage of a comparator with hysteresis as the average voltage between the upper trip point voltage and the lower trip point voltage. This can be expressed as

$$V_{\text{CTR}} = \frac{V_{\text{UTP}} + V_{\text{LTP}}}{2} \qquad \textbf{(12.5)}$$

Thus, for Figure 12.3(a)

$$V_{\text{CTR}} = V_{\text{REF}}\left(\frac{R_i + R_f}{R_f}\right) \qquad \textbf{(12.6)}$$

and for Figure 12.4(a)

$$V_{CTR} = V_{REF} \left(\frac{R_f}{R_i + R_f} \right)$$ (12.7)

where V_{CTR} is the average of the two trip point voltages. (These equations are valid only when $V_{Lsat} = -V_{Hsat}$.)

The difference between the upper and lower trip points is the hysteresis voltage, expressed as

$$V_H = V_{UTP} - V_{LTP}$$ (12.8)

For Figure 12.3(a) this becomes

$$V_H = \frac{R_i}{R_f} (V_{Hsat} - V_{Lsat})$$ (12.9)

and for Figure 12.4(a)

$$V_H = \frac{R_i}{R_i + R_f} (V_{Hsat} - V_{Lsat})$$ (12.10)

where V_H is the hysteresis voltage. As can be seen from these expressions, the center voltage and the hysteresis voltage are both dependent on the ratio of R_f to R_i. Changing this ratio will change both of these voltages.

It is possible to modify the circuit of Figure 12.3(a) so that these two voltages are independent of each other, as shown in Figure 12.5, by adding a third resistor. The equations for this circuit are as follows:

$$V_{UTP} = -\frac{R_i}{R_f} (V_{Lsat}) - \frac{R_i}{R_N} (V_{REF})$$ (12.11)

$$V_{LTP} = -\frac{R_i}{R_f} (V_{Hsat}) - \frac{R_i}{R_N} (V_{REF})$$ (12.12)

where R_N is the resistor from the noninverting input to the negative supply.

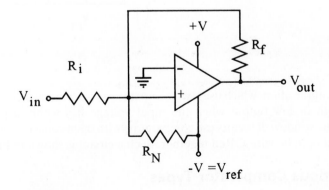

FIGURE 12.5
Comparator with independent hysteresis and center voltage adjustment.

FIGURE 12.6
Window detector with active low output.

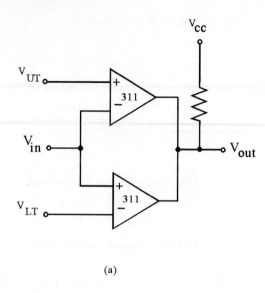

(a)

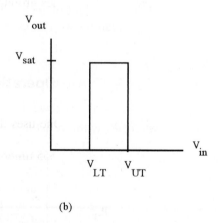

(b)

Another application, which uses two comparators, is a window detector, which provides a high or low output when the input voltage lies within a preselected range. Normally window detectors use comparators with open collector outputs such that the outputs can be wire-ORed together. Such a circuit is shown in Figure 12.6.

12.2.3 Various Comparator Types

In listing the various types of comparators (Table 12.1), note that this is not a comprehensive list nor a totally representative list. Rather, the list shows a sample

TABLE 12.1
Types of comparators

Type	Technology	Features and Comments
311	Bipolar	Industry standard comparator, high output current sinking capability, strobed output
2903	Bipolar	Single power supply operation, two comparators in a package, available four to a package (2901), common mode range includes ground
1017	Bipolar	Dual comparator, very low power
3098	Bipolar	Designed to provide independently adjustable hysteresis points, high output current
3290	Bipolar/MOS	Very high input impedance, single power supply operation, wide common-mode range

of the wide variety of existing devices and makes the reader aware of the many different types of comparators available for different applications. In addition, most comparators come in different temperature ranges and in different packages such as plastic or ceramic DIP and TO-5. Many are also available two or four to a package. Many also will operate from a single voltage supply with a common mode range which includes ground.

12.3 TIMERS

12.3.1 Basic Operation

Two common methods are used in existing integrated circuits to obtain the basic timing function. One uses the time required to charge or discharge a capacitor in an analog mode. The other uses a frequency source and a counter for timing in a digital mode. The 555 timer and its various manifestations use the first method and devices such as the 2240 use the second method.

The basic 555 timer is shown in Figure 12.7. In this diagram, the external components, R and C, are shown connected to provide the monostable function. In the stable state, the output is low and the set/reset flip-flop is reset. The output of the flip-flop in the reset state is high, turning on the discharge transistor and keeping the timing capacitor discharged. The three internal 5-kΩ resistors provide reference voltages for the upper and lower comparators of two-thirds and one-third of the supply voltage, respectively.

When the voltage on the trigger input pin is pulled below $(1/3)V_{cc}$, the lower comparator sets the flip-flop, causing the output to switch high and turning off the discharge transistor. The capacitor then charges up through the external resistor. When the voltage at the threshold input reaches $(2/3)V_{cc}$, the flip-flop is reset. This, in turn, causes the output voltage to go low and also turns on the discharge transistor,

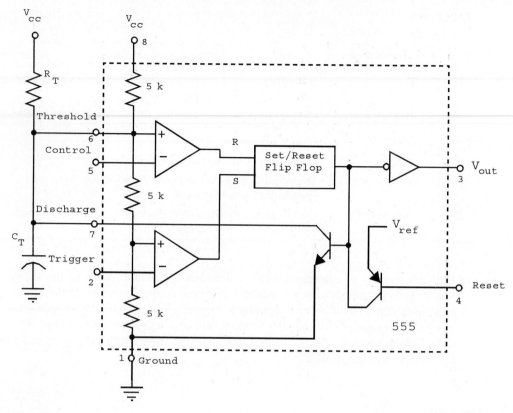

FIGURE 12.7
Basic 555 timer.

which discharges the capacitor. The basic timing equation in the monostable mode is:

$$t_o = 1.1 R_T C_T \tag{12.13}$$

where t_o = width of the output pulse (s)
        R_T = external resistor (Ω)
        C_T = external capacitor (F).

The 555 can also be connected to operate as an astable multivibrator as shown in Figure 12.8. In this connection, the capacitor alternately charges between $(1/3)V_{cc}$ and $(2/3)V_{cc}$. Since charging takes place through the sum of R_A and R_B and discharge only through R_B, we can see that the duty cycle has to be greater than 50%. The following equations apply to the 555 in the astable mode as shown by the waveform of Figure 12.8:

$$t_1 = 0.69(R_A + R_B)C_T \tag{12.14}$$

$$t_2 = 0.69 R_B C_T \tag{12.15}$$

$$f = \frac{1}{T} = \frac{1}{t_1 + t_2} = \frac{1.45}{(R_A + 2R_B)C_T} \tag{12.16}$$

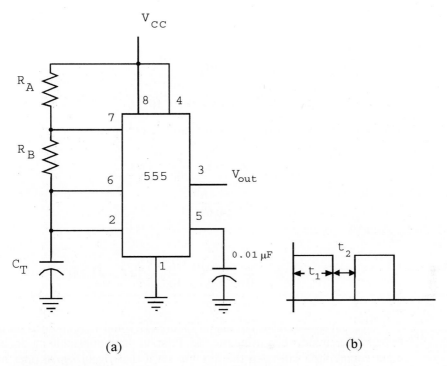

(a)   (b)

FIGURE 12.8
The 555 timer in the astable mode: (a) Circuit and (b) output waveform.

where f = frequency of the output (Hz)
          t_1 = on time of the output (s)
          t_2 = off time of the output (s).

EXAMPLE 12.2

Design a square-wave oscillator with an output frequency of 200 Hz and a duty cycle of 80%.

Solution
There are two design parameters, frequency and duty cycle, and three unknowns, the two timing resistors and the timing capacitor. One of these elements can be arbitrarily chosen. Because the discharge transistor in the 555 has to carry the current through R_A, it is recommended that R_A be larger than 5 kΩ. The capacitor is usually the critical element: too large and it has to be electrolytic with high tolerance and large leakage currents; too small and it approaches the value of the wiring and distributed capacitance. Let us try 0.1 μF for C_T; then the value of R_B can be calculated from Eq. (12.15) as

$$t_2 = 0.2 \times \frac{1}{200} = 1 \text{ ms}$$

so

$$R_B = \frac{t_2}{0.69 C_T} \quad \text{or}$$

$$R_B = \frac{1 \times 10^{-3}}{0.69 \times 0.1 \times 10^{-6}} = 14.5 \text{ k}\Omega$$

Then the value of R_A can be found from Eq. (12.16) as

$$t_1 = 0.8 \times \frac{1}{200} = 4 \text{ ms} \quad \text{or}$$

$$R_A = \frac{t_1}{0.69 C_T} - R_B$$

so

$$R_A = \frac{4 \times 10^{-3}}{0.69 \times 0.1 \times 10^{-6}} - 14.5 = 43.5 \text{ k}\Omega$$

We can see from the circuits of Figures 12.7 and 12.8 that certain things have to be kept in mind when using the 555. First, in the monostable mode, because the input comparator will keep the flip-flop set if the trigger signal does not go high again, the user must make certain that the trigger input goes high before the timing period is over. If this is not possible, then this input should be capacitively coupled. Second, in the astable mode, the first period will be longer than the remaining ones because the capacitor is initially discharged. The maximum frequency in the astable mode is typically limited to 500 kHz. The output in the astable mode may also experience oscillations at the switching points which can usually be eliminated by adequate power supply bypassing. One other point about the 555 is that when driving an inductive load, the load should be diode coupled so that the output cannot go negative when the load is turned off. Otherwise the output stage could possibly latch up.

The reset input of the 555 can be used to reset the internal flip-flop from an external source, thus setting the output low and turning on the discharge transistor. Reset overrides all other inputs to the timer. The control voltage input may be used to vary the reference voltages used by the internal comparators. Thus, by applying the external voltage to the control voltage pin, the reference voltages may be varied from their internally set values of $2/3 V_{cc}$ and $1/3 V_{cc}$.

The 2240 type of timer is shown in Figure 12.9 in simplified form. The 2240 consists of a modified 555 astable circuit with associated control circuitry that drives an eight-bit binary counter. The 2240 may be operated in either the monostable or astable mode. In either mode, the voltage across the timing capacitor, C_T, shown in Figure 12.10, sits at V_{cc} until the device is triggered by a positive signal at the trigger input. The timing capacitor is then discharged very quickly down to $0.27 V_{cc}$ and is then allowed to charge up to $0.73 V_{cc}$. These values give an oscillator period

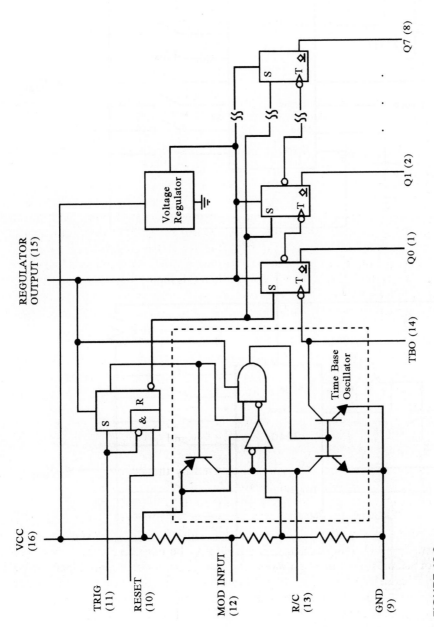

FIGURE 12.9
The 2240 timer.

395

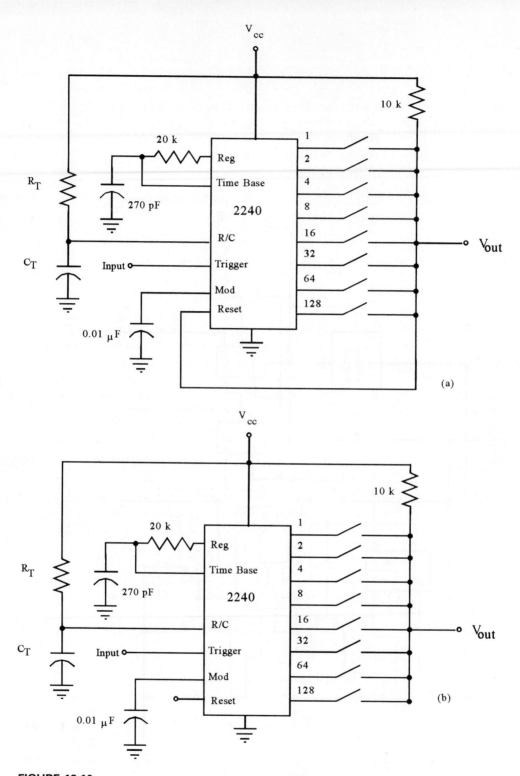

FIGURE 12.10
2240 timer basic operating modes: (a) Monostable connection and (b) astable connection.

equal to $R_T C_T$. During each discharge of the timing capacitor, a narrow pulse appears at the time base output, which is used to drive the binary counter.

The connections for the two operating modes are shown in Figure 12.10. In the monostable mode, the output normally sits high until the device is triggered and then goes low. The duration of the output pulse is determined by which counter outputs are connected together and tied to the reset pin. The total period is the sum of the connected outputs multiplied by the oscillator period or nT where T is the product of the timing resistor and the timing capacitor. In the astable mode, the period of the output is the same as the period of the monostable pulse and thus the frequency is the reciprocal of $2R_T C_T$. In the astable mode, the duty cycle is fixed at 50%.

EXAMPLE 12.3

Use a 2240 in the monostable mode to provide an output pulse of 20-ms duration when triggered.

Solution
Considering that it might be desirable to modify the pulse width after the circuit is designed, advantage can be taken of the flexibility of the 2240 by placing the 20 ms somewhere in the middle of the maximum pulse width capability. For example, if the clock has a 0.1-ms period, a pulse width can be generated from 0.1 to 25.5 ms. Using this criterion as a guide, the clock has to have a 0.1-ms period or a frequency of 10 kHz. Thus, $R_T C_T$ is just 0.1 ms and selecting C_T as 0.001 μF, R_T becomes 100 kΩ.

12.3.2 Typical Applications

The applications that have been developed for the 555 timer are too numerous to list here. Some of the more common ones include a missing pulse detector, darkroom timer, linear ramp generator, pulse width modulator, pulse position modulator, and tone burst generator. As an example, consider the circuit of Figure 12.11, which is a linear ramp generator. In this circuit, transistor $Q1$ serves as a constant current source, which in turn will cause the voltage across C_T to increase linearly. The time required for the capacitor voltage to reach $(2/3)V_{cc}$ determines the width of the output pulse and is given by

$$t = \frac{(2/3)V_{cc}C_T}{I_T} \tag{12.17}$$

The capacitor current is determined by the voltage at the base of $Q1$ minus the V_{BE} drop divided by the emitter resistance or

$$I_T = \frac{[(V_{cc}R_B)/(R_A + R_B)] - V_{BE}}{R_E} \tag{12.18}$$

In general, the current should be held to less than 1.0 mA and $Q1$ should be chosen to provide high gain at that current. Figure 12.11(b) shows the ramp generated.

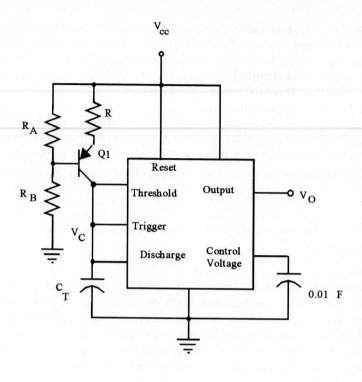

(a)

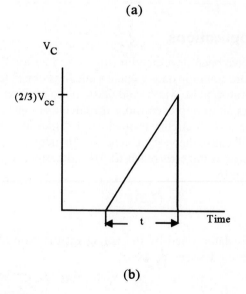

(b)

FIGURE 12.11
The 555 timer used as a linear ramp generator. (a) Circuit and (b) output mode.

Another example, Figure 12.12, shows how the 555 in the astable mode can be modified to provide duty cycles of less than 50%. The modification consists of adding diodes that isolate the two timing resistors so that they are independent of each other as shown in Figure 12.12(a). The associated equations are given in Figure 12.12(b). These equations are only approximate since the diode drops will modify the threshold voltages to some degree.

A final example shows how the 555 may be used as a pulse width modulator. The circuit is shown in Figure 12.13. There are maximum and minimum limits on the range of voltages that can be applied to the control voltage input for proper operation. If the voltage is too high, the capacitor will not be able to charge to a high enough value to trip the upper comparator. If set too low, the input voltage will have to approach zero in order to trip the lower comparator. In this particular

FIGURE 12.12

The 555 timer on astable mode modified for less than 50% duty cycle: (a) Circuit diagram and (b) circuit equations.

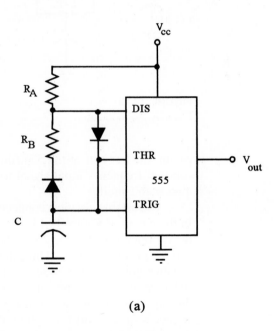

(a)

$$t_{HI} = 0.69 R_A C$$

$$t_{LO} = 0.69\ R_B C$$

$$f = 1/T = 1/(t_{HI} + t_{LO}) = 1.45/(R_A + R_B)\ C$$

(b)

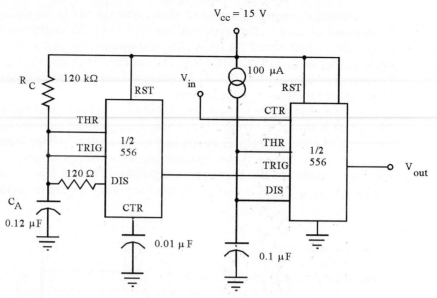

FIGURE 12.13
Pulse width modulator.

circuit, the left half of the 556 (dual 555) is used in the astable mode to provide a constant frequency input signal to the right half of the 556, which is connected in monostable mode.

The astable circuit is connected differently than the one shown in Figure 12.8. In the circuit of Figure 12.13, the on time of the output is given by $0.694 R_C C_A$ and the off time by $0.694 R_D C_A$. For the values given in this figure, the on time is 10 ms and the off time is 10 μs. Thus, the left-hand astable circuit provides a very narrow negative-going trigger pulse to the right-hand monostable circuit. For the monostable circuit, the timing capacitor is charged by a constant current source so that the capacitor voltage is a linear function of time. If the impedance of the input voltage source is much less than the 5-kΩ internal impedances of the timer, then the widths of the output pulses will be directly proportional to V_{in}. For the values shown, a V_{in} of 10 V provides a full-scale output pulse width of 10 ms.

12.3.3   Various Timer Types

As in the case of the comparator, a wide variety of timers are available on the market today. They come in different temperature ranges and in a variety of packages. The 555 timer is also available two to a package (the 556) and four to a package (the 558/559). Analog timers are also obtainable in low-power CMOS versions, which are particularly suitable for battery-powered applications. Various other analog timers besides the 555 are shown in Table 12.2, and examples of digital timers are given in Table 12.3.

TABLE 12.2
Types of analog timers

Type	Technology	Features and Comments
2905	Bipolar	Positive edge triggered, output can be a positive or negative pulse, output transistor is floating and current limited
TLC555	CMOS	CMOS version of the 555, very low power, high input impedance
4047	CMOS	CMOS monostable or astable device, positive or negative trigger, astable mode strobe, can be retriggered in the monostable mode, Q and $\overline{Q}$ outputs available simultaneously

TABLE 12.3
Types of digital timers

Type	Technology	Features and Comments
4060	CMOS	Standard CMOS logic, built-in oscillator followed by a 14-bit ripple counter
7240, 7250, 7260	CMOS	Similar to the 2240, the 7240 has a binary counter to 255, the 7250 has a decimal counter to 99, and the 7260 counts to 59 for seconds, minutes, and hours
7242	CMOS	Fixed-length version of the 2240, counts to 128 or 256

12.4 PHASE-LOCKED LOOPS

12.4.1 Basic Operation

A phase-locked loop (PLL) is a closed-loop feedback control system in which the phase of the reference input signal is compared to the phase of the output signal and the output frequency is adjusted to keep the two signals at the same frequency. Phase-locked loops are available that will work with either analog or digital signals. They are capable of many different functions including signal generation, modulation, demodulation, frequency selection, frequency multiplication, and frequency division.

A PLL is shown in Figure 12.14. The output of the phase detector drives a voltage-controlled oscillator (VCO) through a low-pass filter to keep the loop in lock. The range of frequencies over which the loop will stay in lock is called the *lock-in range.* This range is always equal to or greater than the capture range, which is the range of incoming frequencies onto which the loop is able to lock.

Digital PLLs use one of two types of phase detectors, either a simple exclusive OR gate or a network that looks at the edges of the two signals. Both types have

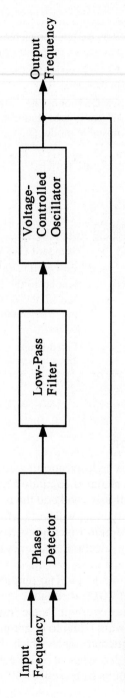

FIGURE 12.14
Phase-locked loop.

their advantages and disadvantages. The primary advantages of the exclusive OR phase comparator are its simplicity and good noise rejection. The edge comparator has the advantages of not being input duty cycle dependent, of not locking onto harmonics of the input signal, and of providing a lock range that covers the full VCO range. Analog PLLs commonly use a four-quadrant analog multiplier to mix the input and output signals for analog phase detection. Because this mixing is actually multiplication, the output of the phase comparator is a function of the amplitudes, frequencies, phase relationships, and duty cycles of the two signals.

The low-pass filter is usually just a simple lag filter consisting of a single RC combination or a lag-lead filter consisting of an R and an RC combination.

12.4.2 Application Considerations

Because the PLL is a reasonably complicated device, no single set of rules is applicable to all applications. However, several factors are common to most applications and normally should be taken into account. These include such things as selection of the VCO free running frequency, the lock range, the capture range, and the input signal amplitude. Application factors also include the loop lock-up time and the tracking rate. The following factors all affect lock-up time: input phase, low-pass filter characteristics, loop damping, input frequency deviation, input amplitude, noise, and VCO center frequency.

12.4.3 Typical Applications

The following applications show the versatility of the PLL and how it can be used to perform a variety of different functions.

A very straightforward application for either an analog or digital PLL is frequency multiplication, as shown in Figure 12.15. In Figure 12.15(a), integer multiplication is performed by dividing down the output of the VCO with a digital counter so that the output frequency of the counter matches the input frequency, which normally comes from a fixed frequency oscillator. For fractional frequency multiplication, two counters are used as shown in Figure 12.15(b) such that the desired output frequency is the input frequency multiplied by the ratio of the two divisors.

Another PLL application is AM demodulation. One AM demodulation technique is shown in Figure 12.16. In this method of demodulation, the PLL locks onto the carrier frequency of the AM signal. The output of the VCO is a nonmodulated signal at the frequency of the carrier. If the input signal is phase shifted 90° and then multiplied by the VCO output, the average dc value of the multiplier output will be directly proportional to the amplitude of the input signal. The advantage of this method is that it offers a higher degree of noise immunity than the conventional peak detection type of AM demodulation.

Motor speed control for small dc motors such as those used in tape drives is another application of digital PLLs that is finding wide use. As shown in Figure

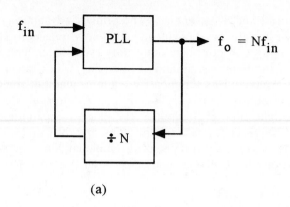

(a)

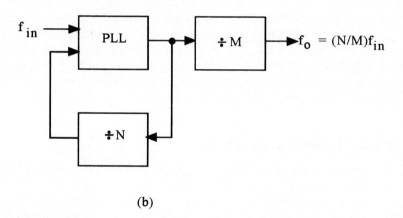

(b)

FIGURE 12.15
PLL frequency multiplication: (a) Integer frequency multiplication and (b) fractional frequency multiplication.

12.17, the motor and associated speed sensor become the VCO portion of the loop. If the input frequency is a very stable source, this method of speed control provides very stable and precise speeds. For good speed control, the speed sensing device should provide high resolution. Either an optical encoder or a magnetic speed sensor could be used for this purpose.

Another common use of the PLL is as a frequency shift keyed (FSK) demodulator. Such a circuit is shown in Figure 12.18. Here the output of the low-pass filter is further filtered before being input to a comparator. The output of the comparator becomes a logic zero or a logic one corresponding to the two input frequencies. The purpose of the additional low-pass filter is to remove the sum frequency components. Several PLLs are available that have been designed specifically for FSK modulation and demodulation.

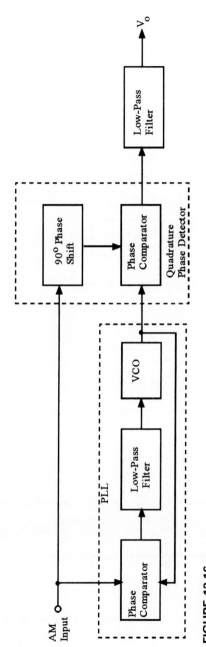

FIGURE 12.16
PLL AM demodulator.

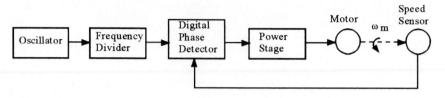

FIGURE 12.17
PLL motor speed control.

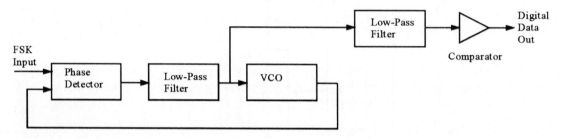

FIGURE 12.18
PLL FSK demodulator.

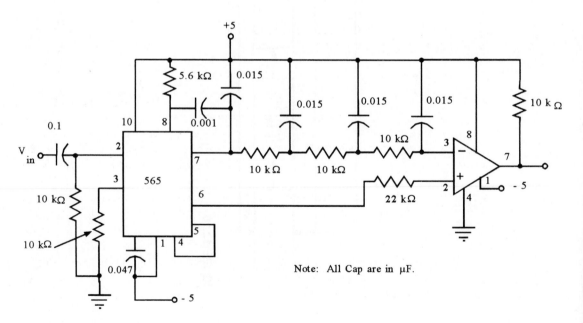

FIGURE 12.19
The 565 PLL used as an FSK demodulator.

406

TABLE 12.4
Various types of phase-locked loops

Type	Part	Technology	Features and Comments
Digital	4046	CMOS	Contains both types of digital phase detectors, 1 MHz maximum VCO frequency
Analog	560	Bipolar	General-purpose PLL, 30-MHz operation
Analog	565	Bipolar	General-purpose low-frequency PLL
Analog	210	Bipolar	Designed as an FSK modulator/demodulator, has built-in comparator
Analog	567	Bipolar	PLL plus a quadrature phase detector used for tone and frequency decoding

Figure 12.19 shows a 565 PLL used as an FSK decoder. In this circuit, the FSK frequencies are 1070 and 1270 Hz. The three-stage RC ladder low-pass filter is designed with a band edge approximately halfway between the transmission frequency of 300 baud (150 Hz) and twice the lower FSK frequency. The free-running VCO frequency is set approximately halfway between the two FSK frequencies.

Various types of phase-locked loops are listed on Table 12.4.

12.5   VOLTAGE-TO-FREQUENCY AND FREQUENCY-TO-VOLTAGE CONVERTERS

12.5.1   Basic Operation of Voltage-to-Frequency Converters

The basic operation of the 566 voltage-to-frequency (V/F) converter or VCO can be described in terms of the block diagram shown in Figure 12.20. The variation of output frequency with input voltage is due to the voltage-controlled current source. The current source charges the timing capacitor, C_T, linearly with time at a rate dependent on the current and the capacitor value as per Eq. 12.17. The capacitor voltage is applied to a Schmitt trigger, which is simply a comparator with hysteresis. The output of the Schmitt trigger causes the current source to switch back and forth between being a current source and being a current sink via the current mirror. This allows the capacitor to alternately charge and discharge linearly. The signals from both the Schmitt trigger and the capacitor are buffered to provide square-wave and triangular-wave outputs, respectively.

The basic conversion equation is

$$f_o = \frac{2(V_{cc} - V_{in})}{V_{cc} R_T C_T}$$

(12.19)

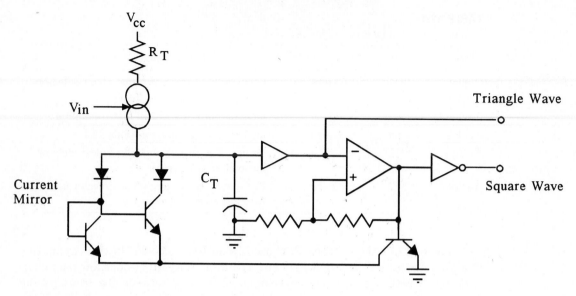

FIGURE 12.20
Voltage-to-frequency converter.

where f_o = output frequency (Hz)
V_{cc} = supply voltage (V)
V_{in} = control voltage (V)
R_T = timing resistance (Ω)
C_T = timing capacitance (F).

This technique of creating a VCO is representative of the way in which voltage-to-frequency conversion can occur. Other types of V/F converters use a monostable multivibrator such that the output pulse width is constant and the duty cycle changes with input voltage to give a varying output frequency.

A typical application of a 566 used as a VCO is shown in Figure 12.21. For best operation, R_T should lie between 2 and 20 kΩ. The allowable range of input voltage is from $0.75 V_{cc}$ to V_{cc}. For example, with the values shown in Figure 12.21, the range of output frequencies is from 833 to 5000 Hz for input voltages from 11.5 down to 9 V. Note that the output frequency is inversely proportional to the input voltage.

12.5.2 Basic Operation of Frequency-to-Voltage Converters

The operation of the 2907 family of frequency-to-voltage (F/V) converters is described in block diagram form in Figure 12.22. The input signal goes into a Schmitt trigger to provide some hysteresis around the switching point, which can be ground or some other voltage. The output of the Schmitt trigger drives a charge pump, which alternately charges and discharges the timing capacitor. The average amount

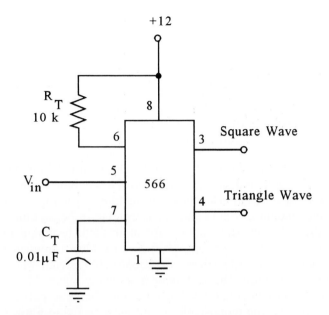

FIGURE 12.21
The 566 used as a voltage-controlled oscillator.

of charge flowing into or out of the timing capacitor is directly proportional to the
frequency of the input voltage. The output circuit of the charge pump then mirrors
the charging current into the load resistor. These pulses of current are integrated
by the timing capacitor, C_T, to give an output voltage proportional to the input

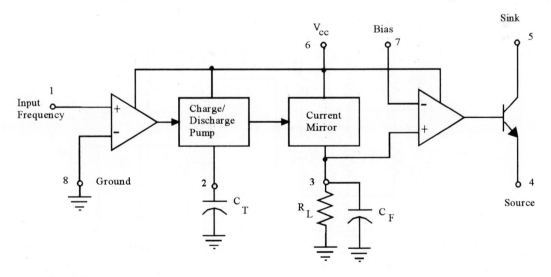

FIGURE 12.22
The 2907 frequency-to-voltage converter.

frequency. The basic conversion equation is given by

$$V_o = V_{cc}C_T R_L f_{in} \qquad (12.20)$$

where
V_o = output voltage (V)
V_{cc} = supply voltage (V)
f_{in} = frequency of the input signal (Hz)
C_T = timing capacitance (F)
R_L = load resistance (Ω).

The filtered output is further buffered by an operational amplifier, which in turn drives an uncommitted output transistor. The amplifier can be used as a comparator or as an amplifier. The choice of the filter capacitor, C_F, depends only on the allowable amount of ripple and the acceptable response time.

Most of the V/F converters that use the monostable technique of conversion may also be used for F/V conversion. When used to convert frequency to voltage, the input frequency is used to trigger the monostable circuit, which provides a constant width output pulse. This technique is shown in Figure 12.23 using a 4151 V/F along with the design equations. Since the time between pulses is a function of the input frequency, the monostable output must be fed to a low-pass filter to obtain an average output voltage proportional to the input frequency. In this circuit, R_B and C_B form the low-pass filter. The input signal is differentiated by the input capacitor and resistor network to trigger the monostable circuit. For the values shown, this circuit will provide a full-scale output of 10 V for an input frequency of 1 kHz.

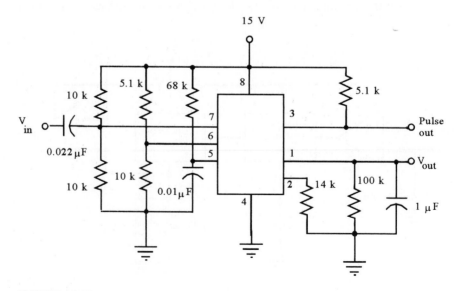

FIGURE 12.23
Monostable F/V conversion using a 4151.

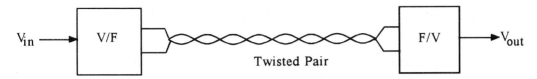

FIGURE 12.24
Analog data transmission using V/F and F/V converters.

12.5.3 Typical Applications

One common application of the V/F converter is to send analog data a reasonable distance using a twisted-pair cable. When used with an F/V at the receiving end as shown in Fig. 12.24, analog voltage can be recreated. Alternatively, the output frequency can be used directly by a microprocessor or a frequency counter.

Another typical application of the F/V is for speed measurement with a magnetic pickup. The output of the F/V can be read directly by a voltmeter to give a measurement of speed. Such an application is shown in Figure 12.25. Using a 2907 type of F/V also allows use of the op amp in the device as a comparator to make a speed switch. The op amp can also be used as an active filter to reduce the output ripple.

Another common use for V/F converters is for FSK generation. In this application, the input voltage levels representing the two logic levels select one of two output frequencies. A typical device used for this purpose is the 2206 as shown in

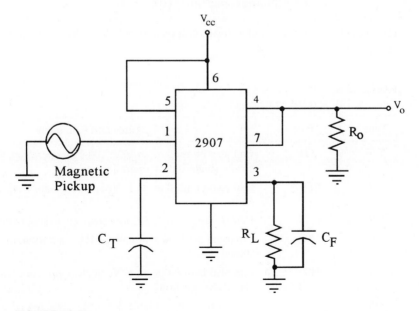

FIGURE 12.25
F/V used for speed measurement.

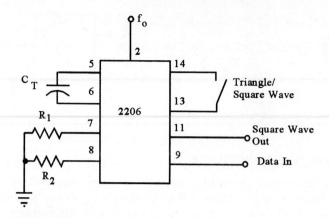

FIGURE 12.26
V/F used for FSK generation.

Fig. 12.26. The frequency of oscillation for this device is given by

$$f_o = \frac{1}{R_1 C_T} \quad \text{HIGH-LEVEL} \atop \text{DATA IN} \qquad f_o = \frac{1}{R_2 C_T} \quad \text{LOW-LEVEL} \atop \text{DATA IN} \qquad \textbf{(12.21)}$$

where
f_o = output frequency (Hz)
$R_{1,2}$ = timing resistances (Ω)
C_T = timing capacitance (F).

High-level data are defined as input voltages of >2 V and low-level data are defined as input voltages of <1 V.

TABLE 12.5
Types of F/V and V/F converters

Type	Part	Features and Comments
V/F	331	Can also be used as an F/V, 100-kHz operation, very high linearity, single supply operation
V/F	VFC32	Can also be used as an F/V, open collector output, high linearity
V/F	4151	Can be used as an F/V, low cost, single supply operation
V/F	2207	Provision for binary inputs, 1-MHz operation, low drift, high stability
V/F	9400	Can also be used as an F/V, single supply operation possible, 100-kHz operation
F/V	2907	Four variations available, low cost, good linearity, internal op amp for different applications

12.5.4   Various Types of F/V and V/F Converters

Many V/F converters can also be used as F/V converters. The choice of a converter for a particular application depends on many factors and can only be determined by careful evaluation of the various parameters of each device. Some of the more common devices are given in Table 12.5.

12.6   OSCILLATORS AND FUNCTION GENERATORS

We can see from the preceding sections that many of the devices previously discussed could also be used as fixed-frequency oscillators. However, most of those provided only square-wave or square- and triangular-wave outputs. In this section, those devices that also produce sinusoidal outputs are discussed. Many of these use the same principles of operation as the square-wave generators previously discussed, but follow the square-wave output with additional circuitry to generate a sine wave. Typical of such a device is the 8038 function generator. A basic application for the 8038 is as a multiwaveform source as shown in Fig. 12.27. In Figure 12.27, the 8038 will supply square-wave, triangle-wave, and sine-wave outputs over a frequency range of 20 Hz to 100 kHz. Switch SW1 is used to select the output frequency range and potentiometer R4 adjusts the frequency within the selected range.

The 8038 must be externally trimmed for both a 50% duty cycle at the square- and triangular-wave outputs and for minimum distortion at the sine-wave output. Potentiometer R3 adjusts for a 50% duty cycle, and pots R7 and R9 are used to adjust for minimum sine-wave distortion.

The square-wave output swings between the supply voltages. The square-wave amplitude may be adjusted via potentiometer R12. The triangular-wave output swings above and below the center point of the supply voltages by one-third of their values. The amplitude of the triangle wave can be set by potentiometer R11. The sine wave swings symmetrically above and below the center point of the supply voltages to a peak of 22% of their values. Potentiometer R10 adjusts the sine-wave amplitude. All three outputs should be buffered if used to supply any appreciable current because of their high output impedances.

The basic frequency equation for the 8038 is

$$f_o = \frac{0.3}{R_A C_T} \tag{12.22}$$

where f_o = output frequency (Hz)
R_A = resistance between V_{cc} and pin 4 or pin 5 (Ω)
C_T = capacitance between pin 10 and $-V_{cc}$ (F).

For optimum performance, the charging current should be kept between 0.001 mA and 1.0 mA. The charging current is found from

$$I_c = \frac{V_{cc} - (-V_{cc})}{5R_A} \tag{12.23}$$

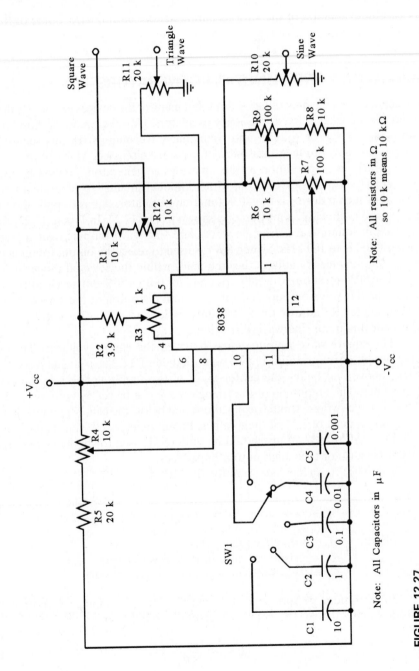

FIGURE 12.27
The 8038 function generator.

where I_c = charging current (mA)
            V_{cc} = positive supply voltage (V)
       $-V_{cc}$ = negative supply voltage (V)
         R_A = resistance between pin 4 or pin 5 and V_{cc} (Ω).

Other devices that will supply sine-wave outputs include the 205 and 2206. Both of these will also provide square, triangular and ramp outputs as well as VCO operation.

12.7 FUNCTION MODULES

A rich variety of other analog devices exist for various applications. This section examines some of the more commonly used types of such circuits.

12.7.1 Amplifiers

Logarithmic Amplifiers Logarithmic amplifiers are constructed such that the output voltage is the log base 10 of the ratio of the input currents times a constant. They are useful for log, log ratio, antilog, data compression, optical density measurements, and data linearization. Typical types include the LOG100, 8048, 8049, and the 755, 757, and 759.

Audio Amplifiers Audio amplifiers are integrated circuits that have been optimized to operate over the audio range of frequencies. In addition, they usually have very low output impedances in order to drive a loudspeaker directly. In general, they are also designed to operate from single-ended supplies and deliver several watts of output power. Typical devices include the 3360, 1306, 380, 2878 (dual), 2002, and 2004. These range in output power from 0.25 to 12 W.

Isolation Amplifiers Isolation amplifiers are used where it is desirable or necessary to isolate the signal source from the output signal and any associated electronics connected to the output. Typical applications would include medical instrumentation and power measurements. The usual techniques for isolation include optical coupling and transformer coupling. Up to several thousand volts of isolation can be achieved with either method. Typical devices include the 204, 277, 290, 293, ISO100, 3650, 3652, and 3656.

12.7.2 Other Functions

Sample and Hold Sample-and-hold devices are used to sample a voltage and hold it until additional circuitry can use the signal. Typical applications include peak voltage measurement and as an input for an analog-to-digital converter. The sampled voltage is held on either an internal or external capacitor.

Important parameters involved with sample-and-hold operation include acquisition time, aperture time, dynamic sampling error, gain error, hold settling time, and hold step. If an external capacitor is used, it should be a good-quality, low-leakage type such as polystyrene, polypropylene, or Teflon. Typical sample-and-hold devices include the LF398, the IH5110-5115, the AD346 and 582, and the SHC76 and 85.

RMS-to-DC Converters   RMS-to-dc converters convert the rms value of a complex signal to dc. This type of conversion is particularly useful for power measurements involving nonsinusoidal signals. Such signals are prevalent in circuits involving phase control of SCRs or TRIACs. When used with isolation amplifiers, rms-to-dc converters provide a convenient method of measuring power without having a direct connection to the power portion of the circuit. Typical rms-to-dc converters available include the AD536A, AD636, 4340, and 4341.

Multipliers   Multipliers are used to perform multiplication and division of analog signals. Typically most multipliers will also square and find the square root of analog signals. Because of their versatility, certain types can also perform such functions as finding the difference of squares, linear AM modulation, voltage-controlled amplification, and function generation. They are either two- or four-quadrant devices meaning they will work with either positive only or bipolar signals, respectively. Typical multipliers include the ICL8013, AD532, AD533, AD534, MPY100, MPY534, MPY634, and 4203-4206.

Multiplexers   Frequently it is convenient to have the ability to switch different analog signals around in a circuit. The multiplexer allows such switching to occur where the signals to be switched are controlled by logic level control signals. Various multiplexers are available—from two line to one line up to sixteen line to one line. They are also available with either normally open or normally closed switches. All use either JFET or MOSFET devices as switches and many are CMOS. Most will switch bipolar signals with on resistances ranging from 10 Ω or less to several hundred ohms. Typical multiplexers include the DG180-191, IH5025-5038, AD7501-7503, CD4016, CD4051-4053, CD4066, LF13331-13333, LF13201, LF202, LF508, LF509, MPC8S, MPC4D, MPC16S, MPC8D, MPC800, and MPC801.

Potpourri   A perusal of the catalogs listed in the Bibliography (Section 12.8) will show a large list of devices that perform a variety of special functions. A partial listing would include such things as peak detectors, voltage references, temperature sensors, modulators, demodulators, servo circuits, compandors, voltage sensors, motor speed regulators, smoke detectors, ultrasonic transceivers, and dot bar display drivers. This list is not meant to be inclusive but rather points out that the circuit needed to do a function might already be available.

12.8 BIBLIOGRAPHY

The following list of books is a cross representation of the literature available regarding the application of various special function integrated circuits.

1. Mitra, S. K., *An Introduction to Digital and Analog Integrated Circuits and Applications,* Harper & Row, New York, 1980.
2. *Applications Manual,* Signetics, Sunnyvale, CA, 1979.
3. *Intersil Application Handbook,* Intersil Inc., Cupertino, CA, 1985.
4. Pippenger, D. E., and E. J. Tobaben, *Linear and Interface Circuits Applications, Volume 1,* Texas Instruments, 1985.
5. *Linear Applications Handbook,* National Semiconductor, Santa Clara, CA, 1978.
6. McMenamin, J. M., *Linear Integrated Circuits,* Prentice Hall, Englewood Cliffs, NJ, 1985.
7. Gayakwad, R. A., *Op-Amps and Linear Integrated Circuits Technology,* Prentice Hall, Englewood Cliffs, NJ, 1983.
8. Hughes, F. W., *Op-Amp Handbook,* Prentice Hall, Englewood Cliffs, NJ, 1986.
9. Coughlin, R. F., and F. F. Driscoll, *Operational Amplifiers and Linear Integrated Circuits,* Prentice Hall, Englewood Cliffs, NJ, 1982.
10. Hill, W., and P. Horowitz, *The Art of Electronics,* Cambridge University Press, Cambridge, England, 1980.

The following data books were also used in the preparation of this chapter. They are a noninclusive list of various linear and analog integrated circuit manufacturers.

1. *1984 Databook,* Analog Devices, 1984.
2. *Integrated Circuits Databook,* Burr-Brown, 1986.
3. *Databook,* Exar, 1985.
4. *Component Data Catalog,* Intersil, 1986.
5. *Linear Databook,* Linear Technology, 1986.
6. *CMOS Data Acquisition Catalog,* Maxim, 1985.
7. *Linear Integrated Circuits,* Motorola, 1979.
8. *Linear Databook,* National Semiconductor, 1980.
9. *Linear Integrated Circuits,* RCA, 1978.
10. *CMOS Integrated Circuits,* RCA, 1983.
11. *Analog Data Manual,* Signetics, 1981.
12. *Linear Circuits Data Book,* Texas Instruments, 1984.

13

DIGITAL ELECTRONICS*

13.1 NUMBER SYSTEMS

13.1.1 Types of Number Systems

In digital electronics numbers are represented by a series of 1's and 0's, i.e., in binary notation. This should imply that, in order to represent numbers with a digital circuit, a binary system must be used, although hexadecimal (base 16) is often used for reasons of space. Let us consider briefly the binary and hexadecimal numbering systems. First any number N can be represented in any base by the following equation:

$$N = d_n B^n + \cdots + d_3 B^3 + d_2 B^2 + d_1 B^1 + d_0 B^0 \qquad (13.1)$$

where N is the number represented in base 10, the d's are the digits in base B, and B is the number base of the number system. A simple example in base 10 is the number 372_{10}, which is equal to 3 times 10^2 plus 7 times 10^1 plus 2 times 10^0, or

$$
\begin{array}{r}
3 * 100 \\
7 * 10 \\
+\,2 \\
\hline
372
\end{array}
$$

---

* This chapter was written by Prof. Roger Hack, Purdue University.

EXAMPLE 13.1

What does 1001_2 equal in base 10?

Solution

The binary number is 1001_2, which is equal to 1 times 2^3 plus 0 times 2^2 plus 0 times 2^1 plus 1 times 2^0,

$$
\begin{array}{ll}
1 * 8 & 8 = 2^3 \\
0 * 4 & 4 = 2^2 \\
0 * 2 & \\
+\,1 * 1 & \\
\hline
10 & \text{in decimal or base 10.}
\end{array}
$$

Another number system used quite often in digital electronics is that of hexadecimal (hex). In hex the base is 16, with the first six letters of the alphabet used to represent the numbers from 10 to 15. Equation (13.1) can be used for converting from the hex number system to another system.

EXAMPLE 13.2

What does 3A2F in hexadecimal equal in base 10?

Solution

The number $3A2F_{16}$ would be equal to

$$
\begin{array}{ll}
3 * 4096 & (4096 = 16^3) \\
10 * 256 & (\ 265 = 16^2) \\
2 * 16 & \\
+\,15 * 1 & \\
\hline
14895 &
\end{array}
$$

Another number system that is used occasionally with computer systems is base 8 (octal). At one time octal was very popular; today, however, it is only used by a few computer systems. Equation (13.1) can also be used to convert from base 8 to decimal.

EXAMPLE 13.3

The technical reference manual of a computer gives the port address of 326_8, and the address is needed in decimal.

Solution

The octal number 326_8 is equal to

$$
\begin{array}{ll}
6 * 1 & (\ 1 = 8^0) \\
2 * 8 & (\ 8 = 8^1) \\
+\,3 * 64 & (64 = 8^2) \\
\hline
214 &
\end{array}
$$

13.1.2   Conversions to Other Number Systems from Base 10

We often find it necessary to convert from decimal to another number system such as binary. The simplest method of doing this is to create a table with the values of each digit location of the target number system. This table starts on the right with the value of the target base raised to the power zero (1), and moves to the left with the target base raised to increasing powers. The power of the base is increased by one with each move to the left. The table is terminated when the value of the position is greater than the number being converted. With this method, the leftmost position will be equal to zero in the final number. The next step is to divide the largest position value into the number being converted to find the value of the new number in the target base. Next subtract this number times that position value from the original number, and repeat this process with the next position to the right until done.

EXAMPLE 13.4

Convert 43 in base 10 to binary.

Solution

The position values are as follows:

2^6	2^5	2^4	2^3	2^2	2^1	2^0	Exponent value
64	32	16	8	4	2	1	Decimal value
0	1	0	1	0	1	1	Binary number

EXAMPLE 13.5

Convert 8951 in base 10 to hex.

Solution

The position values are as follows:

16^4	16^3	16^2	16^1	16^0	Exponent value
65536	4096	256	16	1	Decimal value
0	2	2	F	7	Hex number

If we need to convert from a base other than 10 to another base other than 10 it is often easier to convert the original number to base 10, and then convert this number to the new base. An exception to this rule is converting from hex or octal to binary. The reason hex and octal are used is the ease with which we can convert between these bases and binary.

13.1.3   Converting Hex to Binary

One digit in hex or base 16 can be easily converted to four binary digits. The reason for this is that 2^4 equals 16. Thus, to convert to binary from hex all one needs to do is convert each digit one at a time.

EXAMPLE 13.6

Convert 3AF2H (another way of showing base 16) to binary.

Solution

Starting with the leftmost digit, convert each digit one at a time to binary.

3	A	F	2	Number in hex
0011	1010	1111	0010	Number in binary

13.1.4 Converting Octal to Binary

One digit in octal or base 8 can easily be converted to three binary digits. This is because 2^3 is equal to 8. All we need to do to convert from octal to binary is convert one digit at a time to binary.

EXAMPLE 13.7

Convert 341Q (another way of showing octal) to binary.

Solution

Simply convert one digit at a time to binary:

3	4	1	Number of octal
011	100	001	Number in binary

If you need to do a lot of number conversions, it might be wise to purchase a calculator with built-in number conversion functions. Several such calculators are on the market, and there are also resident software packages that will perform these functions for the operator.

13.2 THE LOGIC SYMBOLS

13.2.1 The AND Gate

The AND gate will produce a logic 1 output if and only if all inputs are logical 1's. The Boolean representation of the AND gate is $C = A * B$, and is read $C = A$ AND B. Electrically, the AND gate can be represented by series-connected switches. The truth table and logic symbol for a two-input AND gate are shown in Table 13.1 and Figure 13.1, respectively.

FIGURE 13.1
Two-input AND gate.

TABLE 13.1
AND truth table

A	B	C = A * B
0	0	0
0	1	0
1	0	0
1	1	1

TABLE 13.2
OR truth table

A	B	A + B = C
0	0	0
0	1	1
1	0	1
1	1	1

TABLE 13.3
NOT truth table

A	$\overline{A}$
0	1
1	0

13.2.2 The OR Gate

The OR gate will produce a logic 1 output if there is a logical 1 on any one of the inputs into the gate. The Boolean representation of the OR gate is $C = A + B$, and is read $C = A$ OR B. Electrically the OR gate can be represented by switches connected in parallel. The truth table and logic symbol for a two-input OR gate are shown in Table 13.2 and Figure 13.2, respectively.

13.2.3 The NOT Gate or Inverter

The NOT gate, or inverter, as it is often called, will have a logic 1 output if the input is at logic level zero. The NOT gate has only one input and one output. The Boolean representation of the inverter is $C = \overline{A}$, and is read $C = A$ NOT, $C =$ NOT A. The truth table and logic symbol are shown in Table 13.3 and Figure 13.3, respectively.

13.2.4 The NAND Gate

The NAND gate has a logic 1 output in all cases except when all inputs are at logic level 1. It is the same as an AND gate with an inverter on the output. The Boolean representation of the NAND gate is $C = \overline{A * B}$, and is read $C = A$ NAND B. The truth table and logic symbol for the NAND gate are shown in Table 13.4 and Figure 13.4, respectively.

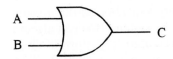

FIGURE 13.2
Two-input OR gate.

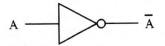

FIGURE 13.3
Inverter.

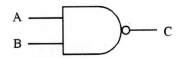

FIGURE 13.4
Two-input NAND gate.

TABLE 13.4
NAND truth table

A	B	$\overline{A * B}$
0	0	1
0	1	1
1	0	1
1	1	0

TABLE 13.5
NOR truth table

A	B	$\overline{A + B}$
0	0	1
0	1	0
1	0	0
1	1	0

TABLE 13.6
XOR truth table

A	B	$A \otimes B$
0	0	0
0	1	1
1	0	1
1	1	0

13.2.5 The NOR Gate

The NOR gate has a logic 1 output if and only if all the inputs to the gate are at logic level zero. The NOR gate is the same as an OR gate with an inverter on the output. The Boolean representation for the NOR gate is $C = \overline{A + B}$, and is read as $C = A$ NOR B. The truth table and logic symbol are shown in Table 13.5 and Figure 13.5, respectively.

13.2.6 The XOR Gate

The XOR gate has a logic 1 output if and only if one of the inputs is at logic level 1, and a zero output if more than one of the inputs are at logic level 1. The XOR gate is often referred to as an exclusive OR gate. The Boolean representation for the XOR gate is $C = A \otimes B$. The truth table and logic symbol for the XOR gate are shown in Table 13.6 and Figure 13.6, respectively.

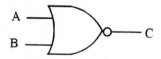

FIGURE 13.5
Two-input NOR gate.

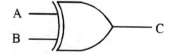

FIGURE 13.6
Two-input exclusive OR gate, XOR.

13.3 RULES OF BOOLEAN ALGEBRA

Boolean algebra is a special form of algebra that deals with the manipulation of TRUE–FALSE, or Boolean expressions. Boolean expressions can have only one of two types of output, either TRUE or FALSE. The output of a Boolean expression depends on the Boolean equation and the inputs to the system. Boolean algebra is used to simplify the amount of hardware that is required to perform a given Boolean expression. The following are standard rules for manipulating Boolean algebra equations. The major use of these rules is to reduce the complexity of a Boolean algebra expression. In many cases the simpler Boolean expression will lead to the smallest hardware solution. In other cases the number of gates used may not be as important as the types of gates that are available in the circuit.

13.3.1 The NOT Operation and the Law of Complementation

First law of complementation:

$$\text{If } A = 0, \text{then } \overline{A} = 1 \tag{13.2}$$
$$\text{If } A = 1, \text{then } \overline{A} = 0 \tag{13.3}$$

Second law of complementation:

$$A * \overline{A} = 0 \tag{13.4}$$

Third law of complementation:

$$A + \overline{A} = 1 \tag{13.5}$$

Fourth law of complementation:

$$\overline{\overline{A}} = A \text{ (double complement law)} \tag{13.6}$$

13.3.2 The Commutative Laws

First law of commutation:

$$A + B = B + A \tag{13.7}$$

Second law of commutation:

$$A * B = B * A \tag{13.8}$$

13.3.3 The Associative Laws

First law of association:

$$A * (B * C) = (A * B) * C \tag{13.9}$$

Second law of association:

$$A + (B + C) = (A + B) + C \tag{13.10}$$

13.3.4 The Distributive Laws

First law of distribution:

$$A * (B + C) = (A * B) + (A * C) \tag{13.11}$$

Second law of distribution:

$$A + (B * C) = (A + B) * (A + C) \tag{13.12}$$

13.3.5 The Laws of Tautology

The six laws of tautology are as follows:

1.	$A * A = A$	**(13.13)**
2.	$A + A = A$	**(13.14)**
3.	$A * 1 = A$	**(13.15)**
4.	$A * 0 = 0$	**(13.16)**
5.	$A + 1 = 1$	**(13.17)**
6.	$A + 0 = A$	**(13.18)**

13.3.6 The Laws of Absorption

The first law of absorption:

$$A * (A + B) = A \qquad \text{(13.19)}$$

The second law of absorption:

$$A + (A * B) = A \qquad \text{(13.20)}$$

Another set of rules can be derived from the ones shown. These rules are given without explanation.

$$A + \overline{A} * B = A + B \qquad \text{(13.21)}$$
$$A * (\overline{A} + B) = A * B \qquad \text{(13.22)}$$
$$A + C * B = (A + B) + (A + C) \qquad \text{(13.23)}$$
$$A * B + \overline{A} * C = (A * C) + (\overline{A} + B) \qquad \text{(13.24)}$$

EXAMPLE 13.8

Simplify $f = x * (\overline{x} + y)$.

Solution

First multiply out the expression

$$f = (x * \overline{x}) + (x * y) \qquad \text{Using Eq. (13.11)}$$

Next use Eq. (13.4), $x * \overline{x} = 0$. Thus, $f = x * y$, which is the simplest expression.

The notation $f = xyz$ is also used to represent $f = x * y * z$; the two expressions are used interchangeably.

EXAMPLE 13.9

Simplify $w = xy + \overline{xy}z$.

Solution

$$w = xy + \overline{xy}z \qquad \text{Using Eq. (13.8)}$$
$$w = xy + z \qquad \text{Using Eq. (13.21)}$$

This form is the simplest form of the original expression.

13.4 DE MORGAN'S THEOREM

De Morgan's laws imply that "any Boolean function can be accomplished using either AND gates and inverters, or OR gates and inverters." This is very important, since this law reduces the number of types of gates required in any digital logic problem. De Morgan's laws are the basis for the use of most families of digital logic gates. First case of De Morgan's theorem:

$$\overline{A * B} = \overline{A} + \overline{B}$$

Second case of De Morgan's theorem:

$$\overline{A + B} = \overline{A} * \overline{B}$$

13.4.1 Minterm and Maxterm Forms

The terms *minterm* and *maxterm* refer to the format in which a logic function is expressed. The minterm format is often referred to as the sum of products (SOP) format. The maxterm format is referred to as the product of sums (POS) format.

Examples of the SOP format:

$$f = (\overline{A} * \overline{B} * C) + (\overline{A} * B * \overline{C}) + (A * B * C)$$
$$f = (\overline{A} * \overline{B} * C * D) + (A * \overline{B} * \overline{C} * D) + (A * \overline{B} * C * D)$$

Examples of the POS format:

$$f = (A + B + C) * (A + B + C) * (A + B + C)$$
$$f = (A + B + C + D) * (A + B + C + D) * (A + B + C + D)$$

By applying De Morgan's law:

1. Any logic operation can be implemented in either a minterm or a maxterm format.
2. Any minterm format equation can be converted into a maxterm equation.
3. Any maxterm format equation can be converted into a minterm equation.

Statements 2 and 3 state that the two formats are complementary. A demonstration of what we mean by *complementary* is accomplished by the following example:

$$f = (\overline{A} * \overline{B} * C) + (\overline{A} * B * \overline{C}) + (A * \overline{B} * C)$$

and is shown in Table 13.7.

By applying De Morgan's theorem the above function can be written

$$f = (\overline{A} + \overline{B} + \overline{C}) * (\overline{A} + B + C) * (A + \overline{B} + \overline{C}) * (A + B + \overline{C}) * (A + B + C)$$

This function is shown in Table 13.8. It can be seen that, in the transformation from minterm format to maxterm format, all terms that were 0 are now 1, and all terms that were 1 are now 0. In other words, the function in minterm format is equal to the NOT, or inverse, of the function in maxterm format. This is important

TABLE 13.7
Truth table

m	A	B	C	f
0	0	0	0	0
1	0	0	1	1
2	0	1	0	1
3	0	1	1	0
4	1	0	0	0
5	1	0	1	1
6	1	1	0	0
7	1	1	1	0

TABLE 13.8
Truth table

M	A	B	C	f
0	0	0	0	1
1	0	0	1	0
2	0	1	0	0
3	0	1	1	1
4	1	0	0	1
5	1	0	1	0
6	1	1	0	1
7	1	1	1	1

to remember when trying to simplify a logic function using Karnaugh maps, as discussed in the next section.

EXAMPLE 13.10

Given the function of five variables

$$f(A,B,C,D,E) = (A + \overline{BC}) * (\overline{D + BE})$$

express the function as a sum of products (SOP).

Solution
We use De Morgan's theorem and the distributive law to find

$$
\begin{aligned}
f(A,B,C,D,E) &= (A + \overline{BC}) * (\overline{C + BE}) \\
&= (A + \overline{B} + \overline{C}) * \overline{D} * (\overline{BE}) \\
&= (A + \overline{B} + \overline{C}) * [\overline{D} * (\overline{B} + \overline{E})] \\
&= (A + \overline{B} + \overline{C}) * (\overline{B} * \overline{D} + \overline{D} * \overline{E}) \\
&= A * \overline{B} * \overline{D} + A * \overline{D} * \overline{E} + \overline{B} * \overline{D} + \overline{B} * \overline{D} * \overline{E} + \overline{B} * \overline{C} * \overline{D} + \overline{C} * \overline{D} * \overline{E}
\end{aligned}
$$

The last line above is clearly in SOP format. This conversion will enable the function to be built with fewer types of gates than the original function.

13.5 THE KARNAUGH MAP

The Karnaugh map, or K-map, is a special truth table designed specifically for simplifying equations. It is designed such that one enters the desired logic function into the map, and reads out the simplest function to implement the logic function.

TABLE 13.9
Two-variable K-map

	$\overline{A}$	A
$\overline{B}$	$\overline{A} * \overline{B}$	$A * \overline{B}$
B	$\overline{A} * B$	$A * B$

(a)

OR

	$\overline{A}$	A
$\overline{B}$	0	2
B	1	3

(b)

The K-map assumes that the function is expressed in minterm format; however, if it is expressed in maxterm format zeros are entered instead of ones in the following instructions. Also, if the simplified expression is desired in maxterm format, read out the zeros instead of the ones in the following instructions. Note that the K-map is an easy method of converting from SOP to POS notation. Shown in Tables 13.9, 13.10 and 13.11 are the maps for two-, three- and four-variable K-maps.

Another notation that is used when working with minterms and maxterms is either the sum of minterms or product of maxterms. To use this notation, every variable must appear in each term, and each term is assigned a number correspond-

TABLE 13.10
Three-variable K-map

	$\overline{A}\overline{B}$	$\overline{A}B$	AB	$A\overline{B}$
$\overline{C}$	$\overline{A}\overline{B}\overline{C}$	$\overline{A}B\overline{C}$	$AB\overline{C}$	$A\overline{B}\overline{C}$
C	$\overline{A}\overline{B}C$	$\overline{A}BC$	ABC	$A\overline{B}C$

(a)

OR

	$\overline{A}\overline{B}$	$\overline{A}B$	AB	$A\overline{B}$
$\overline{C}$	0	2	6	4
C	1	3	7	5

(b)

TABLE 13.11
Four-variable K-map

	$\overline{A}\overline{B}$	$\overline{A}B$	AB	$A\overline{B}$
$\overline{C}\overline{D}$	$\overline{A}\overline{B}\overline{C}\overline{D}$	$\overline{A}B\overline{C}\overline{D}$	$AB\overline{C}\overline{D}$	$A\overline{B}\overline{C}\overline{D}$
$\overline{C}D$	$\overline{A}\overline{B}\overline{C}D$	$\overline{A}B\overline{C}D$	$AB\overline{C}D$	$A\overline{B}\overline{C}D$
CD	$\overline{A}\overline{B}CD$	$\overline{A}BCD$	$ABCD$	$A\overline{B}CD$
$C\overline{D}$	$\overline{A}\overline{B}C\overline{D}$	$\overline{A}BC\overline{D}$	$ABC\overline{D}$	$A\overline{B}C\overline{D}$

(a)

OR

	$\overline{A}\overline{B}$	$\overline{A}B$	AB	$A\overline{B}$
$\overline{C}\overline{D}$	0	4	12	8
$\overline{C}D$	1	5	13	9
CD	3	7	15	11
$C\overline{D}$	2	6	14	10

(b)

ing to the row in the truth table. The function is then written as a product or sum of the terms in the function.

EXAMPLE 13.11

Express the function $f = \overline{A} * \overline{B} * \overline{C} + \overline{A} * B * \overline{C} + \overline{A} * B * C + A * B * C$ in SOP notation.

Solution

The function $f(A,B,C) = \overline{A} * \overline{B} * \overline{C} + \overline{A} * B * \overline{C} + \overline{A} * B * C + A * B * C$ can be expressed by looking at the terms individually. The first term can be expressed by a zero, in other words the function is true if $A\ B\ C = 0\ 0\ 0$. The second term can be expressed by 2, or the function is true if $A\ B\ C = 0\ 1\ 0$, or 2 in binary. Thus, we could write the entire function as $f(A,B,C) = \Sigma(0,2,3,7)$, or $f(A,B,C) = m(0,2,3,7)$.

EXAMPLE 13.12

Express the function $f(A,B,C) = (\overline{A} + \overline{B} + C) * (A + \overline{B} + C)$ as a product of maxterms.

Solution

The first term is true if $A = 0$, $B = 0$, or $C = 1$; thus, it is represented by 1. The second term is true if $A = 1$, $B = 0$, or $C = 1$; thus, it is represented by 101_2 or 5. Therefore, the entire function can be represented by $f(A,B,C) = \Pi(1,5)$ or $f(A,B,C) = M(1,5)$.

Any adjacent pair represents a reducible pair of minterms. To enter a function into the K-map, simply place a one into the location in the map where there is a minterm.

EXAMPLE 13.13

Enter the following function into a K-map:

$$f = (\overline{A} * \overline{B} * C) + (\overline{A} * B * \overline{C}) + (A * \overline{B} * C)$$

or $\quad f = m(1,2,5)$

Solution

See Table 13.12.

TABLE 13.12
K-map for Example 13.13

	$\overline{AB}$	$\overline{A}B$	AB	$A\overline{B}$
$\overline{C}$	0	1	0	0
C	1	0	0	1

After entering the function into the K-map, the next step is to draw loops around the adjacent squares with ones in the square. The following is a list of rules for drawing such loops.

1. In the three-variable map, the first and fourth columns are considered to be adjacent.
2. In the four-variable map the first and fourth columns and rows are considered to be adjacent.
3. Each loop should be drawn around the largest group of two, four, or eight adjacent squares as possible. The number of entries in a loop *must* be an integral power of 2. Note diagonal entries are not considered adjacent.
4. A one can be included in more than one loop; however, do not draw a new loop unless it contains at least one that has not been included in existing loops.
5. When all loops are drawn, inspect the map to check if any loops are redundant. That is, have all the entries in that loop been included in the other loops?

Once the loops have been drawn, the next step is to read out the simplified equation. The following is a set of rules to follow for reading out the simplified equation in SOP format:

1. Each loop represents a simplified minterm for the equation. All minterms are ORed together after each has been read out.
2. Any variable in any given loop that appears in both the complemented and uncomplemented form drops out of the term. The variables left make up that term, and they are ANDed together.

EXAMPLE 13.14

Simplify the function $f(A,B,C) = m(1,2,5)$ using a K-map.

Solution
The map from Example 13.13 would have the loops drawn and read out to make the simplest implication of the function. This K-map with the loops drawn is shown in Table 13.13. The simplest function is then

$$f = (\overline{B} * \overline{C}) + (\overline{A} * B * \overline{C})$$

TABLE 13.13

K - Map

	$\overline{A}\overline{B}$	$\overline{A}B$	AB	$A\overline{B}$
$\overline{C}$	0	1	0	0
C	1	0	0	1

EXAMPLE
13.15

Simplify the function $f = M(0,1,2,4,5,8,9,12,13)$ using a K-map.

Solution

The K-map to read out in POS notation is shown in Table 13.14. *Note:* Read out the zeros. The K-map to read out in SOP notation is shown in Table 13.15. *Note:* Read out the ones.

TABLE 13.14

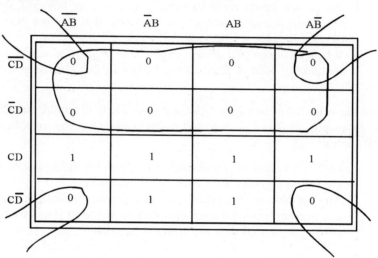

	$\overline{A}\overline{B}$	$\overline{A}B$	AB	$A\overline{B}$
$\overline{C}\overline{D}$	0	0	0	0
$\overline{C}D$	0	0	0	0
CD	1	1	1	1
$C\overline{D}$	0	1	1	0

TABLE 13.15

	$\overline{A}\overline{B}$	$\overline{A}B$	AB	$A\overline{B}$
$\overline{C}\overline{D}$	0	0	0	0
$\overline{C}D$	0	0	0	0
CD	1	1	1	1
$C\overline{D}$	0	1	1	0

The POS simplified equation is $f = \overline{C} * (\overline{B} + \overline{D})$, and the simplified SOP simplified equation is $f = (C * D) + (B * C)$. If we were to choose the simpler of the two logic circuits, it would be the POS notation; however, both circuits will perform exactly the same logic function and are interchangeable.

13.6 SSI IMPLEMENTATION OF BASIC DIGITAL CIRCUITS

13.6.1 The Transistor-Transistor Logic Family

Transistor-transistor logic or TTL is the culmination of several years of improvement in digital small-scale integrated (SSI) circuits. The current TTL families, 74XX and 54XX, are the standard for most digital circuits in use in the mid-1990s. This includes the 74L/54L, 74H/54H, and 74LS/54LS families as well, which are modifications on the original designs. TTL logic is the most widely used SSI and medium-scale integrated (MSI) logic family. Nearly every manufacturer has a TTL product line, and the most common ICs are produced by several different companies. The major difference between the 74 and the 54 families is the operating temperature. The 74 is the commercial family and is rated from 0 to 70°C, whereas the 54 is the military family and is rated from −55 to 125°C. Other than the temperature rating, these families are the same; however, the military version is usually much more expensive, and should not be considered unless the temperature environment requires it. These devices require an input voltage, V_{cc}, of 5 V.

The major dynamic differences between the different families are shown in Table 13.16. As you can see from Table 13.16 the selection of which family to use is a matter of speed, power requirements, and operating frequency. In most cases TTL families are not mixed, because the difference in gate delay can create timing problems. If a device is carelessly replaced by one from another family, a timing problem can become extremely difficult to find. Currently the 54/74 family is being

TABLE 13.16
TTL families

Series	Gates		Flip-Flops
	Gate Propagation Delay (ns)	Power Dissipation (mW)	Clock Input Frequency Range
54LS/74LS	9.5	2	dc to 45 MHz
54L/74L	33	1	dc to 3 MHz
54S/74S	3	19	dc to 125 MHz
54/74	10	10	dc to 35 MHz
54H/74H	6	22	dc to 50 MHz

replaced by the 54LS/74LS as the most popular. However, the 54S/74S is very popular where high speed is a criterion. The 54/74 TTL family is not specified too often anymore except when a 54LS/74LS part is not available. Also the 54S/74S family is being replaced by CMOS family devices.

13.7 FAN-OUT

First, a load is defined as the current drawn at an input pin of a standard gate within the logic family. Fan-out is then defined as the number of loads the output of a particular family can drive without modification. Furthermore, if an input is said to have a fan-in of four, that means it requires four times the current of the standard gate to drive this input. Fan-out and fan-in numbers are unique to a particular TTL family. If families are mixed, these numbers are invalid. In this case the designer must consult the appropriate data sheets to determine output and input current levels and then determine if the output current is below the sum of the loads the device must drive. Each family and subfamily has a few devices with larger fan-outs than the rest of the family. These devices are used to drive large numbers of gate inputs, to couple a low-power family to a higher power TTL device, or to drive outside world devices such as LEDs or displays. The typical fan-out of the 74/54, 74H/54H, and 74S/54S TTLs is 10. The typical fan-out of the 74L/54L and 74LS/54LS TTLs is 20. However, it is wise to consult the specific data sheet for the device in question.

EXAMPLE 13.16

The output on a 74LS00 must be used as the input for 12 other 74LS devices. The fan-out for the 74LS family is 20 standard loads. Determine if the device would drive the 12 loads if the loads were from the 74S family.

Solution
From the *TTL Data Book for Design Engineers* from Texas Instruments, the maximum high-level input current for the 74LS family is 20 μA, and for the 74S family, 50 μA. Thus, the maximum current the 74LS00 can supply is 400 μA. The maximum that could be attempted to be drawn on the device is 12 74S loads, which is 600 μA. Therefore, if the device is used in a circuit that has 12 74S loads, it could fail.

13.8 LOGIC LEVELS

Most TTL devices work with positive logic, that is when the input is high (about 5 V) the input is treated as a *high*. Another term for this is *active high*. Some devices, however, are *active low,* which means that when the input is low (about 0 V) the input is treated as true (a one on the truth table). Active-low devices are said to use negative logic. The truth table, or data sheet for the device, will indicate

TABLE 13.17
Logic levels for TTL families

	Input High (V)	Input Low (V)	Output High (V)	Output Low (V)
Minimum	2.0	0.0	2.4	0.0
Typical	3.3	0.6	3.3	0.2
Maximum	5.0	0.8	3.6	0.4

TABLE 13.18
Some TTL devices

TTL Number	Device Name
7400	Quad two input NAND gate
7420	Quad two input NOR gate
7404	Hex inverter
7410	Triple three input NAND gate
7420	Dual four input NAND gate
7432	Quad two input OR gate

whether the device is positive or negative logic. Many times a negative logic device is shown with inverters on the inputs. Remember also when using a negative logic device that the output is low when the output is true. The logic levels for TTL families are shown in Table 13.17.

Some popular TTL devices are listed in Table 13.18. Please note that the pin-outs are the same within the TTL family for any particular device. That is, a 7400 and a 74LS00 perform the same function and have the same pin-out. Refer to a TTL data book for more information on any particular device.

EXAMPLE 13.17

Design a digital automobile ignition switch such that the car will not start until certain preset conditions are met. The following is a list of the inputs of the system.

1. *Driver seat-belt detector:* Output is high when seat belt is fastened.
2. *Passenger seat detector:* Output is high when a passenger is in seat.
3. *Passenger seat-belt detector:* Output is high when seat belt is fastened.
4. *Ignition switch:* Output is high while switch is closed.

The system is to produce a high output under the conditions listed in the truth table shown in Table 13.19. The output is the starting motor relay.

Solution

$$f(A,B,C,D) = m(9,15)$$

Using the K-map to simplify the equation we get the result shown in Table 13.20. There are no adjacent ones, thus the output equation is read as $f(A,B,C,D) = (A * B * C * D) + (A * \overline{B} * \overline{C} * D)$. The logic circuit for this equation is shown in Figure 13.7(a), followed by a possible circuit wiring diagram as shown in Figure 13.7(b).

TABLE 13.19
Truth table for Example 13.17

A B C D	$f(A,B,C,D)$	
0 0 0 0	0	
0 0 0 1	0	
0 0 1 0	0	
0 0 1 1	0	
0 1 0 0	0	
0 1 0 1	0	
0 1 1 0	0	
0 1 1 1	0	
1 0 0 0	0	
1 0 0 1	1	
1 0 1 0	0	
1 0 1 1	X	Not a likely condition
1 1 0 0	0	
1 1 0 1	0	
1 1 1 0	0	
1 1 1 1	1	

TABLE 13.20

	$\overline{A}\,\overline{B}$	$\overline{A}B$	AB	$A\overline{B}$
$\overline{C}\,\overline{D}$	0	0	0	0
$\overline{C}D$	0	0	0	(1)
CD	0	0	(1)	0
$C\overline{D}$	0	0	0	0

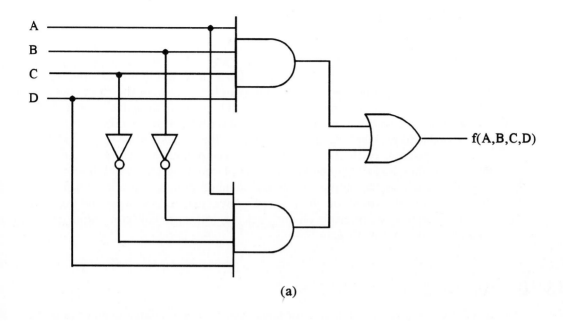

(a)

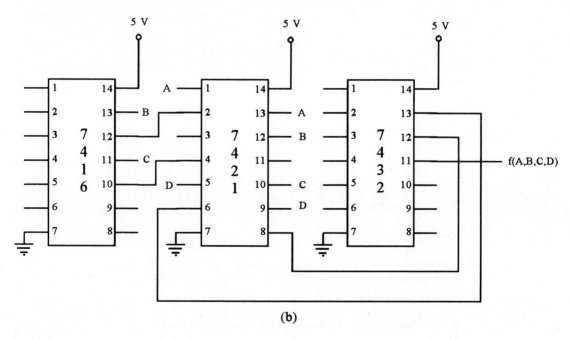

(b)

FIGURE 13.7
(a) Logic diagram and (b) circuit diagram for Example 13.17.

13.9   OPEN COLLECTOR GATES

Most TTL gates use what is called a *totem pole output stage.* The outputs can be tied together only via the inputs of another gate. If the outputs were to be tied together directly and one output was high while another was low, the output would be at some level, not high or low. In addition, one or more of the gates could be damaged as a result of excess current. Some TTL gates are available using a single output transistor with the output collector open (not tied to V_{cc}). In such a case, an external load resistor is required to complete the connection to V_{cc}. Figure 13.8 is a typical example of an open collector gate. The outputs of this type of gate can be tied together, and a common load resistor placed between the outputs and V_{cc}. Then if any one of the outputs is low the common output is at a low level. This allows the use of "wired" or "hard-wired" gates. Figure 13.9 shows the use of a 7405 hex inverter (six inverters on one package, with open collector output gates) wired as a six-input NOR gate.

13.10   TRISTATE OUTPUT GATES

The tristate output was designed for large systems where a large number of gates must tie to a single point, or *bus.* This is often the case in most large digital and microprocessor circuits. Tristate logic uses the standard two high/low logic levels, but they also have a third state. This third state is a high impedance output, which effectively disconnects the output from the common point. Most gates of this type will not have any output other than an apparent open circuit unless it is enabled, usually by a high (or low) on the enable input of the device. In applications that

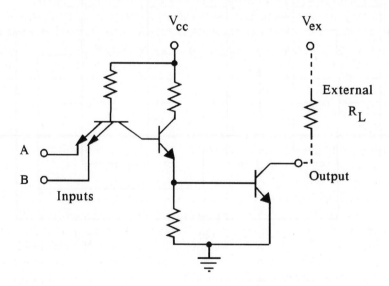

FIGURE 13.8
Typical open collector gate circuit.

FIGURE 13.9
Open collector hex inverters wired for a six-input NOR.

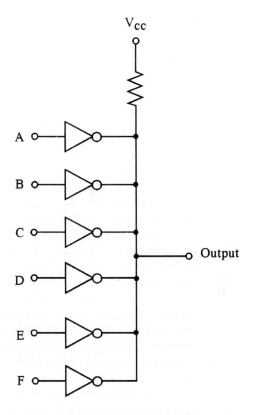

require tristate output, no other type of device can be substituted. Also care must be used in assuring that only one device is enabled at any given time for any given output point. If two devices are enabled at the same time and their outputs are at different levels, the status of that point is not defined. This type of problem is often referred to as "fighting for the bus" or "bus contingency" in microprocessor circuits.

13.11 CMOS FAMILY

CMOS stands for complementary metallic-oxide semiconductors. These devices have several improvements over the TTL or TTL-LS families. The major advantage is that of power consumption. The CMOS family is rated at 10 nW per gate under static conditions, and 10 mW per gate when toggling at a 1-MHz rate. It offers a 50:1 to 100:1 power saving when compared to the TTL-LS family. The noise margin is better than 2 V when operated from a 5-V supply, while TTL-LS will only guarantee a 0.4-V margin. The CMOS family also has a theoretical fan-out of infinity, because the input impedance of the gates is on the order of 10^{12}. However, because of input capacitance and gate delays, the practical fan-out is about 50. The last advantage of CMOS is that of the power supply. CMOS devices will operate over a power supply range of 3 to 15 V. This removes the need for tight power supply regulation. The logic levels for CMOS devices are listed in Table 13.21.

TABLE 13.21

Logic levels for CMOS devices

	% of V_{dd}		With 5-V Supply	
	High (%)	Low (%)	High (V)	Low (V)
Minimum	70	0	3.5	0.0
Maximum	100	30	5.0	1.5

Some final notes must be made, however, about using CMOS family devices. The first is that of static protection. CMOS devices are very sensitive to damage from static electricity. Therefore, care must be used when handling these devices. Grounding straps and other forms of static protection should be used when handling CMOS devices. Although most CMOS ICs are now protected by diode circuits at the inputs, they must still be handled with care. The last item to consider is floating grounds on inputs with CMOS devices. While TTL families may permit the floating of unused inputs, because of the high capacitance of the inputs of CMOS devices this practice will absolutely not work. All inputs that do not connect to gate outputs must be tied to the positive supply or ground. A floating CMOS input can cause problems of the worst kind. The input impedance is so high that time constants are often in terms of hours; therefore, the circuit may work for some time, but when the input drifts high or low the circuit may then fail. Furthermore, a floating input can cause a gate to oscillate and draw abnormally high currents from the supply.

13.12 MSI IMPLEMENTATION OF MORE GENERAL DIGITAL CIRCUITS

As discussed in Section 13.5, digital circuits are often simplified using a K-map. This method leaves the circuit in either POS or SOP notation. It is certainly easier to use an IC that already has the required gate configuration. Table 13.22 presents devices that provide commonly required expressions, along with the function of that device. Please note that several of these devices are expandable, which means they can be used to implement a more complex circuit. The circuit diagrams for these devices can be found in any TTL data book. Table 13.22 is not by any means a complete list of the multiple gate devices on the market. Again it is wise to consult a TTL data book to determine whether or not the logic circuit has already been implemented on a single chip.

Another method of implementing digital logic functions is with the use of a multiplexer. A multiplexer is a device designed to select one of several signals for output. The user selects which input is to be output using the "Select Inputs" pins. The device will then decode the selection and output whatever is on the selected "Data Input" pin. These devices were originally intended to be used for sharing

TABLE 13.22
Some MSI gate combinations

TTL Part Number	Function
7450	$Y = AB + CD + X$, two such gates, one expandable X is the output of a 7460 expander
7451	$Y = AB + CD$, two such gates
7452	$Y = AB + CDE + FG + HI + X$ X is the output of a 7461 expander
7453	$Y = AB + CD + EF + GH + X$ X is the output of a 7460 expander
7454	$Y = AB + CD + EF + GH$
7455	$Y = ABCD + EFGH + X$ X is the output of a 7462 expander
7460	$X = ABCD$ and $X = ABCD$, two gates, both X and X are available for use with 7423, 7450 or 7453
7461	$X = ABC$, three such gates, for use with 7452

data lines with microprocessor circuits; however, they also provide a straightforward method of designing digital logic circuits. In fact, any eight-to-one multiplexer can provide a one-package logic circuit for any three-variable logic function. Any 16-to-1 multiplexer can be used to implement any four-variable function.

EXAMPLE 13.18

Use a 74LS151 Data Selector/Multiplexer to realize the truth table of Table 13.23.

TABLE 13.23
Truth table for Example 13.18

m	C	B	A	f
0	0	0	0	0
1	0	0	1	1
2	0	1	0	1
3	0	1	1	0
4	1	0	0	0
5	1	0	1	1
6	1	1	0	0
7	1	1	1	0

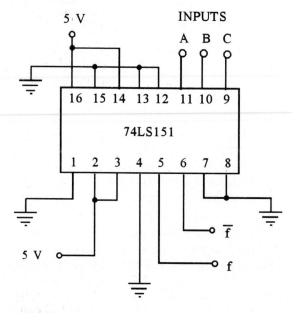

FIGURE 13.10
Solution to Example 13.18.

Solution
The technique is really quite simple. First start with the truth table in Table 13.23. The entire procedure is to connect the signals C, B, and A to the appropriate "Select Inputs" on the multiplexer. Then place on the "Data Inputs" the desired function. In other words, the inputs D_0 through D_7 will be the same as the last column in Table 13.23. This is done by tying each of the "Data Inputs" either to ground or V_{cc}. The solution is shown in Figure 13.10. One added advantage in using a multiplexer is that the inverse output is also available if needed.

13.13 LATCHES AND FLIP-FLOPS

Latches and flip-flops are both sequential logic devices. The only difference between latches and flip-flops is that the term *flip-flop* is generally reserved for devices that trigger on a clock edge. A latch on the other hand is level sensitive. In other words, the output is not only dependent on the present inputs, but also on the previous state of the device. The simplest of such devices is the R-S latch. This device is made up of two NAND gates and is activated by LOW logic levels. A low on the $\overline{S}$ input on the circuit will set Q high, and $\overline{Q}$ low. This output is held until the $\overline{R}$ is pulsed low, at which time outputs are reversed. Shown in Figure 13.11 is an R-S latch from NAND gates, the symbol for an R-S latch, and a timing diagram for the device. Also an R-S latch can be made from NOR gates. This latch works the same, except that a high pulse sets and resets the latch. The R-S latch does have the

FIGURE 13.11
An R-S latch made from two NAND gates.

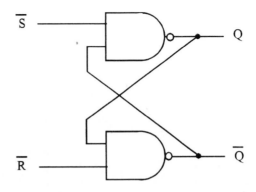

problem that if both inputs are active, it will be driven into an illegal state. When this happens the output is unpredictable, and this is an unacceptable situation.

To correct this problem, the J-K flip-flop is used in most situations. This device is readily available from all IC manufacturers. The simplest way to understand how the J-K flip-flop works is by looking at a state table. Note that the J-K flip-flop only changes state at the trailing edge (for most J-K flip-flops, although some change at the leading edge) of the clock input. The purpose of the clock is to ensure that all events happen at the same time. Table 13.24 shows the truth or state table for a J-K flip-flop. The inputs on a J-K flip-flop labeled S and R are used for setting the initial state of the device. A high on S will set the initial state high, and a high on R will set the initial state low. These inputs are override functions that are independent of the clock and other inputs. Only one of these inputs can be used at a time. They are not normally used as auxiliary data inputs.

One of the most popular uses of flip-flops is in the design of counters. Figure 13.12 shows a counter referred to as a binary ripple counter. Table 13.25 shows the output of the counter for each clock cycle. This counter outputs the number of clock cycles that have been input to the clock of J-K flip-flop A. The outputs are reversed, however: A is the least significant bit, and C is the most significant bit for the counter.

Counters can be designed to count up to any number less than the maximum for any number of flip-flops. The maximum number of unique states that a counter can have is referred to as the *natural modulus* of the counter. The natural modulus of a counter is equal to 2^N, where N is the number of flip-flops in the counter. To

TABLE 13.24
Truth table for a J-K flip-flop

J	K	Q_{n+1}	Result
0	0	Q_n	(No change)
0	1	0	(Reset)
1	0	1	(Set)
1	1	NOT(Q_n)	(Toggle)

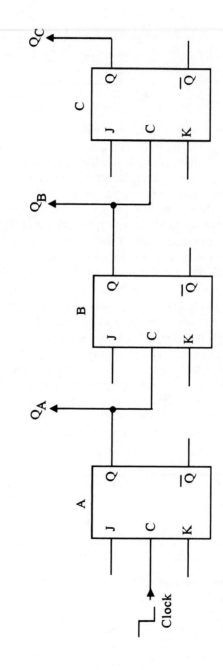

FIGURE 13.12
Binary ripple counter made from J-K flip-flops.

TABLE 13.25
Binary ripple counter output

Count	Q_c	Q_b	Q_a
0	0	0	0
1	0	0	1
2	0	1	0
3	0	1	1
4	1	0	0
5	1	0	1
6	1	1	0
7	1	1	1

design a counter with a modulus less than the natural modulus, one must use feedback. An example of a counter that uses feedback to reduce the modulus of a counter is shown in Figure 13.13. This counter counts from 0 to 4, and resets to zero.

Many commercial counters are available, so the need to design a discrete counter is eliminated. Many of these counters, such as a 74LS90, are designed to

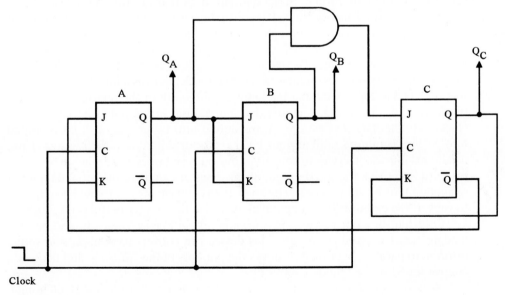

FIGURE 13.13
Counter to implement 0 to 4 and reset.

count a preset sequence. The 74LS90 counter is designed to count up to 9 (binary) and reset to zero. Also, some counters are programmable, such as the DM8555 and DM8556. The DM8555 is a decade counter (will count up to 9 binary and reset) that can be preloaded with a starting count whenever it is reset. Thus, if a counter is desired that will count six unique states, one needs to preload the counter with binary 4. The counter will then have six unique states (4, 5, 6, 7, 8, and 9). The DM8556 behaves the same way, except it is a binary counter (will count to 15 binary and reset). Another important group of commercially available counters are up/down counters. Thus far we have discussed only up counters, that is, counters that will increase the count on every clock pulse. Up/down counters have two clock inputs. Every time the up input is pulsed, the counter will increase the output, and every time the down input is pulsed, the count will decrease by one. An example of an up/down counter is the 74LS192 up/down BCD counter and the 74LS193 up/down binary counter. Both versions are fully programmable and can be used as modulo-N counters. Each output can be set to a high or a low by entering the desired data at the inputs and then pulsing the load input low. This is an asynchronous load since it is independent of the clock inputs. More information is available on all of the mentioned counters in any TTL data book.

13.14 THE SHIFT REGISTER

Shift registers are another category of digital circuits based on sequential logic elements. They are useful for temporary data storage, and for conversion from serial to parallel data format. Serial data are data that are transmitted one bit at a time over a transmission line. Parallel data have all eight bits transmitted at one time (assuming an eight-bit system). Both systems have their advantages and disadvantages. The serial type of system is slower, but only requires one (two with a ground) transmission line. The parallel system requires eight (nine with a ground) or more transmission lines, but it is faster. Figure 13.14 shows a four-bit shift register that uses D flip-flops. Note that the input is shifted out of the register with a four-bit delay from flip-flop D. The first data bit is input into flip-flop A. At the next clock cycle the data that were in A are shifted into B, and the next bit is shifted into A. This continues until all four bits have been shifted into the register. Thus, it requires four clock cycles to convert a four-bit serial word to parallel. If an eight-bit word is desired, the serial output of this register can drive the serial data input of another four-bit shift register. However, then eight clock cycles would be required to convert the eight-bit word to parallel. A device that is very useful is the 74LS194 universal four-bit shift register. It is capable of shift right, shift left, and parallel loading, and has parallel outputs. This device can convert parallel data to serial, or serial to parallel. Figure 13.15 shows the pin-outs of the 74LS194, and the mode control for S1 and S2 mode control pins.

The output of a shift register can be connected back to its input, creating a counter with N unique states, where N is the number of flip-flops in the shift register.

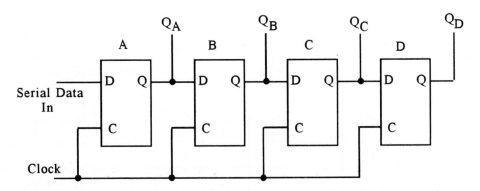

FIGURE 13.14
A four-bit shift register using D flip-flops.

This type of counter is called a *ring counter*. Referring to Figure 13.16, assume an initial condition of A being high, and all others low. On the first negative clock edge, the high bit will shift to the right by one position. It will shift to the right again once each clock pulse, until the fifth, at which time it will move back to A. Then the process will start over again. Ring counters are sometimes used as control sequencers: one operation follows another, followed by another, and so on, until

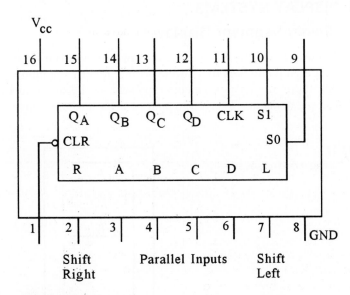

FIGURE 13.15
74LS194 four-bit shift register.

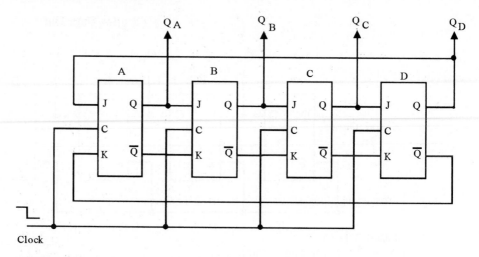

FIGURE 13.16
Ring counter using J-K flip-flops.

the process restarts. Please note in Figure 13.16 that some means of starting with flip-flop A high and all others low needs to be added. This can be done using the PSET and the clear pins of the flip-flops.

13.15 DIGITAL DISPLAY SYSTEMS

13.15.1 Seven-Segment Display

Figure 13.17 shows a typical seven-segment display. The seven-segment display is used to convert digital numbers to a display format in terms of numbers or characters. Note that there are seven segments representing the different LEDs on the display. These segments are referred to as *a* to *g*. To present a given number on

FIGURE 13.17
Segment identification for seven-segment display.

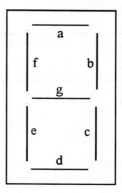

the display, one must light the correct LEDs. The two types of seven-segment displays are *common anode* and *common cathode*. The common anode has all seven anodes tied together, and this common point is tied to V_{cc}. The segments are then lit up by connecting the desired segments to ground. The common cathode seven-segment display has all seven cathodes tied together, and this common point is tied to ground. The segments are then lit up by connecting the desired segments to V_{cc}. Current-limiting resistors are usually required between V_{cc} and the display on the common cathode display, and between ground and the display on the common anode display. This is to prevent the current from becoming large enough to damage either the display or the display driver.

13.15.2 Display Decoder-Drivers

A decoder is a circuit or device having output that is related to its inputs by a preset pattern. Decoders are often used to convert binary, BCD, or ASCII to an uncoded form. An example of a BCD to seven-segment code decoder is the 74LS47 decoder-driver. The 74LS47 converts a BCD number on the four input pins $(A-D)$ to the correct code for driving a seven-segment display (pins $a-g$). The output of pins $a-g$ can be connected directly through a current-limiting resistor to the display, since each output pin can sink 24 mA of current. The 74LS47 is used with common anode displays, and when a given segment is turned on, will take that pin to ground. The 74LS47 also has an RBI input, and a RBO output. If the RBI is low, the BCD code of 0000 will cause the display to be blank, and produce a low on the RBO

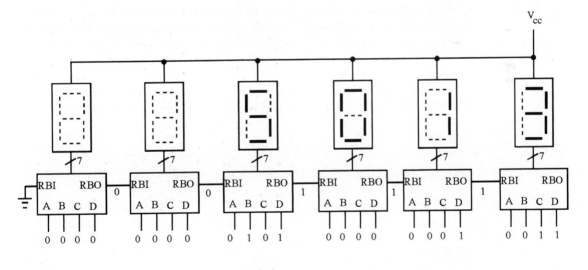

FIGURE 13.18
Using the 74LS47 to drive seven-segment displays.

pin. If the RBI is high, the BCD code 0000 will be displayed as a zero, and the RBO will be high. If the input BCD is anything but zero, the number will be displayed, and RBO will be high. The purpose of this is to blank out leading zeros. Figure 13.18 shows a six-digit display using seven-segment displays and a 74LS47.

13.16 USING DIGITAL DEVICES FOR MEMORY SELECTION IN A MICROCOMPUTER SYSTEM

A memory map is an important item when working with a microprocessor system. In a typical microprocessor system, several memory and I/O devices usually must be accessed. The addresses of these devices are placed on the address lines on the microprocesor's output pins. A typical memory chip may have 4096 (4K) memory locations, whereas the microprocessor with a 16-bit address bus can address 65,535 memory locations. There also may be several such memory devices in the system (up to 16). Therefore we need a method of determining which memory device is to be accessed at any given time. Suppose there are 16 address lines coming out of the microprocessor; then for 4K memories only 12 address lines are required into the memory devices. Table 13.26 shows a possible memory map using six such memory devices. Note in Table 13.26 that the least significant three hexadecimal digits are the 12 bits of the address bits that are input into the memory devices, and the most significant hex digit is the upper four bits of the address. It is these four bits that must be decoded to select which memory device we want to talk to at any given time.

One method to do this would be to use discrete digital logic to decode the upper four bits of the address. This would require a digital circuit for each memory device. However, this is not necessary. The easiest method would be to use a device such as a 74L154 4-line-to-16-line decode/demultiplexer. Figure 13.19 shows how this can be accomplished. Note that the outputs of the decoder are input to the chip-select-not inputs of the memory device. On the 74L154 all the outputs are active low, but this is standard on many microprocessor systems. Also note that

TABLE 13.26

Memory map

Address	Device Number
$0000–$0FFF	1
$1000–$1FFF	2
$2000–$2FFF	3
$3000–$3FFF	4
$4000–$4FFF	5
$5000–$5FFF	6

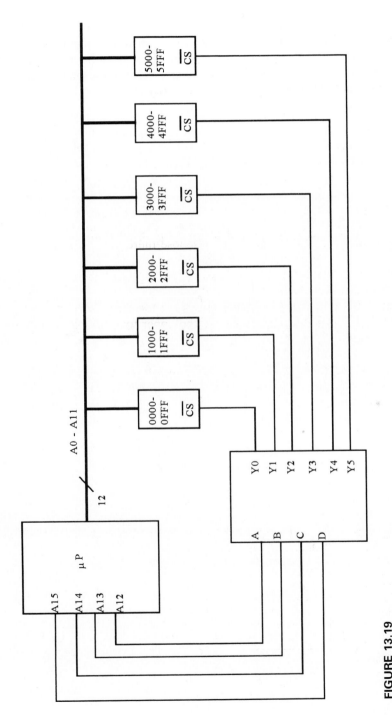

FIGURE 13.19
Decoding memory address using a 74L154 decoder.

there are two select pins on the 74L154 that are active low. These pins must be held low in order for the device to work. Another popular device for memory selection is the 74LS138, a 3-to-8 line decoder. This device has both an active-low and an active-high chip select (it has two active-low pins). Note that with both devices all chip selects must be active for the chip to be active. It is also possible for a 4-to-16 line decoder to drive the chip select on a smaller decoder to decode more than four address lines.

14

LARGE-SCALE INTEGRATION CIRCUITS*

14.1 INTRODUCTION

In the early 1960s, the electronics industry began commercially producing integrated circuits (ICs). With the introduction in 1971 of the first 8-bit general-purpose microprocessor by Intel Corporation, and the rapid development of semiconductor memories, the sale of integrated circuits mushroomed into a multibillion dollar industry. We are now in an age in the electronics industry where entire system functions are being integrated onto silicon chips. We refer to these chips as *large-scale integration (LSI) circuits*. It is becoming rare for a designer in any area of the electronics industry not to be exposed to LSI semiconductor circuits.

This chapter discusses a variety of LSI circuits that engineers and technologists may encounter in their work. A basic understanding of the subject matter is assumed, but we emphasize definitions of terminology and concepts when they are first introduced.

Large-scale integration (LSI) and *very-large-scale integration* (VLSI) are two good examples of terms without standard definitions. VLSI definitions have changed as technology has advanced and can now mean more than 100,000 devices per chip, or more than 1000 equivalent logic gates per chip. In this chapter, no distinction is made between LSI and VLSI circuits.

To aid in the understanding of the actual use of the LSI devices, several types of LSI chips are presented along with examples of their use.

Most of the LSI devices discussed here are silicon. Although devices using gallium arsenide and other materials are in use, GaAs devices pose some unique fabrication problems. Silicon predominates because it has an ideal combination of

* This chapter was written by Prof. Robert J. Borns, Purdue University.

bandgap, oxide stability, and natural abundance as compared to other fabricating materials.

A large part of the chapter is devoted to semiconductor memories and logic, because they comprise the bulk of all LSI devices currently in use. Some special ICs are discussed in Section 14.4, and the remainder of the chapter is devoted to packaging and quality, and how those two factors influence the choice of an appropriate LSI device.

14.2 SEMICONDUCTOR MEMORIES

Semiconductor memories are LSI devices that have the ability to store information in binary form. They are used in conjunction with a microprocessor or some other type of computer that is able to control the storage and retrieval (writing and reading) of information.

Almost every semiconductor company produces some type of LSI memory device. A typical semiconductor memory data book may contain from 20 to well over 100 different memory part numbers. The dynamic RAM semiconductor market is currently in excess of four billion dollars. The strength of the entire semiconductor industry is closely related to the success of memory products.

14.2.1 Technology Types

Depending on the semiconductor process and device technology, memories can be categorized into several technology types.

NMOS, PMOS, and CMOS are the three main types of *field-effect transistor* (FET) memories. *N-channel metal-oxide semiconductors* (NMOS) memories are currently the most widely used, because they are less expensive to fabricate than the other FET technologies, and have some density advantages.

P-channel MOS (PMOS) technology is seldom used exclusively to fabricate LSI memories since *p*-channel transistors have slower operating speeds than comparable *n*-channel transistors.

Complementary MOS (CMOS) memories have both *n*-channel and *p*-channel transistors in their circuitry. This allows designers to design circuits that consume less power than circuits designed exclusively with NMOS or PMOS technology. In addition, CMOS has superior noise immunity and temperature stability characteristics. As CMOS technology improves, it is gradually replacing NMOS as the dominant memory technology.

Bipolar memories are the other main technology type currently in use. Bipolar memories use bipolar transistor circuits, which can be designed to operate faster than FET transistor circuits. The penalty for the increased speed is a dramatic increase in the power requirements.

Bubble memories operate on the principle of storing information via magnetic domains called *bubbles.* Although bubble memories are durable and can store very large amounts of data (1-Mbit devices are currently available), they require support chips to supply a magnetic field. The additional chips add to the memory cost, making bubble memories much more expensive per bit than other types. The main advantages of bubble memories are large storage capability and nonvolatility.

One disadvantage of bubble memories is the inability to access memory bits randomly. Bubble memories, therefore, are a type of *serial* memory.

14.2.2 Comparing Memory Technology Types

The major memory technologies currently in use are the two FET memories, NMOS and CMOS. FET memories can be manufactured economically, have good performance characteristics, and generally have low power requirements. Disregarding cost, CMOS would be the choice over NMOS as the CMOS technology becomes more mature. The speed of CMOS is comparable to NMOS, and the standby power requirements are less. CMOS, however, has an inherent problem with latch-up, a condition in which the memory timing signals latch to a nonfunctioning state. The problems with latch-up are progressively being resolved, and CMOS in most cases is as reliable as NMOS.

If memory speed is the most important design parameter, bipolar memories are the memory of choice. The drawbacks are higher cost per bit, increased power requirements, and lower available densities. Bipolar memories have memory access speeds of around 5 to 25 ns, whereas NMOS and CMOS memories have access speeds in the 100- to 200-ns range.

EXAMPLE 14.1

A design engineer has been asked to choose a memory that will be fast enough to interface with an Intel 8085AH-1 microprocessor. What memory technology should be chosen?

Solution
An Intel 8085AH-1 has a maximum operating frequency of 6 MHz from its specification data. The minimum clock period is 1/6 MHz = 167 ns. Any memory with a cycle speed of less than 167 ns will be faster than any memory operation the microprocessor can accomplish. NMOS or CMOS memory technology would be chosen.

14.2.3 Memory Storage Characteristics

Memory can also be classified by its storage characteristics. *Volatile* memories lose their stored information when power is removed. *Nonvolatile* memories retain the stored information even when power is removed.

Dynamic memories require periodic refreshing of the stored information in order to maintain bit integrity. *Static* memories will maintain stored information (bits) without any type of refreshing. Some memory manufacturers also produce a *pseudostatic* memory, which is actually a dynamic memory that has built-in automatic refresh circuitry, making it appear to be static memory.

14.2.4 Memory Classification

Memories are further classified as *random access memory* (RAM) and *read-only memory* (ROM). The memory content of RAM can be changed by the user. (They are also known as *read/write memories.*) Standard ROM memories cannot be changed by the memory user. The memory contents are programmed into the memory during manufacturing.

A special type of ROM, called a *programmable ROM* or *PROM,* allows the memory user to program the memory if the memory IC is removed from the circuit application. A special device called a memory code "burner" or PROM programmer is used.

If the PROM can be erased and the memory reprogrammed, it is called an *erasable PROM* or *EPROM.* Currently, two types of EPROM are available. *EEPROMs* are electrically erased by applying a high voltage (+20 V) to the memory. An *ultraviolet EPROM* or *UVEPROM* is erased by exposing the memory cells to a concentrated ultraviolet light source. A transparent window in the plastic or ceramic IC packaging is provided for this purpose. EPROMs usually have a limited number of erasures.

EEPROMs have several advantages over UVEPROMs. With EEPROMs, selected bits can be changed. The entire memory content, however, is erased with a UVEPROM. Second, EEPROMs allow for changes to be made *in situ,* that is, in the circuit application.

EEPROMs also have their disadvantages. The reliability of EEPROMs may be less than that of UVEPROMs, and often the cost is higher per memory bit. Figure 14.1 illustrates memory classification.

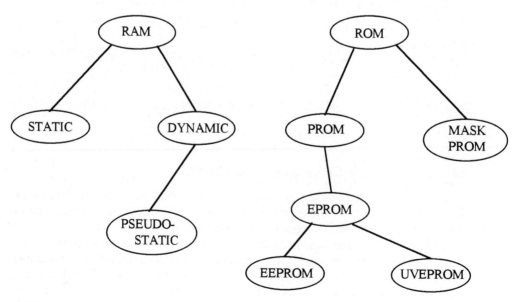

FIGURE 14.1
Memory families.

EXAMPLE 14.2

How is a UVEPROM erased? How long does it take? Will normal room lighting erase the contents?

Solution

Most UVEPROMs are erased by exposure to a high-intensity ultraviolet light source with a wavelength of less than 4000 Å for 15 to 60 min depending on the manufacturer.

The *integrated dose* is defined as:

$$\text{UV intensity (W/cm}^2) \times \text{Exposure time (s)} \qquad \textbf{(14.1)}$$

The minimum integrated dose for erasure is typically 15 W-s/cm^2.

Fluorescent lamps and sunlight contain light wavelengths of less than 4000 Å. To ensure memory integrity, an opaque cover should be placed over the erasure window.

14.2.5 Memory Operation and Associated Terminology

To understand fully the operation of LSI memories, we must first have a basic understanding of the many terms and definitions used to describe memory operation.

The purpose of all types of memories is to store data. The IC memory stores data by means of the binary scheme previously described. The time it takes for a memory device to retrieve stored information is described by means of performance parameters. The two most common performance parameters are *access time* and *cycle time.*

Access time is the time it takes a memory to retrieve a bit of information out of its storage array and take it to the data output.

Cycle time is the total time it takes to get a bit (or bits) of information out of storage until you can begin to get another bit of memory data.

The definitions for access time and cycle time are similar, but not exactly the same. Cycle time includes the additional time it takes in some memories to restore the data that are read and to set up a new memory bit address. So, for many LSI memories, the cycle time will be longer than the access time. Some semiconductor memories, such as static RAMS (SRAM), can have cycle times equal to the access time since the read operation does not require rewriting of the memory bit read. For this reason, the read operation in a static RAM is termed *nondestructive.*

Memory manufacturers like to quote access times for memory chips since they are faster than cycle times and naturally sound better. Often when designing with memories, the designer is just as concerned with cycle time, since overall system performance is typically related to cycle time as well as access time.

Memory *configuration* is a broad term that refers to the format of the memory data output. Various design schemes are used to output memory bits depending on the intended memory application. The type of memory (RAM, ROM, EPROM) and the technology (CMOS, bipolar, etc.) play an important part in determining the memory configuration.

All semiconductor memories have an area of the chip that is used strictly for memory storage. This area is called the memory *cell* area. Each cell is capable of storing 1 bit of information.

Nearly all dynamic RAMS (DRAMs) have cells made up of one FET transistor plus storage area (capacitance). Static RAMs require four or six transistors plus the storage capacitance for each memory cell. In either case, the transistor is used to control the input and output of the cell's data bit.

To define each DRAM memory cell uniquely, a series of control lines called *bit lines* and *word lines* operates the transistors and thereby controls access to the cell. The clock signals that control the word lines and bit lines are known as *row address strobe* or *RAS,* and *column address strobe* or *CAS. RAS* and *CAS* are timed with the address lines to select which memory cells will be accessed for reading or writing. Additional details are discussed in the DRAM chip operation example that follows.

Static RAMs have less complicated timings than DRAMs since the memory size is smaller and not dynamic. An output enable (OE) and write enable (WE) are used in conjunction with the address lines to select memory in an SRAM.

ROMs have the least complicated timings of memory chips, since no write operations occur. An OE and address line timings are used to select memory to read.

A *serial* memory has the data bits coming out one at a time. A 64K $\times$ 1 memory has 2^{16} memory cells, for a total of 65,536 memory cells. The term "64K" is shorthand for the 65,536 memory cells. The "$\times$ 1" designates that the memory is serial, and is called the memory's *organization.*

Typical memory organizations are $\times$ 1, $\times$ 4, $\times$ 8, and $\times$ 16. Organizations of $\times$ 2 and $\times$ 9 are less common. An 8K $\times$ 8 memory chip would still have 65,536 cells for storage, but now each time you access data, 8 bits come out at a time. This of course means that the chip needs to have eight data output lines instead of one. The 8-bit organization is often called *byte organized,* whereas the $\times$4 organization is called *nibble organized.*

Eight-bit organizations are also often referred to as *words.* An 8K $\times$ 8 memory organization can also be referred to as an 8K word with 8 bits per word. Care must be taken when using "word" terminology, because a word is defined as the actual number of bits that are operated on at any one time by a microprocessor system. A word can vary from 4 to 32 bits depending on the microprocessor being used.

Some LSI memories have internal *registers* or *buffers* that temporarily store output data until the device using the memory chip is ready to use the data. A register can increase the *effective* bit access time by allowing the memory chip to go after the next output data while the previous output data are being used. (Recall that part of the cycle time is used for doing things other than accessing data.) Put simply, a register allows for more efficient use of performance time.

A feature that enhances effective DRAM memory performance is *page mode.* With page mode, each time a memory bit (cell) is read, the capability exists to read additional memory bits along the column (changing CAS) with all of the bits having the same row address (fixed RAS). This increases net performance.

EXAMPLE 14.3   Compare the bit access times for a memory with an access time equal to 100 ns, and a cycle time of 250 ns, with and without page mode.

Solution

To read 3 bits without page mode, it would take 250 ns for each of the first 2 bits, and 100 ns for the third bit (since we do not have to wait until the cycle has completed for the last bit) for a total of 600 ns.

Now with page mode, if we were reading 3 bits along the same row, and the page mode cycle time is 140 ns, the total time required to read 3 bits would be 250 ns for the first bit (since all RAS and CAS timings are the same as the normal read); then, the second bit would require the page mode cycle time of 140 ns; and finally the third bit would require 100 ns for the access time. The total time is 490 ns, a savings of 110 ns. If additional memory bits were read along the same row, the net performance gain would be proportionally higher.

Nibble mode is similar to page mode, but only 4 bits at a time can be read while in nibble mode.

Savings in performance can also be achieved during memory write cycles using page or nibble node with DRAMs.

14.2.6 Dynamic RAM Terminology

Some definitions and terminology apply only to dynamic RAMs due to their operating characteristics. A dynamic RAM requires periodic refreshing of its stored memory data. This is due to the gradual leakage of capacitive charge that the DRAM uses to store information. For example, if the storage of charge in a memory cell is defined as a "HIGH" or "1," and the absence of charge is defined as a "LOW" or "0," the HIGHs would eventually lose their charge due to parasitic capacitance charge leakage inherent in DRAM designs.

To prevent the loss of memory bits, all cells in a DRAM are periodically *refreshed* or rewritten to replace any stored charge that has discharged. If a memory chip has a *refresh address counter* (RAC) built in, the memory user need only supply a periodic pulse to a pin that is the input to the RAC. The RAC then handles the addressing to ensure that all cells get refreshed. Other memory chips without an RAC may require that the user periodically cycle all addresses to refresh the memory. Cell data retention is the maximum amount of time any particular memory cell can maintain good data bit integrity. The frequency at which a memory user must supply a refresh pulse is referred to as the *refresh rate*. Two to 6 ms is a typical refresh rate in a DRAM, and will always be specified in the memory specifications.

Availability is a parameter related to refresh rates in DRAMs. Obviously, if a cell is being refreshed, it is not available for any type of normal cycle (read, write, etc.). Availability is defined as 100% minus the percentage of time refresh will interfere with read/write operations. Typical availabilities for DRAMs are 95 to 99%. If a memory chip is going to be used at the minimum access/cycle time, it is advisable to contact the manufacturer's applications engineers to aid in determining how availability will affect the design.

Some manufacturers have the refresh cycle interwoven with a write cycle to give *hidden refresh* capability. This can translate to an availability of 100% in some applications.

EXAMPLE 14.4

An engineer wants to design a low-cost hand-held device with a large memory capability. Dynamic RAMs are cheaper than static RAMs. Should DRAMs be chosen?

Solution

DRAMs have a higher density and are therefore cheaper than SRAMs, but are usually not practical for small consumer products. This is due to the necessity of refresh circuitry and, often, parity circuitry for DRAM designs. These additional circuits consume space and add to cost.

DRAMs are used predominantly in large memory systems where the additional circuitry space and cost are insignificant compared to the volume memory cost savings through the use of DRAMs instead of SRAMs. The hand-held device will probably need only a few memory chips, so CMOS SRAMs should be chosen.

EXAMPLE 14.5

A DRAM has in its specification a memory cell data retention time (refresh rate) of 4.0 ms, and 128 unique RAS addresses. What are two possible refresh methods?

Solution

Method 1: A concentrated refresh can be used by performing 128 consecutive refresh timing cycles every 4.0 ms.

Method 2: A *dispersed* or *deconcentrated* refresh can be used by performing 1 refresh timing cycle every 31.25 μs. (4 ms/128 = 31.25 μs).

14.2.7 Memory IC and Chip Characteristics

In addition to knowing the types of LSI memories and their operation, there are several other characteristics that LSI memory users need to understand in order to choose intelligently the type of memory that will fit their needs. Several of the more important characteristics are density, signal levels, power, and voltage. We also define redundancy and soft error rates. Timing is discussed in the memory chip examples.

The number of memory cells or bits that a memory IC has is called its *density*. Density is closely related to organization, since the organization can be multiplied to produce the density. A chip that has an organization of 256K $\times$ 4 bits would have a density of 1024K bits, which is often referred to as a density of 1 megabit, or 1 Mb. Table 14.1 shows a list of common memory densities.

The term is also used in a broader sense to indicate the number of memory cells per unit area in a memory chip or design. The larger the number of cells in a given chip area, the higher the density.

TABLE 14.1
Common memory
densities

Memory Density	Actual Number of Bits
1K	1,024
2K	2,048
8K	8,192
16K	16,384
64K	65,536
256K	262,144
512K	524,288
1 Mb	1,048,576
4 Mb	4,194,304
8 Mb	8,388,608
16 Mb	16,777,216
4 Gb	4,294,467,296

EXAMPLE 14.6

A 32K × 16 memory system is being designed. (a) How many 8K × 8 chips would be needed for the system? (b) How many *actual* memory bits will the system have?

Solution
(a) In this example, the "width" or "word" is 16 bits. Two 8K × 8 chips will be needed for each 8K × 16 memory block. Since we need four of the 8K × 16 memory blocks to form a 32K × 16 system (4 × 8 = 32), a total of eight 8K × 8 chips is needed.
(b) The total number of bits is found as follows:

$$8K = 8192 \text{ bits}$$
$$8K \times 8 = 8192 \times 8 = 65{,}536 \text{ bits}$$
$$65{,}535 \times 8 \text{ chips} = 524{,}288 \text{ bits.}$$

This is a 0.5-Mb memory system.

Power is another term that has several meanings when applied to semiconductor memories. Normal power consumption when the memory is being exercised (reads or writes) is called *active* power. This is calculated by multiplying the operating current times the supply voltage:

Active power: $\quad P_{av}(\text{active}) = I_{cc}(\text{active}) \times V_{cc}$ **(14.2)**

Memories also have another type of power called *standby* power. Standby is the state the memory is in when it is powered, but not being exercised. Standby power is found by multiplying the standby current times the supply voltage:

$$\text{Standby power:} \qquad P_{av}(\text{standby}) = I_{cc}(\text{standby}) \times V_{cc} \qquad \textbf{(14.3)}$$

In addition, DRAMs will have a *refresh* power related to the amount of current required to perform memory refresh operations. It is calculated in the same manner as active and standby power.

Some manufacturers also have memories with a "low-power" feature that will maintain volatile memory using a minimal amount of power. These memories are useful in applications that use a battery as a source of power.

CMOS is the technology of choice for memory users that have low power requirements. The main difference between CMOS and NMOS power requirements is the standby power. Active power between comparable NMOS and CMOS memories will not differ substantially. Bipolar chips have larger power requirements than NMOS or CMOS.

EXAMPLE 14.7

A TC51425GP CMOS 1-Mbit DRAM is used in a 4-Mbit memory system. What is the maximum active and standby power for the memory system?

Solution

Since the memory chip is 1 Mbit, four chips will be used. From the specifications in a data book, we find

$$V_{cc}(\text{max}) = 5.5 \text{ V}$$
$$I_{cc}(\text{active, max}) = 65 \text{ mA}$$
$$I_{cc}(\text{standby, max}) = 2 \text{ mA}$$

Therefore, using Eqs. (14.2) and (14.3):

$$\text{Active power} = 5.5 \times 65 = 357.5 \text{ mW (per chip)}$$
$$\text{Standby power} = 5.5 \times 2 = 11 \text{ mW (per chip)}$$

For a four-chip system,

$$\text{Active power} = 357.5 \text{ mW} \times 4 \text{ chips} = 1.43 \text{ W}$$
$$\text{Standby power} = 11 \text{ mW} \times 4 \text{ chips} = 44 \text{ mW}$$

Most state-of-the-art LSI memories require only a single 5-V power supply. This has been made possible through semiconductor circuit design innovation. Several older generation LSI memory chips and a few current chips require multiple voltage supplies to operate. A three-supply memory would typically need supply voltages of −5 V, +5 V, and +12 V. The negative voltage is called the *substrate* voltage. A negatively biased substrate is required on all silicon semiconductor memories, but current chips generate the substrate voltage internally, transparent to the user.

In addition to the single or multiple voltage supplies, all LSI memory chips require a ground connection. Ground is usually referred to as V_{ss}, the +5 V voltage as V_{cc}, the +12 voltage as V_{dd}, and the substrate voltage of −5 V as V_{bb} or V_{sub}. Information on supply voltages and tolerances can always be found in the dc operating conditions section of an IC's specification sheets.

Bipolar memories are designed to be compatible with *emitter coupled logic* (ECL) voltage. Bipolar memory chips have two voltage inputs, commonly labeled V_{cc} and V_{ee}. In use, V_{cc} is grounded, and V_{ee} becomes the supply voltage, which has a range for most bipolar memories from 0.5 to −7.0 V, with a typical value of −5.2 V.

Signal levels are a very important design consideration. The input and output voltages must be compatible with the other components used in the design. Fortunately, signal levels have been standardized to avoid voltage mismatches.

The most prevalent voltage convention is *TTL*. TTL stands for *transistor-transistor-logic*. TTL voltage levels have been defined for inputs and outputs as shown in Table 14.2. LSI components that meet the above specifications are called *TTL compatible*. The best LSI components will operate well within the TTL margins specified in Table 14.2. The closer to 0.0 V an input or output LOW is, the better. Similarly, the farther the HIGH voltages are above 2.0 or 2.4 V, the better. The important point here is that not all LSI chip manufacturers will have components with identical I/O voltages, since TTL compatibility falls within voltage ranges. This may have to be taken into consideration if designs are to have good operating margins.

The other major voltage level standard is CMOS. The input LOW voltage for CMOS compatibility is 0.0 to 1.5 V, and the input HIGH voltage is from 3.5 to 5.0 V.

Another logic voltage level standard is ECL. As previously discussed, ECL is a logic level coupled to bipolar transistor operation. Binary 0 is represented by −1.55 V. A binary 1 is represented by −0.75 V. ECL has gone through several evolutions. Two common ECL standards for LSI devices such as memories are the 10,000 (10 kb) and 100,000 (100 kb) ECL.

Another often used term when discussing signal levels on outputs is *tristate* or *three-state*. Tristate outputs have the ability to go into a high impedance mode in addition to the normal TTL high and low voltage levels. For memory chips, this translates into faster allowable logic level changes on the data bus. A tristate output in the high impedance mode is completely off, thereby not adding any load to the data bus. Tristate buffering allows for bidirectional signal flow on a data bus.

TABLE 14.2
TTL voltage levels

Signal	Typical Symbol	Defined Voltage Range (V)
Input LOW	VIL	0.0–5.0
Input HIGH	VIH	2.0–5.0
Output LOW	VOL	0.0–0.4
Output HIGH	VOH	2.4–5.0

Dynamic RAM memory chips are able to store bits of information by storing a capacitive charge in its memory cells. In high-density DRAMs, chip designers must make the memory cells as small as possible to increase the density of the memory. As the size of the storage cell decreases, the amount of charge that represents a bit of memory decreases proportionally.

DRAMs with densities of 64K and larger have memory cells that are small enough to be susceptible to alpha-particle radiation. Alpha particles have sufficient energy to destroy the charge stored in high-density memory cells. When this occurs, it is called a *soft error*. A memory bit that changes its logic level due to a soft error is not permanently damaged. Only the content (charge) of the cell is destroyed. A *hard error* is a memory cell that is permanently damaged and has lost its read/write storage capacity. The number of alpha particles a memory device is exposed to varies depending on the environment and alpha-particle sources present. Also, not all the alpha particles that penetrate a memory chip will destroy cell data. For these reasons, alpha-particle susceptibility data are given statistically, based on experimental data. A typical way of indicating the soft error rate (SER) is "percent fails per thousand hours of operation":

$$SER = X\%/\text{KPOH} \tag{14.4}$$

where X is the percentage of cell failures, and KPOH is 1000 power-on hours.

Another way of specifying SER is in *failures in time,* or *FITs*. Ten FITs would be 10 cell failures in one million operation hours. The acceptable industry standard for SER is 1000 or fewer FITs.

Soft errors can actually occur in almost any type of memory device with high-energy alpha particles. This includes static RAMs and bipolar RAMs. However, the soft error mechanisms are different and occur less frequently.

Most applications that use high-density RAMs have *error correction coding* (ECC) circuitry to correct soft errors that occur. SER is also the reason high-density memories are not used in small low-cost applications. The additional circuitry needed for ECC adds to the cost.

14.2.8 Choosing the Correct Memory for Your Application

Now that we have reviewed the memory technologies, types, and terminology, it is appropriate to look at the criteria for choosing the correct LSI memory based on design requirements. The major design criteria to consider are cost, performance, power consumption, amount, and function.

Quality and packaging are also important criteria, but their influence is considered separately in subsequent sections.

The assumption is that a particular circuit has been defined, and the decision has already been made to use LSI memory as opposed to tape, disc, or other nonsemiconductor memory. One or several different types of memory may be appropriate, and choices must be made.

Performance is how fast you intend to read (or write) data from the memory. If you have a microprocessor-based application, the performance may be dictated by the speed at which the microprocessor can read the memory. Keep in mind that cycle time will often be your *usable* memory speed. In general, the faster the memory, the more it will cost. Bipolar will be the fastest, and CMOS the slowest. Figure 14.2 shows the relative cost/performance of memories.

Function indicates the storage characteristics of the memory. This includes volatility, static or dynamic, and RAM, ROM, or PROM. Function and performance are closely related since the performance is dependent on the specific function type. A static RAM can provide cycle times down to 15 ns, while NMOS DRAMs are limited to about 50-ns cycle times.

Oftentimes memory selection depends on exactly where the design is in the production cycle. EPROMs are most often used in the development stage of products when code changes are often necessary. In mature products, ROMs are used since they are less expensive than EPROMs.

After deciding the performance requirements, the next step is deciding how much memory you need. The memory organization will be one of the decision inputs. Let's say the circuit operates with bytes of data, and the overall system

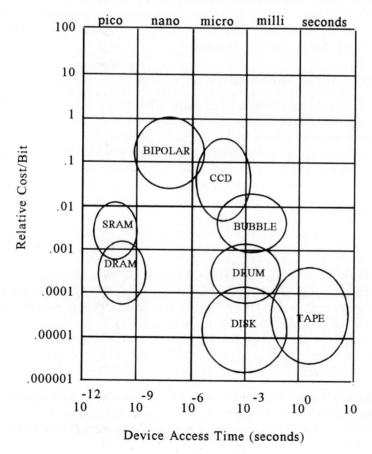

FIGURE 14.2

Cost/time comparison of some memory types.

memory requirement is 1 Mb. The next step would be to look in a memory data book from a semiconductor manufacturer to find out which chips are available with an 8-bit output organization. If 8K × 8 memory chips are chosen, 16 chips would be needed to provide 1 Mb of storage whereas one could also use a single 128K × 8 chip.

Power requirements are dictated by the type of memory IC selected and by the number of memory chips that will be active at any one time.

Memory *cost* is affected by many factors. Most manufacturers have their prices tied to the volume of memory purchased. The higher the volume, the lower the cost. Higher speed, reliability, or density will all add to the cost.

NMOS is usually the cheapest technology. CMOS is second, and bipolar is the most expensive.

DRAMs are cheaper than SRAMs since they have a higher density, but SRAM designs are less complex and costly since they do not require refresh circuitry. However, if memory requirements are large, the large savings on the DRAM cost will more than offset the added circuitry cost.

Another item to consider with cost is the number of support chips necessary to implement your memory scheme. Decoders and drivers may be needed in some applications, and not needed in others.

14.2.9 DRAM Technology

Dynamic RAMs command a large share of all semiconductor sales. Their read/write capability makes them very useful and their high bit density is advantageous for system memory density and low cost per memory bit.

Many new circuit innovations are first used in DRAM circuits, and manufacturers use DRAM density as a benchmark of technical competence.

The largest DRAM in current common use is 1M × 1, although 4M × 1 is available. Within the next few years both the 16M × 1 and then the 256M × 1 will appear.

The following paragraphs outline some of the common characteristics of the DRAM.

Refresh　As noted, the DRAM stores a 1 state as charge on a capacitor. This charge will gradually leak away and memory will be lost. Therefore, it is necessary to refresh the DRAM data periodically. Typical refresh requirements range from 2 to 6 ms.

Refresh is supplied by DRAM controllers connected between the memory and the processor bus. If read or write operations do not occur within the requisite time, the controller automatically initiates a refresh cycle.

Access Time　The time required to access data in the commonly available DRAMs ranges from 60 to 100 ns. This is adequate for a majority of microprocessors but slow for many of the high-speed processors, operating at clock frequencies of 66 to 100 Mhz (15 to 10 ns). In such cases *wait states* must be inserted while memory access occurs, slowing down overall execution.

Multiplexed Address Lines   A 1M × 1 DRAM has one input data line and one output data line, since one bit at a time is accessed. However it must have 20 address lines. To keep the package size and pin number reasonable it has become common to multiplex the address lines. In the case of the 1M × 1 DRAM, there would be only 10 address lines, as shown in Figure 14.3. The control line $\overline{OE}$ is for *output enable* when reading the memory, and $\overline{WE}$ is for *write enable* when writing to the memory. $\overline{RAS}$ and $\overline{CAS}$ are used for multiplexing as described in the next paragraph.

The 1M bits are arranged in a square matrix of 256K × 256K. Thus ten lines are used to address a row in the matrix and another ten are used to address a column. The row and column address select one bit location for input or output. Using multiplexing via the $\overline{RAS}$ and $\overline{CAS}$ lines, the same ten lines are used for both row and column addressing.

The READ cycle is detailed in this list:

1. The row addresses are set (A0 through A9).
2. The active-low $\overline{RAS}$ signal is applied to latch the row addresses.
3. The column addresses are set (A0 through A9).
4. The active-low $\overline{CAS}$ signal is applied to latch the column addresses.
5. The read signal, $\overline{OE}$, goes low for a read operation.
6. The chip places valid data on the D_{out} line.

Manufacturers are very helpful in explaining specifications when asked. Use their resources!

SIMM   Most microprocessor-based computers provide for expansion of RAM via sockets built into the computer motherboard that can accommodate additional memory. In most cases this memory is constructed from DRAM in the

FIGURE 14.3
Typical connections to a 1M × 1 DRAM

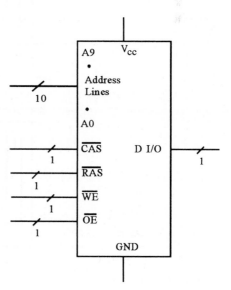

form of single in-line memory modules (SIMMs). These modules consist of a number of DRAMs arranged to provide the requisite memory.

EXAMPLE 14.8

Show how 1M × 1 DRAMs can be connected to provide a 1M × 8 computer memory system.

Solution

Clearly we will need 8 1M × 1 DRAMs connected in parallel as shown in Figure 14.4. Note that the $\overline{CAS}$ and $\overline{RAS}$ lines are connected in parallel as well as the

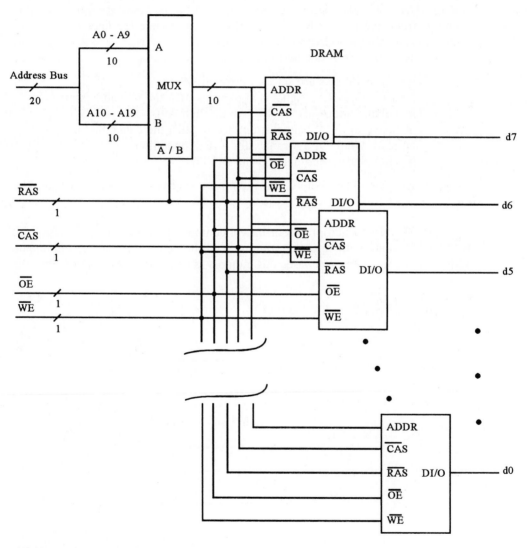

FIGURE 14.4

A 1M × 8 memory constructed from 1M × 8 DRAMs.

$\overline{\text{WE}}$ line for writing to the memory and $\overline{\text{OE}}$ for reading memory. Since all eight DRAMs operate in unison, each holds 1 bit of the 8-bit data. The 20 address lines from A0 through A19 must be passed through a 16-bit multiplexer to select A0 through A9 for the row and A10 through A19 for the column.

14.2.10 Static RAM Technology

Static RAMs (SRAMs) are similar in operation to DRAMs, but the timing waveforms are simpler due to the chip's static nature. The penalty for static memory is a fourfold increase in the number of transistors needed for each memory cell (one transistor in a DRAM versus four transistors in an SRAM). Static RAMs are available up to 1 Mb.

Older SRAMs had access times of over 200 ns, which means that an 80386 operating at 16 Mhz (62.5 ns) would be slowed down by the need for wait states during memory access. Now SRAMs are commonly available with access times down to 15 ns and can be used with processors operating up to 66 Mhz. Thus one of the primary advantages of SRAMs is their speed. For this reason they are often used for level 2 cache, which is external to the microprocessor chip.

The cell size of SRAM is much larger than DRAM so that there is little advantage to 1-bit organization or for multiplexed address lines. A generic 8K × 8 SRAM might then have pin connections as shown in Figure 14.5. The $\overline{\text{WE}}$ and $\overline{\text{OE}}$ lines are used to enable writing or reading memory, respectively, while the $\overline{\text{CS}}$ line is used to enable the SRAM chip. Important features of SRAM are as follows.

Access Time As already noted, SRAM is fast memory with access times of less than 10 ns for special, high-cost units. Typical access times will run from 15 to 50 ns.

Standby/Active Power Power consumption has always been a problem with SRAM. Older units would consume a significant amount of the total computer

FIGURE 14.5
Typical pin connections to an 8K × 8 SRAM.

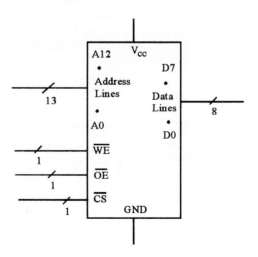

power. During the active state, when read and write operations are occurring, SRAMs may draw 30 to 100 mA at 5 V, or 150 to 500 mW per chip. For this reason many SRAMs provide for a standby mode with significantly reduced current and reduced supply voltage, as necessary, only to maintain the memory content in the chip. This is typically less than 100 μA, or about 500 μW per chip, a significant reduction.

EXAMPLE 14.9

Contrast power consumption of a 64-Kbyte memory composed of eight 8K $\times$ 8 SRAMs drawing 40 mA in the active state and 75 μA in the standby state with a 5-V supply.

Solution

It would take eight SRAMs to make up the 64 Kbytes required. Since they would be in parallel, the total current in active state would be $I = 8(40\text{ mA}) = 320$ mA and the corresponding power would be $P = (5\text{ V})(320\text{ mA}) = 1.6$ W. In the standby state the corresponding values would be $I = 8(75\ \mu\text{A}) = 600\ \mu$A so the power dissipation would be $P = (5\text{ V}) (600\ \mu\text{A}) = 3$ mW. This is a reduction of over 500 in power.

In many cases power consumption can be further reduced by operating the SRAM at reduced voltage levels. Some can operate down to 2 V for another factor of 2.5 reduction.

14.2.11 EPROM Technology

EPROMs are used in the initial design stage of a product when memory needs are still being defined and refined. EPROMs allow the designer to change memory as needed.

EPROMs are available with a variety of densities to suit the many needs of users. Densities in common use range from 16 Kb to 1 Mb. Organization is typically with either 8- or 16-bit words, with 8 bits the most common. Address and data lines are not typically multiplexed. Control lines consist of a chip select ($\overline{\text{CS}}$) and a means for programming ($\overline{\text{PGM}}$). Figure 14.6 shows typical pin connections for a 64-Kb EPROM organized as 8K $\times$ 8.

The following issues are of importance for EPROMs.

Access Time EPROMs are not particularly fast for memory elements. Older 2K $\times$ 8 units that are still in use today have access times of 400 to 500 ns. Clearly wait states are required even in computers with clock speeds below 8 Mhz. Newer EPROMs have access times that vary from 250 ns for 1-Mb (128K $\times$ 8) units to as low as 100 ns for 16-Kb devices organized as 2K $\times$ 8.

Erasure UV erasable EPROMs have a UV-transparent window covering the memory area of the chip. This window is typically kept covered during use to

FIGURE 14.6
Typical pin connections to an 8K × 8 EPROM.

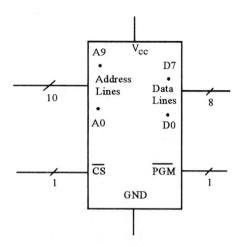

avoid stray UV causing any loss of memory. Erasure by intense UV light typically takes from 15 to 25 min.

Electrically erasable EPROMs appear in the form of any other integrated circuit. A typical time of 1 ms is required to program a bit, i.e., the write command to that bit location would have to be held for at least 1 ms.

In any case the manufacturer provides information on how the memory can be programmed. In most cases programming can be carried out in multiple bits, such as 8-, 16-, or even 32-bit blocks. For a 1-Mb EPROM, such block programming can significantly reduce programming time. Software routines and dedicated computer-based programming units are usually employed to automate the programming process. A typical automated programming time would be 1 s.

Power Consumption   EPROMs are not designed or intended for low power consumption applications and a summary of specifications demonstrates this fact. Typical active current will range from 50 to 100 mA at 5 V. If the EPROM has a standby facility, the current is typically 0.1 to 2 mA at 5 V.

EXAMPLE 14.10   Suppose we wanted to program an 8K × 8 EPROM by hand, i.e., each bit was to be set and the proper programming pulse was to be provided. Suppose it took 3 s per bit. How long would it take to program the unit? What if it was programmed by an automatic unit at a specified write time of 1.5 ms/bit?

Solution
Well, just for fun, because one would never do this by hand, the time would be $(3 \text{ s})(8)(8192) = 196,608$ s, which is about 55 hours, continuous. This is why automated EPROM programming was developed. On the other hand, the automated unit could perform the operation in a time of about $(1.5 \text{ ms})(8)(8192) = 98.3$ s or about 1.6 min. If it was programmed in 8-bit blocks, which is normal, the programming time would be only about 12 s.

14.3   LSI SEMICONDUCTOR LOGIC

LSI semiconductor logic is a technological extension of the digital logic principles discussed in Chapter 13. Although microprocessors and memory are in reality "LSI logic devices," the term "logic" in design circles refers to the logic gates and functions that are used in a design.

The 7400 TTL logic family and the 4000 CMOS logic family are a collection of small-scale integration (SSI) and medium-scale integration (MSI) ICs. Today, semiconductor manufacturers have been able to create LSI logic ICs with thousands of logic gates per IC. *Custom logic* allows designers a new degree of design flexibility unheard of only a few years ago. Logic designers no longer have to work around standard logic since one can literally custom design your own logic circuitry!

LSI semiconductor logic's three main advantages are as follows:

1. Efficient use of board space
2. Reduced power requirements
3. High speed.

LSI logic is used extensively in applications where speed is the critical parameter. A microprocessor has a flexibility advantage over LSI logic, but is over an order of magnitude slower. A microprocessor would have to run at a frequency of several hundred megahertz to compete with the logic available today. Today's high-speed microprocessors operate at or under 100 MHz.

14.3.1   Standard, Custom, and Semicustom Logic

Logic can be classified into two broad categories, *standard* and *custom* logic. *Semicustom* logic is a subset of custom logic.

Standard logic is the 7400/4000 type SSI/MSI logic ICs that have traditionally been used in logic design. Custom logic refers to requesting a semiconductor manufacturer to design and build a specific logic device for a custom application. Standard custom logic has also been around for a number of years. Historically, custom logic was very expensive and had very long turnaround times from design to actual hardware.

As personal computers and computer-aided design (CAD) logic programs became widely available, a whole new area of semicustom logic was created. Semicustom logic is designed by the logic designer with the aid of computer-driven logic design software and sent to a semiconductor company (also known as a foundry) for the manufacture of prototype and volume production parts. Many new "semicustom" foundries went into business to provide a dedicated quick turnaround production service.

Today, most designers refer to custom and semicustom logic as *application-specific logic*. Custom or semicustom ICs are also known as application-specific ICs or ASICs.

14.3.2 Types of Custom and Semicustom Logic

Custom and semicustom logic can be further classified into four main technology types: full custom, standard cell library, gate arrays, and programmable logic devices. Each of these approaches to LSI logic design is in use today.

Full custom logic design is a unique (customized) logic design that is translated initially into a custom LSI logic design, and then fabricated into an LSI chip. A full custom design can be designed by the engineer or technologist requiring the design or by a vendor of semiconductor logic designs.

The advantages of full custom logic are efficient layout, resulting in lower manufacturing costs (higher density), and oftentimes specific functionality that may not be possible with the other design methods.

A drawback to full custom logic is that it is very design intensive, making it time consuming and not ideally suited for quick turnaround times. Also, the testing of design functionality is more complicated with full custom logic.

A less design-intensive approach to ASIC design is the *cell library design methodology*. Simply stated, cells are logic design blocks that will do a predetermined logic function. For example, instead of building a flip-flop with logic gates, the designer would place the "flip-flop" cell in the design. Each custom/semicustom IC manufacturer has design software they use or sell to designers containing hundreds of standard cells. The designer then designs in the high-level cells. The actual final IC layout is usually then completed by the device vendor.

Design time is shortened with the standard cell approach, but flexibility is sometimes lost. Oftentimes you must use a cell that does more than you want, but is the only cell available. This results in some design inefficiency, along with cell interconnect inefficiency.

It is important to choose carefully the device vendor that you would like to use for standard cell design since you will also be using their software and cell library. Some companies have very extensive SSI/MSI/LSI cell libraries; others may not have the flexibility needed for a particular logic design.

A variation of the standard cell approach offered by some vendors is the *cell compiler*. A compiler takes design input and creates a cell for the specific application. It has the increased layout efficiency of full custom design, but it is still a modular cell approach. The software used is much more extensive (and costly) than standard cell software.

Gate arrays are LSI devices that contain a large number of gates and logic functions that are all interconnected by metallization. Gate arrays are programmed by selectively blowing metal fuse interconnects to customize the chip for a specific logic design.

Gate arrays can be programmed in two ways. One way is to give a vendor your logic design and have the vendor program the gate array. Also available are *user-programmable gate arrays* or *UPGAs,* which allow the user to program the gate array with the aid of a development system.

The advantage with any type of gate array over full custom or standard cell design is quicker turnaround time from design to actual hardware. There are no custom parts to manufacture. The standard gate array is programmed and then tested to verify the logic. In addition, once a custom design is defined, volume production parts can be produced with only a level or two of customization, far less than with custom or semicustom ICs.

Programmable logic devices or *PLDs* are application-specific ICs that are customized by the designer to implement a specific logic function. PLDs differ from gate arrays in their logic structure and, in addition, some gate arrays require several layers of customization (processing). PLDs are usually bipolar fuse technology, and gate arrays are often CMOS devices with much larger logic gate capacity.

Monolithic Memories first introduced the programmable array logic (PAL™) as its solution to increasing turnaround time to device hardware. The PAL device has a programmable first array of AND gates, and a fixed second array of OR gates. Simple PLDs are used to implement product of sums (POS) and sum of products (SOP) logic equations. The two levels of gates come in varying widths.

In addition to PAL devices, there are *IFL™* (Signetics) PLDs that have a first level of programmable AND gates and a second level of programmable OR gates. A PROM PLD has a fixed first array of AND gates and a second array of programmable OR gates.

In 1984, Altera Corporation introduced a new type of PLD device, the *erasable PLD* or *EPLD*. An EPLD is user-programmable logic that can be erased with exposure to ultraviolet light. An EPLD used in conjunction with personal computer software allows a logic designer to create a design and implement it on a chip in one day. The EPLD has the additional advantage over PLDs of allowing for real-time logic changes because of its erasable feature.

EPLDs have from 300 to 2000 logic gates, and are between PLDs and gate arrays in terms of cost per function and complexity. EPLDs can replace up to 60 conventional CMOS or TTL logic chips in function, and can be used for applications such as microprocessor memory address decoding. EPLDs offer faster decode times and use less board space than conventional discrete logic.

14.3.3 Gate Array Technology

The Advanced Micro Devices (AMD) Am3550 is a bipolar mixed ECL/TTL I/O mask programmable gate array.

The Am3550 is a typical gate array that has up to 124 selectable I/Os, and the equivalent of 5228 gates. There are 200 customizable cell locations. The 576 internal cells have a fixed array of 19 transistors and 32 resistors that can be interconnected with a metal pattern to create unique logic functions known as *macro cells.*

A list of some of the macro cells available for the Am3550 is shown in Table 14.3. Other manufacturers will have unique cell definitions, but some of the standard logic components are common to all gate array devices. Some common logic components are adders, AND, OR, and NAND gates, and flip-flops.

TABLE 14.3
Samples from the Am3550 gate array macrocell library

MPA-2	Macrocells	AMD	Proprietary Macrocells
AM203	8-input OR/NOR	AM2001	3-input NOR/2-input NOR
AM222	Dual 2-2 OR/AND/XNOR	AM2014	2-bit D flip-flop
AM262	1 of 4 decode with enable	AM2031	Dual 4:1 MUX with enable
AM293	D latch		
AM312	3-3-3 AND/OR	I/O	Macrocells
AM370	Differential line receiver	AF201	PECL-TO-TTL 3-state OR
AM393	D latch with clock enable	AF208	PECL-TO-TTL transceiver
		AOX2002	4-input OR/NOR

There are four phases in the implementation of the Am3550 and they are typical steps in any gate array design:

1. Design acceptance
2. Layout
3. Test generation
4. Fabrication.

Design acceptance is the step where the customer completes a logic design and sends it to the manufacturer. Layout and test generation are performed by the manufacturer for the custom chip. The last step is the fabrication of sample parts, which are sent to the customer for approval. Upon customer acceptance, the gate arrays are fabricated in volume.

Figure 14.7 shows a flowchart of the gate array development flow.

14.3.4 PLD Principles and Applications

The PLD has become an important tool for both prototyping and production. The reason for this is the versatility of this LSI circuit in implementing complex sets of Boolean equations, sequential digital circuits, and complex decoding processes. The following paragraphs present the basic principles of PLD operation and schematic representation.

Schematic Representation   Consider the symbolic representation in Figure 14.8(a) where three vertical lines represented by A, B, and C cross a single horizontal line leading to an AND gate symbol. The horizontal line is called a "product" line because any of the input lines connected to it will show up as an output of the AND gate. A connection is shown as a black dot. In this case the figure represents the Boolean product, BC, i.e., B AND C.

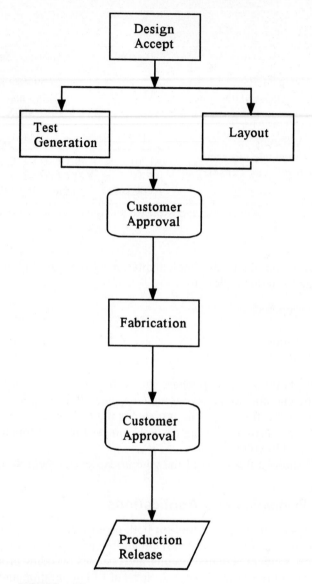

FIGURE 14.7
Gate array code development flow.

By contrast Figure 14.8(b) shows three horizontal lines crossing a single vertical line leading to an OR gate symbol. The vertical line is called a sum line because any line connected to it will be summed by the OR gate. In this case the output of the OR gate will be A + C, i.e., A OR C.

So a PLD consists of a large array of inputs crossing many product lines. The outputs of the product lines in turn cross sum lines. A new PLD has connections

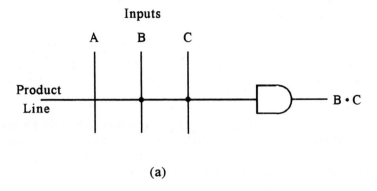

(a)

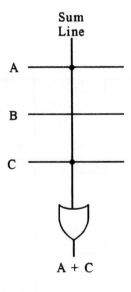

(b)

FIGURE 14.8
Symbolic PLD product and sum representations.

at every crossing via what is called a fuse. The PLD is "programmed" by blowing open fuses at every crossing that should not have a connection.

In addition, each input line actually has both the inverted and noninverted version of the input as product crossing lines.

EXAMPLE 14.11

Show how a PLD with three inputs, four product lines, and two outputs would be programmed to provide the following combination logic:

$$Y = A\overline{B} + \overline{A}C + \overline{B}\,\overline{C}$$
$$Z = A\overline{B}\,\overline{C}$$

Solution

Figure 14.9 shows the solution. Note that connections are made for product line 1 with A and B, which are then ANDed and connected to the Y sum line. This concept is then repeated for the remainder of the equations. Remember that all of the connections to a sum line are ORed together.

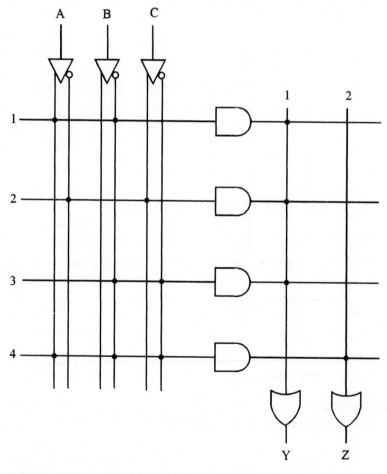

FIGURE 14.9
Combination logic solution to Example 14.11 using a PLD.

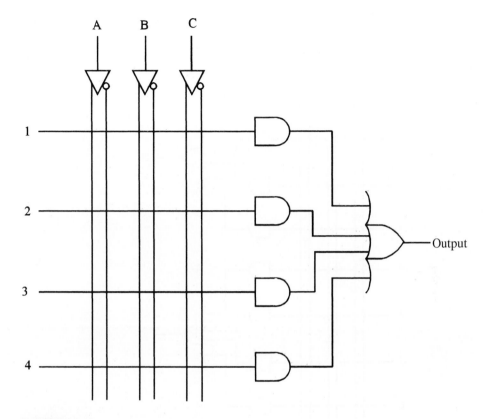

FIGURE 14.10
Basic structure of a PAL with permanent sum lines.

Programmable Logic Array (PLA)   Example 14.11 demonstrates a PLA type of PLD. This is a PLD that is fully programmable; that is, the user can select product and sum connections. This is the most versatile of PLDs, but also the most complicated to use and program. They are also typically slower than other digital technology. For these and other reasons PLAs are not frequently used.

Programmable Array Logic (PAL)   A PAL is a PLD where the sum connections have been made permanent. Thus the user can program only the product lines. The schematic diagram of the PAL is frequently modified as shown in Figure 14.10 to emphasize that certain product line outputs are permanently ORed. The programming of PALs is much easier and they are faster than PLAs.

EXAMPLE 14.12   A PAL has three input lines, A, B, and C. It has two banks of four product lines ORed to produce two outputs. Show how such a PAL can implement

$$X = AB + \overline{A}C + \overline{A}\,\overline{B}C$$
$$Y = BC + A\overline{C}$$

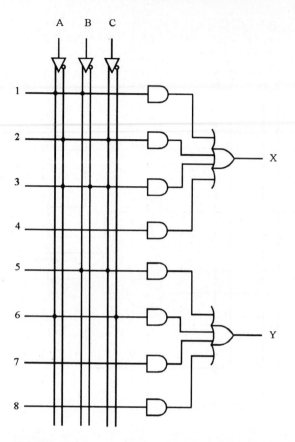

FIGURE 14.11
Combination logic solution to Example 14.12 using a PAL.

Solution
The solution is shown in Figure 14.11. Note that since the PAL comes with all connections made, each of the nonconnections in Figure 14.11 would have to be blown following the chip manufacturer's procedures.

Example 14.12 suggests how the PAL is employed for a very simple example. Actual PALs will commonly be packaged in 40-pin chips with thousands of gates, 10 to 20 inputs, and 8 to 10 outputs. Each output line can result from many product terms.

PAL Enhancements. Many enhancements can be made to the basic PAL architecture presented in Figures 14.10 and 14.11. These include the following:

1. Provide feedback of the sum term back to the product term input. This leads to the development of sequential logic circuits with stable and unstable states. This is called a *combinatorial PAL configuration*.

2. The presence of gated flip-flops on the sum line outputs provides what is called the *registered PAL configuration*. In this case, counters and shifters can be implemented.
3. Some PALs can be erased and reprogrammed, using the same UV technology employed with EPROMs. Unfortunately the entire program is erased so this cannot be used to "fix" a single or a few incorrectly blown fuses.

In summary, the PLD and the PAL, in particular, have become quite common for implementation of functions such as combination logic, multiplexing, counters, memory addressing, and a host of other functions in modern computer systems. In many cases a large number of SSI or MSI logic devices can be replaced by a single PAL.

14.4 SPECIAL LSI CIRCUITS

Many LSI semiconductor circuits are available that do not fall neatly into the categories of memory or logic. Many of the special LSI devices are support chips for memory and logic, or a combination of several analog and digital functions integrated into one circuit.

In general, data books and IC specifications are the best source of detailed information about a particular LSI chip. However, several of the more widely used special LSI circuits are discussed here as an informational aid.

14.4.1 Microprocessor-Based Interface Circuits

Many LSI ICs contain functions that are used to interface with microprocessors. It is useful to review the basic functions of some of the more common circuits.

Buffers are circuits that have the ability to isolate one part of a circuit from another to prevent signal interference. Buffers are often tristate. Buffers that pass data in two directions are called *transceivers*.

Many microprocessor-based systems use a multiplexed address/data bus to achieve maximum IC pinout efficiency. *Latching* is necessary to separate the address and data information. Latches are devices that hold the valid address or data until they are used.

Both memory and I/O ports need address decoding to enable selection of a particular memory location or I/O port. Many designers are turning to PLDs for more efficient decoding.

14.4.2 Speech Synthesis

Several speech synthesis chips are on the market. Most are designed to operate in microprocessor-based applications, using an 8-bit code input. The National Digitalker, for example, has a limited programmed vocabulary that is selected by the 8-bit input code. The Signetics MEA8000 is user programmable, and the speech code needs to be stored in a separate memory.

14.4.3 Communication Circuits

Hundreds of special-purpose communications LSI chips are currently available. Communications controllers, local-area network (LAN) ICs, asynchronous interface adaptors, and audio and video amplifiers are available as specialized LSI ICs, to name a few. Serial communication is dependent on universal asynchronous receiver/ transmitters (UARTs). If the specialized UART can also handle synchronous, it is an industry standard USART. UARTs and USARTs are used with modems to transmit data via telephone lines.

The two basic types of serial communications formats are *synchronous* and *asynchronous.* Both formats require additional coding bits for detection and decoding. These coding bits are called *framing bits.*

Asynchronous formats have framing bits added to each character code. Synchronous formats have framing schemes that code entire blocks of character data. For this reason, synchronous communication is faster than asynchronous, but requires more complex decoding circuitry.

A third type of format, known as *isosynchronous,* occurs when an asynchronous format is used with a synchronous modem.

14.5 PACKAGING

Packaging is the final manufacturing step in the fabrication of LSI devices. Packaging serves two main purposes. First, to protect the semiconductor device from the environment. This includes *electrical* protection (against electrostatic discharge, ESD, for example), *mechanical* protection, and *moisture* or corrosive element protection. LSI chips are usually mounted on a plastic or ceramic carrier and hermetically sealed to isolate the chip from the environment.

The second function of packaging is to ensure that the semiconductor chip is in a usable form. This involves putting the chips into standardized packages. *Dual-in-line packages,* or *DIPs,* shown in Figure 14.12(a), come in standard widths and pin counts, allowing them to be universally compatible within a standard package size.

Today, a vast assortment of packaging types is available to fit almost any type of circuit board requirement. However, the large increase in packaging types has brought with it an increase in the importance of properly choosing the correct LSI package early in the design phase of a project. Many types of automated assembly machines require a certain type of package, and care must be taken to ensure that the LSI components chosen in a design are available in the package needed for manufacturing. Package costs vary tremendously, and once again early engineering effort is necessary to identify the package that will do the job at the right cost.

14.5.1 Packaging Standards

There are two main types of LSI semiconductor packaging standards. The Joint Electron Device Engineering Council (JEDEC) has set packaging standards for most of the semiconductor devices produced. The JEDEC standards are followed by most manufacturers with the exception of the Japanese.

(a)

(b)

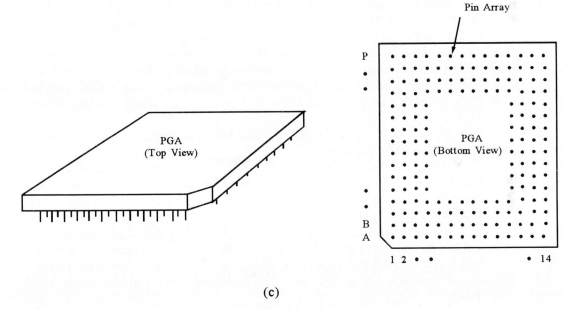

(c)

FIGURE 14.12
Packaging for LSI. (a) DIP. (b) SOIC or flatpack. (c) PGA oblique and bottom views.

The "Japanese" packaging standard is the other widely followed guideline. The differences between the two standards are almost nonexistent with DIP packages, but become significant when working with *surface mount devices* (SMDs). It is important to check the physical packaging dimensions carefully for a particular part as shown in the manufacturer's data book to avoid component compatibility problems. In addition, not all manufacturers follow the JEDEC guidelines!

14.5.2 Types of Packaging

Packaging technology has produced a wide array of package types in the last decade. Until recently, DIPs were the standard LSI component package found in almost every board design. With the advent of larger pinout requirements, and the need for increased card density, the DIP was no longer a suitable package type.

Several of the new types of LSI packaging are discussed, but SMDs appear to be the major next-generation package.

Most of the LSI packages in use are either molded plastic, or ceramic. Plastic technologies tend to be a little less reliable due to lack of complete hermeticity, but they are very cost efficient. Ceramic technologies involve mounting the semiconductor chip on or between ceramic. Ceramic is more expensive, but usually considered more reliable. DIPs are mounted on a metal lead-frame and surrounded by plastic in a standardized rectangular shape. The pins are bent and form two rows, with the pins at 0.1-in. center spacing. They come in a variety of pin sizes (numbers) and DIP widths.

DIPs are usually mounted on circuit boards through holes and either soldered or wire-wrapped. This is often called *pin-through-hole* technology. They can also be mounted into DIP sockets for ease of replacement.

Small outline integrated circuits (SOICs), as shown in Figure 14.12(b), are surface mount devices that have a rectangular shape similar to DIPs, but with smaller length, width, and height dimensions. Instead of having straight pins like DIPs, the SOICs have pins that are bent to provide a flat surface mountable area. They are often called *gull-wing* packages.

As with DIPs, there are JEDEC and Japanese dimension standards for all the SOIC standard pin sizes.

SOICs are surface mounted to a circuit board with one of several soldering techniques. If there are surface mount devices on only one side of the board, it is called a single-sided board. If devices are mounted on both sides, it is called a dual-sided board. This applies to DIPs as well.

Another common SMD package is the *J-lead*. The J-lead package is square and has its pins bent into a "J" shape with the surface mountable part of the pin bent under the package. J-lead packages are used for high pin count packages.

J-lead packages are mounted with techniques similar to that of the SOIC SMDs.

Pin grid arrays (PGAs) shown in Figure 14.12c are packages that have an array of pins located physically under the package. The chip is mounted either between or on top of the pins. The name "pin grid" comes from the arrangement of the pins in equal *x-y* spacing, to form a grid. Ceramic is often the material used for pin grid arrays. Pin grid is another type of package useful for logic or other high pin applications. Pin grid arrays use pin-through-hole mounting technology.

Plastic chip carriers (PCCs) are square packages with leads spaced 0.50 in. apart on four sides. They are used in applications where the chips are I/O intensive. *Tape automated bonding* (TAB) technology is used in place of wire bonding to connect the chip to the pin leads.

Flat packs are really very similar to PCCs, or SOICs. They are surface mountable devices with leads coming out of two or all four sides. They were designed to handle high I/O and for ease of manufacturability with automated equipment. At this time flat packs are not standardized, but JEDEC is addressing the problem.

Several companies place several semiconductor chips onto one package or *module.* Circuit board density is the main thrust behind multichip modules. Several manufacturers stack the chips for increased density. Most multichip modules (MCMs) are produced for captive (in-house) use. The main drawback to MCMs is the cost involved in replacing the entire module if one of the chips fails.

Another method for packaging LSI chips is to attach the die (chip) directly to the circuit board (wire-bonding, for example). This is called *direct chip attach* (DCA). The chips are then covered with a passivation material. This is generally a captive process designed to save costs by eliminating the intermediate packaging connections.

14.5.3 Parameters Influencing Package Selection

Semiconductor companies package their LSI chips in several different types of packaging. The type of packaging chosen is basically up to the designer. The important parameters to consider when choosing packaging are summarized next.

The most important consideration is whether the package will fit into the overall design. If the design is, for example, microprocessor based, and the company you chose does not offer an SMD micro, it does not make sense to choose an SOIC memory since you will have a design mismatch (pin-through-hole and SMD). Such mismatches can be alleviated by the use of pin-through-hole SMD sockets, but in general, it is not advisable to mix packaging types in the same design.

Another important consideration is the availability of the part in the package type chosen. Some parts are not stocked heavily in some of the offered package types. *Lead time* refers to the amount of advance notice a company has to be given in order to deliver parts in time for the design deadline.

Device costs vary depending on the type of packaging. As mentioned, ceramic costs more than plastic. DIPs are usually the cheapest since they are the most mature packaging technology. When in doubt, always get the latest price quote from the manufacturer.

Manufacturability refers to the relative ease of being able to mass produce a product with a given LSI device. If a package has high manufacturability, it is compatible with the automated equipment and will not cause a manufacturing problem.

Most LSI devices chosen for a design are bought or *sourced* from two different companies. This reduces the risk involved with being dependent on one source for a part. This strategy is called *second-sourcing.* It is imperative to choose sources that have complete package dimensional compatibility so that the parts are totally interchangeable.

Serviceability refers to how replaceable a part is in actual use. If component serviceability is needed, the components should be placed in sockets.

Manufacturers keep data on the reliability of different types of packages. Upon request, most companies will supply this information. With data from several companies, one can choose the company with the most reliable package type.

14.6 QUALITY

In today's electronics industry, additional emphasis is being placed on product quality. Companies like Hewlett-Packard and IBM owe a large part of their success to their commitment to high-quality products. Japanese companies have gained a large share of the world semiconductor market by having "zero defect" goals for their components. To compete, LSI component manufacturers for the most part have accepted the zero defect philosophy for their products. However, there is still a wide range in the quality of the LSI components available. For this reason, it is very important for the component user to be familiar with the terms and criteria used to measure quality.

14.6.1 Types of Quality

Quality can be divided into two major types. The initial quality of a component is often referred to as *average outgoing quality level* (AOQL), or *shipped product quality level* (SPQL). The long-term quality of a component is called its *reliability*.

AOQL is the number of defective components in a group of supposedly good components manufactured and shipped by the manufacturer. It is most often expressed as a percentage.

EXAMPLE 14.13

Company X orders 1000 64K DRAMs from Company Q. Company Q claims an AOQL rate of 0.3%. Company X decides to test the DRAM memories before installing them in their application, and finds five defective DRAMs.

Solution

$$AOQL = (\text{Number defective/Number in sample}) \times 100$$
$$= (5/1000) \times 100 = 0.5\%$$

Because 0.5% is greater than 0.3%, the quality target was exceeded.

Reliability is the estimated measure of how reliable a product or component is during its use. Reliability numbers give component users some idea of the component's long-term quality. Customer satisfaction depends on product quality being maintained throughout the product's life.

The reliability of a product must be measured before shipment to assure this goal is met. One could collect accurate, actual reliability information by testing a product at use conditions for a time equal to the projected lifetime of a product. But this is impractical since some components have lifetimes of many years.

LSI component manufacturers use accelerated life testing to evaluate component reliability. This circumvents the problem of waiting for actual field failure data to calculate reliability. Accelerated life tests shorten the time needed to find failure mechanisms by accelerating one or a combination of parameters.

Some of the more common acceleration factors used are temperature, voltage, humidity, electric field, pressure, current density, and radiation. Temperature is by far the most common acceleration variable, because it is somewhat easy to control and model.

Component manufacturers will supply reliability information for a particular component in which you may be interested. If you are purchasing a large volume of components or are particularly interested in achieving high reliability, it is prudent to discuss your particular application with the manufacturer. A particular subset of the reliability test data may be of special interest depending on how you will be using a component.

For example, you may be using a microprocessor in an automobile ignition control circuit. Automobiles during their lifetime will experience thousands of extreme temperature cycles. Temperature cycling is one of the reliability life tests that most manufacturers use. Maybe this particular manufacturer has poor temperature cycling reliablity data but average reliability is good. In-depth investigation can avert serious reliability problems.

14.6.2 LSI Reliability Stress Testing

Several common stress tests are being used in the component industry to evaluate LSI/VLSI reliability:

High-temperature bias (HTB): Components are dynamically operated at an elevated temperature of 100 to 150°C. The operating voltage can also be elevated to provide additional stress.

Static HTB (SHTB): Similar to the HTB test, except components are not dynamically operated. Constant voltages on internal circuitry provide a different type of electrical stress.

Low-temperature bias (LTB): Components are operated at temperatures of 0 to −50°C.

Temperature and humidity (T/H): By applying a high relative humidity of 80 to 90% and a temperature typically 85°C, the corrosion resistance is tested.

Temperature cycling (TC): Parts are subjected to a temperature variation from typically −50 to 150°C, two to three times an hour. This test will stress wirebonds, solder connections, and surfaces with different coefficients of expansion and contraction.

Soft error rate (SER): A radiation source is placed in proximity to the component to subject it to alpha-particle emission. Alpha particles will cause memories to lose their stored data.

In addition to the electrical reliability tests listed here, manufacturers do a series of mechanical tests to verify the packaging integrity of the component.

EXAMPLE 14.14

A manufacturer's reliability data indicates three component failures out of a 100-part sample for a T/H stress test after 2000 hours. The components did not fail in several other stress tests. Is there a problem?

Solution

Carefully examine the reliability data and match it to your particular application. Find out how many application hours 2000 hours of stress corresponds to. Also find out after how many stress hours the parts failed. It may be that 0/100 fails were found after 1000 hours of stress, and 1000 hours of stress corresponds to 10 years of actual use. Then there would be no problem. However, if your application would be in a humid environment, it would be best to avoid components with high T/H failure rates even if the other failure rates are good.

Fortunately, most reliability failure mechanisms follow the *bathtub curve.* The name "bathtub curve" comes from the resemblance of the curve to a bathtub! The reliability bathtub curve is a plot of component failures on the *y* axis versus time on the *x* axis. LSI components have their largest failure rate within the initial hours of operation. This initial high failure section of the curve varies from a few hours to possibly 100 hours for some components. After the initial failures, the failure rate is fairly constant for thousands of hours of operation. Finally, the component

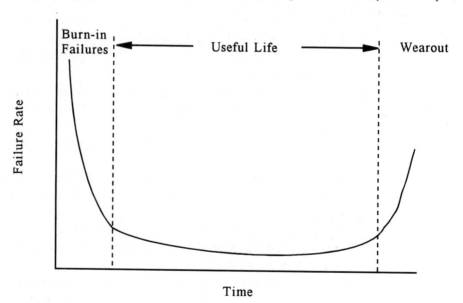

FIGURE 14.13
Bathtub reliability curve for electronic components.

will wear out, and the failure rate curve will begin to rise again. A typical reliability curve for LSI components is shown in Figure 14.13.

Manufacturers take advantage of the high statistical failure rates that occur within the initial hours of operation. A process known as *burn-in* is used to remove the initial failures from a component population prior to shipping. With burn-in, the product the customer sees is the components that have the fairly low constant failure rate.

Burn-in can be static or dynamic. A *static burn-in* subjects the components to a high temperature for a specific amount of time, 2 hours, 24 hours, etc. Then the parts are retested and the failures removed. *Dynamic* or *in-situ burn-in* subjects the components to high temperatures while the component is operated and tested. In-situ burn-in is preferred, but adds cost to the component.

High reliability parts are usually parts that have experienced a longer burn-in time. This, too, adds cost to the component.

EXAMPLE 14.15

Manufacturer Q uses a 2-hour burn-in. Manufacturer Z uses a 12-hour burn-in for the same type of part. All other reliability parameters being equal, which part should you buy?

Solution

If the burn-in conditions and price are equal, always buy the components with the largest amount of burn-in. Additional early life failures will be removed. However, excessive burn-in could be an indication of a processing problem; request the initial and long-term reliability failure data to check for problems.

15

MICROPROCESSORS*

15.1 INTRODUCTION

15.1.1 Basic Microprocessor Structures

A microprocessor unit (MPU) is a very-large-scale (VLSI) or extremely large-scale integrated circuit chip, which contains an arithmetic-logic unit (ALU), a timing and control unit, sets of registers, a program counter, an address bus, a data bus, a built-in processor control mechanism, and an instruction interpreter or decoder.

The MPU is a central processing unit (CPU) whose main functions are to execute given instructions and to control the activities of the bus-oriented microprocessor system.

It is commonly understood that the number of data bits of a microprocessor is mainly determined by the number of the microprocessor's external data lines. Thus, there are 8 external data lines for an 8-bit microprocessor, 16 external data lines for a 16-bit microprocessor, and 32 external data lines for a 32-bit microprocessor.

The microprocessor is packed in one of the following three forms: dual-in-line package (DIP), leadless package, and pin-grid-array (PGA). Figure 15.1 shows general DIP and PGA packages.

15.1.2 Microprocessor-Based Computer Systems

A microprocessor alone is not an operational computer, i.e., it is unable to perform the arithmetic, logic, and control operations of conventional computers. Memory chips, peripheral devices, and input/output devices must be provided and interfaced

---

* This chapter was written by Dr. Paul I-Hai Lin, Purdue University.

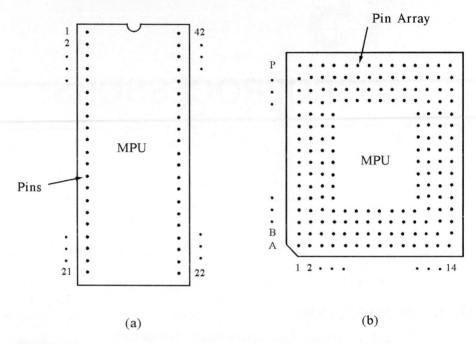

FIGURE 15.1

Two types of microprocessor packages. (a) DIP. (b) PGA.

with the microprocessor to form a microcomputer system. In addition, the necessary software must be provided for running the microprocessor to implement the required functions.

A typical configuration of a microcomputer system is shown in Figure 15.2. This includes an MPU, read-only memory (ROM) and read/write memory (RAM) (for program and data storage, respectively), input devices (a keyboard and switches), output devices (a CRT display, actuators, and indicators), and mass storage such as floppy diskettes, tapes, CD-ROMs, and hard disks.

As expected, microprocessor-based computer systems have taken on effective roles as key components in office and factory automation, instrumentation, engineering, and science laboratories.

15.1.3 Single-Chip Microcomputers

A single-chip microcomputer is a VLSI chip that contains an MPU, programmable ROM (PROM) or erasable PROM (EPROM), data memory RAM, clock, and I/O capabilities. This is shown in Figure 15.3.

Sometimes, a digital-to-analog converter (DAC), and/or an analog-to-digital converter (ADC) is integrated in the single-chip microcomputer for interfacing with analog world.

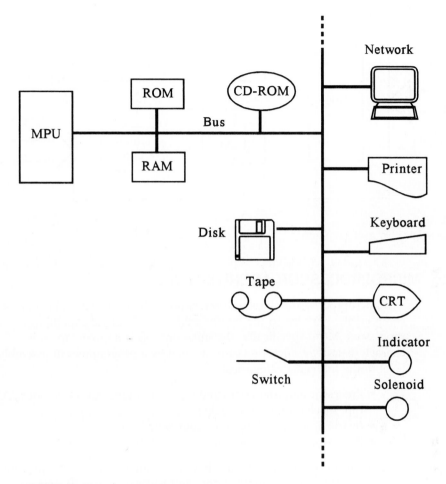

FIGURE 15.2
A typical microcomputer system.

Examples of single-chip microcomputers or single-chip microcontrollers are as follows:

- *8-bit microcomputers:* Intel MCS-48 and MCS-51, Motorola MC6801 and MC68701, and Zilog Z8 families
- *16-bit microcomputers:* Intel MCS-96 family.

Single-chip microcomputers are found in stand-alone and peripheral controllers, refrigerators, dishwashers, washing and drying machines, automobile subsystems, and so on. These applications tend to require small memory space and less computation power, and they have lower system hardware costs.

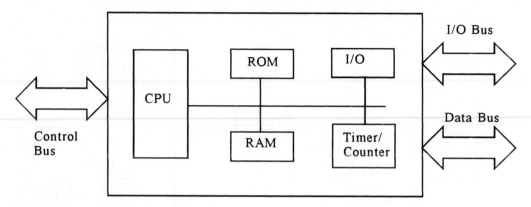

FIGURE 15.3
A typical single-chip microcomputer.

15.2 MICROPROCESSOR ARCHITECTURES

The microprocessor architecture refers not only to the general layout of the major components, but also to the operational and functional capabilities of the microprocessors. More specifically, the microprocessor architecture refers to the basic characteristics of a microprocessor as seen by a programmer at assembly language level in the following categories:

- Processor unit (data sizes and representations, register models, ALU, interrupt mechanisms, and stacks)
- Instruction set and memory addressing modes
- Processor bus structure and input/output mechanisms.

Microprocessor architectures have been influenced by the semiconductor technology, microprocessor applications, and software crisis. Because of technology limitations in the 1970s, early 8-bit microprocessors were limited by the small number of registers, 8-bit data width, 64K memory size, limited addressing modes, and less powerful instruction sets.

Advancements in semiconductor technology enabled microprocessor manufacturers to escape these limits and jump to the development of modern 16-bit, 32-bit, and 64-bit microprocessors. These microprocessors are used in the large-scale systems that require greater addressibility. Also, these faster microprocessors are necessary to support multitasking and multiuser operating systems and modular and high-level language programming.

15.2.1 Data Types and Sizes

Primitive data types that can be found in microprocessors are unsigned integers, signed integers, floating-point numbers, ASCII characters, packed and unpacked binary coded decimal (BCD) numbers, and logical data.

Unsigned Binary Integer Numbers   All microprocessors support this type of data representation with all data bits used to represent the number's magnitude. For an n-bit number, the largest and smallest numbers that can be represented are $2^n - 1$ and 0, respectively. Therefore, with a bit width of 8 bits, 16 bits, or 32 bits as a single word, the equivalent decimal numbers that can be represented range from 0 to 255, 0 to 65,535, or 0 to 4,294,967,295, respectively.

Logical Data   The logical data are treated as unsigned integers in that each word bit can take on binary values of 0 or 1.

Signed Integers   A signed integer is a data representation with the most significant bit (MSB) as a sign bit and the rest of the bits as magnitude. If the sign bit is 1, a number is considered negative and represented by two's complement; otherwise the MSB will be 0 and, hence, considered a positive number with $n - 1$ bits as its magnitude.

ASCII Characters   "ASCII" stands for American Standard Code for Information Interchange. The ASCII character set includes 96 displayed characters and symbols, and 32 nondisplayed control characters. The MSBs of the ASCII characters are normally reserved for parity bits.

Unpacked BCD   In BCD representation, the lower 4 bits of a byte are used to represent decimal digits (0–9).

Packed BCD   A packed BCD uses a byte to represent two BCD digits.

Floating-Point Numbers   Most modern 16- and 32-bit microprocessors support signed 32-, 64-, or 80-bit real number representations via hardware math or floating-point coprocessors that follow IEEE floating-point standards.

15.2.2   CPU Registers

All microprocessors contain registers for local or scratchpad storage of frequently used data, memory referencing, instruction accessing, and so on. These might include general-purpose registers, accumulator(s), a program counter or instruction pointer, stack pointers, index registers, data registers, address registers, segment registers, and status/condition flag registers.

General-Purpose Register   A general-purpose register can be used as a data register, an address register, a counter, or as defined through available instructions.

Accumulator   The accumulator is normally used by the arithmetic and logic operations.

Program Counter/Instruction Pointer   The program counter (PC) is a program address register that usually points to the next available instruction to be executed in the main memory. In general, the program counter has the same number of bits as the address bus.

Instruction pointers are found in the segmented 16- and 32-bit Intel microprocessors. An instruction pointer (IP) in the 16-bit 8086 is very similar to a program counter in the 16-bit MC68000. But the IP points to the next instruction to be fetched by a bus interface unit of a processor.

Stack Pointer   The stack pointer (SP) is used to access a special memory section called *stack* that has been reserved for temporary storage of data, return addresses, and processor status.

Index Register   The index register is used to define the location of an operand and to modify the operand address of an instruction. Index registers are normally used in a program for manipulating a list of data such as tables or arrays.

Segment Registers   The segment registers are special address registers used in segmented memories for identifying different memory segments such as data, program, and stack segments.

Flag/Status/Condition Code Registers   The flag or status registers are used by the processor to hold certain information, such as zero result, negative sign, overflow, and so on, that is caused by the execution of certain arithmetic and logical instructions. The status of the processor and control information is also stored in this register.

15.2.3   Addressing Modes and Instruction Set

Addressing Modes   An instruction must either contain an operand or it must specify where to obtain the data to use as an operand. In general, the operand may be included in the instruction, or in memory, or in the registers. Since the two-address instruction format is used by microprocessors, instructions should carry the addresses of the sources and the destinations of data.

The schemes that are used for addressing main memory and CPU registers are generally called addressing modes. The addressing modes found in microprocessors might include the following:

- *Immediate addressing:* The operand follows op-code in consecutive memory location(s).
- *Direct (absolute) addressing:* The address of the operand is given as part of instruction.
- *Indirect addressing:* The address of the operand is in the register or in main memory locations.
- *Relative addressing:* This type of addressing is normally used in JUMP or

BRANCH instructions that cause program control to shift by a displacement relative to the current address.

- *Index addressing:* The effective address of the instruction is produced by the sum of a displacement and content of the index register.
- *Base-register addressing:* The effective address of the instruction is the sum of a displacement and content of the base register.
- *Segment addressing:* This is a form of base addressing used in the microprocessors with segmented memory. The effective address of the operand and op-code is computed by adding the segment register and an offset register (the instruction pointer in the case of the 8086).
- *Stack addressing:* The stack pointer contains the address of the top of the stack where data items are to be accessed.

Instruction Set   The instruction set refers to a group of fundamental operations that the computer can perform on data. This is also a set of instructions that is available to the programmer at assembly language level. In general, instructions of microprocessors may include the following instruction groups:

- Data movement instructions (MOVE, LOAD, STORE)
- Arithmetic instructions (ADD, SUB, INCREMENT, etc.)
- Logic instructions (AND, OR, EX-OR, etc.)
- Shift and rotate instructions
- Bit manipulation instructions
- Program sequencing and control (JUMP, or BRANCH)
- Subroutine linkages (CALL, RETURN)
- Stack instructions (PUSH, POP)
- Interrupt instructions
- I/O instructions (if supported)
- Processor control instructions (HALT, WAIT, etc.).

Multiplication, division, string manipulation, iteration, conversion, and block manipulation instructions are generally included in advanced microprocessors to form a rich and powerful instruction set.

Instructions that support high-level language programming, such as array boundary checking, stack frames, and so on, are found in some 16- and 32-bit microprocessors.

Multiprocessor communication, system control, and floating-point instructions are available through the 16- or 32-bit microprocessors that support multiprogramming, multitasking, multiprocessing, and coprocessing.

CISC and RISC   Microprocessors have evolved in two approaches with respect to architecture. One type is referred to as a *complex instruction set computer* (CISC) because a great number of instructions are available for performing complex operations on data. Such instructions require many clock cycles for execution. Most of the earlier microprocessors such as the Intel 80XX and Motorola 68XX lines were of the CISC type. Many of the modern processors such as the Intel 80XXX and Pentium lines continue to use this architectural design.

The alternative is called the *reduced instruction set computer* (RISC) for which fewer instructions are built into the processor architecture. Complex operations are performed by programs using the RISC instruction set. The concept is mainly that the RISC instructions are principally executed in a single clock cycle and the instruction decoding is much simpler so that even complex operations can be executed faster than for CISC platforms. The Motorola PowerPC line is an example of a common RISC processor. Debate continues on which approach is better, but good performance is obtained from either.

15.2.4 Microprocessor Buses

The pins on a microprocessor chip can be divided into three types: *address, data,* and *control.* These three types of buses are used by the microprocessor to communicate with external memory chips and I/O peripherals. A three-bus microprocessor system is shown in Figure 15.4.

The Address Bus   The address bus is a unidirectional bus that carries the address information from a microprocessor to memory or I/O devices for the selection of I/O devices, program code, and data in the system.

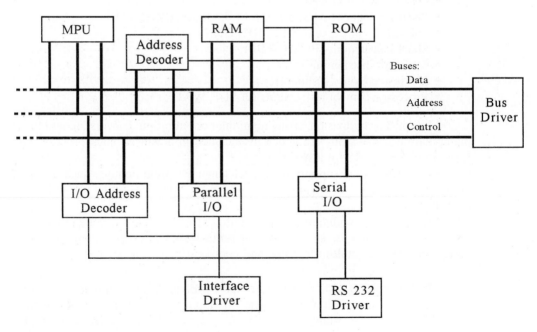

FIGURE 15.4
A three-bus microcomputer system.

The address buses of most 8-bit microprocessors are 16 bits wide. Thus, the memory sizes of such microprocessors as the 8085, the 6800, the 6502, and the Z80 are all equal to 64K or 65,536 bytes.

The external address buses and direct addressing capabilities supported by most 16-bit microprocessors are considerably different; therefore, these alternatives might affect the application suitabilities of the various microprocessors. The variations in memory spaces among various 16-bit microprocessors are shown in Table 15.1.

Most 32-bit microprocessors have full 32-bit address buses that permit a physical address space of 4 gigabytes, except that of the NS32032, which has a 24-bit address bus.

The Intel Pentium and Motorola PowerPC series of microprocessors support 32-bit address buses providing 4-gigabyte (Gbyte) addressing. Some versions of the PowerPC have a 40-bit address bus so that addressing is extended to 1 terabyte (Tbyte), which is 1,073,741,824 Kbytes!

The Data Bus   A data bus is a bidirectional bus used to transfer data to and from memory and I/O devices. The bit width of a microprocessor is defined as the number of parallel wires contained in the data bus. In general, a microprocessor with n data lines is normally called an n-bit microprocessor.

Thus a microprocessor with an 8-bit data bus is called an 8-bit microprocessor, the microprocessor like the 8088 (16-bit internal/8-bit external data bus) is considered an 8-bit microprocessor, and the MC68000 (32-bit internal/16-bit external data bus) is considered a 16-bit microprocessor.

Separated Data Bus and Address Bus   If a microprocessor is designed with a simple separated data bus and address bus, then no demultiplexing circuit is needed. These are typical examples of separated data/address bus microprocessors:

- *8-bit:* 6502, MC6800, MC6809, Z80
- *16-bit:* MC68000, Intel 80286

TABLE 15.1
Memory spaces of various 16-bit microprocessors

Name of Processor	External Address Bits	Physical Memory Size	Memory Management
Intel 8086	20	1 Mbytes	Segmented memory
Intel 80286	24	16 Mbytes	Segmented, virtual memory (1 Gbyte)
MC68000	24	16 Mbytes	N/A
MC68010	24	16 Mbytes	Virtual memory (2 Gbytes)
MC68012	30	1 Gbytes	Virtual memory (2 Gbytes)
Z8001	23	8 Mbytes	Maximum 48 Mbytes using MMU
Z8002	16	64 Kbytes	Maximum 384 Kbytes using MMU
NS32016	24	16 Mbytes	Virtual memory (4 Gbytes)

- *32-bit:* MC68020, Intel 80386, Intel 80486, and AT&T Bell Labs WE32100 (formerly Bellmac-32A)
- *64-bit:* Intel Pentium, Motorola PowerPC.

Multiplexed Data/Address Bus   Because of the pin number limitation, a number of microprocessors are designed with a multiplexed data/address bus. These microprocessors are:

- *8-bit:* 8085, 8088
- *16-bit:* 8086, NS32016, Z8001/8002
- *32-bit:* Z80000, NS32032.

To separate the data/address bus information of an 8-bit microprocessor, an octal latch, such as the 74LS373 or Intel 8282, can be used to capture the address portion and hold it until the microprocessor sends out more address information. Figure 15.5 shows how this can be done.

For 16- or 32-bit address/data multiplexed microprocessors, we would have to add more tristate latches. For example, two 74LS373 or Intel 8282 latches can be used as shown in Figure 15.6 for demultiplexing the address/data bus (AD0 through AD15) of the 8086.

Bus Buffering   Bus buffers can be used to increase the driving capability of the bus and to protect the bus from overload conditions. For a microprocessor system with light loading and relatively short bus length, output buffers are not normally required.

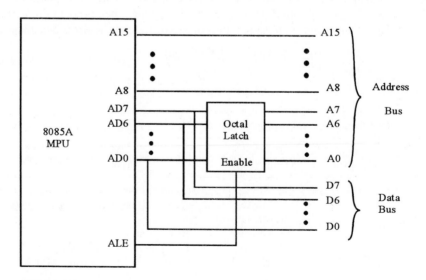

FIGURE 15.5
A data-address bus demultiplexing circuit for the 8085 MPU.

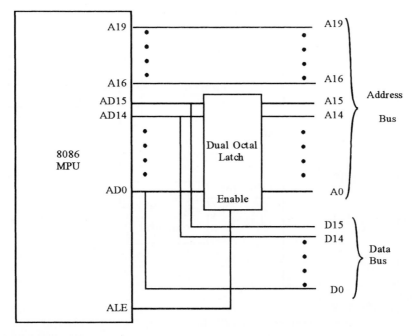

FIGURE 15.6
A data/address bus demultiplexing circuit for the 8086 MPU.

Octal buffer 74LS244 or Intel 8283 can be used for unidirectional address bus buffering. In general, there is no need to add buffers for the demultiplexed address bus because it is already buffered by the demultiplexing circuit. For example, address bus buffering for the 80286 CPU, which has separated data/address buses, is shown in Figure 15.7.

For the bidirectional data bus buffering, we can use bidirectional buffers or transmitter-receivers (transceivers), such as the 74LS245 or Intel 8286.

The Control Bus   The control bus is used to pass control information back and forth among the processor, memory, and I/O devices. Some of the control lines are inputs to the microprocessor, and others are outputs.

For 8-bit microprocessors, the control information might include memory synchronization, I/O synchronization, interrupts, bus request, clock, and reset.

However, some additional control signals are added to modern 16- and 32-bit microprocessors for supporting multiprocessing, multiprogramming, coprocessing, and/or memory management.

15.2.5   Processor States

Modern microprocessors have two operating modes: user and privileged or supervisor mode. When the microprocessor operates in the supervisor mode, the entire

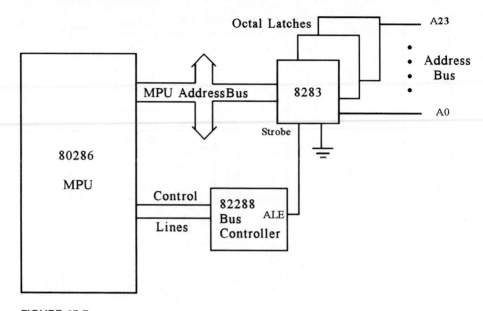

FIGURE 15.7
Address bus buffering for the 80286 MPU.

instruction set is available to the operating system or privileged programs. While it is in the unprivileged or user mode, only a restricted subset of instructions is available to unprivileged or normal user programs.

For example, the 32-bit MC68020 has supervisor/user states and the 32-bit 80386 uses real/protected address modes to distinguish the two operating states.

15.2.6 Interrupts, Exceptions, and Reset

The concept of interrupts supports multiprogramming by allowing more than one program to reside in main memory at the same time. These programs may include the main program and interrupt service routines.

In general, microprocessors provide two types of interrupts: hardware interrupts via the interrupt pins of a microprocessor, and software interrupts through the execution of particular instructions. Interrupt priority levels, and maskable/nonmaskable identification were defined during the design of the microprocessors and they cannot be altered by the user.

Two common hardware interrupts are called nonmaskable interupt (NMI) and a regular interrupt. The NMI is an interrupt request that cannot be refused or disabled by the processor, whereas a regular interrupt is maskable.

The software interrupt or exception is an interrupt issued to the microprocessor through an instruction; thus, it is normally designed as an NMI.

Multiple interrupts are serviced in order of their priority, regardless of the sequence in which they are received.

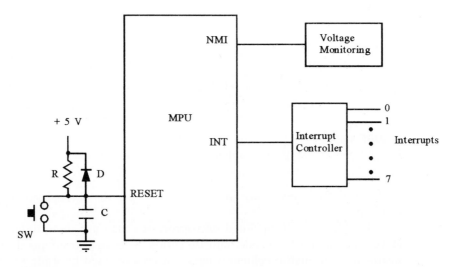

FIGURE 15.8
A typical example of using NMI, INT, and RESET.

To handle interrupts, there are reserved memory locations for storing the addresses of interrupt service routines, as well as special registers or flags needed for reporting information about interrupts or for representing the disabling or enabling of interrupts.

Modern microprocessors are equipped with many interrupt vectors for internal exceptions and external hardware interrupts, and they may work with interrupt controller chips. The exceptions may include divide error, overflow error, invalid op-code, stack under/overflow, coprocessor error, bus error, address error, trace, and so on.

Reset Input All microcomputers require a reset line for system reset or initialization. When the microprocessor senses RESET, an initialization routine is unconditionally invoked with the highest priority.

A typical example of using interrupts and resets is shown in Figure 15.8. The NMI is generally used by the microprocessor's voltage level monitoring subsystem to handle power failures. An interrupt controller connects interrupt requests in an appropriate order to the interrupt input of the microprocessor. Also, an RC time delay circuit provides the microprocessor with power-on reset capability and the push button switch offers manual reset operation. The reader is referred to microprocessor user manuals for optimal values of R and C. Finally, the diode D is used to protect surge voltage from the power supply and prevent microprocessor damage.

15.2.7 Advanced Architectural Features

Pipelining Most higher performance 16- and 32-bit microprocessors are designed with an integration of the well-recognized concurrent operation technique

Time Cycle

Functional Stage	0	1	2	3	4	5	6	7	8	9
Execution			I1	I2	I3	I4	I5	I6	I7	I8
Decode		I1	I2	I3	I4	I5	I6	I7	I8	I9
Fetch	I1	I2	I3	I4	I5	I6	I7	I8	I9	I10

FIGURE 15.9
Ideal overlapped execution of a three-stage pipelined computer.

called *instruction pipelining* in order to increase the throughput of the micro-processors.

For example, a pipelined microprocessor may use three different function stages such as fetch, decode, and execution. The ideal overlapped execution of instructions through pipelining stages can be explained by using the reservation table shown in Figure 15.9. In the reservation table, rows indicate the function stages and the time intervals are shown on the top. From time interval 2 and up, we observe that instruction execution, decoding, and fetching stages perform on different data in parallel fashion.

Dual-Port Memory   As the name implies, this memory subsystem has two ports. This system is designed to be used as a shared memory but not allow simultaneous access by two processors. Thus, an arbitration control logic is used to ensure mutual exclusion.

For example, in a multimicroprocessor system the dual-port memory serves both as local memory to processors on a single local bus, and as system memory to other processors in the global system bus. Thus, this memory system can be commonly used by two processors as shared memory for sharing data and codes and for interprocessor communication.

Intel has presented an example of implementing a dual-port memory by using the 8027 dynamic RAM controller as arbitration and synchronization logic and dynamic RAM memory.

As another example, the dual-port memory may be used in graphics systems to keep the flow of data from overwhelming the bus.

Interleaved Memory   The interleaved memory is a parallel memory or multimemory system that is used to speed memory access by allowing overlapped memory fetches.

Cache Memory   The cache memory is a high-speed memory in between the CPU registers and main memory. It is normally used for storing frequently used parts of the code and data, and hence eliminates the number of fetches from main memory that a CPU must perform.

Some microprocessors, such as the Intel 80386, use high-speed cache memory, which is external to the processor, e.g., an external memory chip. On-chip addressing of this cache is provided. This is called level 2 cache. Level 1 cache is more common on newer processors for which the cache memory is built into the processor chip.

The Intel 486 series has an 8-Kbyte level 1 cache for instruction and data prefetches. The Intel Pentium uses a 16-Kbyte level 1 cache organized as two 8-Kbyte caches, one exclusively for instructions and the other exclusively for data. The PowerPC 620 Motorola microprocessor uses two separate level 1, 32-Kbyte instruction and data caches.

Virtual Memory   The virtual memory technique allows a large memory addressing capability that exceeds the physical size of RAM main memory by using hardware and software to manage the mapping and data transfer between main memory and secondary storage (disc memory system).

This memory management method is implemented in many mainframes and minicomputers. Currently, most 32-bit and some 16-bit microprocessors support this capability either on-chip or through a separated memory management unit (MMU). Currently, the segmentation, the paging and mix of segmentation, and paging schemes are three different implementations of virtual memory in microprocessors.

In the segmentation scheme, a block of memory provides the processor with many, completely independent address spaces, called *segments*. Each segment is a particular portion of memory consisting of a linear sequence of addresses, from relative zero to some maximum of variable size.

In the paging scheme, the address space is split into a number of equal-sized blocks called *pages*. Typical page size is either 2 or 4 Kbytes. An address translation mechanism uses table registers, segment registers, and page tables to relate virtual addresses to physical addresses.

Multiprocessor System   A multiprocessor system could be defined as a system composed of a number of microprocessors operating at the same time to solve a given application, or to run several jobs in parallel on separate microprocessors. Compared to a single microprocessor system, a multimicroprocessor system can significantly speed up processing because of the introduction of parallelism into the basic sequential architecture.

Bus Arbitration   Consider the situation in which a number of processors share a common bus. If more than one processor tries to access the bus at the same time, then a bus contention or conflict is introduced. A *bus arbiter* is a device that uses a given scheme to resolve the bus conflicts by activating only one processor and putting other processors into wait states.

Coprocessing   Special-purpose processors such as math, numeric, or floating-point coprocessors, I/O coprocessors, and direct memory access coprocessors may be incorporated with CPUs to form a multiprocessing computer system.

For example, using a math coprocessor would effectively extend the instruction and register set and the data types of the microprocessors for handling increased computational loads.

Parallel Input/Output (PIO) Interfaces Parallel I/O interface chips such as the Motorola 6821 PIA (peripheral interface adapter), Intel 8255A PPI (programmable peripheral interface), and Z-80 PIO are used to perform simple byte data transfers between I/O devices and the CPU.

For example, two 8255A PPIs connected in parallel can be used by the 8086 microprocessor to process word I/O, and four 8255A PPIs are appropriate for a 32-bit microprocessor, such as the 80386, to handle 32-bit I/O data transfers.

Serial Interfaces The serial interface chips are used for serial communication, in which the transmission of data is in a bit stream. The MC6850 asynchronous communication interface adapter (ACIA), the Z-80 SIO, and the Intel 8251 USART (universal synchronous and asynchronous receiver and transmitter) are peripheral chips that can be used for serial communication applications.

Asynchronous serial communication requires no common clock and is generally used for data rates below 19.2 Kb/s. To transmit an ASCII character, a start bit is issued to inform the receiver to get ready, and then 8 data bits are sent at fixed data rates. Finally, 1 or 2 stop bits are used to indicate the end of a transmission.

Synchronous serial data communication is used for high-data rate and noninterrupted data transmission. There is no start and stop bit in between the transmission of characters, and synchronizing characters are normally sent in the form of uninterrupted strings of bits.

Direct Memory Access (DMA) Controller The DMA controller is a specialized chip including a control register, a bus interface unit, and some independent channels, each of which contains an address register, and a counter. The DMA is programmed via the CPU. After the DMA has been initialized, it may request the use of both the address bus and the data bus and suspend operation of the CPU while performing a block of data transfer directly between an I/O port and main memory.

Superscalar Architecture Many new microprocessors are incorporating a feature called *superscalar operation* wherein the processor has more than one instruction execution unit. This allows more than one instruction to be executed at the same time, as long as the instructions or their results do not depend on each other.

The Intel Pentium has three instruction execution units, two for integer operations and one for floating-point operations. Thus up to three instructions can be executed at the same time. Estimates are that efficient use of this feature can lead to a 40 to 50% improvement in execution speed.

The Motorola PowerPC also includes superscalar operations with six independent instruction units in the PowerPC 620. The design allows for simultaneous execution of up to four instructions per clock cycle.

15.3 EIGHT-BIT MICROPROCESSORS

Popular 8-bit microprocessors such as the MC6800, the 6502, the 8085, and the Z-80 have the following general characteristics: an ALU, a set of programmer accessible registers, a program counter, an instruction decoding unit, a timing and control unit, an 8-bit parallel data and input/output bus, a 16-bit address bus (maximum 64 Kbytes of memory space), and interrupting capability. Enhanced versions of 8-bit microprocessors normally include added instructions, larger memory space than 64 Kbytes, and higher speed.

15.3.1 The MOS Technology 6502 and 65816 Microprocessors

The 6502 has a 16-bit address bus (64K address space), an 8-bit data bus, and two interrupt inputs ($\overline{\text{IRQ}}$ and $\overline{\text{NMI}}$). In addition, the 6502 has a 16-bit program counter, PC, and the following 8-bit registers: accumulator A, index register X, index register Y, stack pointer S, and processor status register P. The flags in the processor status register are carry, zero, IRQ disable, decimal mode, software break or interrupt command, overflow, and negative.

The 6502's registers and the vector locations of reset, interrupt and break command, and nonmasked interrupt are shown in Figure 15.10.

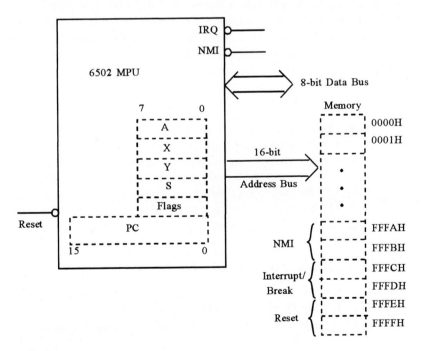

FIGURE 15.10
The 6502 CPU's registers and addresses of the interrupt vectors.

The 6502's instructions include the following groups: loads, stores, arithmetic, logical, shift and rotate, increment, decrement, compare, conditional branch (BCC), jump, jump to subroutine (JSR), stack manipulation (push and pull), flag manipulation, and register transfer.

To speed up instruction execution, we can write an instruction with a lower byte for the address that references to the *zero page* (memory address 00 to FF). For example, the instruction LDA $FOH will move the data from memory location 00F0H into the A register. The upper byte of the address is supplied internally. Therefore, a speed up of instruction execution is obtained.

The stack memory of the 6502 is located on the second page of memory addresses 0100H through 01FFH. The 8-bit stack pointer can be set or read via the X register. For example, to initialize the stack pointer to the top of the stack, we need two instructions:

```
LDX #$FF
TXS            ; transfer X to stack pointer
```

The peripherals for configuring the 6502 microcomputer system can include the 6520 PIA, the 6522 versatile interface adapter (VIA), the 6530 ROM/RAM-I/O timer, and/or the 6532 RAM-I/O timer.

The 65816 This microprocessor is an 8-/16-bit version of the 6502. The internal data path is 16 bits wide, but the external data path is 8 bits wide. The address bus is expanded from 16 to 24 bits to provide a 16-Mbyte addressing capability. Additional 8-bit address lines are multiplexed with the data bus. It was designed to operate in either the 6502 emulation, or the 65816, which supports multiprocessing through the memory lock (ML) control signal. In addition, the widths of internal registers are extended to 16-bit.

The CPU registers of the 65816 include the following:

- X: a 16-bit index register composed of two 8-bit index registers (XH and XL)
- Y: a 16-bit index register composed of two 8-bit index registers (YH and YL)
- S: a 16-bit stack register composed of two 8-bit stacks (SH and SL)
- C: a 16-bit accumulator of two 8-bit A and B accumulators
- PC: a 24-bit program counter
- D: a 16-bit direct register composed of two 8-bit direct registers (DH and DL)
- P: an 8-bit status register.

Six flags of the 6502's status register have the same meaning as in the 65816. Other flags and their definitions in this CPU are:

- X: index register select; 1 = 8 bits and 0 = 16 bits
- E: emulation; E = 1 for the 6502 emulation
- M: memory select; 1 = 8-bit accumulator, and 0 = 16-bit accumulator.

New instructions are introduced for covering additional registers. The MVP and MVN are two new block move instructions that allow us to move a block memory within 16-Mb address space. MVP moves a block of data downward, whereas MVN moves the data upward.

To support multiprocessing, two instructions, TRB (test and reset bits) and TSB (test and set bits), can be used to set and clear semaphores that arbitrate the shared memory and peripherals.

15.3.2 The Motorola MC6800 and MC6809

The MC6800 The 8-bit MC6800 microprocessor uses a 16-bit address bus and a 16-bit program counter to provide a typical 64K address space. Other CPU registers are two 8-bit accumulators (A and B), a 16-bit stack pointer, a 16-bit index register, and an 8-bit condition code register (CCR).

The flag bits of the CCR include zero, negative, interrupt-mask, half-carry, carry, and overflow bits.

The interrupt vectors that hold the locations of interrupt service routines include the reset (FFFF-FFFE), the nonmaskable interrupt NMI (FFFD-FFFC), the software interrupt SWI (FFFB-FFFA), and the maskable interrupt IRQ (FFF9-FFF8).

The MC6821 peripheral interface adapter is for parallel interfacing and the MC6850 asynchronous communication interface adapter is for serial interfacing. Other peripherals such as the MC6840 programmable timer, the MC6843 floppy-disc controller, the MC6828 priority interrupt controller, and so on are also available.

The instruction set of the MC6800 is divided into three general groups: (1) accumulator and memory operations, (2) program control operations, and (3) CCR operations.

All of the load and store, arithmetic and logical, rotate and shift, stack operations (push and pull), test zero or minus, and transfer instructions are included in the first group.

The program control operations contain the conditional branches, jump, and jump to subroutine instructions. Note that the jump to subroutine is the same as a call to a subroutine in other microprocessor.

Finally, the CCR manipulation instructions are bit manipulation operations for modifying processor status.

The MC6809 This is an enhanced version of the MC6800. The MC6809 supports modern programming techniques through the hardware and added instructions.

Special features of this microprocessor are improved stack manipulations, 8×8 unsigned multiply, 16-bit arithmetic, transfer/exchange for all registers, push/pull for all registers, and load effective address instructions.

The CPU registers are shown in Figure 15.11. These include two 8-bit accumulators A and B or a 16-bit D register (pairing A and B), two 16-bit index registers X and Y, two 16-bit stack pointers U and S, a 16-bit program counter, an 8-bit condition code register, and an 8-bit direct page register.

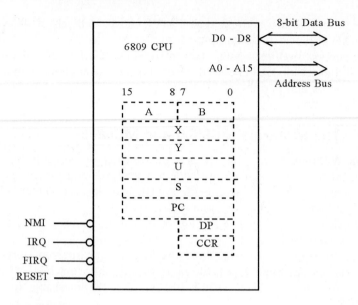

FIGURE 15.11
The MC6809 CPU's registers.

The U stack pointer is designed to be used by the programmer as a stack marker for passing arguments to and from subroutine, but the S stack is used automatically by the 6809 during interrupts and subroutine calls.

The direct page register is designed to be used in direct addressing for holding high byte address information (A8–A15).

The MC6809 instruction groups include the 8-bit and 16-bit accumulator and memory instructions, index register/stack pointer instructions, branch instructions, and miscellaneous instructions.

All the peripherals that are used by the MC6800 can be interfaced to the MC6809.

15.3.3   The Intel 8085A

This microprocessor has an 8-bit data bus multiplexed with the lower 8 bits of the address bus (AD0–AD7), 64-Kbyte addressing capability, an on-chip clock generator (with an external crystal, LC or RC network), a serial in/serial out port, seven 8-bit general-purpose registers (A, B, C, D, E, H, and L), a 16-bit program counter, a 16-bit stack pointer, and a flag register containing the carry, zero, sign, auxiliary, and parity flags. Furthermore, six of the 8-bit registers can be paired as three 16-bit registers BC, DE, and HL.

The 8085 has one nonmaskable interrupt TRAP with associated vector address 24H. In addition there are four maskable interrupts and their related vectors: RST7.5 (3CH), RST6.5 (34H), RST5.5 (2CH), and INTR. The 8085 CPU will need to obtain the RST code from an external hardware circuit in order to respond to the INTR interrupt. These restart codes are the same as the restart instructions. The names and their vector addresses are RST0 (0000H), RST1 (0008H), RST2 (0010H), RST3 (0018H), RST4 (0020H), RST5 (0028H), RST6 (0030H), and RST7 (0038).

The reset will load PC with 0000H and performs a system initialization by executing the instructions that are stored in the memory specified by the PC.

The functional grouping of the 8085 instruction set includes data transfer operations (MOVE, LOAD, and STORE), stack operations (PUSH and POP), arithmetic operations, logical operations (AND, OR, XOR, and COMPARE), shift and rotate operations, conditional and unconditional jumps, call and return operations, I/O operations, restart operations, reset and set interrupt operations, and processor control operations.

For example, the minimum configuration of the 8085A microcomputer system includes the 8085A itself, the 8155 (256 × 8 RAM, I/O timer/counter), and the 8355 (2K × 8 ROM, I/O) or the 8755A (2K × 8 EPROM, I/O).

The special-purpose support devices for the 8080/85 family are the 8255 PPI, the 8251 UART, the 8254 programmable interval timer, the 8259A interrupt controller, and the 8237 DMA controller.

15.3.4 The Zilog Z-80

The frequently noted features of this microprocessor are its on-chip dynamic RAM refresh capability, bit manipulation, block transfer instructions, duplicate sets of general-purpose registers, and index registers. This chip was very popular in the old CP/M operating system applications and also runs 8080/8085-based software.

The Z-80 CPU registers include 8-bit general-purpose registers (A, B, C, D, E, H, and L), a flag register, a set of 8-bit alternate registers (F′, A′, B′, C′, D′, E′, H′, and L′), an 8-bit interrupt register, a 7-bit dynamic memory refresh register, a 16-bit program counter, a 16-bit stack pointer, and two 16-bit index registers (IX and IY). The CPU registers are shown in Figure 15.12.

The alternate register is a duplicate of the main registers used by exchange instructions for faster response in processing interrupt service routines.

The general-purpose registers can be paired as 16-bit registers: FA, BC, DE, HL, F′A′, B′C′, D′E′, and H′L′.

The 8-bit interrupt register is used to store the page address of an interrupt service routine.

The Z-80 instruction set includes 78 instructions of the 8080/8085A as a subset. Also an extensive instruction set contains string, bit, byte, and word operations. In addition, the following powerful instructions are supported: exchange, block search, block transfers, and block I/Os.

The EXX instruction exchanges the register bank and auxiliary bank (BC ↔ B′C′, DE ↔ D′E′, and HL ↔ H′L′). Thus, this instruction can be used in subroutine calls for faster saving of the registers.

The CPIR and CPDR are two block search instructions. Both instructions use the HL register as an index register and the BC register as a byte counter.

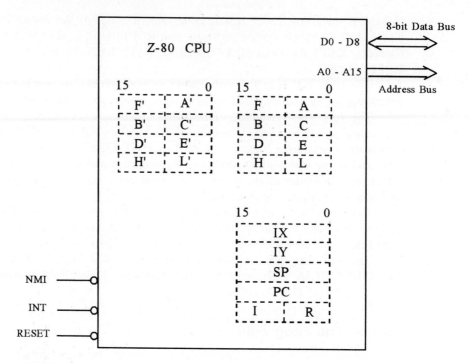

FIGURE 15.12
The Z-80 CPU's registers and interrupt inputs.

Before the execution of the block search instructions, HL must be initialized with the beginning address of the memory block that is to be searched; the BC is loaded with the size of the block; and the pattern should be in the A register.

The CPIR is used to search from low memory to high memory locations with these operations:

- If the contents of the A register and the location pointed to by the HL index are equal, namely, [A] = [HL], then set the zero flag and stop the search.
- Else increment the HL by one, decrement the BC by one, compare again, repeat, until the byte is found or until BC = 0.

The CPDR has the same operations as the CPIR except that the HL is decremented.

Two block transfer instructions are LDIR and LDDR. Here, HL is used as a source index, DE is used as a destination index, and BC is a byte counter.

The LDIR will move the [BC] byte of data from the memory location pointed to by the HL register to the memory location pointed to by the DE register in an upward direction (from low to high). That is, repeat the following operations until the BC is equal to zero:

```
[DE] ↔ [HL], DE ↔ DE + 1, HL ↔ HL + 1, and BC ↔ BC −1
```

The LDDR will move a block of data in a downward direction, namely:

[DE] ↔ [HL], DE ↔ DE − 1, HL ↔ HL − 1, and BC ↔ BC −1

Block I/O instructions are INDR, INIR, OTDR, and OTIR. Here, registers B (byte counter), C (port register), and HL (memory pointer) should be initialized before the execution of these block I/O instructions. The INIR or INDR can be used to input a maximum of 256 bytes of data from an input port to main memory in the upward or downward direction; the OTIR or OTDR is for output memory bytes to output ports.

The interrupt-related instructions are enable and disable interrupt (EI, DI), return from interrupt RETI, return from nonmaskable interrupt RETN, set interrupt mode 0 through 2 (IM0, IM1, IM2), wait for interrupt or reset HALT, and move data from A register to or from interrupt register (LD I, A; LD A, I).

The restart instructions (RST 00H, RST 08H, RST 10H, ,RST 18H, RST 20H, RST 28H, RST 30H, and RST 38H) are like that of the 8085A.

There is a nonmaskable hardware interrupt NMI with associated vector address 0066H. The Z-80 accepts the maskable interrupt as one of three options:

- *Mode 0:* with the IM0 instruction in effect and looking for a 1-byte restart instruction from the data bus
- *Mode 1:* vector address 0038H
- *Mode 2:* using the I register and hardware device code input from data bus to form a vector.

Peripheral devices that support the Z-80 are Z80-DMA/Z8410-DMA (direct memory access), Z80-PIO/Z8420-PIO (parallel I/O ports controller), Z80-SIO/Z8440-SIO (serial I/O controller), and Z80-CTC/Z8430-CTC (counter timer controller).

The Z64180 This is a modified version of the Z-80 designed to include an on-chip memory management unit (MMU) that provides 512K of physical memory space, and an on-chip serial communications interface and timers.

The Z800 The Z800 microprocessor family is designed to provide an upgrade path for the existing Z-80 software. The features of this microprocessor follow: up to 25-MHz operation, on-chip MMU providing logical addressing space up to 16 Mbytes, a 256-byte on-chip cache, four built-in DMA channels, three counter/timers, a UART, an interrupt controller, a dynamic RAM refresh mechanism, and a clock oscillator.

15.4 16-BIT MICROPROCESSORS

Advances in technology, market demands, and high software costs have motivated manufacturers to design high-performance 16-bit microprocessors. The architectural features of these microprocessors might include a 16-bit data bus, the ability to address large physical memory space, a pipeline architecture that allows instructions to be prefetched during spare bus cycles, the support of coprocessing, multitasking and multiuser capabilities, and multiprocessing. In addition, the 16-bit microprocessors normally support multiply, divide, and some other powerful instructions, which are absent in most 8-bit microprocessors.

Representative examples of these 16-bit microprocessors include the Intel 8086 and 80286, Motorola MC68000/68010/68012, Zilog Z8001/8002, and National Semiconductors NS32016. They are found in both general-purpose personal computer systems and in embedded computer systems for such real-time applications as instrumentation and control systems, telecommunication switches, industrial robots, and so on.

15.4.1 The Intel 16-Bit Microprocessors

The Intel 8086 This is a true 16-bit microprocessor designed with a two-stage pipeline, namely, a BIU (bus interface unit) and an execution unit (EU). In addition there is a 6-byte instruction queue between BIU and EU to increase system performance. Furthermore, modular programming support and high-level language orientation are two architectural concepts that are incorporated in the 8086 design.

The 8086's internal registers include four 16-bit general-purpose registers (AX, BX, CX, and DX), two 16-bit pointers (stack pointer SP and base pointer BP), two 16-bit index registers (source index SI and destination index DI), four 16-bit segment registers (code segment CS, data segment DS, stack segment SS, and extra segment ES), a 16-bit instruction pointer, and a 16-bit flag register (carry, parity, auxiliary, zero, sign, overflow, interrupt enable, direction, and trap flags).

The four 16-bit general-purpose registers can be referenced on a byte-wide basis, i.e., they can be identified as either high-byte registers (AH, BH, CH, and DH) or low-byte registers (AL, BL, CL, and DL).

The main memory of the 8086 has a maximum linear space of 1 Mbyte. A memory segmentation technique is used to divide the main memory into code, data, stack, and extra segments of up to 64 Kbytes each, with each segment falling on 16-byte boundaries. Memory locations 0000 through 003F (1 Kbyte) are reserved for 256 four-byte interrupt vectors containing the address of the interrupt service routines.

In addition, I/O instructions and I/O address space of up to 64Kbytes are supported by the 8086.

Two modes of operations (minimum mode for the 8086 itself in the system, and maximum mode for multiprocessing) are supported by the 8086. For example, a typical three-processor system may include a general data processor (8086), a math coprocessor (8087), and an I/O coprocessor (8089) in a maximum mode operation.

The 8086's instruction set can be separated into the following groups:

- *Data transfer instructions:* move, push, pop, exchange, translate, input, output, load effective address, and so on
- *Arithmetic instructions:* ADD, SUB, increment, decrement, ASCII adjust

for addition and subtraction, decimal adjust for addition and subtraction, negate, compare, unsigned and integer multiply (MUL and IMUL), ASCII adjust for multiplication and division (AAM and AAD), unsigned and integer divide (DIV and IDIV), convert byte to word (CBW), and convert word to double word (CWD)

- *Bit manipulation instructions:* NOT, AND, OR, XOR, TEST, shift arithmetic left and right (SAL and SAR), shift logical left and right (SHL and SHR), rotate left and right (ROL and ROR), rotate through carry left and right (RCL and RCR).
- *String instructions:* move byte or word string (MOVS), compare string (CMPS), load byte or word string (LODS), store byte or word string (STOS), and repeat (REP, REPE/REPZ, REPNE/REPNZ)
- *Program transfer instructions:* jump on conditions, unconditional transfers (CALL, RET, JMP), iteration controls (LOOP, LOOPE/LOOPZ, LOOPNE/LOOPNZ, JCXZ), and interrupts (INT, INTO, IRET)
- *Processor control instructions:* flag operations, halt until reset or interrupt (HLT), wait for test pin active (WAIT), escape to external processor (ESC), lock bus during next instruction (LOCK), and NOP.

Interrupts of the 8086 include external hardware interrupts (INTR and NMI), reset, software interrupts (INT n), and internal interrupts (divide error, single step, breakpoint, and overflow).

The INTR line is usually connected and driven by an Intel 8259A interrupt controller, which accepts up to eight interrupt requests from external devices.

The 80186 This is a highly integrated version of the 8086 microprocessor, effectively combining two independent DMA channels, three 16-bit timer/counters, a programmable interrupt controller, and an on-chip clock generator. Note that the maximum memory space (1 Mbyte) and the CPU registers are the same as the 8086.

The new instructions added to the 80186 include push immediate data (PUSH immediate), push all registers on stack (PUSH A), pop all registers from stack (POP A), block input and output byte or word string (INS and OUTS), signed integer multiply, immediate boundary checking for data array, enter and leave for build, and tear down of stack frames for high-level blocked-structured languages.

The 8088 This is an 8-bit data bus version of the 16-bit 8086. The 8-bit data bus is multiplexed with the lower portion of the 20-bit address bus. Thus, an external demultiplexing circuit is still needed in order to separate data and address information. The instruction queue of the 8088 is 4 bytes long instead of the 6 bytes of the 8086. The 8088 has an instruction set, memory addressing capability, and CPU registers that are identical to the 8086.

Two examples of the 8088-based microprocessor system are (1) a minimum system including the 8088 CPU, the 8155(s) (RAM, I/O, and timer), and the 8755A (EPROM, and I/O), which is shown in Figure 15.13, and (2) a typical 8088-based

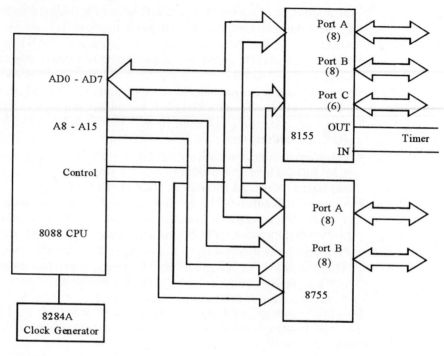

FIGURE 15.13
A minimum system for the Intel 8088-based microprocessor system.

personal computer including the 8088 CPU, the single-chip microcomputer 8048 (performing keyboard scanning, power-on self test, and system requested tests), the 8259A interrupt controller, the 8255A PPI, the 8253 timer, the 8337A DMA controller, ROM, and RAM. This is shown in Figure 15.14.

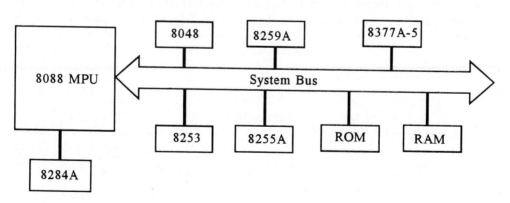

FIGURE 15.14
A typical 8088-based microprocessor system.

The 80188   This processor is the 8-bit data bus version of the 80186. All the features in the 80186 are applied to the 80188.

The 80286   This 16-bit CPU has a four-stage pipeline (bus, instruction, execution, and address unit), hardware integration of task switching and the memory protection mechanism for supporting multitasking and multiuser systems, and a built-in MMU for supporting virtual memory.

Furthermore, the 80286 is capable of executing either binary-level 8086 code in a 1-Mbyte real address mode, or the 80286 code in protected mode of 16-Mbyte real address space (24-bit address) and up to 1 Gbyte of virtual address space. Note that the 24-bit address bus and 16-bit data bus are separated.

A segmented virtual memory is implemented in the 80286, giving a logical address space of 16K times 64K segments, which is 1 Gbyte.

The 80286 system might use such support devices as the 80287 math coprocessor, the 8089 I/O coprocessor, the 82289 bus arbiter, the 82288 bus controller, the 8259 programmable interrupt controller, the 8254 programmable timer, the 8274 multifunction USART, the 8255 PPI, the 8207 advanced dynamic RAM controller, the 802586 Ethernet coprocessor, and the 802730 text/graphics coprocessor.

A full register set of the 80286 includes all of the registers of the 8086. Also, a 1-bit nested task flag (NT), and two bits for I/O privileged levels (IOPL) are added to flag registers. Plus, a 16-bit machine status word (MSW) contains four status bits: task switch (TS), processor extension emulated (EM), monitor processor extension (MP), and protection enable (PE).

Finally, a set of descriptor table registers and hidden descriptor caches is introduced for translating virtual addresses of programs to real addresses. These are four 48-bit hidden descriptor caches for CS, DS, ES, and SS selectors; two 40-bit system address registers, namely, global and interrupt descriptor table registers (GDTR and IDTR); and a 56-bit local descriptor table register LDTR including a 16-bit visible selector and a 40-bit hidden descriptor cache.

The LDTR is used to identify and isolate the segments that belong to each task. The GDTR defines global address space available to all the tasks in the system.

The 80286's instructions include all of the 8086's plus some extended instructions:

- *Push all instructions:* PUSHA (push all registers on stack) and POPA (pop all registers from stack)
- *Block I/O instructions:* INS (input byte of word string) and OUTS (output byte string)
- *Execution environment control instructions:* LMSW (load machine status word) and SMSW (store machine status word)
- *High-level language instructions:* ENTER (create the stack frame for procedure entry) and LEAVE (restore the stack frame for procedure exit)
- *Array index checking instruction:* BOUND.

Reserved interrupts and exceptions for the 80286 are divide error, single step, NMI, breakpoint, INTO overflow exception, BOUND range exceeded exception, invalid op-code, processor extension not available, double exception detected, processor extension segment overrun interrupt, invalid task state segment, segment not present, stack segment overrun or not present, and general protection.

The NMI and INTR are two hardware interrupts to the 80286. Again, the 8259A interrupt controller is generally connected between the interrupt input and external devices that request the interrupts.

15.4.2 The Motorola MC68000 Family

The MC68000   Basic features of the MC68000 include separated address (24-bit, 16-Mbyte address space) and data (16-bit) buses, 32-bit internal registers, memory mapped I/O, eight priority levels of interrupts, two states of operation, namely, user and supervisor, and structured modular programming support through instructions.

The MC68008   This is an 8-bit data bus and 20-bit address bus version of the MC68000. Thus, the address space of the MC68008 is 1 Mbyte. Note that both the MC68000 and the MC68008 have identical CPU registers.

The MC68010   This is a virtual memory/virtual machine version of the MC68000, with 16-bit data bus and 24-bit address bus. The user state programming model (CPU registers) is the same as the MC68000. But the supervisor state of the MC68010 has three more registers, namely, a 32-bit vector base register (VBR) and two 2-bit alternate function registers (SFC and DFC).

The 68012   This is an extended address version of the 68010 equipped with a 16-bit data bus and 30-bit address bus.

These different versions of processors are equipped with identical data registers and address registers, except that the supervisor state registers are different in order to meet the requirements of different applications.

Eight data registers (D0 through D7), all 32 bits wide, can be used to store data. They can also be used as index registers to hold offset addresses.

Seven address registers (A0 through A6) are all 32 bits wide. They are not used for data storage but for address information storage such as base addresses, address values, and pointer addresses.

Address register A7 is called the *user stack pointer,* which is used to store the return addresses, data, and other parameters that are required for handling a subroutine call.

When the 68000/68008/68010/68012's are set in the user mode, all the data and address registers, program counters, CCRs, and user stack pointer A7 are available to user program.

In the supervisor mode, the supervisor state pointer (SSP) or A7' is used to point to the supervisor stack, which is used by supervisory calls such as interrupts, internal exceptions, and software exceptions. The status register includes a user byte and a system byte. The user byte is the same as the CCR. The system byte contains five bits: T for trace, S for supervisor state, and three bits called I2, I1, and I0 or interrupt mask. The privileged instructions can be executed only in this mode.

The 68000 family's instruction set includes the following instruction groups:

- *Data movement instructions:* MOVE data, MOVEA address, MOVEP peripheral data, MOVEM multiple registers, MOVEQ quick, EXG for exchange registers, LEA for load effective address, PEA for push effective register, LINK for link stack, and UNLK for unlink stack
- *Integer arithmetic instructions:* ADD, SUB, multiply (MUL), divide (DIV), arithmetic compare (CMP), test (TST), clear (CLR), negate (NEG), and sign extension (EXT)
- *Logical instructions:* AND, OR, NOT, and exclusive OR (EOR)
- *Shift and rotate instructions:* arithmetic shift left or right (ASL, or ASR), logical shift left or right (LSL or LSR), rotate instructions (ROL and ROR), rotate with extend bit (ROXL and ROXR), and SWAP for exchanging two 16-bit halves of a register
- *Bit manipulation instructions:* bit test (BTST), bit test and set (BSET), bit test and clear (BCLR), and bit test and change (BCHG)
- *Binary coded decimal instructions:* add decimal with extend (ABCD), subtract decimal with extend (SBCD), and negate decimal with extend (NBCD)
- *Program control instructions:* branch on condition (Bcc), test condition, decrement and branch (DBcc), branch always (BRA), branch to subroutine (BSR), jump (JMP), jump to subroutine (JSR), no operation (NOP), return and deallocate parameters (RTD), return from exception (RTE, this is a privileged instruction), return and restore condition codes (RTR), and return from subroutine (RTS)
- *System control instructions:* privileged, trap, and CCR modified instructions
- *Multiprocessor instruction:* test and set an operand (TAS) for synchronization of several processors.

High-level language programming is supported by the MC68000 through simple architecture, multiple registers and stacks, and high-level-language-oriented instructions.

Among them, LINK, UNLINK and MOVE multiple registers are designed to reduce the overhead of subroutine calls. Software TRAP vectors can be used to support operating system routine calls; PEA, LEA, RTR, RTE, JSR, BSR, and RTS can be used to support structured programming.

Hardware and software traps can be used to improve software testability and thus reduce the chance of errors. The exception-generating instructions include unconditional TRAP, TRAPV (on overflow), CHK (check register against bound), DIVS (sign divide), and DIVU (unsigned divide). Seven hardware interrupt requests are encoded and input to IPL2 through IPL0 pins.

One kilobyte of memory from 000H to 3FCH is reserved for exception vectors but some of it is unassigned and reserved for future expansion. These exceptions are grouped and handled on a priority basis as follows:

- *Group 0* (*highest priority*): reset, bus error, and address error
- *Group 1:* trace, interrupts, illegal instruction, and privileged violation
- *Group 2* (*lowest priority*): trap, trap on overflow, check register against bound, and divide by zero.

The support chips of the MC68000 family include the MC68451 MMU, the MC68230 parallel interface/timer, the MC68450 DMA controller, and the MC6800's peripherals such as the MC6850 ACIA, the MC6821 PIA, and so on.

15.4.3 The National Semiconductor NS32016

The NS32016, previously named NS16032, is a 16-bit high-level-language-oriented microprocessor from National Semiconductor, with an internal data bus and an ALU 32 bits wide, multiplexed data and address lines (AD0–AD15), memory-mapped I/O, an 8-byte prefetch queue, and a slave processor interface. A 24-bit external address bus provides up to 16 Mb of uniform addressing.

The NS32016 has three hardware interrupts: nonmaskable interrupt, maskable interrupt, and ABORT. ABORT can be used to abort a bus cycle and any associated instructions. In addition, there are eight internally generated traps or exceptions. These include a slave processor trap, illegal operation trap, supervisor call trap, divide-by-zero trap, flag trap, breakdown trap, trace trap, and unidentified instruction trap.

The NS32081 FPU provides floating-point instructions for single (32-bit) and double (64-bit) precision. The 32082 demand-paged MMU is a slave processor available to the NS32016 CPU to provide operating system support functions such as virtual memory management, memory protection, and dynamic address translation.

Also, the system peripherals such as the NS32201 TCU/timing control unit, the 32202 ICU/interrupt control unit, the DMA controller, and the NS32450 UART/universal asynch are available to this CPU.

The NS32016 has the following registers in the programming model:

- Eight 32-bit general-purpose registers: R0–R7
- Program counter: 24-bit (16 Mbytes of memory)
- SP1: a 24-bit user stack pointer
- SP0: a 24-bit interrupt stack pointer
- FP: a 24-bit frame pointer used by a procedure to access parameters and local variables on the stack
- SB: a 24-bit static pointer points to the global variables of a software module
- INTBASE: a 24-bit interrupt base register contains the address of the dispatch table for traps and interrupts
- MOD: a 16-bit module register contains the module descriptor address of the currently executing software module
- PSR: a 16-bit processor status register
- CFG: a 4-bit configuration register is used to declare the presence of the slave processors such as MMU, FPU, and custom slave processors.

The instruction set of the NS32016 includes move, integer arithmetic, BCD arithmetic, logical and Boolean, shift and rotate, bit field, string, jumps and linkage, CPU register manipulation, floating point, memory management, arrays, custom slave, and miscellaneous instruction groups.

15.4.4 The Zilog 16-Bit Microprocessors

The Z8000 16-bit microprocessor family includes the Z8001, which has a segmented architecture: the Z8002 which has a nonsegmented CPU with a 64K address size; and the Z8003/Z8004, with a virtual memory processing unit.

The Z8001 This CPU has the following features: 8 Mb of address space, sixteen 16-bit (R0–R15) or paired eight 32-bit (RR0, RR2–RR10) general-purpose registers, system and normal execution modes, separated segments of code, data, and stack, instructions for compiler support, special hardware and instructions for multiprogramming support, hardware protection unit for operating system support, three hardware interrupts, and five traps.

The special registers of this microprocessor include a program counter (PC) consisting of a 7-bit segment number and a 16-bit segment offset, a 16-bit flag and control word, a system stack pointer, a normal stack pointer, and a program status area pointer.

The Z8002 This is a 64-Kbyte address space version of the Z8001. This is a nonsegmented CPU with a typical memory space of 64 Kbytes. Thus a 16-bit PC, a 16-bit system stack pointer, and a 16-bit normal stack pointer are equipped for addressing purposes.

The types of Z8000 instructions include load and exchange, arithmetic, logical, program control, bit manipulation, rotate and shift, block transfer and string manipulation, input and output, and CPU control.

The Z8000 is equipped with a nonmaskable interrupt NMI, two maskable interrupts (vectored and nonvectored), and four traps (system call, unimplemented instruction, privileged instruction, and segment trap).

The peripheral chips available to the Z8000 family include the Z8010 MMU, the Z8015 paged MMU, the Z8016 DMA controller, the Z8030 serial communications controller, the Z8031 asynchronous serial communications controller, the Z8036 counter/timer and parallel I/O, and the Z8038 FIFO input/output interface.

15.5   32-BIT AND ABOVE MICROPROCESSORS

32-bit microprocessors such as the Intel 80386, Intel 80486, Motorola 68020, National Semiconductor NS32032, AT&T WE32100, and Zilog Z80000 were designed for applications like CAD/CAM engineering workstations, multiuser office systems, factory automation, robotics, communication switching systems, and high-resolution graphics.

A 32-bit microprocessor generally has full 32-bit internal and external data buses. In addition there is integration of some major features of mainframe computer architecture such as cache memory, virtual memory, pipelining, multiprocessing, multitasking, coprocessing capabilities, system programming, and high-level language programming support.

15.5.1 The Motorola MC68020

The MC68020, a general-purpose microprocessor, has separated 32-bit data and address buses (4 Gbytes of addressing space), memory-mapped I/O, a three-stage pipeline, a coprocessor interface supporting up to eight coprocessors, support of virtual memory through the MC68851 paged MMU, floating-point support through the MC68881 math coprocessor, two processing states (user and supervisor), and on-chip 256-byte cache memory.

The user state programming model of the MC68020 shown in Figure 15.15 is the same as that of the MC68000. But the interrupt stack pointer (ISP) or A7′ of the MC68020 was called the supervisor state pointer (SSP) in the MC68000. Also, the cache control register (CACR), the cache address register (CAAR), and the vector base register (VBR) have been added to the MC68020. Two of three reserved bits of the system byte of the status register are defined as the trace enable and master/interrupt states, respectively.

All the data movement, integer arithmetic, logical, shift and rotate, and bit manipulation instructions of the MC68020 are identical to those of the MC68000.

Variable-length bit field instructions are introduced to enhance the bit manipulation instructions.

Two instructions called PACK and UNPACK are added to the BCD operation group.

CALLM and RTM instructions are new to the program control group. The CALLM, or call module, saves the current module state on the stack and loads a new module state from a specified address. RTM, or return from module, retrieves a previously saved module from the top of the stack.

TRAPcc, trap on condition, has also been added to the group of system control.

There is one instruction called TAS in the MC68000 for supporting multiprocessor operations. New instructions for modify operands as well as coprocessor operations are available to support multiprocessor operations in the MC68020 environment.

The internally generated exceptions and externally generated interrupts are grouped according to their priorities, in which 0.0 is the highest priority and 4.2 is the lowest. The exception groups are as follows:

- *Group 0:* 0.0—reset
- *Group 1:* 1.0—address error
  1.1—bus error
- *Group 2* (*instruction traps*): BKPT #n, CALLM and RTM, DIV, TRAP #n, TRAP cc, TRAPV, CHK, CHK2, and coprocessor-related instructions (cp TRAPcc, cp protocol violation, and cp mid-instructions)

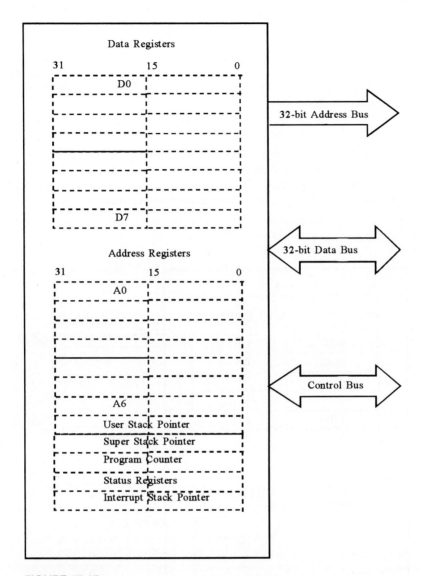

FIGURE 15.15
The Motorola 68020 architecture.

- *Group 3:* 3.0—illegal instructions, line A emulator, unimplemented line F emulator, privilege violation, and cp preinstruction
- *Group 4:* 4.0—cp postinstruction
  4.1—trace
  4.2—interrupt.

15.5.2 The Intel 80386 and 80486

The Intel 80386 This chip includes a true 32-bit CPU with a six-stage pipeline architecture, a memory management unit, and a bus interface. Memory-mapped peripherals and 64K of I/O space are supported by the 80386. It has two modes: real and protected.

The real mode operation carries all the restrictions of the 8086: only 1 Mbyte of memory is directly addressable. The programming model is shown in Figure 15.16.

In the protected mode, the physical addressing size is 4 Gbytes (separated 32-bit address bus and data bus) and the logical or virtual address is 64 terabytes (trillion bytes), which can be decomposed into 16,381 4-Gbyte segments for memory partitioning. Each segment is divided into one or more 4K pages.

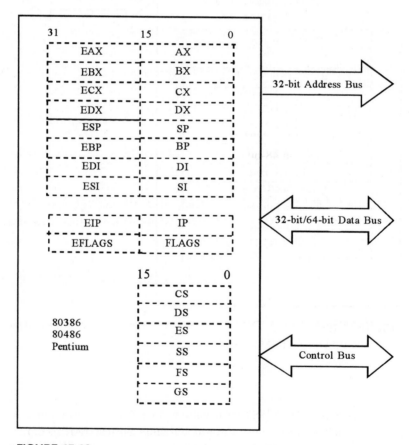

FIGURE 15.16
The Intel 80386/486 architecture.

The registers of the 80386 include the 80286's registers and are divided into the following categories:

- *General-purpose registers (32-bits each)*: EAX, EBX, ECX, EDX, ESI, EDI, EBP, and ESP
- *Segment registers (16 bits each)*: code segment CS, stack segment SS, and data segments DS, ES, FS, and GS
- *A 32-bit flag register EFLAGS:* with 18 bits defined as carry, parity, auxiliary, zero, sign, trap, interrupt enable, direction, overflow, I/O privileged level, nested task flag, resume flag, and virtual mode
- *A 32-bit instruction pointer:* EIP
- *Three control registers (32-bits each)*:

    CR0—machine control register (includes 80286 machine status word) with defined bits: paging enable (PG), processor extension type (ET) for differentiating 80287 (ET = 0) or 80387 (ET = 1), task switch (TS), emulate coprocessor (EM), monitor coprocessor (MP), and protection enable (PE)

    CR1—reserved for future versions

    CR2—a 32-bit page fault linear address register

    CR3—a 32-bit page directory base register with bits 0–11 set to zero, and bits 12–32 reserved for a directory base

- *System address registers:*

    GDTR—a 48-bit global descriptor table register holding a 32-bit linear base address and a 16-bit limit for the global descriptor table (GDT)

    IDT—a 48-bit interrupt descriptor table register holding a 32-bit linear base address and a 16-bit limit for the interrupt descriptor table (IDT)

    TR—a 16-bit selector for the task state segment (TSS)

    LDTR—a 16-bit selector for the local descriptor table (LDT)

- *Debug test registers (32 bits each)*:

    DR0 through DR3—linear breakpoint address 0 through 3

    DR4 and DR5—Intel reserved

    DR6—breakpoint status

    DR7—breakpoint control

- *Test registers (32-bits each for page cache)*:

    TR6—a test control register (command testing register) for the testing of the content addressable memories (RAM/CAM) in the translation lookaside buffer

TR7—test status register (data testing register), containing the data of the translation lookaside buffer test.

The 80286 instructions are all included in the 80386's instruction set. In addition, some new instructions are added for different word widths and new registers. Finally, two groups of new instructions are introduced in the 80386 for bit manipulation and protection model environment.

The 80386 supports two types of hardware interrupts (maskable and nonmaskable) and three classes of exceptions (faults, traps, and aborts).

A fault is an exception that is detected and serviced before the execution of the faulting instruction. A trap is invoked by the execution of the instruction that caused the exception. Aborts are exceptions that are used to report double fault or severe errors, such as a hardware error, or illegal values in the system table.

The faults include divide error, array bound check, invalid op-code, device not available, invalid TSS, segment not present, stack fault, general protection fault, page fault, and coprocessor error.

The traps include debug exception, 1-byte interrupt, interrupt on overflow, and 2-byte interrupt.

The Intel 40486   This is an enhanced version of the 80386. It is compatible with systems that were designed for, and use, the 80386 but offers several improvements when used as a replacement.

The 80486 executes many of its instructions in one clocking period compared to the two required by the 80386. This leads to a significant increase in speed.

The 80486 has an internal 8K cache memory. The cache is used to store both data and instructions. The memory management and cache management units provide for the cache to contain data and instructions that are most likely to be subsequently executed in a running program. Use of the cache provides the one clocking period execution time of instructions contained in the cache.

The 80486 comes in essentially three versions, which are identified by an attached suffix. Thus the 80486DX also contains the equivalent of an 80487 numeric coprocessor. This also leads to more efficient and faster program execution since bus cycles are not necessary. The 80486SX is a version that does *not* contain the coprocessor and is therefore cheaper. Both of these processors are available in 25-, 33-, and 50-MHz versions.

The 80486DX2 is a double-clocked version of the 80486DX, identical in every other respect. This unit is available in 50- and 66-MHz versions. The double clocking refers to the fact that the microprocessor operates at the quoted speed internally but still uses speeds of 25 or 33 MHz for the external bus, as appropriate.

15.5.3   The National Semiconductor NS32032 Family

The NS32032   This is a full 32-bit data bus, virtual memory CPU with a direct address space of 16 Mb (24-bit address lines). The general-purpose and dedicated registers, instruction set, interrupt mechanism, and architectural concepts of the NS32032 are the same as for the NS32016. The coprocessors and the system peripherals of the NS32016 can be used by the NS32032. Note that the lower 24 bits of the data bus (AD0–AD23) are multiplexed with the 24-bit address lines.

The NS32332   This is an enhanced version of the NS32032 virtual memory CPU. Instead of the 24-bit address line in the NS32032, 32-bit multiplexed data and address lines (AD0–AD31) are used by the NS32332. Thus, a 4-Gb uniform addressing space is available.

15.5.4   The Zilog Z80000

The Z80000 utilizes a 32-bit multiplexed address/data system bus, an on-chip MMU, a 254-byte cache, a six-stage pipeline, and other mainframe architectural features that support operating systems and high-level language programming. Sixteen general-purpose registers are RR0, RR2, . . . , RR30.

Two execution nodes, normal and system, with separated stacks called normal and system stacks, are provided for operating system protection. In the system mode, the CPU can execute unprivileged and privileged instructions.

The Z8070 APU arithmetic processing unit is a math coprocessor that can be used to enhance the Z80000.

15.5.5   The Intel Pentium

The Intel Pentium was introduced in 1993 and was originally to be called an 80586 but problems with copyright law moved the company to adopt the literal name. This processor is similar to the 80486, and therefore also the 80386, but with many enhancements. The instruction set of this CISC processor is essentially the same as those of its predecessors. The enhancements include the following:

- Improved technology allows speeds from 60 Mhz to more than 100 Mhz.
- The data bus is increased from 32 bits to 64 bits. This enhances applications using video so that motion approaches the resolution of normal television.
- Cache memory is increased from the joint instruction and data 8K system in the 80486 to two 8K caches, one for instructions and the other for data.
- Superscalar technology is employed to provide three instruction execution systems. Two are for integer data operations and one is for floating-point operations. This means that, in principle, three independent instructions can be executed at the same time.
- The internal floating-point processor has been significantly improved over the version in the 80486.

The Pentium is essentially software compatible with the 804XX Intel series but is not hardware compatible because of the 64-bit data bus. A version of the Pentium provides for a 32-bit multiplexed data bus to make the processor essentially plug compatible with the 80486.

The Pentium is physically packaged in a 237-pin pin grid array (PGA). The chip draws about 3 A, dissipates about 12 W, and must be operated with a good heat sink and high airflow fan.

15.5.6 The Motorola PowerPC

The Motorola PowerPC is the result of a consortium of Apple, IBM, and Motorola combining to enhance and improve the IBM Power microprocessor architecture. The result is a series of RISC processors that are compatible with the Power systems but much improved. The particular features that set the processor apart from others are as follows:

- A 64-bit data bus so that floating-point calculations can be interfaced directly with memory and I/O devices.
- The address bus is 32 bits for a 4-Gbyte available memory space.
- Separate cache memory of at least 8 Kbytes for data and another 8 Kbytes for instructions.
- Special provisions are provided in the instruction set and operating modes that allow it to be used in the 32-bit environment of predecessor processors.

The PowerPC 620 has features that go much further than any other member of the PowerPC family. These enhancements include:

- Separate 32-Kbyte instruction and data caches. This allows fast manipulation of large data systems. Use of level 2 cache interface allows extension of the external cache from 1 to 128 Mbytes.
- This processor has a 128-bit data bus, which allows extension of floating-point precision operations.
- The address bus is 40 bits so that a total of 1 Tbyte is directly addressable by the processor.
- Superscalar architecture is implemented with six instruction execution units. Up to four independent instructions can be executed simultaneously.
- The processor can operate at a clock speed of 133 MHz.

The PowerPC 620 operates from 3.3-V power but draws as much as 9 A for a 30-W dissipation! The processor must be operated with a good heat sink and with good airflow from a fan system.

16

FILTERS*

16.1 INTRODUCTION

Electronic filters are frequency selective circuits that limit the frequency content of the signals they pass. Ideally they pass some frequencies and block others as illustrated in Figure 16.1.

Filters are used in virtually every phase of electronics. They remove unwanted frequencies from radio and television broadcasts, where they are used to pass the frequencies that contain the selected station and block frequencies of all the other stations being simultaneously transmitted. They are used to separate a single telephone call from the thousands being made at any one time. They are used to analyze returning radar signals as well as to assure that only the correct radar signal is transmitted. Filters are used in dc power supplies to remove power line and higher frequencies from the dc voltage. The list of filter uses is far too long to list here, but almost every electronic instrument, communication system, control system, entertainment system, power supply, or computing system will contain at least one, and frequently dozens of filters.

This chapter presents an intuitive description of how several useful passive and active filters operate, as well as simple design procedures for most of the filters presented. Section 16.8 provides a list of references that go into the theory behind the results presented here. The example filter designs are accompanied by simulations of their performance.

---

* This chapter was written by Dr. Luces Faulkenberry, College of Technology, University of Houston.

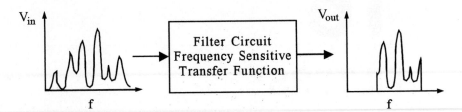

FIGURE 16.1
Filter concept.

16.2 FILTER CLASSIFICATION

Filters are classified according to the range of frequencies they pass. As illustrated in Figure 16.2, *high-pass filters* pass high and block low frequencies, *low-pass filters* pass low and block high frequencies, *bandpass filters* pass a specified range of frequencies, and *band-reject filters* pass all except a specified range of frequencies. Band-reject filters that block only a small range of frequencies are called *notch filters*. Ideally, the transition between the frequencies blocked (stopband) and those passed (passband) would be a straight vertical line in the responses shown in Figure 16.2, but in practice the transition between the stopband and the passband may take several octaves (an octave is a twofold change) or decades (tenfold changes)

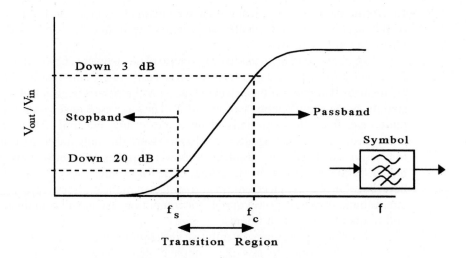

(a)

FIGURE 16.2
Filter classifications: (a) High-pass. (b) Low-pass. (c) Bandpass. (d) Band-reject.

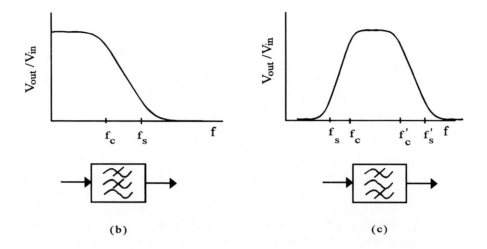

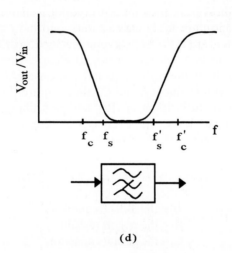

FIGURE 16.2
(*continued*)

in frequency. Different filter types are classified by the steepness of the transition region attenuation (TRA), which is usually given in dB/octave or dB/decade. A decibel (dB) is defined as:

$$dB = 20 \log \left(\frac{V_{out}}{V_{in}} \right) \tag{16.1}$$

where V_{out} = output voltage and V_{in} = input voltage.

Filters are subclassified by the amount of ripple in the passband and the stopband. The frequency at which the filter output is 3 dB less than the input is considered the *cutoff frequency* (f_c or f_{3dB}) of the filter, and is the point at which the transition region begins. Bandpass and notch filters have a middle frequency that is the geometric mean of the upper and lower 3-dB frequencies.

Another major classification criterion of filters is based on whether or not they contain amplifying devices. Those that do not are called passive, and those that do are called active filters.

Passive filters, such as the RC filter of Figure 16.3a, contain only resistors, capacitors, and/or inductors. The advantages of passive filters are (1) ruggedness, (2) small size in higher frequencies, (3) ability to operate at very high frequencies, (4) no need of a power supply, (5) reliability, and (6) relatively low cost at moderate to high frequencies. Their major disadvantages are (1) little isolation between input and output, (2) no gain ($V_{out} \le V_{in}$), (3) difficult tuning because of lack of input/output isolation, (4) difficult to cascade for the same reason, and (6) they are large, bulky, and expensive at low frequencies.

Active filters use resistors, capacitors, and amplifiers to shape the frequency response of a circuit. The lack of inductors in active filters tends to make them smaller and less expensive than passive filters at moderate and low frequencies. Active filters have high input to output isolation, are easy to cascade for higher performance, are fairly easy to tune, and have gain. The major disadvantages of active filters are (1) active components tend to drift more than passive components, causing the filter characteristics to change with time and temperature; (2) the amplifier limits the high-frequency response of the filter, causing the filter cost to increase with frequency; and (3) they require power supplies.

Figure 16.3 shows an RC low-pass filter. As the frequency increases, the capacitive reactance, X_c, decreases causing the output voltage to decrease. The 3-dB frequency is the half power frequency given by

$$f_c = \frac{1}{2\pi RC} \tag{16.2}$$

where f_c = the 3-dB frequency
                R = the circuit resistor
                C = the circuit capacitor.

The change in transition region attenuation is 6 dB/octave (20 dB/decade) for a single RC or RL high-pass or low-pass circuit, and the change in phase response, the delay in phase between V_{in} and V_{out}, is 13.5°/octave (45°/decade).

16.3   POLES (FILTER ORDER)

16.3.1   Poles and Transition Region Attenuation

Each RC or RL network comprises a first-order (also called a single-pole) high-pass or low-pass filter. Each cascaded RC stage contributes an additional pole, thus

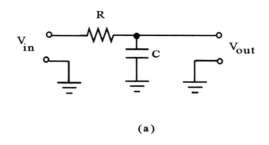

(a)

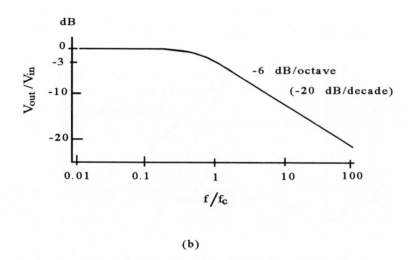

(b)

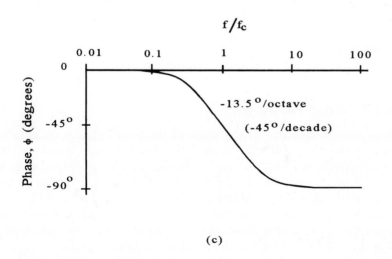

(c)

FIGURE 16.3
Basic RC low-pass filter: (a) Circuit. (b) Frequency response, semilog plot. (c) Phase response, semilog plot.

an additional amount of transition region attenuation and phase response change. For the RC filter of Figure 16.3 each cascaded RC stage would result in an increase in the TRA of 20 dB/decade and an additional 45°/decade of phase change. Thus, for cascaded single-pole filters the total TRA and phase (θ) change is, if n = number of poles,

$$\Delta TRA = n(20 \text{ dB/decade}) \tag{16.3}$$

$$\Delta\theta = n(45°/\text{decade}) \tag{16.4}$$

As shown in Figure 16.4, two poles of RC network will result in a TRA of 40 dB/decade. The more poles, the steeper the filter TRA.

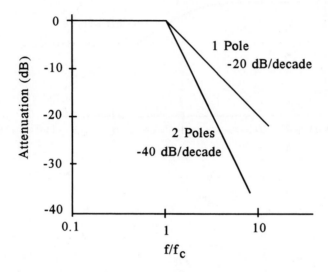

FIGURE 16.4
Poles and transition region attenuation (idealized drawing).

16.3.2 Filter Factor (α)

Filters are also classified by the shape of the passband near the transition region as well as the TRA. The filter factor, α, is related to both. The characteristic shape of a filter frequency response can be changed by varying only α, while keeping all other circuit characteristics the same. The filter factor is related to the damping factor (δ) of a resonant circuit by $\alpha = 2\delta$. A filter must be an RCL or active filter for the filter factor to change the shape of its frequency response. The effect of varying the filter factor is illustrated in Figure 16.5. The lower the filter factor, the lower the damping factor and the more peaked the response.

16.3.3 Cascading Filters

To obtain higher order filters, first- and second-order filters are cascaded as shown in Figure 16.6. The orders of the cascaded filters are added to obtain the order of

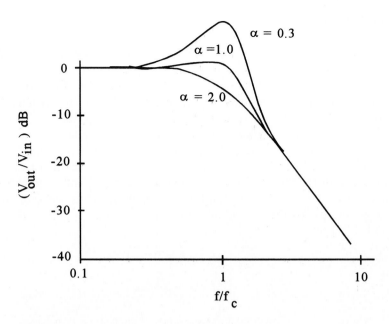

FIGURE 16.5
Filter factor and transition region response shape for second-order low-pass filter.

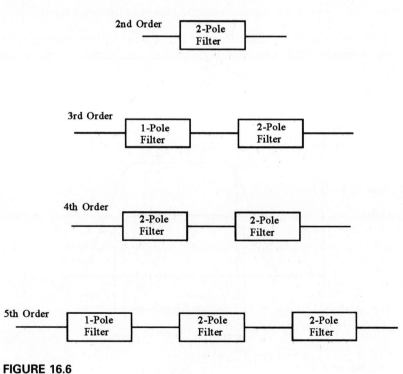

FIGURE 16.6
Cascading to achieve a higher filter order.

the final filter. However, to obtain the desired filter shape, the f_c and filter factor of each of the cascaded filters may have to be altered from that of an uncascaded second-order filter. To understand why, refer to Figure 16.7, which shows the frequency response of two identical bandpass filters with gain. When cascaded, the transfer function of the total filter is the product of the transfer function of each stage. The Bode plot shows the transfer functions in decibels so the total transfer function is the sum of those of the individual stages. The combined filter skirts get steeper, because $(-6 \text{ dB/octave}) + (-6 \text{ dB/octave}) = -12 \text{ dB/octave}$. This causes the bandwidth to shrink. The 0.707 amplitude points, f_1 and f_2, now become half-amplitude points because $(0.707)(0.707) = (0.4949)$. Thus the f_c of the cascaded filter is not that of the individual stages; $f_1 < f'_1$ and $f_2 > f'_2$. The amount of bandwidth shrinkage for n identical stages connected in cascade is

$$f'_2 = f_2 \sqrt{2^{1/n} - 1} \tag{16.5}$$

$$f'_1 = \frac{f_1}{\sqrt{2^{1/n} - 1}} \tag{16.6}$$

where $f_1 =$ lower corner frequency of each stage

$f_2 =$ upper corner frequency of each stage

$f'_1 =$ lower corner frequency of cascaded set

$f'_2 =$ upper corner frequency of cascaded set.

Table 16.1 shows the corrections in f_c and α for various combinations of first- and second-order filters commonly cascaded to obtain higher order responses.

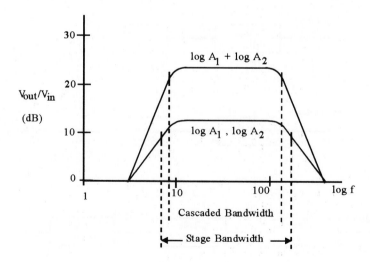

FIGURE 16.7
Bandwidth shrinkage.

TABLE 16.1

Cascaded filter table

For low-pass filters: $f_c = f_{3dB}(f_c \text{ factor})$
For high-pass filters: $f_c = f_{3dB}/(f_c \text{ factor})$

Butterworth Filters

Poles	Stage 1 α / f_c Factor	Stage 2 α / f_c Factor	Stage 3 α / f_c Factor	Stage 4 α / f_c Factor	Stage 5 α / f_c Factor
2	1.414214 1.000000				
3	Real pole 1.000000	1.000000 1.000000			
4	1.847759 1.000000	0.765367 1.000000			
5	Real pole 1.000000	1.618034 1.000000	0.618034 1.000000		
6	1.931852 1.000000	1.414214 1.000000	0.517638 1.000000		
7	Real pole 1.000000	1.801934 1.000000	1.246980 1.000000	0.445042 1.000000	
8	1.961571 1.000000	1.662939 1.000000	1.111140 1.000000	0.390181 1.000000	
9	Real pole 1.000000	1.879385 1.000000	1.532089 1.000000	1.000000 1.000000	0.347296 1.000000
10	1.985377 1.000000	1.782013 1.000000	1.414214 1.000000	0.907981 1.000000	0.312869 1.000000

(*continued*)

Bandpass filters with narrow bandwidths are usually set for a particular Q instead of α ($Q = 1/\alpha$). Q is related to bandwidth by the following equation:

$$Q = \frac{f_0}{f_2 - f_1} = \frac{\sqrt{f_1 f_2}}{f_2 - f_1} \tag{16.7}$$

The change in filter response with Q is shown in Figure 16.8.

16.3.4 Filter Sensitivity

Filter sensitivity is the amount one parameter changes as another varies. The sensitivity of a filter cutoff frequency with the change in a filter resistor, R_1, would be

TABLE 16.1 (*Continued*)

Chebyshev Filters, 0.5-dB ripple, $\varepsilon = 0.34931$

Poles	Stage 1 α / f_c Factor	Stage 2 α / f_c Factor	Stage 3 α / f_c Factor	Stage 4 α / f_c Factor	Stage 5 α / f_c Factor
2	1.157781 1.231342				
3	Real pole 0.626456	0.586101 1.068853			
4	1.418218 0.597002	0.340072 1.031270			
5	Real pole 0.362320	0.849037 0.690483	0.220024 1.017735		
6	1.462760 0.396229	0.552371 0.768121	0.153543 1.011446		
7	Real pole 0.256170	0.916126 0.503863	0.388267 0.822729	0.113099 1.008022	
8	1.478033 0.296736	0.620857 0.598874	0.288544 0.861007	0.086724 1.005948	
9	Real pole 0.198405	0.943041 0.395402	0.451865 0.672711	0.223313 0.888462	0.068590 1.004595
10	1.485045 0.237232	0.651573 0.487765	0.345860 0.729251	0.178208 0.908680	0.055595 1.003661

expressed as $S_{R_1}^{f_c}$, which indicates that if R_1 increases by 1%, f_c decreases by 0.5%. Equations for the sensitivities of filters are given in the bibliography (Section 16.8).

16.4   COMMON FILTER TYPES

16.4.1   Butterworth Filters

The Butterworth filter is characterized by a flat passband and stopband, and a TRA of $20n \log \omega$, where n = number of poles and ω is the radian frequency, which is 6 dB/octave/pole (20 dB/decade). The transition region phase change is 13.5°/octave/pole. Figure 16.9 shows a low-pass Butterworth response for four different orders. The Butterworth filter is a good general-purpose filter.

TABLE 16.1 (*Continued*)

Chebyshev Filters, 1-dB ripple, $\varepsilon = 0.50885$

Poles	Stage 1 α f_c Factor	Stage 2 α f_c Factor	Stage 3 α f_c Factor	Stage 4 α f_c Factor	Stage 5 α f_c Factor
2	1.045456 1.050005				
3	Real pole 0.494171	0.495609 0.997098			
4	1.274618 0.528581	0.280974 0.993230			
5	Real pole 0.289493	0.714903 0.655208	0.179971 0.994140		
6	1.314287 0.353139	0.454955 0.746806	0.124942 0.995355		
7	Real pole 0.205414	0.771049 0.480052	0.316871 0.808366	0.091754 0.996333	
8	1.327947 0.265068	0.511120 0.583832	0.234407 0.850613	0.070222 0.997066	
9	Real pole 0.159330	0.793624 0.377312	0.368610 0.662240	0.180942 0.880560	0.055467 0.997613
10	1.334229 0.212136	0.536341 0.476065	0.280859 0.721478	0.144161 0.902454	0.044918 0.998027

(*continued*)

16.4.2 Bessel Filters

The Bessel filter has a linear phase delay. Thus, nonsinusoidal waveforms are minimally distorted as they pass through the filter, and overshoot and ringing in response to step voltage changes, such as in square waves, are low. Bessel filter TRAs are less than 6 dB/octave/pole, and the transition region phase change is less than 13.5°/octave/pole. The phase response versus frequency plot is almost a straight line, getting straighter as the number of poles increases.

Filters similar to the Bessel are the Gaussian, which has less stopband selectivity (lower TRA), and the parabolic, which has shorter delay times and less selectivity. The catenary and elliptic contour are two similar constant delay flat stopband filters. Table 16.1 contains information on the Bessel filter.

TABLE 16.1 (*Continued*)

Chebyshev Filters, 2-dB ripple, $\varepsilon = 0.76478$

	Stage 1	Stage 2	Stage 3	Stage 4	Stage 5
Poles	α f_c Factor	α f_c Factor	α f_c Factor	α f_c Factor	α f_c Factor
2	0.886015 0.907227				
3	Real pole 0.368911	0.391905 0.941326			
4	1.075906 0.470711	0.217681 0.963678			
5	Real pole 0.218308	0.563351 0.627017	0.138269 0.975790		
6	1.109145 0.316111	0.351585 0.730027	0.095588 0.982828		
7	Real pole 0.155340	0.607379 0.460853	0.243009 0.797114	0.070027 0.987226	
8	1.120631 0.237699	0.394841 0.571925	0.179098 0.842486	0.053512 0.990141	
9	Real pole 0.120630	0.625114 0.362670	0.282589 0.654009	0.137959 0.874386	0.042225 0.992168
10	1.125921 0.190388	0.414283 0.466780	0.214523 0.715385	0.109773 0.897590	0.034169 0.993632

16.4.3   Chebyshev Filters

The Chebyshev filter is an equiripple filter. It has ripples of equal amplitude in the passband (number of ripples = half of the number of poles) as shown in Figure 16.10. The stopband of the Chebyshev filter is flat and monotonic. The reason for the popularity of a filter with ripples in the passband is that the TRA change is greater than 6 dB/octave/pole, and the greater the ripple amplitude, the greater the TRA. Thus, a Chebyshev filter with a given TRA will require fewer stages than a Butterworth of the same TRA. The increase is illustrated in Figure 16.11. The Chebyshev TRA is

$$\text{TRA} = 20 \log \varepsilon + 6(n - 1) + 20 \log \omega \qquad [\text{dB}] \qquad \textbf{(16.8)}$$

where n = number of poles
ε = ripple depth descriptor
ω = radian frequency.

TABLE 16.1 (*Continued*)

Chebyshev Filters, 3-dB ripple, $\varepsilon = 0.99763$

Poles	Stage 1 α f_c Factor	Stage 2 α f_c Factor	Stage 3 α f_c Factor	Stage 4 α f_c Factor	Stage 5 α f_c Factor
2	0.766464 0.841396				
3	Real pole 0.298620	0.325982 0.916064			
4	0.928942 0.442696	0.179248 0.950309			
5	Real pole 0.177530	0.467826 0.614010	0.113407 0.967484		
6	0.957543 0.298001	0.298173 0.722369	0.078247 0.977154		
7	Real pole 0.126485	0.504307 0.451944	0.199148 0.791997	0.057259 0.983099	
8	0.967442 0.224263	0.324695 0.566473	0.146518 0.838794	0.043725 0.987002	
9	Real pole 0.098275	0.519014 0.355859	0.231548 0.650257	0.112754 0.871584	0.034486 0.989699
10	0.972004 0.179694	0.340668 0.462521	0.175474 0.712614	0.089664 0.895383	0.027897 0.991638

(*continued*)

The ripple amplitude is normally given in decibels. Solving Eq. (16.8) for a decade frequency change for second-order Chebyshev filters with ripple amplitudes of 0.5, 1, 2, and 3 dB yields TRAs of 36.8, 40.14, 43.67, and 45.99 dB/decade, respectively. The change in phase angle of a Chebyshev filter is greater than 13.5°/octave/pole, and the phase delay is less linear than that of the Butterworth.

16.4.4 Inverse Chebyshev Filters

Figure 16.12 shows the response of a third-order inverse Chebyshev filter. It has equal ripples in the stopband and a flat passband. Its TRA is greater than 6 dB/octave/pole and its transition region phase change is greater than 13.5°/octave/pole. It is generally more complex to implement than a Chebyshev, and is seldom used.

TABLE 16.1 (*Continued*)

Bessel Filters

Poles	Stage 1 α f_c Factor	Stage 2 α f_c Factor	Stage 3 α f_c Factor	Stage 4 α f_c Factor	Stage 5 α f_c Factor
2	1.732051 1.732051				
3	Real pole 2.322185	1.447080 2.541541			
4	1.915949 3.023265	1.241406 3.389366			
5	Real pole 3.646738	1.774511 3.777893	1.091134 4.261023		
6	1.959563 4.336026	1.636140 4.566490	0.977217 5.149177		
7	Real pole 4.971785	1.878444 5.066204	1.513268 5.379273	0.887896 6.049527	
8	1.976320 5.654832	1.786963 5.825360	1.406761 6.210417	0.815881 6.959311	
9	Real pole 6.297005	1.924161 6.370902	1.696625 6.606651	1.314727 7.056082	0.756481 7.876636
10	1.984470 6.976066	1.860312 7.112217	1.611657 7.405447	1.234887 7.913585	0.706570 8.800155

16.4.5 Elliptic Filters

The elliptic filter, also called a Cauer or Cauer–Chebyshev, has equal ripple in the stopband and the passband, and a very high change in TRA. The frequency response of an elliptic filter is shown in Figure 16.13. If the number of poles is even, elliptic filters have TRAs much greater than 6 dB/octave/pole, but the phase delay is quite nonlinear. The design procedures for elliptic filters are given in several of the bibliography references.

16.5   COMPONENT SELECTION CONSIDERATIONS

Stable, low-noise components are necessary to obtain stable, low-noise filters. Components drift in value with age and temperature, generate noise, and are imprecise. Precision components help solve the last factor. Even with 1% resistors (fairly

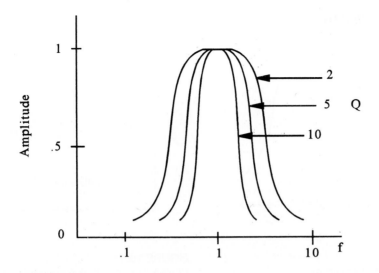

FIGURE 16.8
Bandpass response and Q.

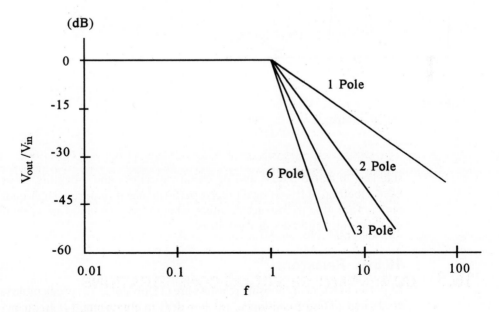

FIGURE 16.9
Low-pass Butterworth filter response as order increases.

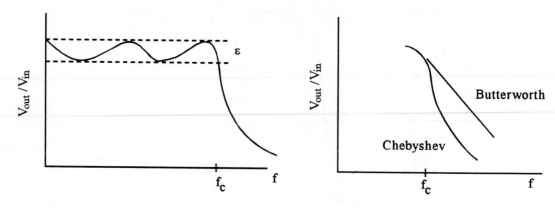

FIGURE 16.10
Chebyshev low-pass filter response.

FIGURE 16.11
Chebyshev filter TRA compared to that of a Butterworth filter.

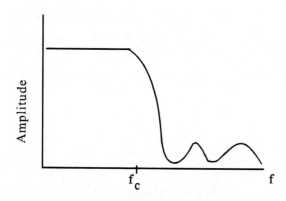

FIGURE 16.12
Inverse Chebyshev third-order low-pass filter.

standard in active filters) and precision capacitors and inductors, most filters will need a method of trimming if precise performance is required. High-quality potentiometers are used to trim active filters, whereas passive filters are normally trimmed with variable inductors or capacitors.

16.5.1 Resistors

Important factors in choosing resistors are (1) low noise, (2) predictable temperature coefficient (TC) of resistance, (3) low drift due to aging, (4) frequency response, (5) values available, and (6) precision. Wire-wound resistors have the lowest noise and low, around 10 ppm/°C (parts per million per degree centigrade), predictable

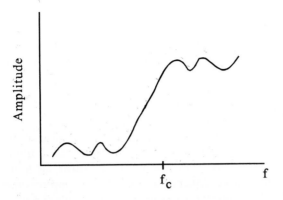

FIGURE 16.13
Fifth-order high-pass elliptic filter response.

TCs. They have limited frequency response and are not available in high resistance values, although they are available with very high precision.

Metal film resistors are available in a very wide range of precisions and values, are low in noise, and have low to moderate TCs (10 to 50 ppm/°C) with both positive and negative TCs available. They are the most used type of resistor in active filters. The thin-film resistors used in hybrid circuitry are metal film and can perform as well as discrete metal film resistors. Metal film potentiometers are good choices for trimming.

Carbon film resistors are available in a wide range of values and in high precisions, but not as high as metal film. Carbon film resistor frequency response is good, but the noise and TCs are not quite as low as those of metal film. The thick-film resistors used in some hybrid circuitry can perform at a level just below discrete carbon film resistors, but are very dependent on production variables.

With higher noise and TCs that range from 200 to 500 ppm/°C, carbon composition resistors are used only in the most forgiving active filter applications. They are available in a wide range of values and have good frequency response, but tolerances tighter than ±2% are not generally available.

16.5.2 Capacitors

Silver mica and ceramic capacitors, with TC designation NPO, have the lowest TCs available, 0 to 70 ppm/°C for mica and 0 to ±30 ppm/°C for NPO, but are only available in low values, generally less than 1 nF. They have very low power factors and dielectric absorption (the dielectric structure holds voltage after the capacitor has been discharged).

Teflon and polystyrene are good dielectrics for middle value capacitors with very low dielectric absorption and power factors, and moderate TCs, around −50 to −100 ppm/°C for polystyrene and −250 ppm/°C for Teflon. Mylar is a good, low-cost, moderate-performance dielectric with TCs around 250 ppm/°C. Polycarbonate capacitors are similar to Mylar in performance except for moderately higher TCs.

The small, high dielectric constant ceramic capacitors are to be avoided for filter use because their values drift with temperature, voltage, frequency, and sometimes humidity.

16.5.3 Op Amps

The choice of operational amplifiers available today is wide; choosing one is a difficult task. For low-frequency, moderate-performance active filters, one of the 741 variants usually suffices. For very low noise, low drift, or higher frequency filters, the choice becomes more difficult because the op-amp parameters must be chosen to optimize the particular filter requirements. In general field-effect transistor (FET) input amplifiers have, along with higher input impedances, higher slew rates (S) than bipolar input op amps, so they can provide higher output voltages at higher frequencies. The relationship between peak output voltage (V_p), frequency (f), and slew rate is $S_{max} = 2\pi f V_p$.

Many companies provide active filters in monolithic or hybrid IC form. To obtain a chosen filter response, the user adds resistors and/or capacitors whose values are selected from nomographs, equations, or software supplied by the filter manufacturer. The additional cost of commercial filters is frequently justified by the design time saved, and reputable manufacturers provide filters with good, predictable performance because the components are matched to the application.

Switched Capacitors   Switched capacitors are frequently mentioned in filter literature and used by commercial manufacturers of filters. Switched capacitors are capacitors that are partially charged and discharged at a frequency higher than the highest filter frequency. The rapid partial charge and discharge causes the capacitors to act like resistors to the applied signals. This is illustrated in Figure 16.14(a). Their advantage is that on-ICs capacitors can be built with very precise ratios and more ideal properties than can resistors. Their disadvantage is that the switching frequency must be filtered out of the output. Figure 16.14(b) shows an inverting amplifier using switched capacitors to set gain.

16.6 PASSIVE FILTERS

16.6.1 RC Filters

RC filters are always Butterworth in response and they have no gain. When $f \ll f_c$ the passband attenuation is 0 dB, and for $f > f_c$ the TRA is 6 dB/octave/pole and the transition region phase change is 13.5°/octave/pole, to a maximum angle of 90°/pole for $f \gg f_c$. Both the high- and low-pass filters, shown in Figure 16.15(a) and (b), have the same Eq. (16.2) for f_c:

$$f_c = \frac{1}{2\pi RC} \tag{16.2}$$

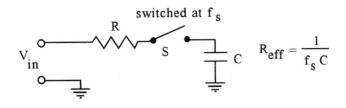

(a)

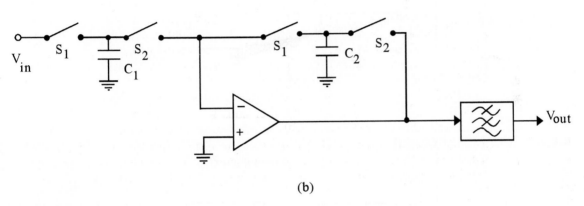

(b)

FIGURE 16.14
Switched capacitors: (a) Switched capacitor as resistor. (b) Switched capacitor used to set gain, $f_s \gg f_{in\,max}$.

The low-pass output voltage, a function of frequency, $V_o(f)_{LP}$, is

$$V_o(f)_{LP} = \frac{V_{in}}{\sqrt{1 + \left(\frac{f}{f_c}\right)^2}} \angle -\tan^{-1}\left(\frac{f}{f_c}\right) \qquad (16.9)$$

The high-pass response is $\sqrt{1 - V_o^2(f)_{LP}}$. The bandpass response, shown in Figure 16.15(c), is obtained by overlapping the responses of a low-pass and a high-pass filter.

EXAMPLE 16.1   Calculate R and C for a low-pass RC filter such that $f_c = 10$ kHz and the filter input resistance is 10 kΩ.

Solution
Let $R = 10$ kΩ. Then

$$C = \frac{1}{2\pi fR} = \frac{1}{2\pi\ 10\ \text{kHz}\ 10\ \text{k}\Omega} = 1.59\ \text{nF}$$

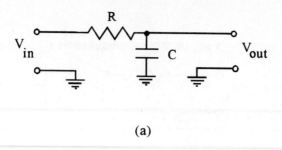

(a)

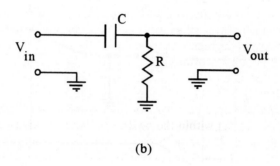

(b)

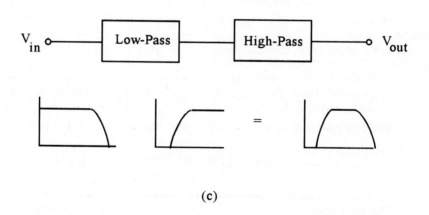

(c)

FIGURE 16.15
First-order RC filters: (a) Low-pass. (b) High-pass. (c) Bandpass.

The closest C is 1.5 nF, so R is recalculated for this value of C:

$$R = \frac{1}{(2\pi fR)} = \frac{1}{2\pi \ 10 \ \text{kHz} \ 1.5 \ \text{nF}} = 10.6 \ \text{k}\Omega$$

The simulation of Example 16.1 is shown in Figure 16.16.

16.6.2 Simulation Plot Information

In Figure 16.16 and the other example simulation plots, the solid line in the plot is the gain (or attenuation) response, and the line with dots is the phase response. The gain is given in decibels with the scale markings on the left side of the graph and the phase in degrees with the scale on the right. The information on the bottom left of the simulation printout is the frequency, phase angle, and gain slope at the far right-hand side of the plot. The information on the bottom right is the gain at the far right side of the plot, the group delay (equal to zero because it was not plotted), and the peak gain of the gain response curve. The range of the simulation variables can be set within the program. The calculated values of the resistors are given to the nearest standard 1% value in the examples, but the actual calculated values were used in the simulations to obtain accurate plots. Even with 1% components, trim potentiometers (pots) must be used in the filters to obtain their calculated responses. The ambient temperature used in the simulations is 27°C, as listed on the top left of the printouts. Many simulations (or cases) can be done as component tolerances are varied. Only one case was used, with each component at its nominal value, in the simulations for this chapter. The op amps used in the active filter section are the LM741, LM309, or LM725.

EXAMPLE 16.2

Calculate R and C for a high-pass RC filter in which $f_c = 500$ Hz, and the minimum resistance the source sees is 2 kΩ. The input resistance of the amplifier the filter feeds is 5 kΩ.

Solution
Solving the parallel resistance equation,

$$R_{\text{total}} = \frac{R_{\text{filter}} R_{\text{in amp}}}{R_{\text{filter}} + R_{\text{in amp}}}$$

from which

$$R_{\text{filter}} = \frac{5 \ \text{k}\Omega \ 2 \ \text{k}\Omega}{5 \ \text{k}\Omega - 2 \ \text{k}\Omega} = 3.32 \ \text{k}\Omega$$

and from Eq. (16.2)

$$C = \frac{1}{2\pi \ 3.33 \ \text{k}\Omega \ 500 \ \text{Hz}} = 95.6 \ \text{nH}$$

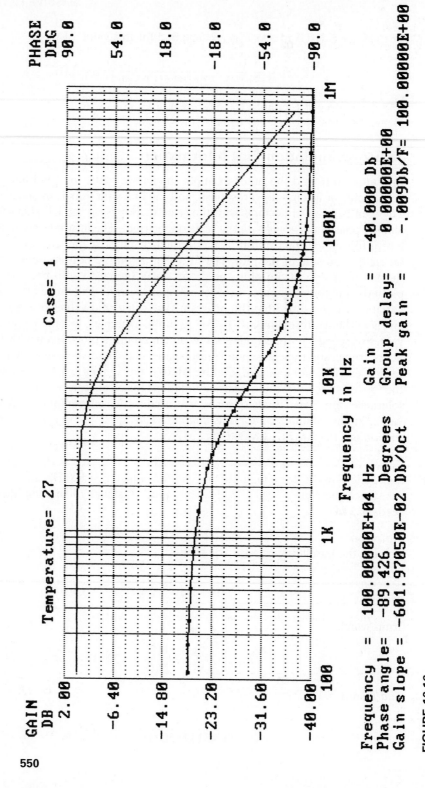

FIGURE 16.16
Example 16.1 simulation.

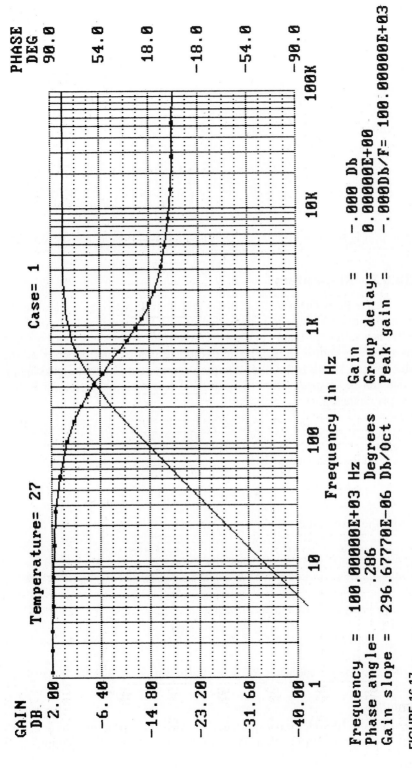

FIGURE 16.17
Example 16.2 simulation.

Use 0.1 μF.

The simulation of Example 16.2 is shown in Figure 16.17.

EXAMPLE 16.3

Calculate the components for an RC bandpass filter in which $f_1 = 1$ kHz and $f_2 = 10$ kHz. Let $C_1 = C_2 = 0.01$ μF.

Solution

Using Eq. (16.2) twice,

$$R_{HP} = \frac{1}{2\pi\, 0.01\ \mu F\, 1\ kHz} = 15.9\ k\Omega$$

$$R_{LP} = \frac{1}{2\pi\, 0.01\ \mu F\, 10\ kHz} = 1.59\ k\Omega$$

The plot is shown in Figure 16.18. The filter response is flat on top only if $f_2 \gg f_1$.

16.6.3   RLC Filters

Low-Pass Filters   RLC filters are inherently two pole because they contain two components whose impedance varies with frequency. A two-pole low-pass RLC filter is a series resonant circuit with the output voltage taken across the capacitor as shown in Figure 16.19. The amplitude response peak depends on the value of the resistor, which is the filter damping component. Since the peaking at the TRA edge can be changed, the filter can have responses other than a Butterworth by simply changing α:

$$f_c = \frac{1}{2\pi\sqrt{LC}} \tag{16.10}$$

The peaking frequency, f_p, is

$$f_p = \frac{1}{2\pi\sqrt{LC}}\sqrt{1 - \frac{R^2 C}{2L}} \tag{16.11}$$

From the resonant circuit relationship and $\alpha = 1/Q$,

$$Q = \frac{1}{R}\sqrt{\frac{L}{C}} \tag{16.12}$$

$$\alpha = \frac{R}{\sqrt{\dfrac{L}{C}}} \tag{16.13}$$

EXAMPLE 16.4

Calculate the components for a low-pass Butterworth RLC filter such that $f_c = 1500$ Hz.

Solution

Select L at 10 mH from experience or trial and error. From Table 16.1 look up $\alpha = 1.414$ for a Butterworth filter. Solve Eq. 16.10 for C to obtain

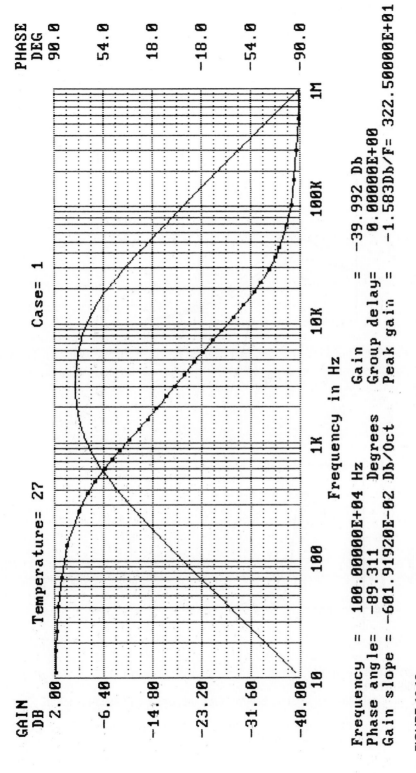

FIGURE 16.18
Example 16.3 simulation.

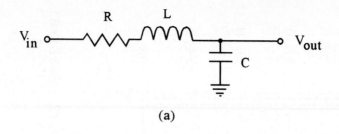

(a)

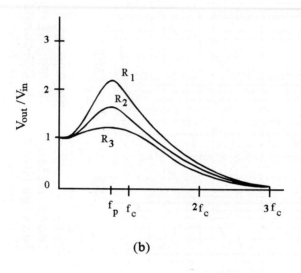

(b)

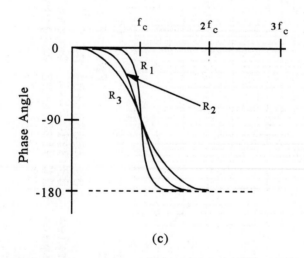

(c)

FIGURE 16.19
RLC low-pass filter: (a) Circuit. (b) Amplitude response. (c) Phase response.

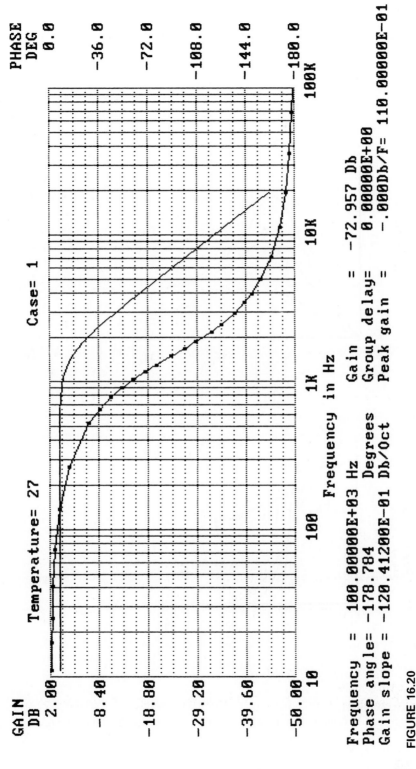

FIGURE 16.20
Example 16.4 simulation.

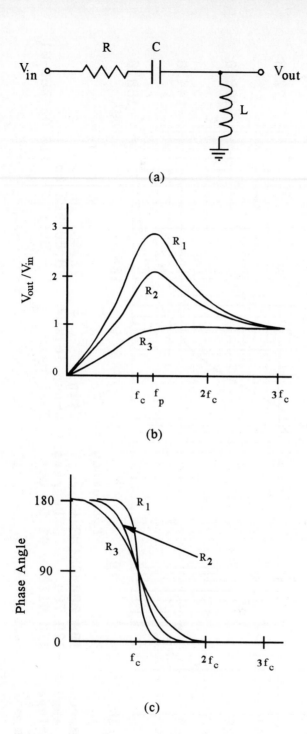

FIGURE 16.21
High-pass RLC filter: (a) Circuit. (b) Amplitude response. (c) Phase response.

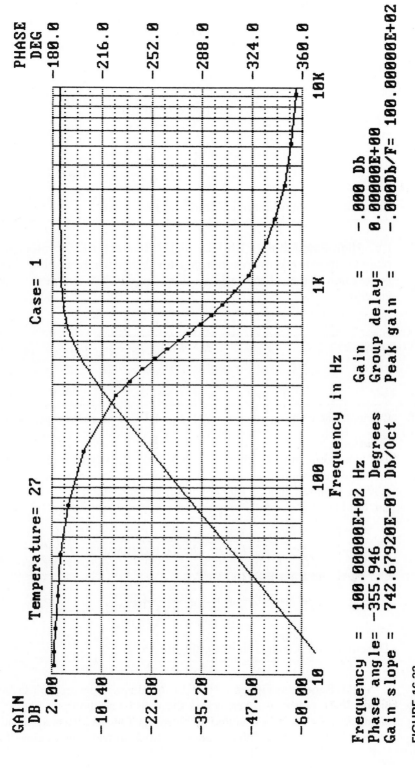

FIGURE 16.22
Example 16.5 simulation.

$$C = \frac{1}{(2\pi f)^2 L} \qquad \text{(16.14)}$$

from which

$$C = \frac{1}{(2\pi\,1.5\text{ kHz})^2\,10\text{ mH}} = 1.12\ \mu\text{F}$$

and from Eq. (16.13),

$$R = \alpha\sqrt{\frac{L}{C}} = 1.414\ \sqrt{\frac{10\text{ mH}}{1.12\ \mu\text{F}}} = 130\ \Omega$$

The plot of this filter output is shown in Figure 16.20.

High-Pass Filters The RLC high-pass filter is a series resonant circuit whose output voltage is taken across the inductor as shown in Figure 16.21. The f_c equation is the same as that of the low-pass filter, Eq. (16.10). The peak response point is given by

$$f_c = \frac{1}{2\pi\sqrt{LC}}\left(\frac{1}{\sqrt{1 - \dfrac{R^2C}{2L}}}\right) \qquad \text{(16.15)}$$

EXAMPLE 16.5

Calculate the components for a high-pass RLC Butterworth filter with $f_c = 500$ Hz.

Solution
The value of L must be chosen so that $X_L \gg R$ to avoid a filter that will not give the required response because it is overdamped. Several iterations of the problem may be required before this condition is satisfied. If $L = 100$ mH then from Eq. (16.14),

$$C = \frac{1}{(2\pi f_c)^2 L} = \frac{1}{(2\pi\,500\text{ Hz})^2\,100\text{ mH}} = 1.01\ \mu\text{F}$$

Use 1 μF, and from Eq. (16.13),

$$R = \alpha\sqrt{\frac{L}{C}} = 1.414\ \sqrt{\frac{100\text{ mH}}{1.0\ \mu\text{F}}} = 453\ \Omega$$

The circuit simulation is shown in Figure 16.22.

RLC Bandpass Filters The RLC bandpass filters are series resonant circuits in which the output is taken across the resistor as shown in Figure 16.23. These bandpass filters are best suited to fairly narrow bandpass applications. Several cascaded stagger-tuned filters (tuned to different frequencies close to each other)

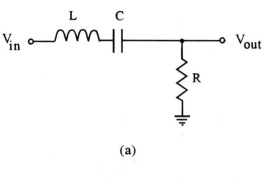

(a)

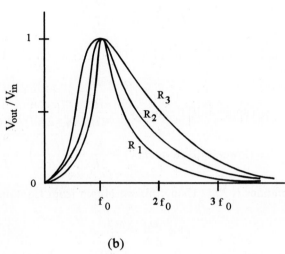

(b)

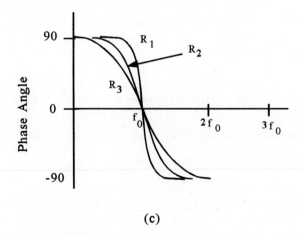

(c)

FIGURE 16.23
Series RLC bandpass filter: (a) Circuit. (b) Amplitude response. (c) Phase response.

are needed for low-Q RLC filters with flat passbands. The resonant frequency of a series RLC filter (f_0) is the frequency at which the inductive and capacitive reactances are equal. This frequency is given by

$$f_0 = \frac{1}{2\pi \sqrt{LC}} \tag{16.16}$$

The quality factor, or degree of peakedness of the response, is

$$Q = \frac{1}{\alpha} = \frac{f_0}{\text{BW}} = \frac{f_0}{f_2 - f_1} \tag{16.17}$$

where BW is the bandwidth of the filter. The upper and lower 3-dB frequencies in terms of ω are

$$\omega_2 \text{ (upper 3-dB frequency)} = \frac{R}{2L} + \sqrt{\left(\frac{R}{2L}\right) + \frac{1}{LC}} \tag{16.18}$$

$$\omega_1 \text{ (lower 3-dB frequency)} = \frac{R}{2L} - \sqrt{\left(\frac{R}{2L}\right)^2 + \frac{1}{LC}} \tag{16.19}$$

Thus, BW in radians is R/L. Using this we obtain from Eq. (16.17) the following:

$$Q = \frac{\omega_0}{\text{BW}} = \frac{\omega_0 L}{R} = \frac{X_L}{R}$$

At series resonance, the circuit impedance is minimum and equal to R. The Q rise in voltage ($V_c = V_L = QV_{\text{applied}}$) must be taken into account when selecting the voltage withstand rating of the circuit components. The series resonant bandpass filter can be connected as a notch filter in parallel with a line to filter out unwanted frequencies.

EXAMPLE 16.6

Calculate the components for a series RLC filter for which $f_1 = 1700$ Hz and $f_2 = 1900$ Hz.

Solution
From Eq. (16.17) we see that the Q required is

$$Q = \frac{\sqrt{f_1 f_2}}{f_2 - f_1} = \frac{\sqrt{1700 \text{ Hz } 1900 \text{ Hz}}}{200 \text{ Hz}} = \frac{1797.2 \text{ Hz}}{200 \text{ Hz}} = 8.99 \approx 9$$

The value of R must be chosen such that the source can drive it. Let $R = 500 \ \Omega$:

$$X_L = QR = 9(500 \ \Omega) = 4500 \ \Omega$$

The simulation results are shown in Figure 16.24.

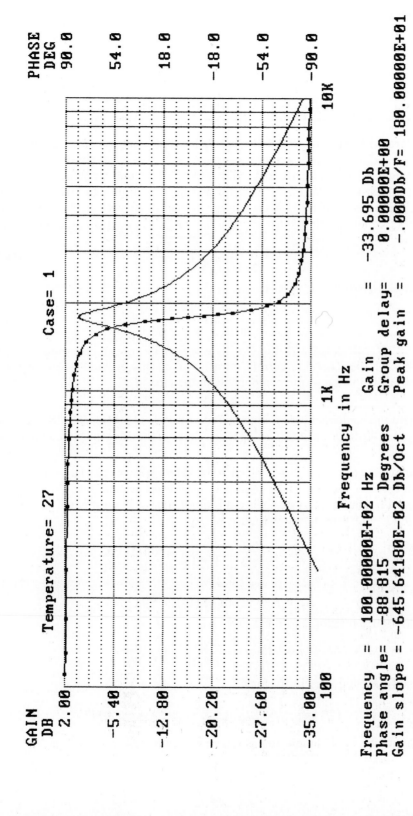

GAIN
DB

Temperature= 27   Case= 1

PHASE
DEG

2.00

-5.40

-12.80

-20.20

-27.60

-35.00

90.0

54.0

18.0

-18.0

-54.0

-90.0

100   1K   10K

Frequency in Hz

Frequency = 100.00000E+02 Hz   Gain   = -33.695 Db
Phase angle= -88.815   Degrees   Group delay= 0.00000E+00
Gain slope = -645.64180E-02 Db/Oct Peak gain = -.000Db/F= 180.00000E+01

FIGURE 16.24
Example 16.6 simulation.

$$L = \frac{X_L}{2\pi f_0} = \frac{4500\omega}{2\pi \, 1800 \text{ Hz}} = 399 \text{ mH}$$

$$C = \frac{1}{(2\pi f_0)^2 L} = \frac{1}{(2\pi \, 1800 \text{ Hz})^2 \, 398 \text{ mH}} = 0.2 \, \mu\text{F}$$

Parallel Resonant RLC Filter   Parallel RLC circuits can be connected as bandpass and notch filters as shown in Figure 16.25. The basic parallel RLC circuit can be modeled as a perfect L and C in parallel with a shunt resistance R_p replacing the series resistance, R_s, of the inductor as shown in Figure 16.25(b). Resonance occurs when $I_c = I_L$, which means that at resonance $X_c = Z_{R+L}$. Thus, the parallel resonant frequency is shifted from that of the same components connected in series:

$$f_0 = \frac{1}{2\pi \sqrt{LC}} \sqrt{1 - \frac{R_s C^2}{L}} \tag{16.20}$$

which can be written as

$$f_0 = \frac{1}{2\pi \sqrt{LC}} \sqrt{\frac{Q^2}{1 + Q^2}} \tag{16.21}$$

If $Q \geq 10$, f_c calculated with Eq. (16.16) is within 1%. Q can be expressed as X_L/R_s or R_p/X_L. The Q rise in current must be taken into account when calculating the power dissipation of the components in the circuit.

EXAMPLE 16.7

Calculate the components of a notch filter with $f_1 = 17.5$ kHz, $f_2 = 18.5$ kHz, and an output of 40 dB down at f_0. The input resistance of the stage the filter feeds is 1 kΩ.

Solution
The required Q is $f_0/(f_1 - f_2)$, which is 18 kHz/1 kHz = 18. This value is high enough that Eq. (16.16) can be used with little error to obtain C. First the value of R_s must be found. Forty decibels is a ratio of 100, so R_p must equal 99 kΩ. Since $R_p = R_s (1 + Q^2)$ the series resistance is

$$R_s = \frac{R_p}{1 + Q^2} = \frac{99 \text{ k}\Omega}{1 + 18^2} = 301 \text{ k}\Omega$$

Thus,

$$X_L = R_s Q = 301 \, \Omega \, (18) = 5.48 \text{ k}\Omega$$

The closest 1% resistor is 5.36 kΩ, from which

$$L = \frac{X_L}{2\pi f} = \frac{5.36 \text{ k}\Omega}{2\pi \, 18 \text{ kHz}} = 47.4 \text{ mH}$$

and finally, solving Eq. (16.16) for C,

$$C = \frac{1}{(2\pi f)^2 L} = \frac{1}{(2\pi \, 18 \text{ kHz})^2 \, 47.4 \text{ mH}} = 1.6 \text{ nF}$$

The simulation for Example 16.7 is shown in Figure 16.26.

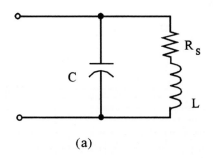

(a)

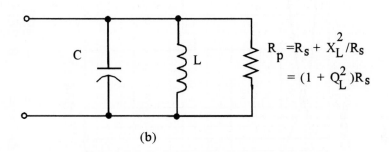

(b)

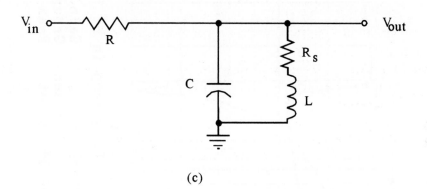

(c)

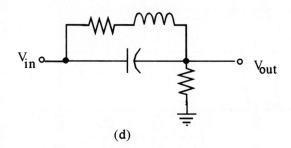

(d)

FIGURE 16.25
Parallel RLC filter: (a) Circuit. (b) Equivalent circuit. (c) Connection for bandpass filter.
(d) Connection for band-reject filter.

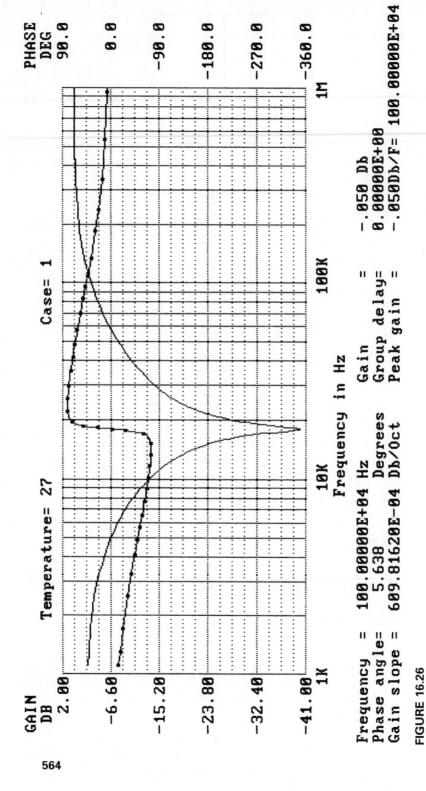

GAIN
DB

PHASE
DEG

Temperature= 27   Case= 1

Frequency in Hz

Frequency = 100.00000E+04 Hz	Gain = -.050 Db
Phase angle= 5.638 Degrees	Group delay= 0.00000E+00
Gain slope = 609.81620E-04 Db/Oct	Peak gain = -.050Db/F= 100.00000E+04

FIGURE 16.26
Example 16.7 simulation.

RLC filters can be cascaded to obtain higher order filters. The end-of-chapter bibliography deals with this topic.

16.7 ACTIVE FILTERS

This section provides the circuit configurations and equations for the more common types of active filters. Because of their use in cascading and low TRA applications, the first-order op-amp low-pass, high-pass, and bandpass filters shown generically in Figure 16.27(a) are covered.

16.7.1 First-Order Filters

Low-Pass Filters The compensated integrator based op-amp first-order low-pass filter shown in Figure 16.27 has improved performance over an RC low-pass filter because it has gain and better input/output isolation. The passband gain and f_c of the circuit are

$$A_p = -\frac{R_f}{R_1} \tag{16.22}$$

$$f_c = \frac{1}{2\pi R_f C} \tag{16.23}$$

EXAMPLE 16.8 Design a first-order low-pass filter with $f_c = 350$ Hz, $A_p = 4.99$ (~14 dB), and $R_{in} = 10$ kΩ. Use the circuit of Figure 16.27(b).

Solution
From the input resistance specification, $R_1 = 10$ kΩ, and from the passband gain equation, $R_f = A_p R_1 = 49.9$ kΩ. From this,

$$C = \frac{1}{2\pi R_f f_c} = \frac{1}{2\pi\, 49.9\text{ k}\Omega\, 350\text{ Hz}} = 9.1\text{ nF}$$

The value of R_s, which compensates for offset voltage generated by the op-amp bias current flowing through the feedback network, is set to the value of R_1 and R_f in parallel, 8.25 kΩ.
The simulated response is shown in Figure 16.28.

High-Pass Filters The compensated differentiator of Figure 16.29 is connected as a first-order high-pass active filter. Equation (16.22) is the passband gain, and $f_c = 1/(2\pi R_1 C)$. All of the high- and low-pass filter cutoff frequency equations of this section are of the form

$$f_c = 1/[2\pi(\text{applicable time constant})]$$

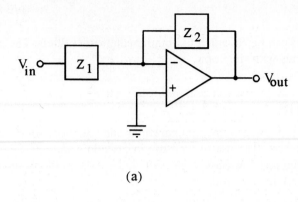

(a)

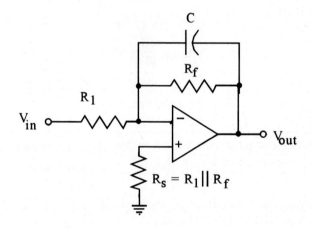

(b)

FIGURE 16.27
First-order low-pass active filter. (a) Generic model. (b) Low-pass filter.

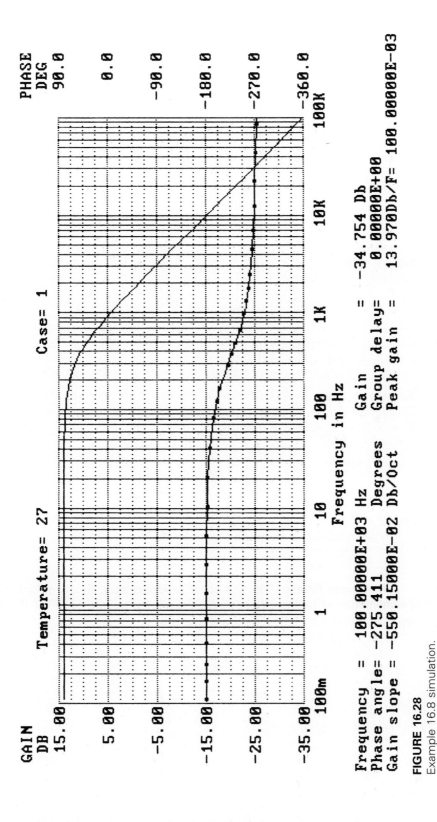

FIGURE 16.28
Example 16.8 simulation.

567

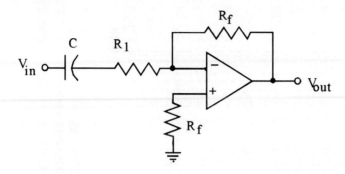

FIGURE 16.29
Differentiator high-pass filter.

EXAMPLE 16.9 Calculate the components for a first-order high-pass filter such that $R_{in} \geq 4.99$ kΩ, $f_c = 12$ Hz, and $A_p = 12$ (21.6 dB).

Solution
To meet the input resistance requirements let $R_1 = 4.99$ kΩ. From the gain equation, $R_f = A_p R_1 = 12(4.99$ kΩ$) = 60.4$ kΩ.

$$C = \frac{1}{2\pi R_1 f} = \frac{1}{2\pi \, 4.99 \text{ k}\Omega \, 12 \text{ Hz}} = 2.66 \ \mu\text{F}$$

and $R_s = R_f$.
The simulation is shown in Figure 16.30.

Bandpass Filters The first-order bandpass filter, shown in Figure 16.31, is a combination of the previous two circuits. Unless the filter $f_2 \approx 6f_1$ the passband response will not look flat. The equations are

$$A_p = \frac{R_f}{R_1}$$

$$f_1 = \frac{1}{2\pi R_1 C_1}$$

$$f_2 = \frac{1}{2\pi R_f C_f}$$

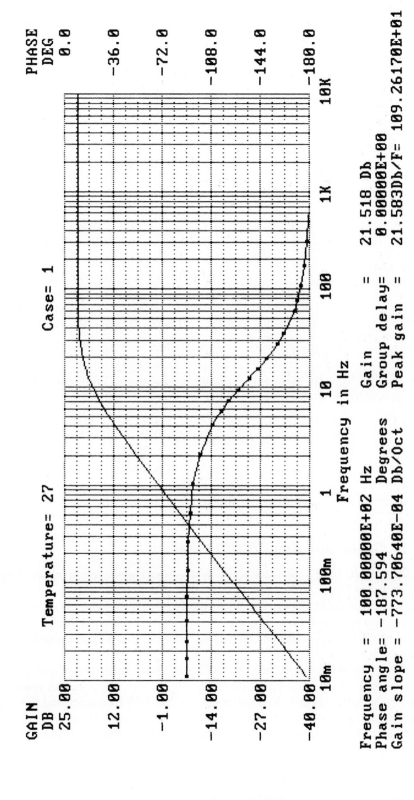

GAIN
DB   Temperature= 27   Case= 1   PHASE
                                                      DEG
25.00   0.0

12.00   -36.0

-1.00   -72.0

-14.00   -108.0

-27.00   -144.0

-40.00   -180.0
    10m   100m   1   10   100   1K   10K
                        Frequency in Hz

Frequency   =   100.00000E+02 Hz   Gain   =   21.518 Db
Phase angle =   -187.594   Degrees   Group delay=   0.00000E+00
Gain slope =   -773.70640E-04 Db/Oct   Peak gain =   21.583Db/F= 109.26170E+01

FIGURE 16.30
Example 16.9 simulation.

569

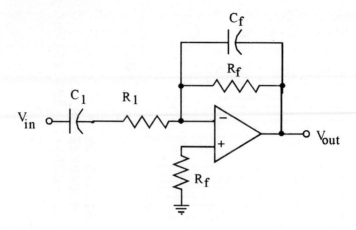

FIGURE 16.31
First-order bandpass filter.

EXAMPLE 16.10

Design a first-order bandpass filter with $A_p = 10$ (20 dB), $R_{in} = 10 \text{ k}\Omega$, $f_1 = 30$ Hz, and $f_2 = 3$ kHz.

Solution
Let $R_1 = 10 \text{ k}\Omega$ to meet the input resistance requirements. From the gain requirements $R_f = A_p R_1 = 10$ (10 kΩ) = 100 kΩ.

$$C_1 = \frac{1}{2\pi \text{ 10 k}\Omega \text{ 30 Hz}} = 530 \text{ nF}$$

$$C_2 = \frac{1}{2\pi \text{ 100 k}\Omega \text{ 3 kHz}} = 0.53 \text{ nF}$$

The simulation results are shown in Figure 16.32.

16.7.2   Sallen and Key (VCVS) Filters

Operation   The Sallen and Key second-order high-pass and low-pass active filters, shown in Figure 16.33, are stable, inexpensive, popular, easy-to-adjust active filters. The two RC circuits provide a cascaded low-pass or high-pass network, thus providing two poles (one per RC), and R_A and R_B set α (which controls the shape function near f_c) to determine the filter type. Since the gain setting resistors provide α, the gain is not variable.

Up to this point the actual and calculated 3-dB frequencies have been the same because we have used either first-order filters with real poles (as opposed to complex), or Butterworth second-order filters. The 3-dB frequency and the f_c used in calculations may not be the same for other types of filters because of ripples in the passband (Chebyshev) or TRA less than 6 dB/octave/pole (Bessel).

Component Calculation Procedure   Generally $R_1 = R_2$ and $C_1 = C_2$ to simplify the calculations and provide better performance; similar components of the same value tend to drift at the same rate, which helps prevent ratio errors. Let f_c equal the value of the corner frequency used in the calculation and f_{3dB} the desired 3-dB frequency. If in Table 16.1 the f_c/f_{3dB} ratio, called the f_c factor, is not one, then calculate f_c. For low-pass filters,

$$f_c = f_{3dB}(f_c \text{ factor}) \qquad \qquad (16.24)$$

and for high-pass filters,

$$f_c = \frac{f_{3dB}}{f_c \text{ factor}} \qquad \qquad (16.25)$$

Select $C = C_1 = C_2$ and let $R = R_1 = R_2$. Then from the general form of Eq. (16.16),

$$R = \frac{1}{2\pi f_c C}$$

Choose an appropriate R_A, normally R, and calculate R_B:

$$R_B = (2 - \alpha)R_A \qquad \qquad (16.26)$$

Now the passband gain is given by

$$A_p = \frac{R_B}{R_A} + 1$$

from the noninverting op-amp amplifier gain equation.

To tune f_c, R_1 and R_2 are varied together, and to change α, R_B is varied.

EXAMPLE 16.11   Calculate the components for a 0.5-dB ripple Chebyshev Sallen and Key high-pass filter with $f_{3dB} = 10$ kHz.

Solution
The circuit of Figure 16.33(b) will be used. From Table 16.1 we obtain f_c factor = 1.21342 and $\alpha = 1.155781$. From Eq. (16.24)

$$f_c = 10 \text{ kHz}/1.21342 = 8.24 \text{ kHz}$$

If the capacitors are set to 0.01 μF, then

$$R_1 = R_2 = R_A = \frac{1}{2\pi f_c C} = \frac{1}{2\pi\ 8.24 \text{ kHz}\ 0.01\ \mu\text{F}} = 1.91 \text{ k}\Omega$$

From Eq. (16.26),

$$R_B = (2 - \alpha)\,R_A = (2 - 1.157781)\,1.91 \text{ k}\Omega = 1.58 \text{ k}\Omega$$

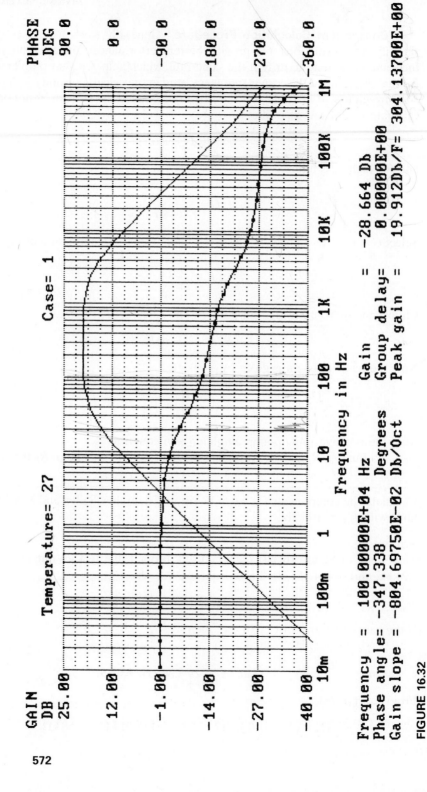

FIGURE 16.32
Example 16.10 simulation.

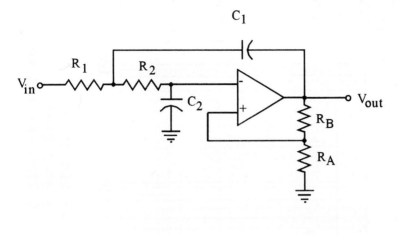

(a)

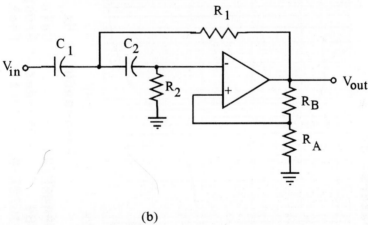

(b)

FIGURE 16.33
Sallen and Key (VCVS) active filters: (a) Low-pass. (b) High-pass.

The passband gain is approximately

$$A_p = \frac{R_B}{R_A} + 1 = \frac{1.58\ \text{k}\Omega}{1.91\ \text{k}\Omega} + 1 = 1.83\ (5.3\ \text{dB})$$

The simulation is shown in Figure 16.34. Notice the dropoff at the far right-hand side of the gain response as the filter output begins to roll off because the op-amp ran out of frequency response.

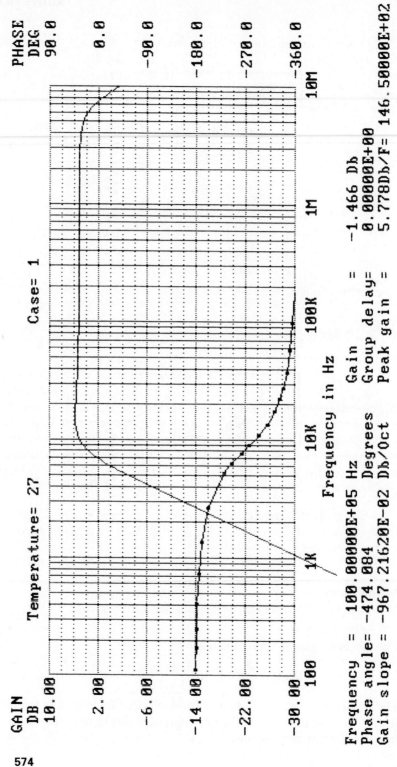

FIGURE 16.34

Example 16.11 simulation. Note the falloff of the gain at the far right side of the plot because the op-amp frequency range was exceeded.

**EXAMPLE
16.12**

Design a second-order 2-dB ripple Chebyshev low-pass filter using a Sallen and Key circuit for $f_{3dB} = 1$ kHz.

Solution

Look up the f_c factor and α from Table 16.1. They are 0.907227 and 0.886015, respectively. Set $C_1 = C_2 = C = 0.1$ μF, as is our convention.

$$f_c = f_{3dB}(f_c \text{ factor}) = 1 \text{ kHz } (0.907227) = 907.227 \text{ Hz}$$

Now let $R_1 = R_2 = R_A = R$ and calculate

$$R = \frac{1}{2\pi f_c C} = \frac{1}{2\pi\, 907.227 \text{ Hz } 0.1\, \mu F} = 1754\, \Omega$$

Use 1.74 kΩ (nearest 1%). R_B and the gain are

$$R_B = (2 - \alpha)R_A = (2 - 0.886015)1.74 \text{ k}\Omega = 1.96 \text{ k}\Omega$$

$$A_p = \frac{R_B}{R_A} + 1 = \frac{1.96 \text{ k}\Omega}{1.74 \text{ k}\Omega} + 1 = 1.13 + 1 = 2.13 \,(6.55 \text{ dB})$$

The simulation is shown in Figure 16.35.

16.7.3 Multiple Feedback Filter

Operation   The multiple feedback bandpass filter is good for low to moderate Q's, up to about 15. The circuit name comes from the two feedback paths, C_1 and R_3 in Figure 16.36. It can be used as a low-pass filter but it works best in bandpass applications. R_1 and C_1 provide the low-pass response, and R_3 and C_2 provide the high-pass response. At frequencies far removed from f_0 the TRA is 6 dB/octave because only one RC network provides each of the high-pass and low-pass responses. The feedback resistors provide the peaking for Q. R_2 increases the input resistance and provides the circuit with a settable gain for $A_p < 2Q^2$. R_4 is set equal to R_3 to compensate for input offset voltage generated by bias current.

Component Calculation Procedure   After selecting f_1 and f_2, make sure the op-amp gain is much greater than Q^2 at f_2. The center frequency is the geometric mean, so f_0 and Q are found from

$$f_0 = \sqrt{f_1 f_2}$$

$$Q = \frac{f_0}{f_2 - f_1}$$

Set $C_1 = C_2 = C$ and calculate:

$$R_1 = \frac{Q}{2\pi f_0 C A_p} \tag{16.27}$$

$$R_2 = \frac{Q}{2\pi f_0 C(2Q^2 - A_p)} \tag{16.28}$$

$$R_3 = \frac{2Q}{2\pi f_0 C} \tag{16.29}$$

Finally, check the passband gain with

$$A_p = \frac{R_3}{2R_1} \tag{16.30}$$

The frequency is tuned with R_1 and R_2 simultaneously, the gain with R_2, and the Q with the R_3/R_1 ratio.

EXAMPLE 16.13

Design a multiple feedback bandpass filter such that the passband gain is 20 (26 dB), $f_1 = 900$ Hz, and $f_2 = 1100$ Hz.

Solution
First calculate Q:

$$f_0 = \sqrt{f_1 f_2} = \sqrt{900 \text{ Hz } 1100 \text{ Hz}} = 994.95 \text{ Hz}$$

$$Q = \frac{f_0}{f_2 - f_1} = \frac{994.95 \text{ Hz}}{200 \text{ Hz}} = 4.975$$

Now check to assure that $A_p < 2Q^2$; $20 < 49.5$, so continue. Set all capacitors equal to 0.01 μF. If the capacitor is poorly chosen the resistor values will not be reasonable. Now use Eqs. (16.28), (16.29), and (16.30) to find the resistor values:

$$R_1 = \frac{Q}{2\pi f_0 C A_p} = \frac{4.975}{2\pi \, 994.95 \text{ Hz } 0.01 \, \mu\text{F } 20} = 3978 \, \Omega$$

$$R_2 = \frac{Q}{2\pi f_0 C (2Q^2 - A_p)} = \frac{4.975}{2\pi \, 994.95 \text{ Hz } 0.01 \, \mu\text{F } (24.975^2 - 20)} = 2697 \, \Omega$$

$$R_3 = \frac{2Q}{2\pi f_0 C} = \frac{2(4.975)}{2\pi \, 995.95 \text{ Hz } 0.01 \, \Omega} = 159.2 \text{ k}\Omega$$

Use 4.02, 2.67, and 158 kΩ for R_1, R_2, and R_3, respectively. Check to see the gain is as desired:

$$A_p = \frac{R_3}{2R_1} = \frac{159.155 \text{ k}\Omega}{2(3978.87 \, \Omega)} = 19.98$$

Figure 16.37(a) shows the overall filter response and Figure 16.37(b) shows a close-up of the passband response.

16.7.4 Biquadratic Filters

Operation The biquadratic filter is shown in Figure 16.38. Called biquad for short, it is a very stable, easily cascadable, bandpass filter capable of Q's of more than 100. Additionally the biquad bandwidth stays constant as f_0 is varied (over a modest range) so f_0 can be tuned without changing the bandwidth. Stage 1 is a summing integrator that provides a low-pass response, stage 2 provides

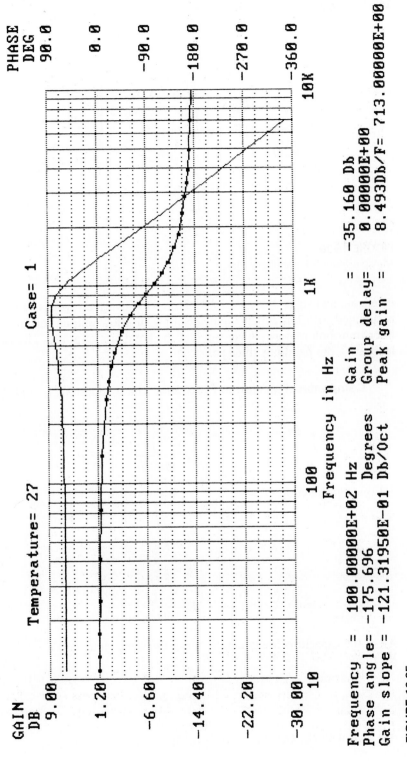

FIGURE 16.35
Example 16.12 simulation.

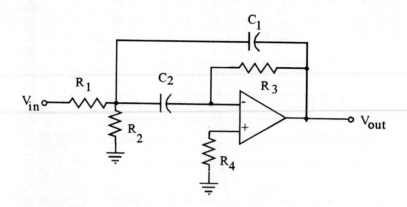

FIGURE 16.36
Multiple feedback bandpass filter.

inversion, and stage 3 is a low-pass filter. The inverted low-pass output of stage 3 is summed with the circuit input, resulting in a low output voltage except in the filter transition region, which provides the bandpass response. The cancellation is least at f_0, the bandpass filter peak.

Component Calculation Procedure First choose f_1, f_2, and A_p. Then

$$f_0 = \sqrt{f_1 f_2}$$

$$Q = \frac{f_0}{f_2 - f_1}$$

Set the gain function G:

$$G = \frac{Q}{A_p} \tag{16.31}$$

$$R_1 = \frac{G}{2\pi f_0 C} \tag{16.32}$$

$$R_f = \frac{Q}{2\pi f_0 C} \tag{16.33}$$

All of the capacitors are generally set to the same value and $R = R_2 = R_3 = R_4 = R_5$, then

$$R = \frac{1}{2\pi f_0 C} \tag{16.34}$$

Adjust f_0 with R_2, Q with R_f, and A_p with R_1.

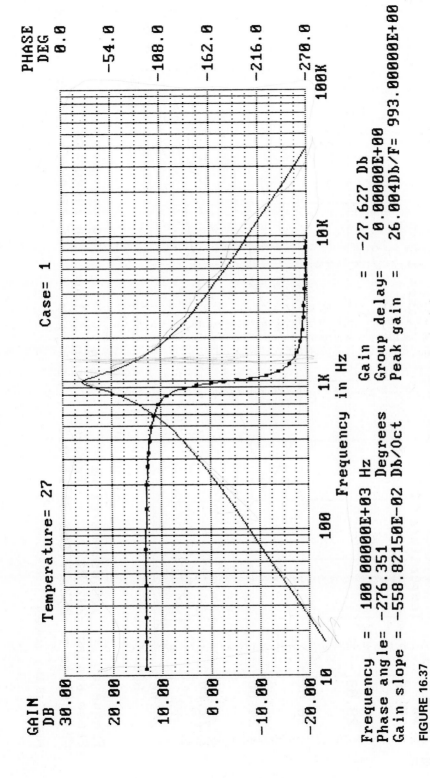

GAIN
DB
30.00

Temperature= 27   Case= 1

20.00

10.00

0.00

-10.00

-20.00

10   100   1K   10K   100K

Frequency in Hz

PHASE
DEG
0.0

-54.0

-108.0

-162.0

-216.0

-270.0

Frequency = 100.00000E+03 Hz   Gain   =   -27.627 Db
Phase angle= -276.351   Degrees   Group delay=   0.00000E+00
Gain slope = -558.82150E-02 Db/Oct   Peak gain = 26.004Db/F= 993.00000E+00

FIGURE 16.37
Example 16.13 simulations: (a) Wide frequency range. (b) Expanded scale simulation.

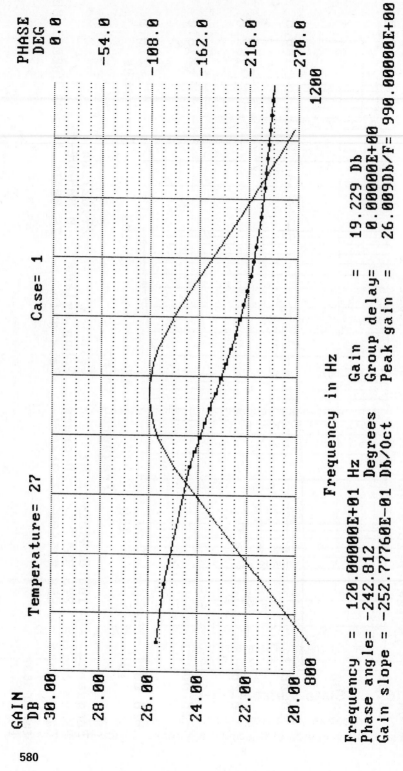

FIGURE 16.37 (continued)

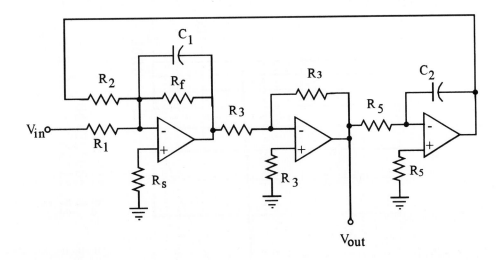

FIGURE 16.38
Biquadratic bandpass filter.

EXAMPLE 16.14

Set up a biquad filter so that $f_1 = 10$ kHz, $f_2 = 10.2$ kHz, and $A_p = 30$ (29.5 dB).

Solution
Using the now familiar equations, $f_0 = 10099.5$ Hz and $Q = 50.5$. The gain function $G = Q/A_p = 50.5/30 = 16.68$. Choosing $C = C_1 = C_2 = 0.01$ μF and using Eqs. (16.32) and (16.33),

$$R_1 = \frac{G}{2\pi f_0 C} = \frac{1.68}{2\pi\, 10099.5\ \text{Hz}\, 0.01\ \mu\text{F}} = 2650\ \Omega$$

$$R_f = \frac{Q}{2\pi f_0 C} = \frac{50.5}{2\pi\, 10099.5\ \text{Hz}\, 0.01\ \mu\text{F}} = 79.6\ \text{k}\Omega$$

The nearest 1% values are $R_1 = 2.67$ kΩ and $R_f = 80.6$ kΩ. Setting $R = R_2 = R_3 = R_4 = R_5$ and solving Eq. (16.34), we get

$$R = \frac{1}{2\pi f_0 C} = \frac{1}{2\pi\, 10099.5\ \text{Hz}\, 0.01\ \mu\text{F}} = 1.58\ \text{k}\Omega$$

Finally $R_s = R_1$, R_2 and R_f in parallel, 976 Ω, and $R_3/2 = 787.9$ Ω. The simulation is shown in Figure 16.39.

16.7.5 State-Variable Filters

Operation The state-variable block diagram is drawn in Figure 16.40, and one implementation of the circuit is shown in Figure 16.41. It is often called the

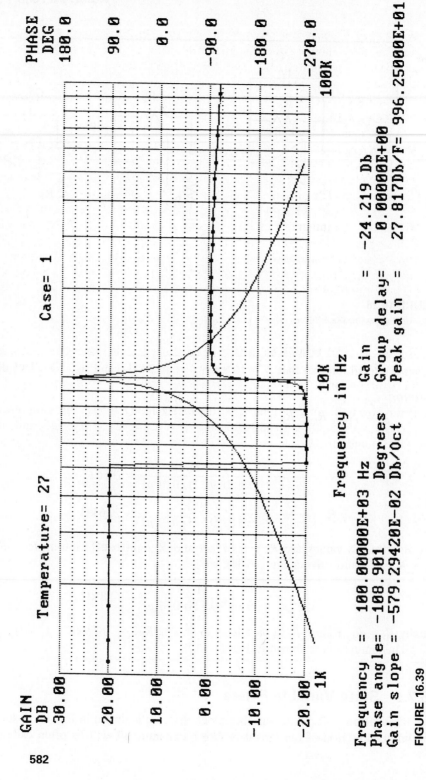

FIGURE 16.39
Example 16.14 simulation.

582

universal filter because it produces a high-pass, low-pass, and bandpass output at the same time, though either the bandpass or the high-/low-pass functions must be optimized—not both at once. The state-variable filter is stable, has low-Q values and α sensitivities, and can provide Q's to 100 as a bandpass filter. It is often used in commercial active filters. We will look at both the unity gain and the variable gain realizations.

The state-variable second-order filter consists of a summer, two integrators (for two poles), and an α setting network. The unity gain version uses the summer to set α so the passband gain cannot be set by the user, while the variable gain version uses a separate amplifier to set α, allowing the summer to be used to set passband gain. The variable gain state-variable is shown in Figure 16.42.

The high-pass function results from the out-of-phase summing of the input and the double integrator output. These signals cancel when $f < f_c$, but when $f > f_c$ the input is allowed through. The low-pass output results from the double integrators, and the bandpass function results from the overlap of the high-pass and low-pass responses. The peaking of the bandpass filter cannot be accomplished if the high- and low-pass filters are optimized, and vice versa.

Component Calculation Procedures—Unity Gain

High- and Low-Pass Filters The procedure for calculation of a high- or low-pass filter is the same except for the calculation of f_c from the f_c factor, except for a Butterworth. First select f_c and α from Table 16.1 for the filter type desired. Calculate f_c if $f_c \neq f_{3dB}$, that is, if the f_c factor is not 1.

Normally, $R_1 = R_2 = R_3 = R_4 = R_f = R'_f = R$, and $C_1 = C_2 = C$. This will be assumed in the rest of the examples. R and R_5 are calculated from

$$R = \frac{1}{2\pi f_c C} \tag{16.35}$$

$$R_5 = R'_f[(3/\alpha) - 1] \tag{16.36}$$

Adjust f_c with R_1 and R_2 simultaneously, and α with R'_f.

Bandpass Select f_1, f_2, and calculate f_0 and Q from

$$f_0 = \sqrt{f_1 f_2}$$

$$Q = \frac{f_0}{f_2 - f_1}$$

The passband gain is equal to Q. Let $C = C_1 = C_2$, and $R = R_1 = R_2 = R_3 = R_4 = R_f = R'_f$.

$$R = \frac{1}{2\pi f_0 C} \tag{16.37}$$

$$R_5 = R'_f(3Q - 1) \tag{16.38}$$

Adjust f_0 as in the high-/low-pass filter, and adjust Q with R'_f.

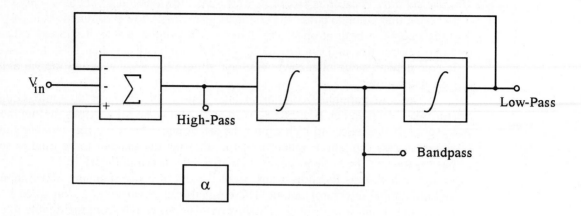

FIGURE 16.40
State-variable block diagram.

EXAMPLE 16.15

Design a unity-gain state-variable low-pass Butterworth filter for a cutoff frequency of 20 kHz.

Solution

From Table 16.1, $\alpha = 1.414$ and f_c factor $= 1$. Let the capacitors equal 0.001 μF; then for all of the resistors except R_5

$$R = \frac{1}{2\pi f_c C} = \frac{1}{2\pi \, 20 \text{ kHz} \, 0.001 \, \mu\text{F}} = 7957 \, \Omega$$

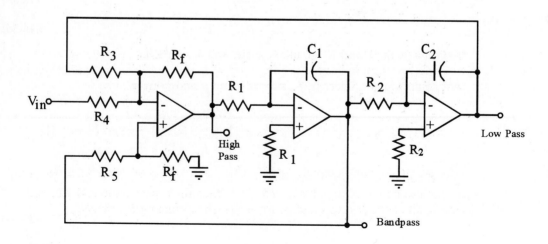

FIGURE 16.41
Unity gain state-variable active filter.

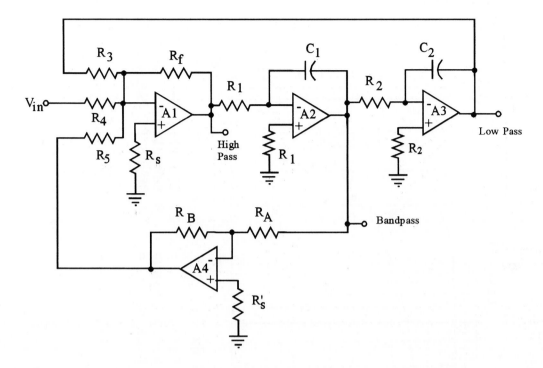

FIGURE 16.42
Variable gain state-variable active filter.

Use the nearest 1% resistor, 7.87 kΩ:

$$R_5 = R'_f[(3/\alpha) - 1] = 8.25 \text{ k}\Omega$$

The simulations are shown in Figure 16.43(a), which shows the ideal response, and Figure 16.43(b), which shows what can happen when the op amp is operated beyond its maximum operating frequency and parasitic effects predominate.

EXAMPLE 16.16

Set up a unity-gain state-variable bandpass filter for $f_1 = 59$ Hz and $f_2 = 61$ Hz.

Solution
Using the appropriate equations, we find that $f_0 = 59.99$ Hz, and $Q = 29.99$, which will also be the passband gain, 29 dB. Setting the capacitors to 0.47 μF, and calculating all of the resistors except R_5, and then calculating R_5, we see

$$R = \frac{1}{2\pi f_0 C} = \frac{1}{2\pi\, 59.99 \text{ Hz}\, 0.47\, \mu\text{F}} = 5.62 \text{ k}\Omega$$

$$R_5 = R'_f(3Q - 1) = 5.62 \text{ k}\Omega[3(29.99) - 1] = 487 \text{ k}\Omega$$

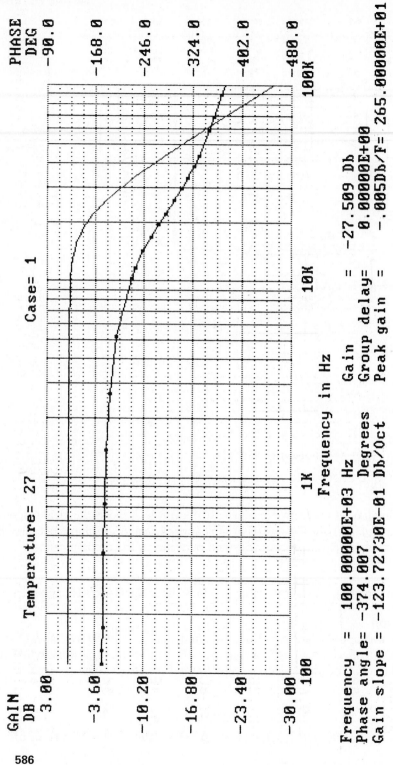

FIGURE 16.43

Example 16.15 simulations: (a) Idealistic simulation. (b) Extended frequency range. Note the nonideal behavior from exceeding the op-amp frequency range.

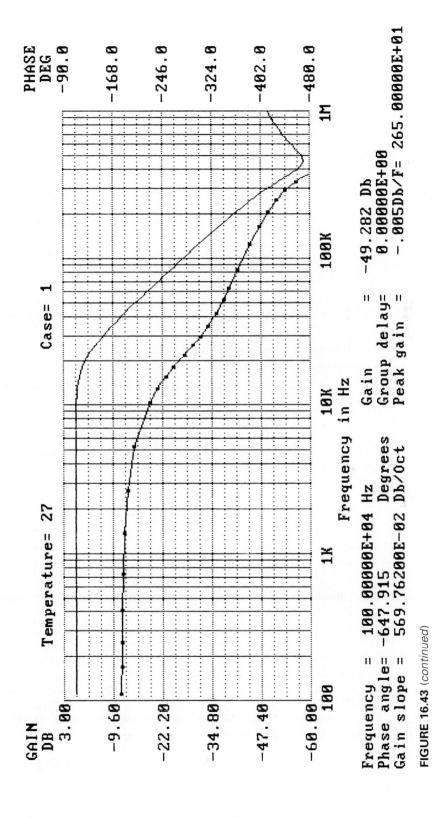

FIGURE 16.43 (continued)

The simulation is shown in Figure 16.44. Figure 16.44(a) is an overview, and Figure 16.44(b) is expanded in scale.

Component Calculation Procedures—State-Variable and Variable Gain

The variable gain state-variable active filter, shown in Figure 16.42, has a separate amplifier to set α (or Q) leaving R_f of the summer available to set the passband gain.

High- and Low-Pass Obtain α and the f_c factor from Table 16.1. Calculate the new f_c if necessary using Eq. (16.24) or (16.25). Set $C_1 = C_2 = C$, and $R_1 = R_2 = R_3 = R_5 = R_f = R_A = R$. Then calculate $R = 1/(2\pi fC)$ and

$$R_4 = \frac{R_f}{A_p} \qquad (16.39)$$

$$R_B = \alpha R_A \qquad (16.40)$$

Adjust α with R_A, A_p with R_4, and f_c with R_1 and R_2 simultaneously.

Bandpass From the desired A_p, f_1 and f_2 calculate

$$f_0 = \sqrt{f_1 f_2} \quad \text{and} \quad Q = \frac{f_0}{f_2 - f_1} \leq 150$$

$$G = \frac{A_p}{Q} \qquad (16.41)$$

Set $C_1 = C_2 = C$, and $R_1 = R_2 = R_3 = R_5 = R_f = R_A = R$. Then calculate $R = 1/(2\pi fC)$, R_4, and R_B as follows:

$$R_4 = \frac{R_f}{G} \qquad (16.42)$$

$$R_B = \frac{R_A}{Q} \qquad (16.43)$$

The adjustments are the same as for the high-/low-pass filter, except they are now Q and f_0 instead of α and f_c.

EXAMPLE 16.17

Calculate the components for a variable gain state-variable 1-dB ripple Chebyshev high-pass filter with a passband gain of 10 (20 dB) and f_{3dB} of 4.7 kHz.

Solution

Let the capacitors equal 0.0022 μF. Look up the f_c factor, 1.050005, and α, 1.045456. Calculate f_c using Eq. (16.25):

$$f_c = f_{3dB}/f_c \text{ factor} = 4.7 \text{ kHz}/1.050005 = 4.476 \text{ kHz}$$

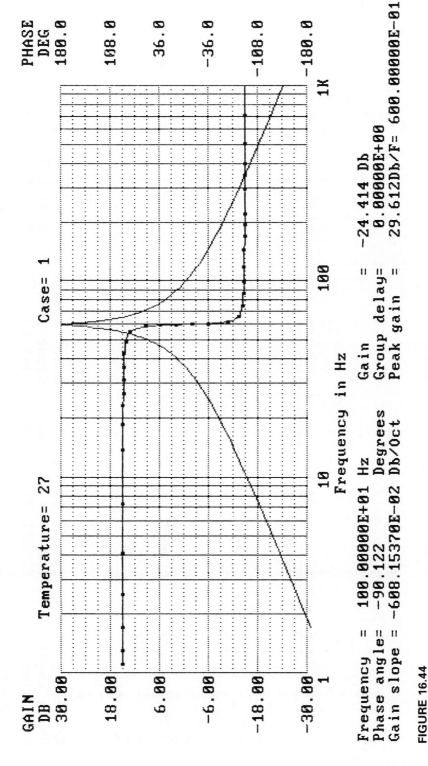

GAIN
DB

PHASE
DEG

Temperature= 27   Case= 1

Frequency in Hz

Frequency   = 100.00000E+01 Hz   Gain   = −24.414 Db
Phase angle= −90.122   Degrees   Group delay=   0.00000E+00
Gain slope = −600.15370E−02 Db/Oct   Peak gain = 29.612Db/F= 600.00000E−01

FIGURE 16.44
Example 16.16 simulations: (a) Wide frequency range simulation. (b) Expanded scale
simulation.

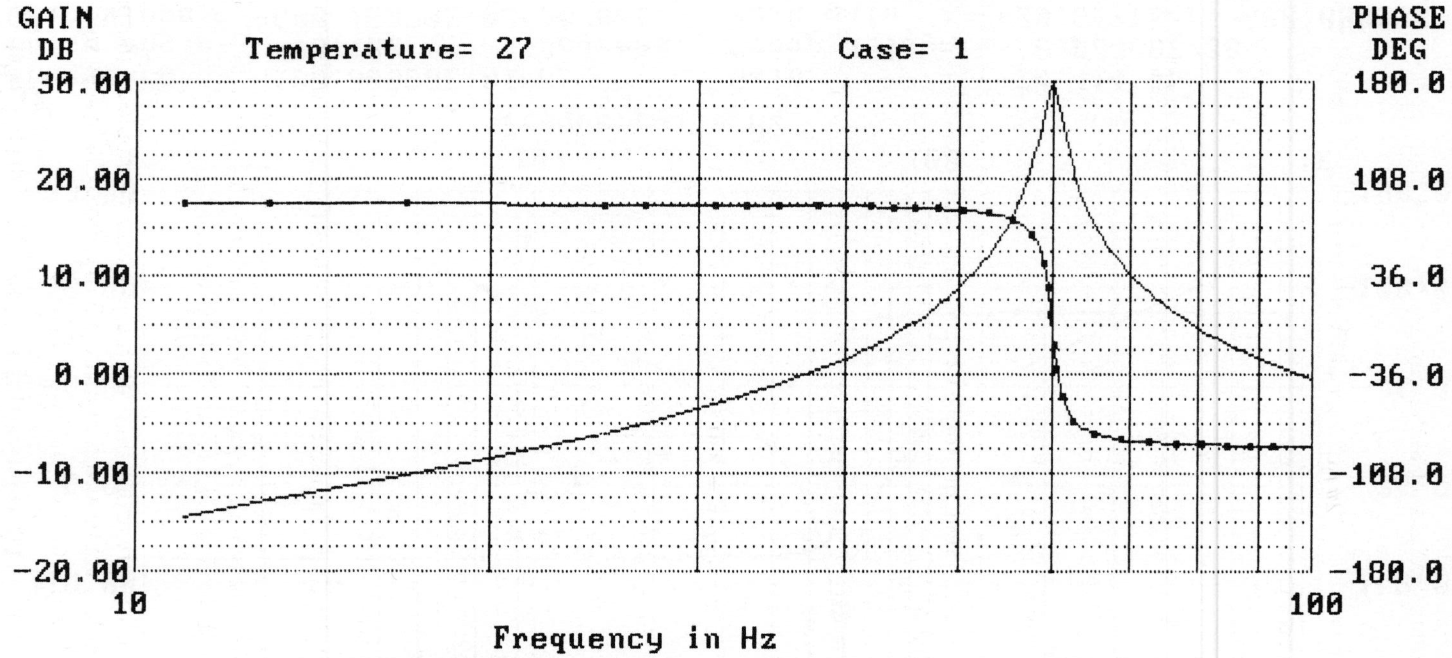

GAIN
DB
30.00   Temperature= 27   Case= 1

PHASE
DEG
180.0

20.00   108.0

10.00   36.0

0.00   -36.0

-10.00   -108.0

-20.00   -180.0
      10   100

                    Frequency in Hz

Frequency   = 100.00000E+00 Hz   Gain   =   -.576 Db
Phase angle= -88.250   Degrees   Group delay= 0.00000E+00
Gain slope = -131.87930E-01 Db/Oct   Peak gain = 29.612Db/F= 600.00000E-01

FIGURE 16.44 (continued)

Now calculate the bulk of the resistors

$$R = \frac{1}{2\pi f_0 C} = \frac{1}{2\pi\, 4.476\ \text{kHz}\, 0.0022\ \mu\text{F}} = 16.2\ \text{k}\Omega$$

Now the rest of the resistors

$$R_4 = \frac{R_f}{A_p} = \frac{16.2\ \text{k}\Omega}{10} = 1.62\ \text{k}\Omega$$

$$R_B = \alpha R_A = 1.045456(16.2\ \text{k}\Omega) = 16.9\ \text{k}\Omega$$

$$R'_s = R_A \| R_B = 8.25\ \text{k}\Omega$$

$$R_s = R_4 \| R_3 \| R_5 \| R_f = 1.24\ \text{k}\Omega$$

The simulation is shown in Figure 16.45.

EXAMPLE 16.18

Calculate the components for a variable gain state-variable bandpass filter with $A_p = 40$ (32 dB), $f_1 = 18$ kHz, and $f_2 = 19$ kHz.

Solution
Calculate f_0 and Q to obtain $f_0 = 18.49$ kHz and $Q = 18.49$. Gain function G is obtained from

$$G = A_p/Q = 40/18.49 = 2.163$$

If we let the capacitors = 330 pF, then the resistors are

$$R = \frac{1}{2\pi f_0 C} = \frac{1}{2\pi\, 18.49\ \text{kHz}\, 330\ \text{pF}} = 26.1\ \text{k}\Omega$$

$$R_B = R_A/Q = 26.1\ \text{k}\Omega/18.49 = 1.40\ \text{k}\Omega$$

$$R'_s = R_A \| R_B = 1.33\ \text{k}\Omega$$

$$R_s = R_4 \| R_3 \| R_5 \| R_f = 4.99\ \text{k}\Omega$$

The simulation is shown in Figure 16.46.

16.7.6 Cascaded Stages

Two examples illustrate the cascading of stages for higher order filters.

EXAMPLE 16.19

Design a high-pass 3-dB ripple Chebyshev filter with $f_{3dB} = 10$ kHz, and an attenuation of 90 dB by 100 kHz. (This is 90 dB/decade.)

Solution
A five-pole Butterworth would do the job with no ripple in the passband, so at least one pole must be saved to make a 3-dB ripple Chebyshev worthwhile. Recall from Eq. (16.8) that the Chebyshev TRA is

$$\text{TRA}(f) = 20 \log \varepsilon + 6\,(n - 1) + 20n \log(2\pi f)$$

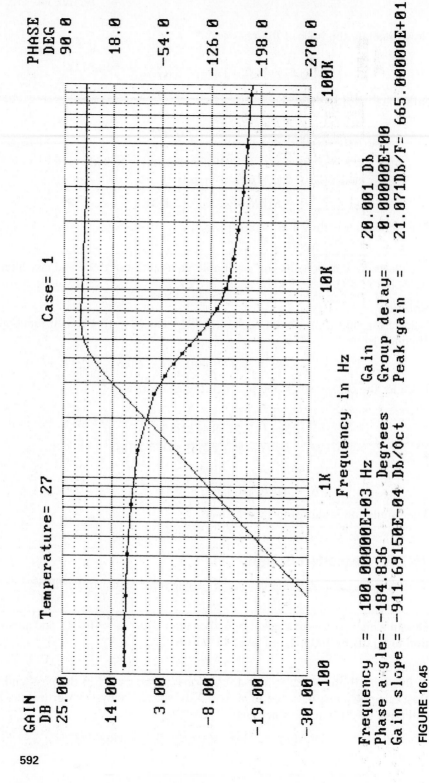

GAIN
DB

25.00

14.00

3.00

-8.00

-19.00

-30.00

Temperature= 27

Case= 1

PHASE
DEG

90.0

18.0

-54.0

-126.0

-198.0

-270.0

100   1K   10K   100K

Frequency in Hz

Frequency   =   100.00000E+03 Hz   Gain   =   20.001 Db
Phase angle= -184.836   Degrees   Group delay=   0.00000E+00
Gain slope = -911.69150E-04 Db/Oct   Peak gain   =   21.071Db/F= 665.00000E+01

FIGURE 16.45
Example 16.17 simulation.

592

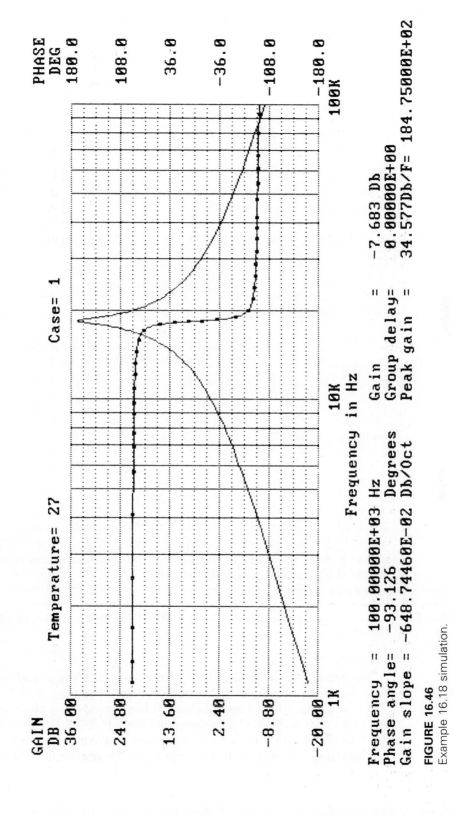

FIGURE 16.46
Example 16.18 simulation.

The TRA for a decade change in frequency is

$$\text{TRA}_{\text{max}} = 20n + 6(n-1) + 20 \log \varepsilon$$

which, evaluated at $n = 4$ and using ε from Table 16.1, is

$$\begin{aligned}\text{TRA}_4 &= (20 \text{ dB/decade})(4) + (6 \text{ dB/decade})(4) + 20 \log (0.99763) \\ &= 80 \text{ dB/decade} + 18 \text{ dB/decade} - 8.868 \text{ dB/decade} \\ &= 97.9 \text{ dB/decade}\end{aligned}$$

which means that a four-pole 3-dB ripple Chebyshev filter, made from two cascaded two-pole stages, will do the job.

From Table 16.1 obtain the α and f_c factor for each stage of the filter: $\alpha_1 = 0.928942$, f_c factor 1 $= 0.442696$, $\alpha_2 = 0.179248$, and f_c factor 2 $= 0.95039$. Thus,

$$f_{c1} = 10 \text{ kHz}/0.442696 = 22.55 \text{ kHz}$$
$$f_{c2} = 10 \text{ kHz}/0.95039 = 10.55 \text{ kHz}$$

Let $C = 0.001 \ \mu\text{F}$. Completing the calculations for stage 1

$$R_{\text{stg1}} = \frac{1}{2\pi f_c C} = \frac{1}{2\pi \, 22.59 \text{ kHz} \, 0.001 \ \mu\text{F}} = 7.05 \text{ k}\Omega$$

Use the nearest 1% resistor, 6.98 kΩ:

$$R_{B\text{stg1}} = (2 - \alpha)R_A = (2 - 0.928942)6.98 \text{ k}\Omega = 7.50 \text{ k}\Omega$$

Similarly, for stage 2,

$$R_{\text{stg2}} = \frac{1}{2\pi \, 10.55 \text{ kHz} \, 0.001 \ \mu\text{F}} = 15.0 \text{ k}\Omega$$
$$R_{B\text{stg2}} = (2 - 0.179248)15.0 \text{ k}\Omega = 27.4 \text{ k}\Omega$$

The simulation is shown in Figure 16.47(a). Gain calculation of the gain with the simple stage gain equation is inaccurate because of the peaking of the individual stages, as shown in Figures 16.47(b) and (c). More complex transfer functions must be solved for accurate gain determination.

EXAMPLE 16.20

Calculate the components for a unity-gain low-pass active filter with a flat passband, 100 dB/decade TRA, no passband ripple, and $f_c = 100$ Hz.

Solution

A five-pole Butterworth filter with two second-order stages and one first-order stage meets the requirements. Normally the first-order stage is first in the circuit. The example was worked using Sallen and Key second-order filters and the simple first-order filter with a real pole shown in Figure 16.27(b). For Butterworth filters, the f_c factor is 1 and the α's for the second and third stages are 1.618034 and 0.618034, respectively. Let all capacitors equal 0.1 μF. For the first stage,

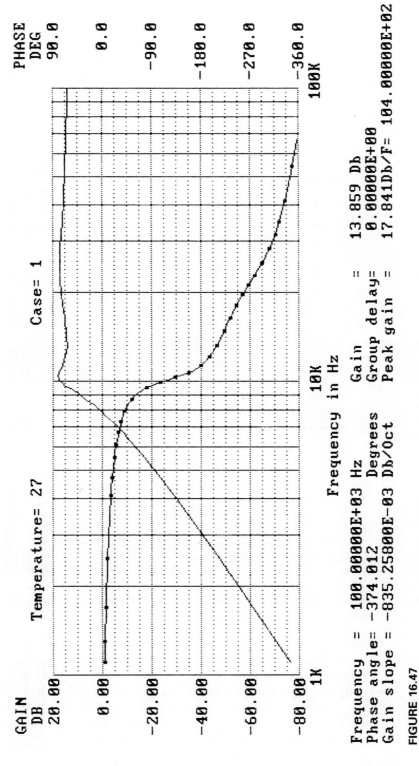

GAIN
DB   Temperature= 27   Case= 1
20.00

PHASE
DEG
90.0

0.00   0.0

-20.00   -90.0

-40.00   -180.0

-60.00   -270.0

-80.00   -360.0
    1K   10K   100K
                  Frequency in Hz

Frequency = 100.00000E+03 Hz   Gain   =   13.859 Db
Phase angle= -374.012   Degrees   Group delay=   0.00000E+00
Gain slope = -835.25800E-03 Db/Oct   Peak gain   = 104.00000E+02

FIGURE 16.47
Example 16.19 simulations: (a) Four pole response. (b) Stage 1 simulation. (c) Stage 2
simulation.

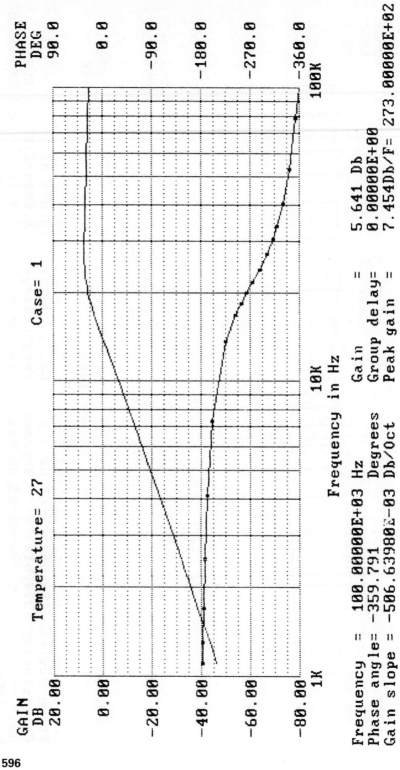

FIGURE 16.47 (continued)

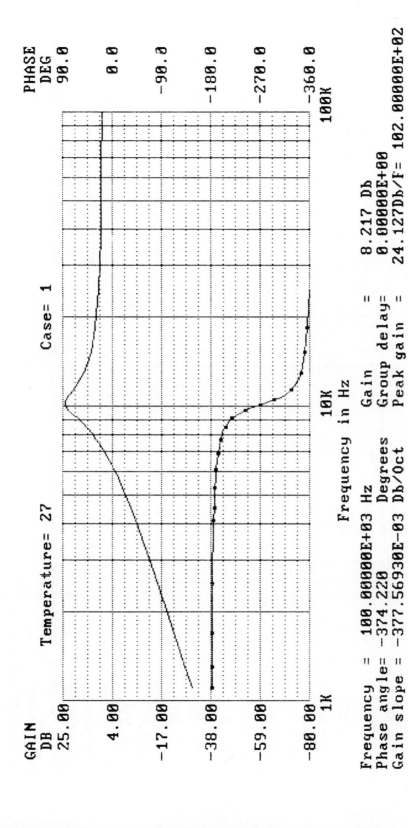

FIGURE 16.47 *(continued)*

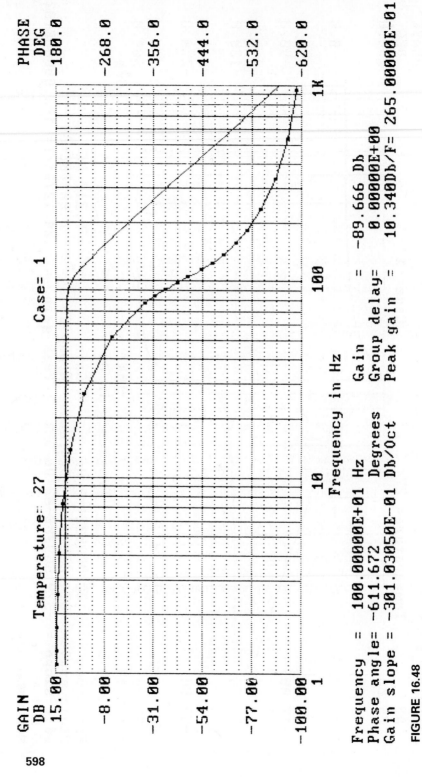

FIGURE 16.48
Example 16.20 simulation.

$$R_1 = R_f = \frac{1}{2\pi \, 100 \text{ Hz } 0.1 \, \mu\text{F}} = 15.8 \text{ k}\Omega$$

and $R_s = R/2 = 7.96$ kΩ.

All of the resistors for the next two stages are the same in value as for the first stage. They use the same equation, except for the values of R_B. Thus, for stage 2,

$$R_{B\text{stg2}} = R_A(2 - \alpha)$$
$$= 15.8 \text{ k}\Omega(2 - 1.618034) = 6.04 \text{ k}\Omega$$

and for stage three

$$R_{B\text{stg3}} = 15.8 \text{ k}\Omega(2 - 0.618034) = 21.5 \text{ k}\Omega$$

The simple gain equations resulted in stage gains of 0, 2.8, and 7.54 dB for stages 1, 2, and 3, respectively, for a total gain of 10.34 dB. The simple equations are accurate because no peaking occurs in the response of any of the filter stages.

The simulation of this filter is shown in Figure 16.48.

16.8 BIBLIOGRAPHY

The short list of books that follows will allow the reader to dig deeper into the fascinating world of filters.

1. Huelsman, L. E., *Basic Circuit Theory with Digital Computations,* Englewood Cliffs, NJ: Prentice Hall, 1972.
2. Johnson, D. E., *Introduction to Filter Theory,* Englewood Cliffs, NJ: Prentice Hall, 1976.
3. Johnson, D. E., and Hilburn, J. L., *Rapid Practical Designs of Active Filters,* New York: John Wiley & Sons, 1975.
4. Lindquist, C. S., *Active Network Design with Signal Filtering Applications,* Long Beach, CA: Steward & Sons, 1977.
5. *Monolithic Filter Handbook, 1990.* Milpitas, CA: Linear Technology, 1990.
6. Oppenheim, A. V., and Willsky, A. S., *Signals and Systems,* Englewood Cliffs, NJ: Prentice Hall, 1983.
7. Tobey, G. E., Graeme, J. G., and Huelsman, L. P., *Operational Amplifiers, Design and Applications,* New York: McGraw-Hill, 1971.

17

MEASUREMENT*

In electricity and electronics, *measurement* refers to finding the value of a variable or component. A variable is a dynamic quantity like current, voltage, and frequency. A component is a static quantity like resistance, capacitance, and inductance. Measurement techniques and instruments are not perfect and so there will be error between an actual value and the value found. It is important to define *how much we trust* the value determined by a measurement.

In this chapter we summarize how to interpret the result of measurements (how good they are) and various techniques and instruments used for making measurements.

17.1 MEASUREMENT INTERPRETATION

Anyone working in electricity and electronics must learn to be critical of measured values. Always interpret the value with respect to what instrument was used to make the measurement. The following terms are used to define measurement.

17.1.1 Measurement Terms

Error and Uncertainty A variable or component has an exact value. The objective of measurement is to try to determine that value. There will virtually always be some difference or *error* between the value determined by measurement and the actual value, even if it is only in the number of digits used to represent the value. This error can never be known, but limits can be placed on the uncertainty. This means that we can know that the measured value is within some range of the actual value.

---

* This chapter was written by Dr. Curtis D. Johnson, College of Technology, University of Houston.

Error limits are usually expressed in percent of either the reading itself or the full-scale (FS) reading possible with the instrument.

EXAMPLE 17.1

A voltmeter measures from 0 to 100 V. The meter measures with an uncertainty of $\pm 3\%$ FS. The meter indicates 45 V. What do we really know about the voltage?

Solution

The value indicated by the meter is uncertain by $\pm 3\%$ of the full scale of 100 V. This is a limit or uncertainty of $(\pm 0.03)(100\ \text{V}) = \pm 3\ \text{V}$. So, our measurement really only tells us that the voltage is between 42 and 48 V!

It is very tempting to ignore the uncertainty and simply assume that the value is, after all, really 45 V. This may cause very serious consequences. Always interpret the measurement.

Sources of Error   Typical sources of measurement error in electricity and electronics include:

1. *Human reading error:* This is primarily of concern with analog instruments. It results from the ability of the human to estimate a value from an analog display. Thus, if a current meter pointer is positioned between 20 and 21 mA, one reader may say it is 20.6 mA and thus closer to 21 mA. Another reader may claim it is 20.4 mA and therefore closer to 20 mA.
2. *Thermal effects:* Many electrical components are affected by temperature, including those used in the measurement instruments themselves. These unpredictable effects can change the apparent value measured from one moment to the next.
3. *Aging effects:* Many electrical components will change value with age. This may mean that there is *actual* change in the value measured in time or that the measurement instrument uncertainty changes in time. In either case it is an unknowable error.
4. *Measurement interference:* Whenever the value of a variable or component is measured, the measurement itself may cause an error. For example, if a voltmeter is connected to measure the voltage between two points in a circuit, its connection may change the voltage. There will then be an added uncertainty (error) in knowledge about the value.

17.1.2   Definitions

To interpret measurements properly it is necessary to understand the meaning of words used to describe the measurement.

Accuracy   Accuracy expresses the *total* uncertainty of a measurement. It is often expressed in percent of full scale for analog instruments, as in Example 17.1.

Digital measurement instruments often express the accuracy as a percent of reading with an added uncertainty of the least significant digits of the readout.

EXAMPLE 17.2

An ac digital voltmeter with a four-digit readout measures 1-kHz signals with an accuracy of

$$\pm(0.75\% \text{ of reading} + 2 \text{ digits}).$$

What is the uncertainty of a reading of 12.33 Vac?

Solution
From the stated accuracy, we have

$$\pm[(0.0075)(12.33) + 0.02] = \pm0.11$$

The 0.02 V is added because each digit represents 0.01 V. From this we conclude the voltage actually lies in the range of 12.22 to 12.44 Vac.

Avoid the tendency to take a digital readout as the absolute truth.

Precision The precision of a measurement instrument defines the effect of *random* errors, which may cause reading-to-reading variations. An instrument with precision will provide very nearly the same value with repeated measurements of the same quantity. Precision is often stated in terms of the statistical *standard deviation* resulting from many measurements of the same quantity. Note that even though an instrument with precision gives a very repeatable value, there may still be significant error and uncertainty between that measured value and the actual value. Precision places a limit on accuracy.

Resolution The smallest change in value that can be detected by a measurement is called the *resolution*. In analog instruments this is hard to define. It may be a matter of opinion on how finely the position of a pointer between scribe marks can be described.

For digital instruments the resolution is very well defined as the value change that will cause the least significant digit to change. A four-digit voltmeter measuring 0 to 199.9 mV can obviously resolve 0.1 mV. This does *not* mean that it is accurate to 0.1 mV! In Example 17.2, the instrument could resolve 0.01 Vac, but the reading accuracy was shown to be ±0.11 Vac.

Significant Figures Interpretation of the results of a measurement also depends on the number of digits that can be justified to represent the number.

Suppose a four-digit digital voltmeter indicates a measurement of 10.45 V. We *cannot* say that the voltage is 10.45000000 V. Apart from accuracy we still only know the reading is not 10.44 and it is not 10.46 but it could be any value in between. Significant figures tell us how many digits can be reliably reported for the measurement. In digital instruments it is simply the number of digits.

For analog instruments the number of significant figures is, as with resolution, hard to define. Suppose an analog meter pointer reads between lines scribed 8 and 9 mA as shown in Figure 17.1. To one significant figure, as the pointer seems closer to 8, it would be reported as 8 mA, but not 8.0 mA because that would be two significant figures. We might estimate that the pointer was a quarter of the way and report 8.25 mA, but that would be three significant figures, which is certainly beyond real readability. As you can see, there is considerable uncertainty in the last digit.

FIGURE 17.1
Estimating analog meter significant figures.

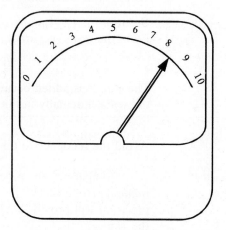

Note that the meter could be very accurate, but the problem of estimating values between scribed lines remains. In many cases a company or the government will simply define what resolution is possible (acceptable) with an analog instrument. Often it is the "closest scribed value" and would thus be 8 mA for Figure 17.1.

The number of significant figures plays an important role in how the results of a measurement may be used. It is the issue again of how a measurement is interpreted, as the following example shows.

EXAMPLE 17.3

A three-digit multimeter is used to measure the current through a resistor and the voltage across the resistor. Assuming the instrument has no error, what is the resistance if the current is 6.01 mA and the voltage is 10.5 V?

Solution
From Ohm's law we have,

$$R = V/I = 10.5/0.00601 \ \Omega$$

A 10-digit calculator gives the result as,

$$R = 1747.088186 \ \Omega$$

But because the number of significant figures of the measurement is only three digits, we can *only* report the result to three digits. Thus, the result is reported as 1750 Ω (rounding up).

Calibration   The accuracy of a measurement instrument is not static in time. Aging effects, mentioned above, can cause the accuracy of an instrument to degrade over time. This new uncertainty is not predictable. An uncalibrated instrument is only good for indications of approximate value.

This situation can be remedied by periodic calibration of the instrument. Calibration is accomplished by measuring well-known standard values and adjusting the instrument response to correspond to the known values. Accuracy can thereby be returned to the original specification.

It is possible to improve the manufacturer's reported accuracy of an instrument by calibration. A manufactured line of meters may be sold as ±3% FS instruments, meaning that all of them will have an accuracy within that limit. A *particular* meter can be calibrated by comparison to known standards and found to be accurate to ±1% FS. Henceforth, that meter could be used for measurements to that accuracy.

Hysteresis   For some measurement instruments, the accuracy depends on how the measurement is approached. This is an error that is different depending on whether the value to be measured is above or below the present instrument reading. For example, suppose the signal input to a frequency meter drops from about 20 to 10.0 kHz. The instrument may indicate a frequency of 10.4 kHz for an error of +4%. Now suppose the measurement process is repeated but for the input signal frequency rising from about 5 to 10.0 kHz. The measured value may now show a value of 9.7 kHz for a −3% error. So there is an error whose value depends on how the measurement is approached. This is the effect of hysteresis in the instrument.

17.1.3   Statistics in Measurement

In most cases of electrical and electronic measurement we are interested in making a single measurement. In that case, as described in the previous section, an interpretation of the measurement depends on the accuracy of the instrument.

If many measurements can be made of the variable or component value, it is possible to use statistics to obtain more information than the accuracy alone can provide. The two most common statistical measures are the average value, or *arithmetic mean,* and the *standard deviation.*

Limitations   Statistical analysis can only improve knowledge of a value whose measurement variations are due to *random* effects. Statistics cannot improve on fixed bias errors. For example, suppose a manufacturer produces meters with an accuracy of ±3% FS. This means that any meter picked from the stock will have an accuracy within that specification. However, the actual accuracy of a particular meter will be some fixed bias. Thus, a particular meter may have an accuracy of −2.3% FS, and another +1.6% FS, and so on. Of course, we will not know what the particular bias accuracy is, only that it is within ±3% FS. Statistics cannot improve on the accuracy of a particular meter with such a fixed bias. Statistics can only improve on knowledge for errors that are random. Thus, if repeated measurements of a variable give a range of readings, then statistics can help us "see through" the randomness to get a better estimate of the actual value, but still limited by the basic accuracy of the instrument.

Arithmetic Mean   The arithmetic mean is simply the average of a set of measurements of some variable or component value. The mean tends to eliminate random error effects on knowledge of the value, i.e., it provides knowledge about the "most likely" value. The arithmetic mean is

$$\langle v \rangle = \frac{v_1 + v_2 + \cdots + v_N}{N} \tag{17.1}$$

where $v_1, v_2, \ldots, v_N$ = the measured values
$\langle v \rangle$ = the arithmetic mean or average
N = the number of measurements.

EXAMPLE 17.4

To determine the value of a frequency, 15 measurements are made. The results, all in kilohertz, are 1.23, 1.28, 1.25, 1.27, 1.26, 1.20, 1.25, 1.24, 1.28, 1.25, 1.23, 1.25, 1.26, 1.24, and 1.24. The instrument has an accuracy of ±1% of reading. What is the arithmetic mean of the frequency and the uncertainty in kHz?

Solution
The arithmetic mean is determined from Eq. (17.1):

$$\langle f \rangle = \frac{18.73}{15} = 1.248666667 \text{ kHz}$$

But, of course, since the readings are only to three significant figures we can report a result of $\langle f \rangle$ = 1.25 kHz. The uncertainty due to instrument accuracy is ±1% of 1.25 kHz or ±0.01 kHz. The frequency is then 1.25 ± 0.01 kHz. Random variations were from −0.05 to +0.03 kHz, so you can see that the average has improved our information about the most likely frequency even though accuracy limitations remain.

Standard Deviation   Another factor of importance when dealing with random error is the spread of values about the mean. Consider the diagrams in Figure 17.2, which show the distribution of voltage readings for two different circuits. Both of these sets of data *have the same arithmetic mean!* The second set of data is much more spread out than the first and so our confidence that the actual value is the

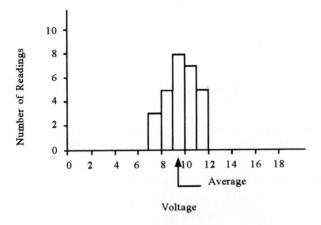

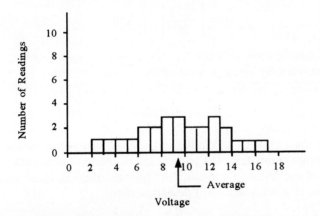

FIGURE 17.2
Sets of readings with identical means but different spreads.

mean is shaken. Also our confidence that another reading will give a value anywhere near the mean in the second case would be low, whereas in the first case it would be high. This kind of information can be described by the standard deviation, which is the root-mean-square (rms) value of the deviations of the readings from the arithmetic mean. It is defined by the following equation:

$$\sigma = \left[\frac{(v_1 - \langle v \rangle)^2 + (v_2 - \langle v \rangle)^2 + \cdot \cdot \cdot + (v_N - \langle v \rangle)^2}{N} \right]^{1/2} \qquad \textbf{(17.2)}$$

where σ = the standard deviation, $\langle v \rangle$ = the arithmetic mean, and the other terms were defined for Eq. (17.1). The standard deviation will be large when the spread of data is large and small when the data are quite peaked.

EXAMPLE 17.5

For the data of Figure 17.2, find the arithmetic mean and standard deviation.

Solution

The arithmetic mean for the two cases is found using Eq. (17.1) and the data shown in Figure 17.2:

$$\langle V \rangle_a = \frac{286}{28} = 10.2$$

$$\langle V \rangle_b = \frac{267}{26} = 10.2$$

From the point of view of the arithmetic mean, the two sets of data seem the same. The standard deviation is found from Eq. (17.2), using the mean calculated earlier:

$$\sigma_a = \left[\frac{42.71}{28} \right]^{1/2} = 1.24$$

$$\sigma_b = \left[\frac{332.6}{26} \right]^{1/2} = 3.58$$

These numbers clearly show that the "a" set of data is much less spread since its standard deviation is only about a third of that of the "b" set.

There is an added statistical interpretation of the standard deviation that is valid when there are many samples, perhaps more than 30. It is an advanced topic in statistics that must be very carefully used so that incorrect conclusions are not drawn. This concept estimates what percentage of the measurements fall within a specified range of the arithmetic mean. For your information, these ranges can be stated as follows. If σ is the standard deviation then:

50.0% of the measurements fall within $\pm 0.675\sigma$ of the mean

68.3% of the measurements fall within $\pm\sigma$ of the mean

95.5% of the measurements fall within $\pm 2\sigma$ of the mean

99.7% of the measurements fall within $\pm 3\sigma$ of the mean.

17.2 ELECTRICAL VARIABLE MEASUREMENT

Measurement of electrical variables means the measurement of ac and dc current, voltage, and power and ac frequency. Although analog instruments are still employed in many cases, most of these measurements are now made using digital instruments.

17.2.1 Current

Current is measured by inserting the current measurement instrument in series with the line through which the desired current is flowing. It is possible to measure the current without breaking the circuit by measuring the magnetic field generated by the flow of current, but this method is far less common. The following sections describe the common instruments and methods for current measurement.

Moving-Coil Meter The most common analog technique for measurement of dc current is based on interaction between a permanent magnetic field and the magnetic field developed as a current passes through a coil of wire. The instrument is called the D'Arsonval or Weston movement.

Figure 17.3 shows the basic structure of the instrument. A coil is mounted on bearings so that it can rotate, but this rotation is resisted by spiral springs. The current to be measured passes through the coil, which produces a magnetic field. This field interacts with the field of the permanent magnet and, like a motor, produces a torque, which causes the coil assembly and attached pointer to rotate until the opposing force of the spring prevents further rotation. The degrees of rotation of the coil are proportional to the current.

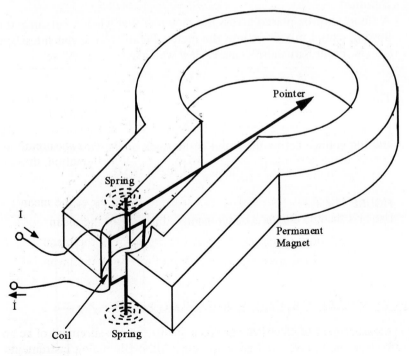

FIGURE 17.3
Basic structure of a moving-coil meter.

The basic movement has some *full-scale (FS) current,* which is the current that will cause the pointer to move to the maximum position. Most current meters are designed to measure from zero to the maximum and must be connected with the correct polarity to drive the pointer up-scale. The *galvanometer* is designed so that the pointer is at the center with no current and can measure bipolar current, i.e., current flow in either direction. Standard FS currents, also called *movements,* are 20 μA, 50 μA, 100 μA, 1 mA, and 5 mA.

The coil of wire has electrical *resistance.* In some cases an additional internal resistance is also placed in series with the coil. It is called the *calibrating* or *swamping resistance.* The swamping resistor is used so that a specific voltage drop (often 50 mV) across the meter will produce FS current deflection. For the standard ranges listed earlier, this would mean net meter resistances of 2500, 1000, 500, 50, and 10 Ω, respectively.

Change of Full Scale   A current meter of any scale can be used to measure a larger current by placing a *shunt resistor* across the meter. This effectively makes the instrument act like a meter with a higher FS current.

EXAMPLE 17.6

How can a 1.0-mA FS movement current meter with a 50-Ω resistance be used to measure up to 100 mA?

Solution

A shunt must be placed across the meter so that when a net current of 100 mA is flowing, only 1 mA flows into the meter. Figure 17.4 shows the schematic for this system. From Kirchhoff's current law we have,

$$I = I_S + I_M$$

Thus,

$$I_S = 100 - 1 = 99 \text{ mA}$$

and the voltage across the shunt is the same as the voltage across the meter:

$$V = I_M R_M = (1 \text{ mA})(50 \ \Omega) = 50 \text{ mV}$$

FIGURE 17.4
Use of a shunt to change a meter range.

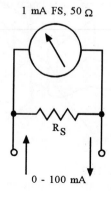

1 mA FS, 50 Ω

R_S

0 - 100 mA

Then from Ohm's law,

$$R_S = V/I_S = (50 \text{ mV})/(99 \text{ mA}) = 0.51 \ \Omega$$

Measurement of Meter Resistance   The internal resistance of a meter movement cannot be determined by simply using an ohmmeter. This is because the ohmmeter passes a current through the resistance to be measured (a meter in this case), which may *exceed* the FS meter current and meter damage may result.

One way to measure meter resistance without damage is shown in Figure 17.5. The procedure is as follows:

1. With S1 open, R is varied until the meter, M, reads FS. The net current, I, is noted on meter M1.
2. S1 is closed and R_S is varied until meter M reads one-half of FS.
3. R is varied to bring the net current back to that noted in step 1. It may be necessary to alternately adjust R_S and R until M1 reads the noted current and M is at one-half of FS.
4. The value of R_S is now equal to the internal resistance of the meter, M. The circuit can be disconnected and R_S measured with a standard ohmmeter.

The maximum value of R_S must be greater than the expected meter resistance. R and V can be any convenient values to assure that the FS current can be produced. For example, R can be selected so that at one-half its maximum the current is one-half FS:

$$R = 2V/I_{FS} \tag{17.3}$$

Alternating Current   It is very important in making ac measurements to account for *frequency* dependence and *waveform* dependence of the measuring instrument. AC instruments will have an accuracy that depends on frequency, becoming less accurate as the frequency increases. There will be some frequency above which they will not work at all.

FIGURE 17.5
Measuring the resistance of a meter.

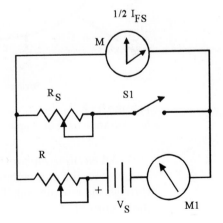

In general, alternating current is expressed as the *rms value* of the current, assuming a sinusoidal waveform. If some other waveform is being measured, a conversion factor will be necessary to convert the readout to the appropriate rms value. In most cases the actual instrument measures the average of the absolute current. This is the same as the average of the full-wave rectified current. The meter display is then marked off in rms voltage. For sinusoidal waveforms the scale factor between average and rms current is $2\sqrt{2}/\pi$, i.e., you multiply the rms meter reading by this quantity to find the average current. When this meter is used to measure waveforms other than sinusoidal a correction factor must be multiplied times the reading. These scale factors are defined in Table 17.1 in the section on ac voltage measurement, although they can be used for current as well.

The following are two instruments commonly used for analog ac measurements.

The *moving-iron movement*, shown in Figure 17.6, can be used for ac. Here the ac is fed through a coil or solenoid. The resulting oscillating magnetic field induces magnetic fields in the soft iron vanes. A repelling force results, which drives the vanes apart until matched by the spiral spring. A common ac movement has a 0- to 5-A range. Shunts can be used, as with the moving coil, to change the scale.

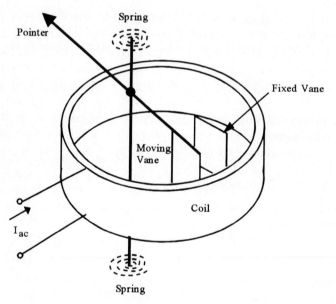

FIGURE 17.6
Moving vane ac meter movement.

An *electrodynamometer movement,* shown in Figure 17.7, uses a moving coil but replaces the permanent magnet by coils excited by the same current that passes through the moving coil. Thus this instrument will respond to either alternating or direct current.

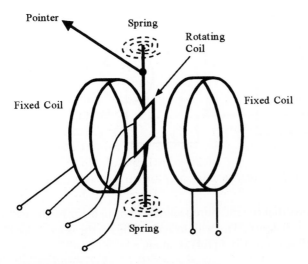

FIGURE 17.7
Electrodynamometer meter movement.

Digital Measurement   The measurement of current by digital instruments typically involves using a digital voltmeter to measure the voltage dropped across a resistor. For ac measurements, the voltage drop is rectified and then measured as a dc voltage. See the discussion of digital voltmeters for more information on how the measurement is made.

17.2.2   Voltage

Voltage measures the potential difference between two points in a circuit. A *floating-ground* voltmeter can be connected to any part of a circuit, although polarity and range must be correct to get a reading. A *fixed-ground* voltmeter can measure only with respect to the common ground. Attempting to connect the meter ground to another part of the circuit could cause a short circuit. Hand-held, battery-operated meters are always floating-ground.

Loading Error   It is possible to introduce a measurement error because of the internal resistance of the voltmeter. Figure 17.8 shows the basic elements of a voltage measurement. The source has been *modeled* by Thévenin's theorem as a voltage to be measured, V_S, and a series resistance, R_S. The voltmeter can be modeled as simply an internal resistance, R_m. The value of V_m is the voltage from a to b in Figure 17.8, with the meter connected. This can be found from Ohm's law:

$$V_m = V_S - \frac{R_S}{R_m + R_S} V_S \qquad \textbf{(17.4)}$$

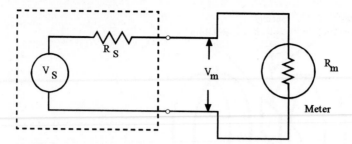

FIGURE 17.8
Loading error due to meter resistance.

This equation shows that the measured value will be in error by the voltage dropped across the source resistance. The way to reduce this error is to make R_m very large so that the second term in Eq. (17.4) is small.

EXAMPLE 17.7

A voltmeter has a 10-kΩ resistance and measures from 0 to 10 V. It is connected to a voltage source with a 2200-Ω source resistance. What percent loading error will result?

Solution
The error is found by dividing the error voltage in Eq. (17.4) by the voltage measured. This gives

$$\text{Error}\ (\%) = \frac{R_S}{R_m + R_S}\ 100 \tag{17.5}$$

So, in this case we get,

$$\text{Error}\ (\%) = [2200/(10{,}000 + 2200)]100 = 18\%!$$

In this case you can simply add 18% back to the reading but in most cases the source resistance is not known. The general conclusion: Make R_m as large as possible.

Analog Measurement   Current meter movements, discussed in the previous section, are used to make voltage measurements. A *multiplier resistor* is used in series with the meter so that the current is kept within the range of the meter, as shown in Figure 17.9.

The value of the multiplier resistance is chosen so that the maximum voltage to be measured results in the FS current of the meter, i.e.,

$$I_{FS} = \frac{V_S}{R_m + R_{MX}} \tag{17.6}$$

where I_{FS} = meter full-scale current
V_S = maximum source voltage (to be measured)

FIGURE 17.9
Voltage measurement with a current meter.

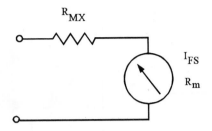

R_{MX} = multiplier resistance
R_m = meter movement resistance.

Solving Eq. (17.6) for R_{MX} gives

$$R_{MX} = \frac{V_S}{I_{FS}} - R_m \qquad (17.7)$$

EXAMPLE 17.8

How can a 50-μA meter movement with a 1000-Ω internal resistance be used to measure 0 to 20 V?

Solution
We need a series multiplier resistance, R_{MX}, so that when V_S is 20 V the current is 50 μA. From Eq. (17.7),

$$R_{MX} = \frac{20 \text{ V}}{50 \text{ } \mu\text{A}} - 1 \text{ k}\Omega = 399 \text{ k}\Omega$$

Voltmeter Sensitivity The sensitivity of a voltmeter is found by dividing the FS voltage by the resistance of the meter. This gives the ohms per volt (Ω/V) of the instrument. The larger this number, the less loading effect the instrument will have on the measurement. For Example 17.8, the sensitivity would be 400 kΩ/ 20 V = 20,000 Ω/V. By contrast, the meter of Example 17.7, which was a 1-mA movement, has a sensitivity of only 1000 Ω/V.

Electronic Voltmeters Electronic voltmeters use electronic circuits to detect the voltage to be measured. This gives them very large sensitivity. The effective resistance of the meter can range from 1 to 100 MΩ so that loading effects are very small. The first versions used vacuum tubes for the circuit and were called vacuum tube volt meters (VTVMs) but modern versions use field-effect transistors and other solid-state circuits.

AC Voltage Measurement AC voltage cannot be measured directly by moving-coil current movements. However, if a rectifier circuit converts the ac voltage to dc then such meters can be used. Often a full-wave bridge rectifier is used for this purpose as shown in Figure 17.10. The meter is marked for rms voltage.

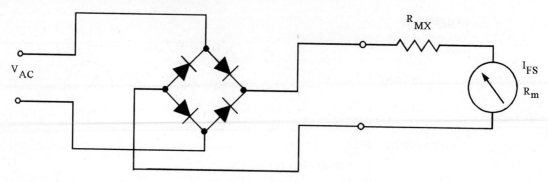

FIGURE 17.10
Use of full-wave bridge to measure ac voltage.

The moving-iron and electrodynamometer movements discussed in the previous sections can also be used for ac voltage measurements. Again, a multiplier resistance is used for a scale shift so that various voltage ranges can be provided.

Most voltmeters, both analog and digital, actually measure the average voltage. If the signal is periodic, this is the average voltage over one cycle, and if aperiodic it is simply the average over the measurement period. The average and rms values are not the same. Thus, for the readout to be in rms, which is desirable, the average is multiplied by an average-to-rms conversion factor. The conversion factor assumes that the ac input voltage is sinusoidal. For such a waveform the conversion factor is $\pi/\sqrt{8} \approx 1.11$.

If the waveform is not sinusoidal, other conversion factors must be used to convert the reading to the rms value appropriate for that waveform. Table 17.1 shows conversion factors for a variety of waveforms which allow determination of the rms, average, peak, and peak-to-peak voltages from the meter readout. This table is good for any meter, analog or digital, which measures the *average* voltage or current and displays the value as the rms for a sinusoid.

EXAMPLE 17.9

Suppose a voltmeter readout, V_{RO}, is 13.5 V rms. What is the rms voltage and peak-to-peak voltage if the input waveform is (a) a sawtooth or (b) a 30% duty cycle square wave?

Solution

The conversion factors of Table 17.1 must be used to multiply the 13.5-Vac reading. (a) For the sawtooth case the answers are

$$V_{rms} = \frac{4}{\pi}\sqrt{\frac{2}{3}}\,V_{RO} = (1.0396)(13.5) = 14.0\ V_{rms}$$

$$V_{pp} = \frac{8\sqrt{2}}{\pi}\,V_{RO} = (3.601)(13.5) = 48.6\ V_{pp}$$

TABLE 17.1
AC Voltage and current conversion factors

Waveform	RMS	Absolute Average	Peak	Peak-to-Peak
1. Sinusoid	1	$\dfrac{2\sqrt{2}}{\pi}$	$\sqrt{2}$	$2\sqrt{2}$
2. Half-wave rectified	$\sqrt{2}$	$\dfrac{2\sqrt{2}}{\pi}$	$2\sqrt{2}$	$2\sqrt{2}$
3. Full-wave rectified	1	$\dfrac{2\sqrt{2}}{\pi}$	$\sqrt{2}$	$\sqrt{2}$
4. Square, $\pm$ symmetry	$\dfrac{2\sqrt{2}}{\pi}$	$\dfrac{2\sqrt{2}}{\pi}$	$\dfrac{2\sqrt{2}}{\pi}$	$\dfrac{4\sqrt{2}}{\pi}$
5. Square, + only	$\dfrac{4}{\pi}$	$\dfrac{2\sqrt{2}}{\pi}$	$\dfrac{4\sqrt{2}}{\pi}$	$\dfrac{4\sqrt{2}}{\pi}$
6. Pulse, $D\%$, ON duty cycle	$\dfrac{\sqrt{2D}}{5\pi}$	$\dfrac{2\sqrt{2}}{\pi}$	$\dfrac{200\sqrt{2}}{\pi D}$	$\dfrac{200\sqrt{2}}{\pi D}$
7. Triangle or sawtooth	$\dfrac{4}{\pi}\sqrt{\dfrac{2}{3}}$	$\dfrac{2\sqrt{2}}{\pi}$	$\dfrac{4\sqrt{2}}{\pi}$	$\dfrac{8\sqrt{2}}{\pi}$

(b) For the 30% duty cycle square wave the answers are

$$V_{rms} = \frac{\sqrt{2D}}{5\pi}\, V_{RO} = (0.4931)(13.5) = 6.66\ V_{rms}$$

$$V_{pp} = \frac{200\sqrt{2}}{\pi D}\, V_{RO} = (3.001)(13.5) = 40.5\ V_{pp}$$

There will be a limitation on the frequency range over which the meter will be accurate, with accuracy falling off for increasing frequency.

Digital Voltmeters A digital voltmeter (DVM) uses an analog-to-digital converter (ADC) to convert a voltage into a multibit digital signal. The digital signal is processed and displayed on a multidigit readout. Figure 17.11 shows the essential features of a digital voltmeter. The input resistance is commonly 10 MΩ.

An input circuit scales the input voltage so that the resulting voltage to the ADC is within the proper range. An ac voltage is converted to a proportional dc voltage by an electronic rectifier circuit.

The ADC produces a digital output proportional to the magnitude of the input voltage. This process takes some time, which is controlled by a clock and timing circuit.

The ADC output is captured by a latch and provided to the display unit. Every time the ADC completes another conversion of input voltage to digital signal,

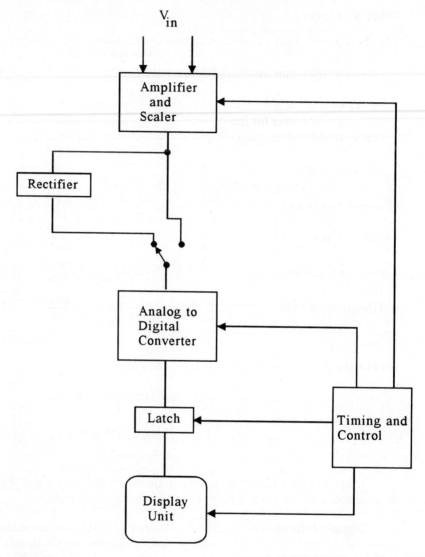

FIGURE 17.11
Basic features of a digital voltmeter.

the latch and display are updated. A typical measurement cycle, from update to update, takes about 0.5 s.

The DVM is also used to measure dc or ac by measuring the voltage dropped across an internal precision resistor and scaling the readout appropriately.

There is a lower and an upper limit to the ac frequency for which the unit can be used. For typical DVMs this is 45 Hz to 1 kHz. The conversion factors of Table 17.1 also apply for the DVM.

17.2.3 Power

Electrical power determines the rate at which energy is being consumed or delivered by an electrical and electronic circuit. The proper unit of this power is the watt (W), which represents an energy delivery of one joule per second (J/s).

DC Power Consider some electrical element or circuit, as shown in Figure 17.12. Electrical power for this object is found by a product of the voltage measured between the two wires and the current into and out of the wires:

$$p = IV \qquad\qquad (17.8)$$

If power is being *consumed* by this object, then the current will be directed into the wire which is more positive, as shown in Figure 17.12(a). If power is being *generated,* then current will be directed out of the wire which is more positive, as shown in Figure 17.12(b).

FIGURE 17.12
Power is consumed by a sink and delivered from a source.

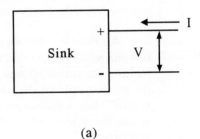

(a)

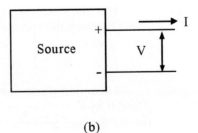

(b)

DC power measurement usually involves separate measurement of the voltage and current.

EXAMPLE 17.10 A transistor radio is found to draw 97 mA from a 9.0-V battery. What power is consumed by the radio?

Solution

From Eq. (17.8) the power is,

$$p = (97 \text{ mA})(9.0 \text{ V}) = 870 \text{ mW}$$

EXAMPLE 17.11

A solar cell is loaded by a resistor. The current through the resistor is found to be 178 mA and the voltage across the resistor is measured to be 0.566 V. What power is the solar cell generating?

Solution

The power generated by the cell is equal to that consumed by the resistor. Therefore, from Eq. (17.8),

$$p = (178 \text{ mA})(0.566)\text{V} = 0.100 \text{ mW}$$

It is possible to use the electrodynamometer shown in Figure 17.7 to measure dc power directly. To do this the moving coil is connected across the load (voltage sensitive) and the fixed coils are connected in series with the load (current sensitive).

AC Power   The measurement of ac power is the same as dc, i.e., a product of voltage across an object and current through the object. The problem here is that the current and power may not be in phase. If they were exactly in phase then the power would be the product. If they were exactly out of phase then the power is zero! Thus, a measurement of ac power requires measurement of the ac voltage across the object, the ac through the object, *and* the phase between the voltage and current.

The simple product of current and voltage is called the *apparent power, p_a*:

$$p_a = iv \qquad\qquad \textbf{(17.9)}$$

The *true power* is given by multiplying apparent power by the cosine of the phase between the voltage and current. The cosine of the phase is also called the *power factor.*

$$p = p_a \cos(\phi) = iv \cos(\phi) \qquad\qquad \textbf{(17.10)}$$

The out-of-phase part of the apparent power is called the *reactive power.* It is found by multiplying the apparent power by the sine of the phase angle between the voltage and current:

$$p_r = iv \sin(\phi) \qquad\qquad \textbf{(17.11)}$$

Although reactive power still has units of watts, it is referred to by the unit name VAR (for volt-amperes-reactive) to distinguish it from real power.

EXAMPLE 17.12

The ac voltage across a load is found to be 56 V and the current is 3.4 A. An oscilloscope shows that there is a 42° phase shift. What are the apparent, true, and reactive powers?

Solution
From Eq. (17.9) we find,

$$p_a = (3.4)(56 \text{ V}) = 190 \text{ W}$$

The true power is found from Eq. (17.10) as

$$p = 190 \cos(42°) = 140 \text{ W}$$

The reactive power is given from Eq. (17.11),

$$p_r = (190) \sin(42°) = 127 \text{ VAR}$$

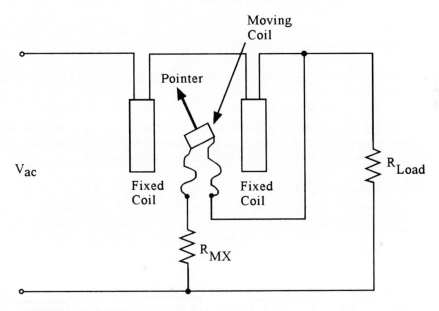

FIGURE 17.13
Use of an electrodynamometer to measure ac power.

A *wattmeter* can be constructed from an electrodynamometer as shown in Figure 17.13. Notice that the moving coil responds to the voltage across the load, whereas the stationary coils respond to the current through the load. This measures true power because only the in-phase components of current and voltage will produce torque to cause coil and pointer motion. The multiplier resistor is used for range calibration of the meter.

17.2.4 Frequency

Frequency is measured as the number of oscillations per second of some periodic signal. The unit for measurement is the hertz (Hz), which represents one oscillation per second.

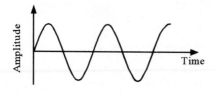

FIGURE 17.14
A pure sine wave is a simple periodic time signal

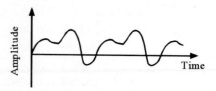

FIGURE 17.15
A complex time signal.

Measurement Interpretation   If a signal has a pure frequency, such as the perfect sinusoidal oscillation shown in Figure 17.14, then a measurement of frequency simply gives the number of cycles per second of the oscillation in hertz. The instrument which performs this measurement is called a *frequency meter.*

Suppose the signal were more complicated, such as that shown in Figure 17.15. A simple measurement by a frequency meter would most likely give as the result only the lowest or *fundamental frequency* determined by the zero crossings of the signal. All of the superimposed higher frequency oscillations would not be determined. Another instrument, the *spectrum analyzer,* is used to display all of the frequencies that make up such a complex signal. The resulting measurement of frequency content is called the *spectrum* of the signal.

Frequency Meter   Measurement of fundamental frequency is accomplished most commonly by a counter circuit such as that diagrammed in Figure 17.16. Here a high-gain amplifier and limiter circuit convert the incoming signal to pulses marking the zero crossings of the signal. These pulses are counted for a fixed time interval and the resulting count is digitally displayed as the equivalent "counts per second" or frequency in hertz.

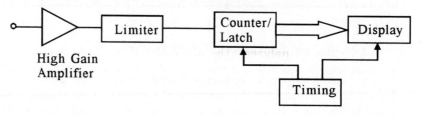

FIGURE 17.16
A frequency meter based on zero crossings.

The display is held by the latch while the next counting interval occurs, and then the display is updated. In this way the display always shows the latest count.

Selection of the time interval required for counting, also called the *gating time,* depends on the range of measurement frequency. A low frequency requires more time to accumulate sufficient counts for a meaningful display.

The presence of noise or superimposed higher frequencies can cause error in the frequency displayed by the frequency meter. For example, the signal in Figure

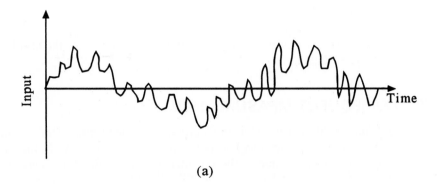

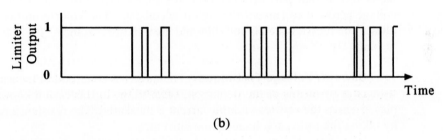

FIGURE 17.17
Using a zero-crossing frequency meter to measure a complex signal.

17.17(a), after the high-gain ac amplifier and pulse forming circuit would provide an input to the counter as shown in Figure 17.17(b). A count will include these erroneous extra pulses and thus not give the correct fundamental frequency of the input signal.

Spectrum Analyzer The alternative to a simple fundamental frequency measurement is to measure the spectrum of the signal. A spectrum analyzer usually presents the results on a CRT, such as that shown in Figure 17.18. Note that the

FIGURE 17.18
Basic display of a spectrum analyzer.

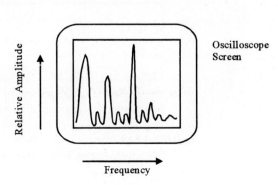

horizontal axis is the frequency, and the vertical is the amplitude as a function of frequency. The amplitude is usually a relative quantity, i.e., it is used just to show the distribution of frequency amplitudes and not to give the absolute amplitude of a given frequency.

17.3 COMPONENT MEASUREMENT

Component measurement in electricity and electronics refers to finding the value of resistance, capacitance, or inductance by measurement. In the case of combined components the value of the impedance is determined by measurement.

17.3.1 Resistance

Resistance is that part of a component's characteristic for which the ratio of the voltage across it to current through it is constant. The voltage and current are in phase. The ratio is called Ohm's law and the unit of resistance is the volt per ampere or ohm (Ω).

Current/Voltage Technique The most basic method of measuring the resistance is by means of the definition, Ohm's law. In this case a known voltage is placed across the resistor and the current is measured. The resistance is then found by taking the ratio of voltage to the current.

$$R = v/i \tag{17.12}$$

Ohmmeter It is possible to construct a meter for direct measurement of resistance by using the basic current movement presented in Section 17.2.1. The most common method is shown in Figure 17.19. The measurement circuit is not dependent on exact knowledge of the voltage source, V. The measurement is made in the following way:

1. The measurement terminals, a and b, are shorted, i.e., R_x is set equal to zero.
2. R_2 is adjusted until the meter reads the maximum FS current. This establishes the zero resistance point as the FS reading.

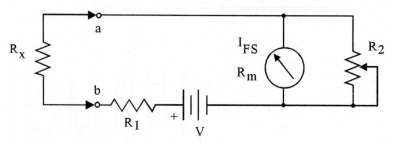

FIGURE 17.19
Use of a current meter to measure resistance.

3. Now as R_x varies from 0 Ω to ∞ (infinity) Ω the meter will vary from FS down to zero.
4. The meter can be recalibrated for variations of the source voltage V by adjusting R_2 to give I_{FS} when $R_x = 0$. There is some error introduced by this procedure as shown later by Example 17.14.

The purpose of R_1 is to define how the resistance is spread between zero and FS. Usually it is selected to define the resistance for one-half FS.

The meter scale in resistance is nonlinear, being much compressed at the lower currents. The instrument is usually designed so that the resistances of interest are in the top half of the scale (from current of half FS to FS). An ohmmeter for which an unknown of R_0 will result in a current of half the FS current requires

$$R_1 = R_0 - \frac{I_{FS} R_m R_0}{V} \tag{17.13}$$

and

$$R_2 = \frac{I_{FS} R_m R_0}{V - I_{FS} R_0} \tag{17.14}$$

where
I_{FS} = the meter full-scale current
R_m = the meter internal resistance
R_0 = the unknown for one-half I_{FS}
V = the ohmmeter supply voltage.

EXAMPLE 17.13

An ohmmeter will use a 50-μA movement with a 3000-Ω internal resistance. The resistance for half scale is to be 10 kΩ and a 3-V battery will be used for a supply voltage. What values of R_1 and R_2 should be used?

Solution
The values of R_1 and R_2 will be found from Eqs. (17.13) and (17.14):

$$R_1 = 10^4\,\Omega - \frac{(50 \times 10^{-6}\,\text{A})(3 \times 10^3\,\Omega)(10^4\,\Omega)}{3\,\text{V}} = 9.5\,\text{k}\Omega$$

and

$$R_2 = \frac{(50 \times 10^{-6}\,\text{A})(3 \times 10^3\,\Omega)(10 \times 10^3\,\Omega)}{3\,\text{V} - (50 \times 10^{-6})(10 \times 10^3\,\Omega)} = 600\,\Omega$$

We use a 1-kΩ pot for R_2 so that it can be adjusted to calibrate the meter as the battery voltage changes.

EXAMPLE 17.14

Suppose the supply in Example 17.13 drops to 2.7 V, a 10% reduction. If the meter is recalibrated by adjusting R_2 to 687 Ω, what is the error in the half I_{FS} resistance indication of 10 kΩ?

Solution
When $I_M = 25 \ \mu A$ the voltage across the meter is 0.075 V. Thus the current through R_2 is 109.2 μA and the total current is 134.2 μA. This current requires $R_x = 10.06$ kΩ. Thus, if the meter reads half FS, the actual unknown resistance is 0.6% higher.

Wheatstone Bridge Figure 17.20 shows a resistance bridge that can be used to determine an unknown resistance. The device labeled ND is a *null detector*. It is used to detect a condition that the voltage across its terminals is zero, i.e., a null.

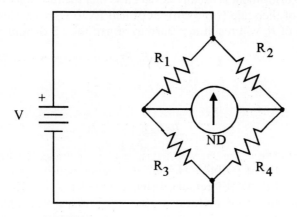

FIGURE 17.20
The Wheatstone bridge.

For any particular set of resistors, R_1, R_2, R_3, and R_4, the *offset voltage*, ΔV, of the bridge is the voltage across the ND:

$$\Delta V = V \left(\frac{R_3}{R_1 + R_3} - \frac{R_4}{R_2 + R_4} \right) \qquad \textbf{(17.15)}$$

If the resistors are varied until a null exists, $\Delta V = 0$, the resistors must satisfy the *null condition*:

$$R_1 R_4 = R_2 R_3 \qquad \textbf{(17.16)}$$

Then if one resistor is unknown, it can be calculated from the others by this equation. The accuracy of this measurement is dependent on two factors:

1. The accuracy of the known resistors
2. The resolution of the null detection.

EXAMPLE 17.15 A Wheatstone bridge uses a null detector with a resolution of 1000 μV, i.e., it resolves zero to within $\pm 500 \ \mu V$. A null condition occurs for $R_1 = 3.000$ kΩ, $R_2 = 12.00$ kΩ, and $R_3 = 457.6 \ \Omega$. What is the value of R_4? What is the uncertainty because of the null detector resolution if the supply voltage is $V = 10$ V?

Solution
The value of R_4 is found by solving Eq. (17.16):

$$R_4 = R_2 R_3 / R_1$$
$$= (12.00 \text{ k}\Omega)(457.6 \ \Omega)/(3.000 \text{ k}\Omega)$$
$$= 1830 \ \Omega \text{ (to four significant figures)}$$

The uncertainty due to resolution is found by finding the values of R_4 that will give an offset voltage of $\pm 500 \ \mu V$:

$$\pm 500 \ \mu V = 10 \frac{457.6}{(3 \text{ k}\Omega + 457.6 \ \Omega)} - \frac{R_4}{(12 \text{ k}\Omega + R_4)}$$

This equation is solved for R_4 using the $+$ and $-500 \ \mu V$ to give a result of

$$R_{4+} = 1831 \ \Omega \text{ and } R_{4-} = 1829 \ \Omega$$

So there is an uncertainty of $\pm 1 \ \Omega$ because of the detector resolution.

Digital Ohmmeter   Digital ohmmeters can be implemented using digital techniques by a system as shown in Figure 17.21. The unknown resistor, R_x, and a reference resistor, R_{ref}, are placed in series with a source voltage, V_s. The instrument then essentially measures the voltage drops across the two resistors and computes

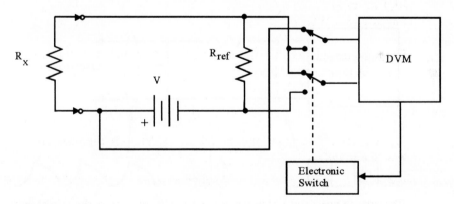

FIGURE 17.21
Use of a digital voltmeter (DVM) to measure resistance.

the ratio. This will be independent of source voltage variation and proportional to the ratio of the resistances.

$$\text{Ratio} = V_x / V_{ref} = (IR_x)/(IR_{ref}) = R_x / R_{ref}$$

Accuracy is dependent on knowledge of the reference resistor and the accuracy of the digital voltage measurement system.

17.3.2 Capacity

Historically, measurement of capacity has been difficult because only analog techniques were available. Digital instruments are now available that make capacity measurements as easy and convenient as resistance measurements. The following paragraphs summarize the methods of capacity measurement.

Charge/Discharge Time The time for a capacitor C to charge through a resistor R to 63.2% of a source voltage is given by

$$\tau = RC \qquad\qquad\qquad (17.17)$$

where τ = time to charge to 63.2% (s)
R = series resistance (Ω)
C = capacity (F)

This is also the time for a charged capacitor to discharge to 36.8% of its initial charge through a resistor. Capacity can be measured by measuring this time and knowing the value of the resistance. If the capacity is large (many microfarads), then this measurement can be made directly with a resistor, voltage source, and stopwatch. For smaller values of capacity the measurement can be made using a resistor, square-wave source voltage, and an oscilloscope to measure the time, as illustrated in Figure 17.22.

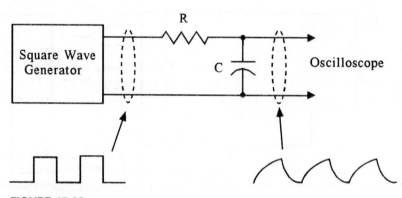

FIGURE 17.22
Use of RC time constant methods to measure capacity.

EXAMPLE 17.16 An 8.2-kΩ resistor is used in series with an unknown capacitor to produce the oscilloscope signal shown in Figure 17.23. The time base is set for 2 ms/cm. What is the value of the capacitor?

Solution
There are ≈ 19 divisions to the peak of the signal vertical trace. Using the discharge portion of the signal, one time constant would be when the signal had decreased

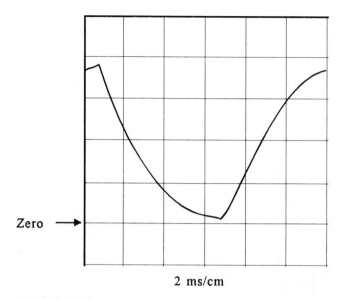

FIGURE 17.23
RC time constant curve for Example 17.16.

to a vertical value of 36.8% of 19 ≈ 7 divisions. From Figure 17.23 you can see that this occurs at ≈1.1 cm from the start of the decay. Thus, the time constant of Eq. (17.17) is $\tau = (1.1 \text{ cm})(2 \text{ ms/cm}) = 2.2$ ms and the capacity if $C = \tau/R = 2.2$ ms/ 8.2 kΩ = 0.27 μF.

Resonance If a variable frequency signal generator and known values of inductance are available, then capacity can be measured by finding the resonant frequency of the inductor/capacitor combination. This frequency is given by

$$f = 1/(2\pi L C_x) \tag{17.18}$$

For the parallel resonant circuit in Figure 17.24(a), resonance will be found as that frequency for which the current registers a minimum as a function of frequency. Current can be measured as the voltage dropped across a series resistance.

In the case of a series resonant circuit as in Figure 17.24(b), resonance is indicated as a maximum of current as a function of frequency.

Accuracy is dependent on the accuracy of frequency measurement, current minimum/maximum detection, and the known value of the inductance.

AC Bridge A common method of impedance measurement is based on an ac bridge circuit as shown in Figure 17.25. The bridge components are adjusted until a null occurs. The null is indicated by a zero of the amplitude of ac voltage across the null detector. When this occurs the bridge arm *impedances* must satisfy

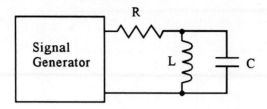

(a)

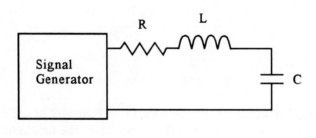

(b)

FIGURE 17.24
(a) Parallel and (b) series resonance to determine capacity.

the relation

$$Z_1 Z_4 = Z_2 Z_3 \qquad\qquad (17.19)$$

Both the real (resistive) and imaginary (reactive) parts of this equation must be satisfied. This gives two null condition equations, from which both the resistive and reactive parts of the unknown impedance may be found.

The bridge of Figure 17.26 is often used for capacity measurement, although many other types are used also. This instrument measures both the capacity, C_x, and the series resistance of the capacitor, R_x. To achieve a true null of both the resistive and reactive parts of the bridge elements, it is necessary to make two adjustments, R_1 and R_3. Often it is necessary to "juggle" back and forth to obtain a null. When the null occurs, the unknowns are given by

$$C_x = \frac{R_1}{R_2} C_s \qquad\qquad (17.20)$$

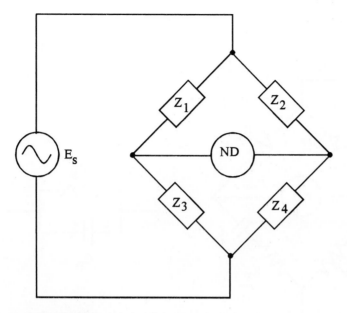

FIGURE 17.25
The ac bridge.

$$R_x = \frac{R_2}{R_1} R_3 \tag{17.21}$$

One of the advantages of the bridge of Figure 17.26 for capacitance measurement is that it does not depend on the signal generator frequency. Thus, accuracy is determined by the accuracy with which the bridge components are known and the resolution used to define the null.

EXAMPLE 17.17

The bridge of Figure 17.26 nulls with the following values: $R_2 = 9.85$ kΩ, $R_1 = 1.76$ kΩ, $R_3 = 0.13$ Ω, and the standard capacitor is $C_s = 97.5$ nF. What is the capacity and resistance of the unknown capacitor?

Solution
The values are found from direct application of Eqs. (17.20) and (17.21):

$$C_x = (1.76 \text{ k}\Omega/9.85 \text{ k}\Omega)(97.5 \text{ nF}) = 17.4 \text{ nF}$$

and

$$R_x = (9.85 \text{ k}\Omega/1.76 \text{ k}\Omega)(0.13 \text{ }\Omega) = 0.72 \text{ }\Omega$$

The series resistance of Example 17.17 is not to be confused with the parallel or *leakage* resistance of a capacitor. Figure 17.27 shows the series and parallel models of a capacitor with resistances. By writing the impedance of each model

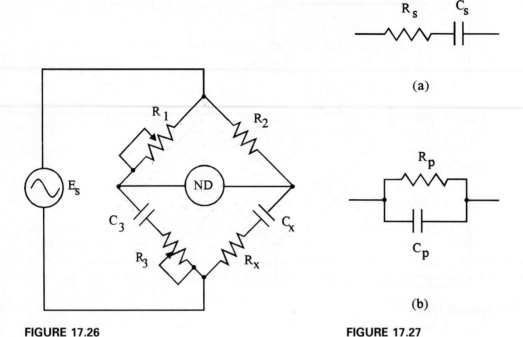

FIGURE 17.26
A bridge for capacity measurement.

FIGURE 17.27
(a) Series and (b) parallel models of a real capacitor.

you can show that the relations between series resistance and capacity to the parallel model are given by

$$R_s = \frac{R_p}{1 + (\omega R_p C_p)^2} \tag{17.22}$$

$$\omega C_s = \frac{1 + (\omega R_p C_p)^2}{\omega C_p (R_p)^2} \tag{17.23}$$

For $(\omega R_p C_p)^2 \gg 1$ or $(\omega C_s R_s)^2 \leq 1$ these equations reduce to

$$R_s \approx \frac{1}{(\omega C_p)^2 R_p} \tag{17.24}$$

$$C_s \approx C_p \tag{17.25}$$

EXAMPLE 17.18

If the bridge of Example 17.17 is operating at 20 kHz, what are the parallel resistance and capacitance?

Solution
Since we solved for C_s and R_s, we first compute the factor

$$(\omega C_s R_s)^2 = [(20 \text{ kHz})(17.4 \text{ nF})(0.72 \text{ }\Omega)]^2$$
$$= 6.27 \times 10^{-8} \ll 1$$

so we use Eqs. (17.24) and (17.25). Thus, $C_p = C_s = 17.4$ nF. From Eq. (17.24) we solve for R_p:

$$R_p = 1/[(20 \text{ kHz})(17.4 \text{ nF})]^2(0.72 \text{ } \Omega) = 11.5 \text{ M}\Omega$$

Digital Capacitance Meters Advances in digital technology have resulted in digital meters that can measure capacitance as easily as voltage, resistance, or current. These devices are based on integrating ADCs for which the unknown capacitor forms a part of the integration process.

It is important to be sure the capacitor is uncharged before connecting the capacitance meter.

17.3.3 Inductance

Since all practical inductors are made of real wire, the series resistance of an inductor is always present and often quite large. Thus, the practical model of an inductor, as shown in Figure 17.28, always includes this resistance. The dc resistance of an inductor can be measured with an ohmmeter. Often the relationship between an

FIGURE 17.28
Circuit model for the inductor.

inductor's inductance and resistance is expressed through the *quality* or Q of the inductor, which is the ratio of the inductive reactance and the series resistance:

$$Q = \frac{\omega L}{R_L} \qquad \textbf{(17.26)}$$

where $\omega = 2\pi f$ and $f = $ the frequency (Hz)
$L = $ inductance (H)
$R_L = $ resistance (Ω).

Resonance Inductance can be measured using the resonance technique described in Section 17.3.2 for capacitors. In this case capacitors with known values are used and the series or parallel resonance frequency with the unknown inductor is found. The inductance can be found from Eq. (17.18).

Bridges The inductance can also be measured using the ac bridge approach outlined in Section 17.3.2 and shown by Figure 17.25 and Eq. (17.19). A null will require both the real and imaginary parts of Eq. (17.19) to be satisfied. These two equations will allow for determination of both the resistance and inductance.

For nulling it is necessary to use either precision variable inductors or capacitors. Because precision capacitors are easier to fabricate, they are often used in inductance measuring bridges. There are two commonly employed bridges, one for higher Q inductors and the other for inductors with lower Q's.

FIGURE 17.29
The Maxwell bridge.

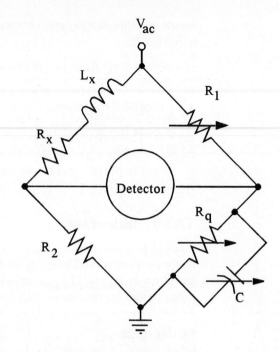

Maxwell Bridge This bridge, shown in Figure 17.29, is used for low-Q (high-resistance) inductors. Satisfaction of Eq. (17.19) requires

$$L_x = R_1 R_2 C \tag{17.27}$$
$$R_x = R_1 R_2 / R_q \tag{17.28}$$

or, in terms of the Q,

$$Q_x = \omega C R_q \tag{17.29}$$

Note that the coil Q depends on R_q but the inductance does not. Thus, the R_q scale of a commercial bridge can be marked directly in Q values if the frequency at which the bridge operates is fixed. R_1 is often a switched set of precision resistors, which determine the inductance measurement range for smooth variations of C.

A null is defined by variations of R_1 and C for a minimum detector reading and then variation of R_q for a null or the best minimum possible.

EXAMPLE 17.19

A Maxwell bridge is needed to measure inductors whose inductance is less than 30 mH and whose series resistance is on the order of 20 to 100 Ω. If a capacitor that varies from 0 to 360 pF is available, what other bridge values can be used for a measurement frequency of 1 kHz?

Solution
To begin we assume that an inductance of 30 mH will correspond to the maximum capacity; then, from Eq. (17.27), the value of $R_1 R_2$ is found to be

$$R_1 R_2 = L_x / C = 30 \text{ mH}/360 \text{ pF} = 83.3 \times 10^6$$

For $R_x = 20\ \Omega$ this means the value of R_q can be found from Eq. (17.28):

$$R_q = R_1 R_2/R_x = 83.3 \times 10^6/20 = 4.16\ \text{M}\Omega$$

and for $R_x = 100\ \Omega$ we have

$$R_q = 83.3 \times 10^6/100 = 0.83\ \text{M}\Omega$$

To use practical values, let's make $R_1 = R_2 = 10\ \text{k}\Omega$, and for R_q we will use a 500-kΩ resistor in series with a 5-MΩ pot. Using these values gives a circuit that can measure L_x from 0 H to a maximum of

$$L_x(\text{max}) = (10\ \text{k}\Omega)(10\ \text{k}\Omega)(360\ \text{pF}) = 36\ \text{mH}$$

and R_x can be measured in the range of

$$R_x(\text{max}) = (10\ \text{k}\Omega)(10\ \text{k}\Omega)/500\ \text{k}\Omega = 200\ \Omega$$

to

$$R_x(\text{min}) = (10\ \text{k}\Omega)(10\ \text{k}\Omega)/5.5\ \text{M}\Omega = 18.2\ \Omega$$

At the maximum inductance these resistances correspond to Q's on the order of 1.1 to 12.4.

Hay Bridge For inductors with a high Q, which means a low value of resistance, the Maxwell bridge is inconvenient. You can see from the preceding example that if the resistances had been on the order of 1 to 5 Ω, which are Q's of 226 to 45, the value of R_q would have to have been 16.6 to 83.3 MΩ! The Hay bridge, shown in Figure 17.30, is much more suitable for high-Q inductance measurements.

FIGURE 17.30
The Hay bridge.

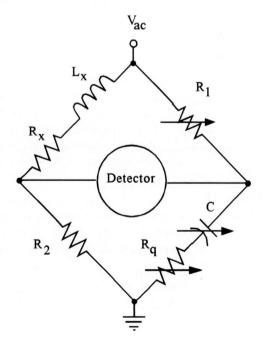

For this bridge the relations for L_x and R_x at null are given by

$$L_x = \frac{R_1 R_2 C}{1 + (R_q \omega C)^2} \tag{17.30}$$

$$R_x = \frac{R_1 R_2 R_q (\omega C)^2}{1 + (R_q \omega C)^2} \tag{17.31}$$

The Q can be shown to be given by

$$Q = 1/(\omega C R_q) \tag{17.32}$$

Many commercial bridges can be switched to either the Maxwell or Hay configuration.

EXAMPLE 17.20

A Hay bridge nulls with $R_1 = 2$ kΩ, $R_2 = 1$ kΩ, $C = 140$ pF, and $R_q = 440$ Ω, and it operates at 50 kHz. What are L_x, R_x, and the Q of the inductor?

Solution
This is just an application of the preceding equations. We find

$$L_x = \frac{(2 \text{ k}\Omega)(1 \text{ k}\Omega)(140 \text{ pF})}{1 + [(440 \text{ }\Omega)(2\pi 50 \text{ kHz})(140 \text{ pF})]^2} = 280 \text{ }\mu\text{H}$$

$$R_x = \frac{(2 \text{ k}\Omega)(1 \text{ k}\Omega)(440 \text{ }\Omega)(2\pi 50 \text{ kHz} \times 140 \text{ pF})^2}{1 + [(880 \text{ }\Omega)(2\pi 50 \text{ kHz})(140 \text{ pF})]^2} = 1.7 \text{ }\Omega$$

$$Q = (2\pi 50 \text{ kHz})(280 \text{ }\mu\text{F})/(1.7 \text{ }\Omega) = 51.7$$

18

COMMUNICATION SYSTEMS*

18.1 INTRODUCTION

The purpose of a communication system is to transmit information from one point to another. We can describe the quality of the system in terms of the information unavoidably lost compared to the information transmitted. Wireless communication systems make use of an rf carrier for transmitting the information. The carrier is of a higher frequency than the message and permits the information to be transmitted in a portion of the frequency spectrum best suited to the medium through which transmission must take place.

All carrier communication systems have in common the fact that they are composed of the basic functional components illustrated in Figure 18.1. The modulator in Figure 18.1 imposes the information to be transmitted on the carrier signal. A transducer (antenna) couples the modulated carrier to the transmission medium and a small portion of this energy is intercepted by the distant receiver. At the reception point another transducer (antenna) couples the now attenuated and distorted carrier signal to a frequency selective receiver. The receiver filters the desired modulated carrier signal and the demodulator separates the modulating signal from the carrier. This computes recovery of the message.

The relative advantages and disadvantages of the different radio frequencies for specific communication links are well understood. Today the entire electromagnetic spectrum up through visible light is being utilized in communication systems (Figure 18.2). A comparison of light and rf communications is illustrated in Figure 18.3. In the radio transmitter the carrier is generated as a periodic electrical signal at a specific frequency, whereas in the light transmitter the carrier is generated directly as electromagnetic radiation. The bandwidth can vary depending on the

---

* This chapter was written by Dr. Harold Killen, College of Technology, University of Houston.

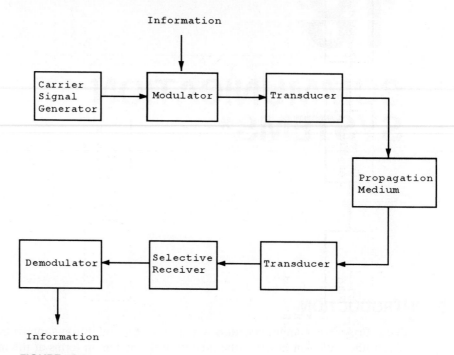

FIGURE 18.1
Basic functional components of a carrier communication system.

type of radiation source. Modulation in the radio system is accomplished electronically with the output being an electrical signal. In the light transmitter, modulation is accomplished by intensity modulation of the light. Traditional angle modulation techniques are still in the laboratory stage. At the light receiver, demodulation is accomplished directly by the receiving photodetector. The radio receiver accomplishes demodulation after the antenna transforms the electromagnetic signal to an electrical signal. Because of the difference in wavelength between a radio communi-

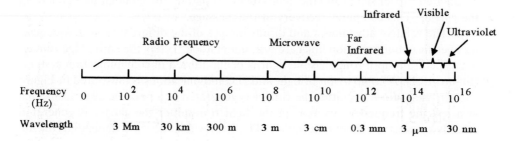

FIGURE 18.2
The electromagnetic spectrum.

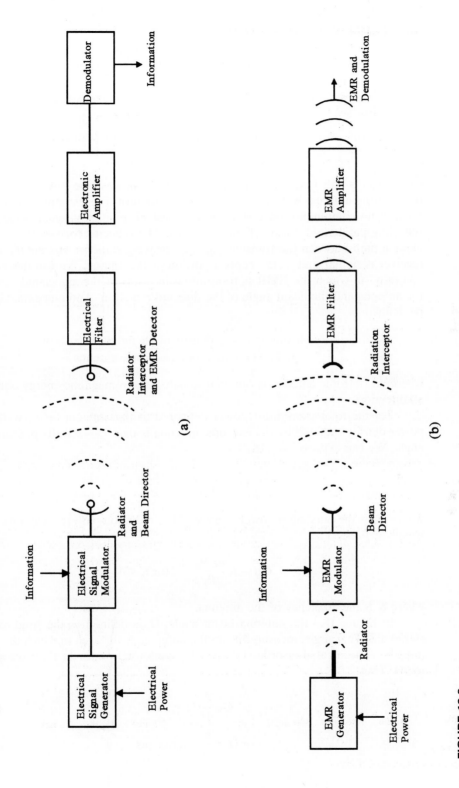

FIGURE 18.3
Comparison of point-to-point (a) radio and (b) light frequency communication systems. EMR, electromagnetic radiation.

cation system and a light communication system, the greater practical difference in design lies in the antennas.

In both systems the wavelength, λ, of the electromagnetic radiation (EMR) is given by

$$\lambda = \frac{300}{f_{MHz}} \text{ meters} \tag{18.1}$$

where f_{MHz} is the frequency in megahertz. We can describe any type of EMR transmitting antenna in terms of its radiation pattern. For example, the antenna for a radio communication system may consist of a small antenna backed by a reflecting paraboloid "dish" [Figure 18.3(a)]. The reflector focuses the radiation along a path between the transmitting and receiving stations. The antenna at the receiver is similar and it intercepts a portion of the energy. We can describe the "pattern" in which the EMR is transmitted in terms of the associated "gain" of the antenna and the beam angle of the directed radiation. Antenna gain, Ga, may be defined as

$$Ga = \frac{\text{Maximum power density of the radiation pattern}}{\text{Power density of an isotropic antenna}} \tag{18.2}$$

where an isotropic antenna is one that radiates electromagnetic energy equally in all directions.

Let the total transmitter power delivered to the antenna be P_T watts. The power density radiated by the isotropic antenna is then $P_T/4\pi$ watts per unit solid angle. We can express Eq. (18.2) as

$$Ga = \frac{P_M}{P_T/4\pi} = \frac{4\pi P_M}{P_T} \tag{18.3}$$

where P_M is the maximum power density of the radiation pattern. Antenna gain is usually expressed as

$$Ga = K\frac{4\pi P_M}{P_T} \quad \text{for } K < 1 \tag{18.4}$$

where K is the efficiency of the antenna.

In Figure 18.4 the antenna beam angle, Ω, is defined as the solid angle in steradians. This angle, measured from the antenna, is the area in which the power must be concentrated in order to have the same total power as the isotropic antenna. Thus,

$$Ga\frac{P_T}{4\pi}\Omega = P_T \tag{18.5a}$$

$$\text{or } \Omega = \frac{4\pi}{Ga} \text{ steradians} \tag{18.5b}$$

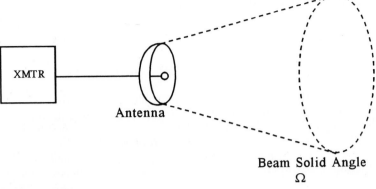

XMTR

Antenna

Beam Solid Angle
Ω

FIGURE 18.4
Transmitting antenna at radio frequencies.

For the paraboloid antenna mentioned earlier, the antenna gain can be expressed as

$$Ga = \left(\frac{4D}{\lambda}\right)^2 \tag{18.6}$$

where D is the diameter of the "dish." We see from this expression that for a given gain, the diameter of the antenna decreases with the square of the carrier frequency. As the carrier frequency is increased (smaller wavelength), a proportionate increase in manufacturing tolerances is experienced. This must be adhered to if the predicted gain of Eq. (18.6) is to be realized.

For very high frequencies, the beam becomes highly directional. The paraboloid antenna becomes impractical due to the small dimensions and power density increases. At optical frequency the antenna becomes a lens. The transmission efficiency remains directly proportional to the square of the frequency for a given diameter and Eq. (18.6) still applies.

18.2   MODULATION TECHNIQUES

The output of the carrier generator in Figure 18.1 is a time-varying sinusoid, which is described by

$$e(t) = E_c \cos(\omega_c t + \theta) \tag{18.7}$$

Three of these parameters can be varied: the amplitude, E_c, frequency, ω_c, and phase, θ. Operating singly on these three parameters produces amplitude (AM), frequency (FM), or phase modulation (PM), respectively. In addition, we can combine the functions to produce a more complex modulation. For example, in digital communications it is often advantageous to combine amplitude and phase modulation. In the following sections we provide specifics.

18.2.1 Amplitude Modulation

One of the earliest forms of modulation was amplitude modulation. We can rewrite Eq. (18.7) as

$$e_{AM}(t) = [E_c + g(t)] \cos \omega_c t \qquad (18.8)$$

where $g(t)$ is the message. To examine the basic characteristics of AM, it is necessary to choose a specific form for the messages. Thus, let the message consist of a term:

$$g(t) = E_m \cos \omega_m \qquad (18.9)$$

Equation (18.8) then becomes

$$e_{AM}(t) = (E_c + E_m \cos \omega_m t) \cos \omega_c t$$

$$= E_c \cos \omega_c t + m \frac{E_c}{2} \cos(\omega_c + \omega_m)t + m \cos(\omega_c - \omega_m)t \qquad (18.10)$$

where $m = E_m/E_c$ is the modulation index. A plot of Eq. (18.10) along with the frequency spectrum is shown in Figure 18.5. From this figure we see that the

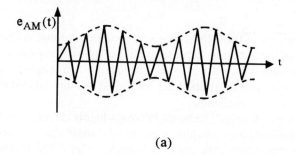

(a)

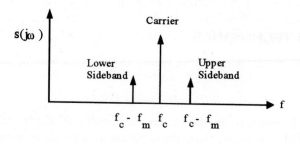

(b)

FIGURE 18.5
(a) Amplitude modulation and (b) frequency spectrum.

spectrum of the AM waveform consists of a carrier with both upper and lower sidebands. Note that the rf bandwidth is twice the frequency of the modulation tone. In fact, since amplitude modulation is a linear process, each modulation process will produce a pair of sidebands. The rf bandwidth may be expressed as:

$$BW = 2 \cdot f_{m(max)} \qquad (18.11)$$

where $f_{m(max)}$ is the highest frequency in the modulating message. An example will illustrate typical calculations.

EXAMPLE 18.1 A 2-MHz rf waveform is modulated with a 5-kHz tone. The peak carrier voltage is 10 V and the minimum is 5 V. What is (a) the unmodulated carrier voltage, E_c; (b) the modulation index, M_a; and (c) the upper (USB) and lower sideband (LSB) frequencies?

Solution

(a) $\quad E_c = \dfrac{V_{c(peak)} + V_{c(min)}}{2} = \dfrac{10 + 5}{2} = 7.5 \ V_{peak}$

(b) $\quad E_m = E_{peak} - E_c = 10 - 7.5 = 2.5 \ V$

$\quad\quad m = \dfrac{E_m}{E_c} = \dfrac{2.5}{7.5} = 0.33$

Alternatively,

$$m = \frac{E_{max} - E_{min}}{E_{max} + E_{min}} = \frac{10 - 5}{10 + 5} = 0.33$$

(c) $\quad f_{USB} = f_c + f_m = (2 \times 10^6) + (5 \times 10^{-3}) = 2.005 \ MHz$

$\quad\quad f_{LSB} = f_c - f_m = (2 \times 10^6) - (5 \times 10^3) = 1.995 \ MHz$

Note from Eq. (18.10), that the carrier is present with or without the modulating tone. That is, carrier power is independent of modulation. The power contained in an AM waveform may be determined from conventional circuit analysis techniques. Assume that the waveform described by Eq. (18.10) is developed across a resistor R. Thus,

$$P_T = \frac{1}{T} \int_0^T \frac{e_{AM}^2(t)}{R} \, dt \qquad (18.12)$$

where T is the period of the modulating tone. Substituting Eq. (18.10) into Eq. (18.12) results in

$$P_T = \frac{1}{RT} \int_0^T E_c^2 \sin^2 \omega_c t \, dt + \frac{1}{RT} \int_0^T m^2 \frac{E_c^2}{4} \cos^2(\omega_c - \omega_m)t$$

$$+ \frac{1}{RT} \int_0^T m^2 \frac{E_c^2}{4} \cos^2(\omega_c + \omega_m)t \, dt \qquad (18.13)$$

We identify this expression as

$$P_T = P_c + P_{LSB} + P_{USB} \qquad \text{(18.14)}$$

where $\quad P_c$ = carrier power
P_{LSB} = lower sideband power
P_{USB} = upper sideband power.

Recognizing that $\int_0^T \sin^2 \omega t \, dt = \int_0^T \cos^2 \omega t \, dt$, Eq. (18.13) reduces to

$$P_T = P_c + \frac{m^2}{4} P_c + \frac{m^2}{4} P_c$$
$$= P_c \left(1 + \frac{m^2}{2} \right) \qquad \text{(18.15)}$$

AM power calculations are illustrated in Example 18.2.

EXAMPLE 18.2

For Example 18.1, determine (a) carrier power and (b) upper and lower sideband powers. Assume a 200-Ω lamp impedance.

Solution

(a) $\quad P_c = \dfrac{1}{RT} \int_0^T E_c^2 \sin^2 \omega_c t \, dt = E^2/2R$

$\qquad = (7.5/\sqrt{2})^2/2(200) = 0.070 \text{ W}$

(b) $\quad P_{USB} = P_{LSB} = \dfrac{m^2}{4} P_c = \left(\dfrac{0.33^2}{4} \right)(0.070) = 0.0019 \text{ W}$

Modulation with nonsinusoidal waveforms produces a more complicated AM waveform as shown in Figure 18.6. Specifically, this is an example of amplitude modulation using digital data. The previous analysis for analog modulation readily transfers to this situation. Consider the limiting case of a 101010 pattern, that is, a square wave. This waveform can be expressed as a Fourier series. From our previous

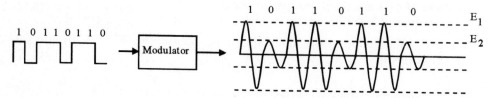

FIGURE 18.6
Amplitude modulation using digital signals. A binary 1 is represented by carrier level E_1, whereas the binary 0 is represented by amplitude E_2.

discussion we know that each frequency in the series produces a pair of sidebands with an associated modulation index depending on the amplitude of the harmonic.

As a matter of generalizing, let the modulating waveform be represented by the following series:

$$g(t) = E_1 \sin \omega_m t + E_2 \sin 2\omega_m t + E_3 \sin 3\omega_m t + \cdots \qquad (18.16)$$

Then Eq. (18.10) becomes

$$e_{\mathrm{AM}}(t) = E_c \left[1 + \frac{E_1}{E_c} \sin \omega_m t + \frac{E_2}{E_c} \sin 2\omega_m t + \cdots \right] \sin \omega_c t$$
$$= E_c[1 + m_1 \sin \omega_m t + m_2 \sin 2\omega_m t + \cdots] \sin \omega_c t \qquad (18.17)$$

The general expression for average power of the modulated waveform is

$$P_T = P_c[1 + (m_1^2 + m_2^2 + m_3^2 + \cdots)]$$
$$= P_c \left(1 + \frac{m_{\mathrm{eff}}^2}{2} \right) \qquad (18.18)$$

where the effective modulation index is given by

$$m_{\mathrm{eff}} = \sqrt{m_1^2 + m_2^2 + m_3^2 + \cdots} \qquad (18.19)$$

Typical frequency spectra for amplitude modulation by complex waveforms are shown in Figure 18.7.

18.2.2 Angle Modulation

Referring to Eq. (18.7),

$$e(t) = E_c \cos(\omega_c t + \theta)$$
$$= E_c \cos[\phi(t)] \qquad (18.20)$$

If ω_c is varied in accordance with the information, frequency modulation occurs. If phase θ is varied by the information, phase modulation occurs. Note that either operation varies $\phi(t)$, and the two techniques are fundamentally the same. This variation of $\phi(t)$ is termed *angle modulation.*

Frequency is the time rate of change $\phi(t)$. That is,

$$\omega(t) = \frac{d}{dt} \phi(t). \qquad (18.21)$$

For the unmodulated wave this becomes

$$\omega(t) = \frac{d}{dt}[\omega_c t + \theta] = \omega_c. \qquad (18.22)$$

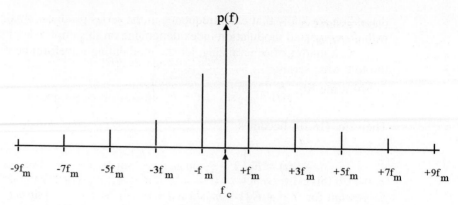

(a)

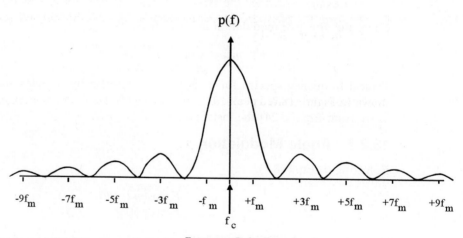

(b)

FIGURE 18.7
(a) Square-wave amplitude modulation. (b) Periodic rectangular wave modulation spectrum.

Next, assume a frequency modulating signal of the form $g(t)$. The instantaneous frequency, $\omega(t)$, becomes

$$\omega(t) = \omega_c[1 + g(t)] \tag{18.23}$$

Substituting Eq. (18.23) into Eq. (18.21) results in

$$\begin{aligned}
\phi(t) &= \int_0^t \omega_c[1 + g(t)] \, dt \\
&= \omega_c t + \omega_c \int_0^t g(t) \, dt
\end{aligned} \tag{18.24}$$

Finally, if this expression for $\phi(t)$ is substituted into Eq. (18.20), we have the classical expression for frequency modulation:

$$e_{\mathrm{FM}}(t) = E_c \cos\left[\omega_c t + \omega_c \int_0^t g(t) \, dt\right] \tag{18.25}$$

We see from this expression that in an FM system, the message, $g(t)$, is integrated before it is transmitted.

To examine further the basic principles of FM it is necessary to assume a specific form for the message. As in the case of AM, the simplest form is a tone. Let $g(t)$ be expressed by

$$g(t) = \frac{\Delta\omega}{\omega_c} \cos \omega_m t \tag{18.26}$$

where $\Delta\omega$ is the desired peak-frequency change and ω_m is the modulating frequency. Thus, from Eq. (18.24) the instantaneous phase, $\phi(t)$, is

$$\begin{aligned}
\phi(t) &= \omega_c t + \omega_c \int_0^t \frac{\Delta\omega}{\omega_m} \cos \omega_m t \, dt \\
&= \omega_c t + \frac{\Delta\omega}{\omega_m} \sin \omega_m t \\
&= \omega_c t + m_f \sin \omega_m t
\end{aligned} \tag{18.27}$$

where $m_f = \Delta\omega/\omega_m$ is the modulation index.

The expression for FM [Eq. (18.25)] may now be written as

$$e_{\mathrm{FM}}(t) = E_c \cos(\omega_c t + m_f \sin \omega_m t) \tag{18.28}$$

This expression is recognizable as the trigonometric form $\cos(A + B)$. The frequency spectrum of the FM wave can readily be found by expanding the expression into a series. This expansion is not difficult and is available in many basic communication texts. The end result is

$$\begin{aligned}
e_{\mathrm{FM}}(t) = {}& E_c J_0(m_f) \cos \omega_c t \\
& + E_c J_{2n}(m_f) \cos[\omega_c t + 2n\omega_m t + \cos(\omega_c t - 2n\omega_m t)] \\
& + E_c J_{2n+1}(m_f)\{\cos[(\omega_c t + (2n + 1)\omega_m t] \\
& - \cos[\omega_c t - (2n + 1)\omega_m t]\}
\end{aligned} \tag{18.29}$$

where the J's are Bessel functions. In this expression n takes on all integer values from 0 to ∞.

An examination of Eq. (18.29) reveals that a sinusoidally modulated FM wave consists of a carrier of amplitude $E_c J_0(m_f)$ and sideband frequencies that are represented by the remaining terms in the series. Since n assumes all integer values, the spectrum is infinitely wide. Fortunately, the energy in the higher order sidebands decreases and the bandwidth is given approximately by

$$\text{BW} \approx 2(\Delta f + f_m) \text{ Hz} \tag{18.30}$$

The frequency spectrum is illustrated in Figure 18.8. Typical calculations are illustrated in Example 18.3.

EXAMPLE 18.3

An FM waveform is described by

$$e_{FM}(t) = 20 \cos(3 \times 10^7 \pi t + 10 \cos 2000\pi t)$$

Find the bandwidth.

Solution

Comparing the expression to Eq. (18.28) we see that $\omega_m = 2000$ rad/s. The instantaneous frequency, $\omega_i(t)$, is given by

$$\omega_i(t) = \frac{d}{dt} \phi(t) = \frac{d}{dt} [3 \times 10^7 \pi t + 10 \cos 2000\pi t]$$

$$= 3 \times 10^7 \pi t - 20{,}000\pi \sin 2000\pi t$$

Thus, $\Delta\omega = 20{,}000\pi$ and

$$\begin{aligned}
\text{BW} &= 2(\Delta\omega + \omega_m) \\
&= 2(20{,}000\pi + 2000\pi) \\
&= 44{,}000\pi \text{ rad/s} = 22{,}000 \text{ Hz}
\end{aligned}$$

Digital data are also transmitted using frequency modulation (Figure 18.9). This technique, known as frequency shift keying (FSK), is widely used. In this example, a binary 1 is transmitted as frequency F_1 and a binary 0 is transmitted as frequency F_2. Two frequencies are the maximum usually used because of the time needed to shift from one frequency to the other.

18.2.3 Phase Modulation

Refer to Eq. (18.20). For phase modulation, the information, $g(t)$, varies the phase $\phi(t)$. Thus,

$$e_{PM}(t) = E_c \cos[\omega_c t + m_p g(t)] \tag{18.31}$$

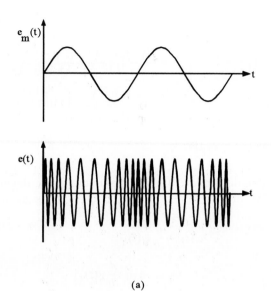

(a)

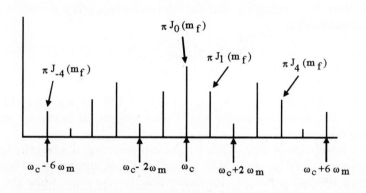

(b)

FIGURE 18.8

Frequency modulation by a sinusoidal waveform and (b) the FM spectrum with sinusoidal modulation.

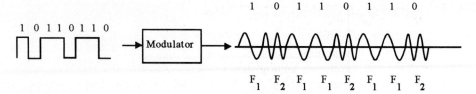

FIGURE 18.9
Frequency shift keying.

where m_p is the modulation index. It is the maximum phase deviation in radians. As in the case for FM, the instantaneous frequency, $\omega_i(t)$, is given by

$$\omega_i(t) = \frac{d}{dt}\phi(t)$$

$$= \omega_c + m_p\frac{d}{dt}g(t) \tag{18.32}$$

Let the message, $g(t)$, be $\cos \omega_m t$. The instantaneous frequency, $\omega_i(t)$, is

$$\omega_i(t) = \omega_c + \omega_m m_p \sin \omega_m t \tag{18.33}$$

It is clear from this expression that the maximum frequency deviation is $\omega_m m_p$. Thus, the bandwidth is

$$\begin{aligned} BW &\approx 2(\omega_m m_p + \omega_m) \\ &\approx 2(f_m \cdot m_p + f_m) \end{aligned} \tag{18.34}$$

Comparing Eq. (18.25) for FM to Eq. (18.31) for PM, we see that FM and PM are basically the same. Frequency modulation utilizes an integrator and phase modulation does not.

The transmission of digital data by phase modulation is a favorite technique. One technique, known as phase shift keying (PSK), is illustrated in Figure 18.10. Note that in this technique, the frequency remains the same while the phase is shifted 180° to denote a logic 0 or 1. The fundamentals of digital communications are discussed in the next section.

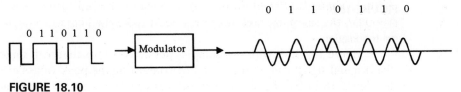

FIGURE 18.10
Phase shift keying.

18.3 DIGITAL COMMUNICATIONS

When information is represented as a finite set of discrete values, the result is referred to as a *digital signal*. Digital data can occur in several ways. For example, we can sample an analog signal and then digitize the resulting pulse. The sample is now represented by a digital word. The operation of analog-to-digital conversion is referred to as *quantization*. As a second example, data may already occur in digital form. A sequence of numbers representing the time at which various events occurred is digital data. If we choose to replace the digital data by another set of symbols, we introduce an additional process—that of *encoding*.

FIGURE 18.11
Block diagram of a digital transmitter.

The various processes mentioned are illustrated in Figure 18.11. The combination of these operations constitutes a process known as *pulse code modulation* (PCM). To complete the system, we need a receiver (Figure 18.12). Generally speaking, the transmission of information as indicated by these two figures is called *digital communications*. Prior to discussing system aspects of digital communications further, we need to consider basic principles such as the sampling theorem.

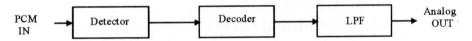

FIGURE 18.12
Block diagram of a PCM receiver.

18.3.1 Sampling Theorem

The process of sampling and digitalizing an analog signal results in the creation of signal elements. The requirements for transmitting N signal elements per second were investigated by Nyquist in 1928. One of the most noteworthy results of this investigation was that it is theoretically feasible to transmit N signal elements per second over a bandwidth of $N/2$ Hz. This bandwidth has come to be referred to as the *Nyquist bandwidth* in his honor. Nyquist showed further that if a signal is sampled, the sampling rate must be at least twice the highest frequency contained in the signal.

Figure 18.13(a), with two samples per hertz, shows that with some smoothing, the original signal can be recovered by connecting the peak values of the samples. In Figure 18.13(b), the signal is undersampled and the recovered signal is a sine wave of lower frequency. We refer to this result as *aliasing*. We see that in a PCM system the sampling rate is an important system consideration.

FIGURE 18.13
Illustration of sampling theory predictions.

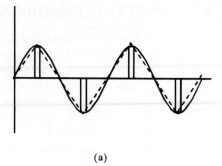

(a)

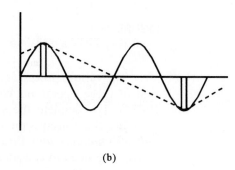

(b)

18.3.2 Quantizing Noise

The end results of digitizing a signal are illustrated in Figure 18.14. The original signal is sampled and an analog-to-digital converter (ADC) produces the digital code shown to the right. If the process is reversed (that is, the digital words converted back to analog), the decoded signal has the appearance shown. The difference between the original signal and the decoded signal is the quantization noise. It can be shown that the rms value of this noise is given by

$$\text{rms} = \frac{1}{\sqrt{12}} \frac{1}{2^n - 1} \tag{18.35}$$

where n is the number of bits in the digital word. The signal-to-quantizing noise ratio can be very important in applications such as speech. A weak signal exhibits a poorer signal-to-quantizing noise ratio than a strong signal. Typical calculations are illustrated in Example 18.4.

EXAMPLE 18.4 A 5-V signal is sampled and digitized to an accuracy of 5 bits. (a) What is the quantizing rms error voltage? (b) What is the signal-to-quantizing noise ratio (SNR)?

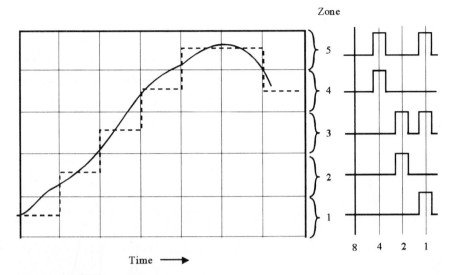

Zone

FIGURE 18.14
Pulse code modulation.

Solution

(a) $\text{rms} = 5 \left[\dfrac{1}{\sqrt{12}} \dfrac{1}{2^5 - 1} \right] = 0.0451 \text{ V}$

(b) $\text{SNR} = \dfrac{5}{0.0451} = 110.85 \rightarrow 40.89 \text{ dB}$

18.3.3   PCM Bandwidth Requirements

We noted earlier that it is theoretically possible to transmit N pulses over a bandwidth of $N/2$ Hz. This is feasible under the ideal conditions postulated by Nyquist. A more practical number is arrived at by considering the frequency spectrum of a single pulse (Figure 18.15). The bandwidth is often taken to be the distance to the first zero crossing. That is

$$\text{BW} = \frac{2\pi}{\tau} \text{ rad/s} \qquad (18.36)$$

where τ is the pulse width. In practice, a PCM wavetrain is composed of a sequence of one's and zero's. Regardless of the sequence, however, the bandwidth can be no greater than that of a single pulse. Suppose, for example, that we sample a signal at the Nyquist rate. The time, T, between pulses is then

$$T = \frac{1}{2f_m} \qquad (18.37)$$

FIGURE 18.15
A single pulse as in (a) has the frequency
spectrum shown in (b).

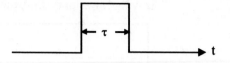

(a)

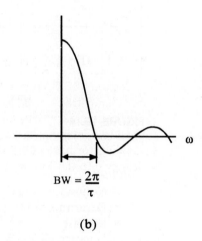

$$BW = \frac{2\pi}{\tau}$$

(b)

where f_m is the highest frequency in the signal. Furthermore, if we extend this to
n time multiplexed signals, the width of the sampled pulse will be

$$T_n = \frac{T}{n} = \frac{1}{2nf_m} \tag{18.38}$$

We now wish to digitize the pulses as illustrated in Figure 18.14. Each digital word
occupies a time of T/n seconds. For an m-bit PCM word, the width of each bit
is T/mn seconds. Consequently, using Eq. (18.36), the PCM bandwidth may be
expressed as

$$BW = \frac{2\pi}{t} = \frac{2\pi mn}{T}$$
$$= 2mn\omega_m \tag{18.39}$$

Now for binary purposes, the number of levels (steps) in the quantization process is

$$L = 2^m \tag{18.40}$$

Solving for m and substituting for m in Eq. (18.39) results in

$$BW_{PCM} = (2 \log_2 L)n\omega_m \qquad\qquad \textbf{(18.41)}$$

EXAMPLE 18.5 A PCM system transmits 24 time multiplexed signals. Each signal is bandlimited to 8 kHz. What bandwidth is required if an 8-bit PCM word is used?

Solution

$$\begin{aligned} BW &= 2 \log_2 Ln\omega_m \\ &= (2)(\log_2 256)(24)(2\pi)(8000) \\ &= 3.072 \text{ MHz} \end{aligned}$$

18.3.4   Digital Encoding Techniques

In Section 18.3.2 we saw how data are converted to digital words. Now the data are actually transmitted as digital signals. For example, binary data are transmitted by encoding each data bit into a signal element. Thus, the encoder performs the function of mapping from data bits to signal elements. Some of the more common encoding formats are illustrated in Figure 18.16. We can group these encoding formats into the following categories:

- Nonreturn to zero (NRz)
- Return to zero (Rz)
- Biphase
- Delay modulation
- Multilevel binary.

The digital signal encoding formats of Figure 18.16 are all of the baseband variety. Basically, the advantage of a digital encoding scheme over analog signaling is lower signal power for a fixed signal quality. Digital signal quality (average signal power to noise power ratio) increases exponentially with bandwidth for elementary encoding schemes. On the other hand, quality increases linearly for nonencoded schemes such as FM. After encoding the binary data, the next logical sequence is to perform digital modulation. This applies specifically to those instances where baseband transmission is not used. We consider techniques related to this in the next section.

18.3.5   Digital Modulation Techniques

We discussed in Section 18.2 the basics of modulation. If a continuum of values of the modulated parameter is allowed, the modulation is said to be analog. If only specific values are allowed, the modulation is said to be digital. Thus, digital modulation refers to techniques in which amplitude, frequency, or phase assumes specific values. These techniques are referred to as amplitude shift keying (ASK), frequency

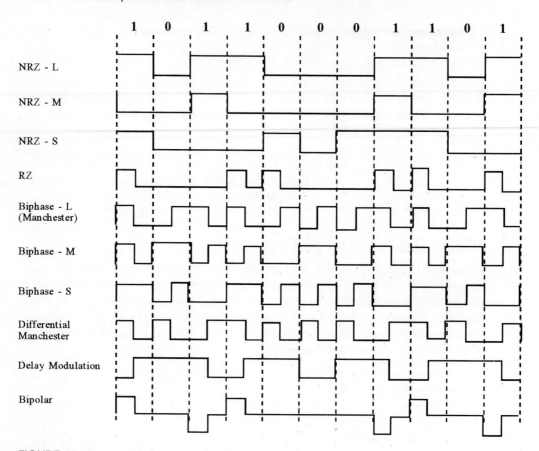

FIGURE 18.16
Digital signal-encoding formats.

shift keying (FSK), and phase shift keying (PSK). Examples are shown in Figures 18.6, 18.9, and 18.10, respectively.

Of these techniques, one of the most popular is PSK. Phase shift keying was developed during the early days of the space program. PSK is considered to be an efficient form of data transmission because it provides the lowest probability of error for a given received signal level when measured over a 1-bit period. In PSK, the incoming binary signal shifts the phase of the output to one of a fixed number of states. Mathematically, the signal is described as

$$e_{PSK}(t) = E \sin \left[\omega_c t + \frac{2\pi(i-1)}{M} \right] \qquad (18.42)$$

where $\qquad i = 1, 2, \ldots, M$
$M = 2^N$, the number of allowable phase states
$N =$ the number of data bits needed to specify the phase state, M.

Three common versions of PSK are binary (BPSK), where $M = 2$; quadraphase (QPSK), where $M = 4$; and $8\phi = $ PSK, where $M = 8$. The phase states are illustrated in Figure 18.17. PSK systems with up to 64 phase states are under development. We refer to these systems as M-ary PSK. Referring to Figure 18.17, if a given amount of information is to be transmitted in a given time interval, the signaling

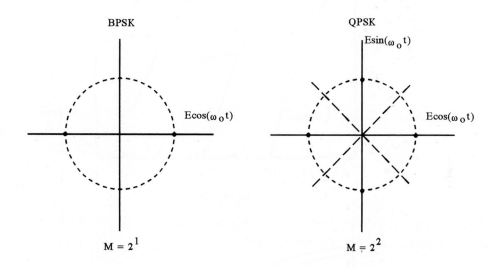

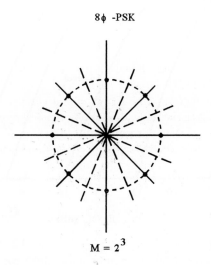

FIGURE 18.17
PSK modulation phase states.

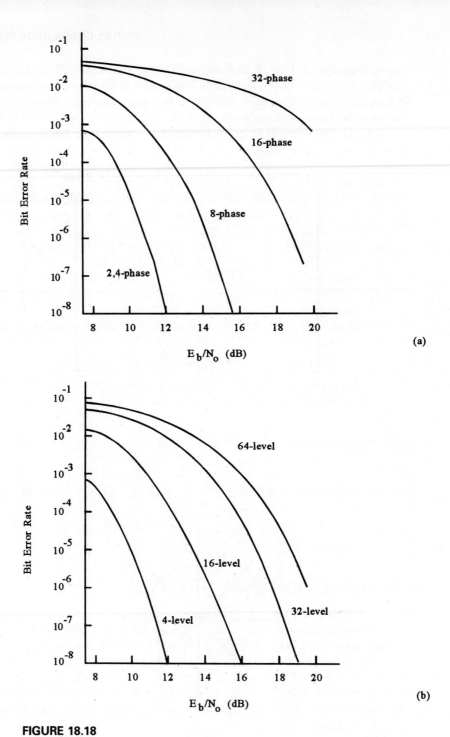

FIGURE 18.18
(a) Increase in noise-induced bit error in the MPSK system. (b) The MAPK system is less susceptible to noise errors.

rate can be reduced in an *M*-ary system by a factor of *N*. Note that in the 8ϕ system, $N = 3$. That is, each phase represents 3 bits of binary information. Note that as the order of the system increases, the phases are closer together. Thus, the system probability of error increases.

Figure 18.9 represents binary FSK. *M*-ary FSK is an extension of binary FSK. Each symbol represents $\log_2 M$ data bits where *M* is equal to a power of two. *M*-ary FSK, however, occupies a large bandwidth and offers no performance advantages. A combination of ASK and PSK does offer performance advantages. This technique is referred to as APSK. We can, for example, use two distinct voltage levels and four phase states. This yields eight distinct states, each representing 3 binary bits. The voltages can be selected to increase the distance between phase states. We begin to see that in digital modulation, a larger number of schemes from which to choose are available. In selecting a modulation technique, it is necessary to make a comparison of the relative performance. One comparison point is the probability of error (Figure 18.18). Note, for example, for a bit-error rate (BER) of 10^{-6}, 16-PSK requires an $E_b/N_0 = 18$ dB versus 14.5 dB for a 16-level MAPK system. Probability of error curves for other modulation schemes are available in the literature.

EXAMPLE 18.6

A 8ϕ PSK modum operates at 4.8 kbit/s. What is the probability of error for this system if the E_b/N_0 is 14 dB? What is the average time between errors?

Solution
From Figure 18.18(a), the probability of error is

$$P_e = 10^{-6}$$

There is one error in 1,000,000 bits. The average time between errors is

$$T_{avg} = \frac{1,000,000}{4800} = 208.33 \text{ s}$$

18.4 DATA COMMUNICATION

Data communications may be defined as a subset of telecommunications. Historically speaking, telecommunications has been defined as the "art and science" of communicating at a distance, as is particularly exemplified by the use of radio waves in radars, radio, and television. Computer communications has extended this array. We must add lines belonging to common carriers such as the Bell systems, AT&T, etc. The machines must be interfaced to the communication system and this has given rise to protocols that are necessary to permit an orderly and efficient transfer of data.

The process of data communication requires at least five elements (Figure 18.19). Specifically, we see that the process consists of a message, a transmitter, a binary serial interface, a transmission channel, and a receiver. Although it is not

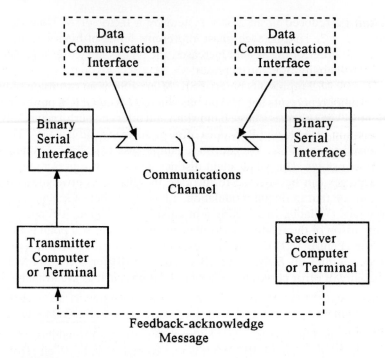

FIGURE 18.19

Elements of the data communication process.

discernible in the figure, we also need intelligent terminals that provide the human/machine interface, modems to convert digital data into a modulated carrier and then demodulate the carrier at the receiver, and test equipment for troubleshooting.

The communication channel may be defined as the path used for electrical or electromagnetic transmission between two or more stations. This could consist of fiber optic lines, coaxial cables, pairs of wires, free space, etc. Obviously the capacity of each channel to carry data will be different. In fact, determining channel capacity is an important part of the design problem.

We noted in Section 18.3.1 that it is theoretically possible to transmit N signal elements per second over a bandwidth of $N/2$ Hz. We infer from this that an X-hertz bandwidth can carry $2X$ separate voltage values at the rate of $2X$ per second. Binary data (bits) require two voltage levels. Transmitting 2 bits simultaneously requires four distinct voltage levels. In one time slot, then, using 2^n signal levels, we can encode n bits; thus, we can transmit $2nX$ bits per second over an X-hertz bandwidth. Let the number of signal levels be $L = 2^n$. The theoretical capacity of the line may be expressed as

$$C = 2nX = 2X \log_2 L \text{ bit/s} \qquad (18.43)$$

EXAMPLE 18.7   Binary data are to be transmitted over a 2.7-kHz line using four signal levels. What is the theoretical capacity of the line?

Solution

$$C = 2X \log_2 L = 2(2700) \log_2 4 = 10.8 \text{ kbit/s}$$

Obviously, a data communication system requires the design of an effective transmission channel. After the channel has been designed, the next step is to select the transmission scheme best suited to the channel. Basically, three transmission schemes are available (Figure 18.20):

- Simplex transmission from A to B but never from B to A
- Half-duplex transmission from A to B during one interval, then from B to A during another interval
- Full-duplex simultaneous transmission from B to A and A to B.

Half-duplex and simplex transmission can be supported with two wire lines. Full duplex requires four lines or else a two-wire channel with different transmission frequencies.

18.4.1 Modems

The actual signaling on the channels is performed by modems (modulators/demodulators). Modems provide the interface between digital signals and analog lines (Figure 18.21). The simplest modems for use are the Bell System 108 series. The outputs of the modems in the transmit mode are 1270 Hz for a binary 1 (mark) and 1070 Hz for a binary 0 (space). The incoming tone in the receive mode is 20025 Hz for a space signal and 2225 Hz for a mark signal. The use of FSK and PSK modulation is common in modems. Generally speaking, PSK is favored for the higher data rates. In terms of bandwidth efficiency, PSK is more efficient.

We can subdivide modems into asynchronous and synchronous classifications. Broadly speaking, *asynchronous* implies a data system that does not transmit data at a constant rate. For example, an interaction computer terminal with its "start" and "stop" characteristics would require an asynchronous modem. On the other hand, a *synchronous* system transmits data at a constant rate. An example of its use would lie in the process of downloading a computer program.

It should be clear at this point that the design of a data communication system is dependent on many factors. A worldwide effort is under way to standardize operational features for the many different systems available. Basically, a set of protocols is in preparation. A protocol is a set of rules governing the exchange of messages between two or more communication devices, as discussed in the next section.

18.4.2 Protocols

The seven-layer Open Systems Interconnection (OSI) reference model (Figure 18.22) being developed by the International Standards Organization can theoretically be used to define every aspect of data communication. The bottom three layers of the OSI model—the physical data link, control, and network layers—apply specifically to the architecture of a local network. The protocols for these levels

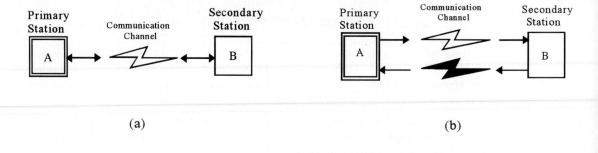

(a)   (b)

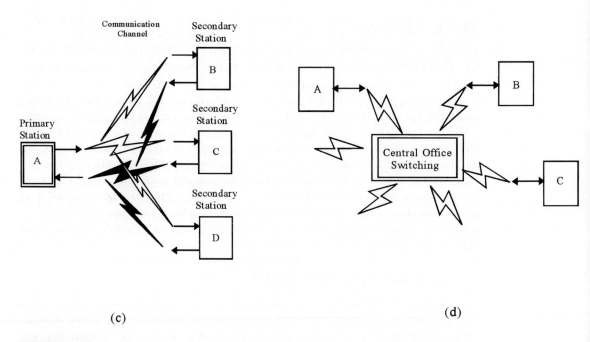

(c)   (d)

FIGURE 18.20

(a) Half-duplex transmission connects A to B over a two-wire channel. Either end can send or receive. (b) Duplex transmission allows simultaneous transmission and reception by using four wire circuits. (c) In nonswitched multipoint transmission, several secondary stations share one duplex channel. The primary station polls its tributaries.
(d) Point-to-point switched systems establish a new transmitter-to-receiver path for each call; thus volatility is the price of versatility.

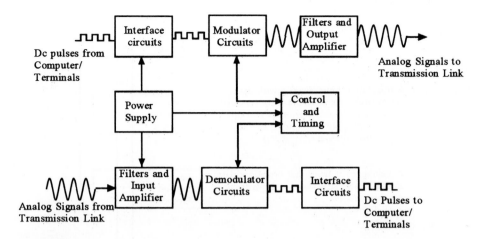

FIGURE 18.21

Modems make up two basic functional sections: The modulator accepts digital inputs from computers or terminals and converts the baseband pulses into analog audio-frequency pulses for transmission. A demodulator in a second modem at the other link end reconverts the audio-frequency analog to baseband digital.

encompass a variety of technologies such as fiber optics, satellites, short- and long-haul techniques, and different accessing methods for different network topologies. The functions and rules are specific to the network type. By way of contrast, the top three layers of the model—session, presentation, and application—are the same to all networks. The transport layer (layer 4) lies between the technology-dependent lower layers and the technology-independent upper layers. The essential function of this layer is to resolve differences between different physical layers and it serves as the boundary between data communications and data processing environments. Thus, all computers, hardware devices, and basic networking software must be able to interface at this layer.

18.4.3 Local-Area Networks

A standard data communication link generally uses a long-distance network consisting of interconnect facilities in different parts of the country or world. A local-area network (LAN) is a data communication system that allows several independent devices to intercommunicate and also confines communications to a moderate-sized area such as a plant, office, campus, etc. Three popular LAN configurations are shown in Figure 18.23. Typically, where many terminals desire access to a dominant central node, the star configuration is applicable. The obvious disadvantage is that if the central node (computer) fails, the entire network is down. By way of contrast, the ring and bus configurations avoid central-node failures. The best known LAN utilizing a bus is Ethernet, developed by Xerox Corporation. IBM has recently chosen a ring configuration as its LAN. In IEEE-802 standard specifies several types of LAN topologies.

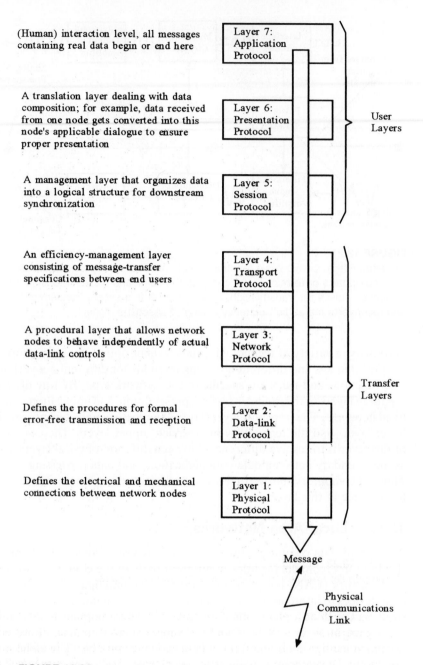

(Human) interaction level, all messages containing real data begin or end here

A translation layer dealing with data composition; for example, data received from one node gets converted into this node's applicable dialogue to ensure proper presentation

A management layer that organizes data into a logical structure for downstream synchronization

An efficiency-management layer consisting of message-transfer specifications between end users

A procedural layer that allows network nodes to behave independently of actual data-link controls

Defines the procedures for formal error-free transmission and reception

Defines the electrical and mechanical connections between network nodes

Layer 7:
Application
Protocol

Layer 6:
Presentation
Protocol

Layer 5:
Session
Protocol

Layer 4:
Transport
Protocol

Layer 3:
Network
Protocol

Layer 2:
Data-link
Protocol

Layer 1:
Physical
Protocol

User
Layers

Transfer
Layers

Message

Physical
Communications
Link

FIGURE 18.22
Seven-level ISO reference protocol.

Nodes (users)

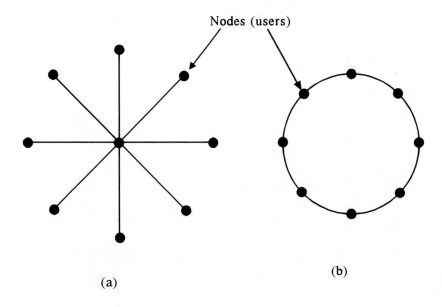

(a)

(b)

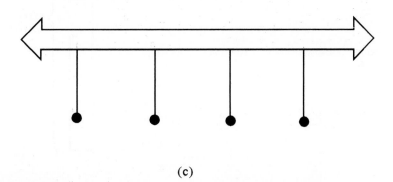

(c)

FIGURE 18.23
Three popular LAN configurations: (a) Star, (b) ring, (c) bus.

In the 1990s, voice communications, for the most part, are digital. In fact, the next generation of private branch exchanges (PBXs) is being designed to switch digital voice signals. Integration of data and voice processing facilities are a natural evolution. Development engineering for integrated office systems is well under way. The "age of information" is surely before us riding on the crest of the communication principles discussed in this chapter. Those countries with technologies advanced enough to take advantage of it will be the beneficiaries.

APPENDIX 1

UNITS, CONVERSIONS, AND CONSTANTS

The international system of units, called the SI system for the French expression Système International d'Unites, was developed by an international congress of representatives from all over the world. The system is based on eight elemental units that describe all of nature. Further descriptions of nature are facilitated by a set of derived units, which are built up from this basic set or can be defined in terms of this set.

1.1 ELEMENTAL SET OF UNITS

Physical Quantity	Elemental Unit	Standard Symbol
Length	meter	m
Mass	kilogram	kg
Time	second	s
Electric current	ampere	A
Temperature	kelvin	K
Luminous intensity	candela	cd
Plane angle	radian	rad
Solid angle	steradian	sr

1.2   PREFIXES

The SI elemental and derived units are used with a set of standard, named prefixes to express large and small numbers.

Quantity	Prefix Name	Prefix Symbol
10^{12}	tera	T
10^{9}	giga	G
10^{6}	mega	M
10^{3}	kilo	k
10^{2}	hecto	h
10	deka	da
10^{-1}	deci	d
10^{-2}	centi	c
10^{-3}	milli	m
10^{-6}	micro	μ
10^{-9}	nano	n
10^{-12}	pico	p
10^{-15}	femto	f
10^{-18}	atto	a

1.3   PARTIAL LIST OF DERIVED SET OF UNITS

Physical Quantity	Derived Unit Name	Definition	Standard Symbol
Acceleration		m/s^2	
Area	square meter	m^2	
Capacitance	farad	A-s/V	F
Charge	coulomb	A-s	C
Density		kg/m^3	
Electric field		V/m	
Energy	joule	$kg\text{-}m^2/s^2 = N\text{-}m$	J
Force	newton	$kg\text{-}m/s^2$	N
Frequency	hertz	s^{-1}	Hz
Illumination	lux	lm/m^2	lx
Inductance	henry	V-s/A	H
Kinematic viscosity		m^2/s	
Luminance		cd/m^2	
Luminous flux	lumen	cd-sr	lm
Magnetic field		A/m	
Magnetic flux	weber	V-s	Wb
Magnetic flux density	tesla	Wb/m^2	T
Power	watt	J/s	W
Pressure	pascal	N/m^2	Pa
Resistance	ohm	V/A	Ω
Stress		N/m^2	
Velocity		m/s	
Viscosity		$N\text{-}s/m^2$	
Voltage	volt	W/A	V
Volume		m^3	

1.4   UNIT CONVERSIONS

The following table shows conversion factors between some other units in common use in the industrial world.

Unit	SI Equivalent
acre	4046.86 m^2
angle (degree)	$(\pi/180)$ rad
atmospheres (atm)	101,325 Pa
bar	10^5 Pa
Celsius (°C)	T(°C) + 273.15 K
cubic inch (in^3)	16.3871 cm^3
dynes	10^{-5} N
electron-volt (eV)	1.602×10^{-19} J
ergs	10^{-7} J
Fahrenheit (°F)	5[T(°F) − 32]/9 °C
foot-pounds (ft-lb)	1.356 J
gallon (gal)	3.785×10^{-3} m^3
gauss (G)	10^{-4} T
horsepower (hp)	0.7457 kW
inch (in.)	2.54 cm
inch squared (in.^{2})	6.4516 cm^2
liter (l)	1000 cm^3
mile (mi)	1.60934 km
ounce (oz)	0.02835 kg
pound (lb, force)	4.448 N
pound (lb, weight)	0.4536 kg
pounds per square inch (psi)	6895 Pa
quart (qt)	9.464×10^{-4} m^3
torr = 1 mm Hg	133.32 Pa
watt-hour (Whr)	3.6×10^3 J

1.5 CERTAIN UNIVERSAL CONSTANTS

The following table contains the descriptions and approximate values of some constants that are found to occur frequently in the electrical and electronics field.

Description	Symbol	Value
Acceleration of gravity	g	9.807 m/s^2
Charge of the electron	e	$1.602 \times 10^{-19} \text{ C}$
Mass of the electron	m_e	$9.109 \times 10^{-31} \text{ kg}$
Mass of the proton	m_p	$1.672 \times 10^{-27} \text{ kg}$
Permeability of vacuum	μ_0	$4\pi \times 10^{-7} \text{ H/m}$
Permittivity of vacuum	ε_0	$8.849 \times 10^{-12} \text{ F/m}$
Pi	π	$3.141592653 \ldots$
Planck's constant	h	$6.626 \times 10^{-34} \text{ J-s}$
Speed of light in vacuum	c	$2.9978 \times 10^{8} \text{ m/s}$

APPENDIX 2

SOLID COPPER WIRE PROPERTIES

The following table gives the American Wire Gauge (AWG) number, diameter in mils (thousands of an inch = 0.001 in.), and resistance in ohms per 1000 feet for soft copper wire.

AWG	Diameter (mil)	Resistance (Ω/1000 ft)	AWG	Diameter (mil)	Resistance (Ω/1000 ft)
0	324.9	0.09825	22	25.3	16.2
2	257.6	0.1563	24	20.1	25.7
4	204.3	0.2485	26	15.9	41.0
6	162.0	0.3952	28	12.6	65.3
8	128.5	0.6281	30	10.0	104
10	101.9	0.9988	32	8.0	162
12	80.8	1.59	34	6.3	261
14	64.1	2.52	36	5.0	415
16	50.8	4.02	38	4.0	648
18	40.3	6.39	40	3.1	1080
20	32.0	10.1			

APPENDIX 3

MATHEMATICAL FORMULAS AND RELATIONS

3.1 PROPERTIES OF OBJECT FIGURES (Figure A3.1)

Object	Circumference or Perimeter	Plane Area	Surface Area	Volume
Box			$2(ab + ac + bc)$	abc
Circle	$2\pi r$	πr^2		
Cone			πrs (regular)	$\pi r^2 h/3$
Cylinder			$2\pi rh + 2\pi r^2$	$\pi r^2 h$
Parallelogram	$2a + 2b$	bh		
Pyramid			$ps/2$ (regular)	$Ah/3$
Rectangle	$2a + 2b$	ab		
Sphere			$4\pi r^2$	$4\pi r^3/3$
Trapezoid	$a + b + c + d$	$(a + b)h/2$		
Triangle	$a + b + c$	$bh/2$		

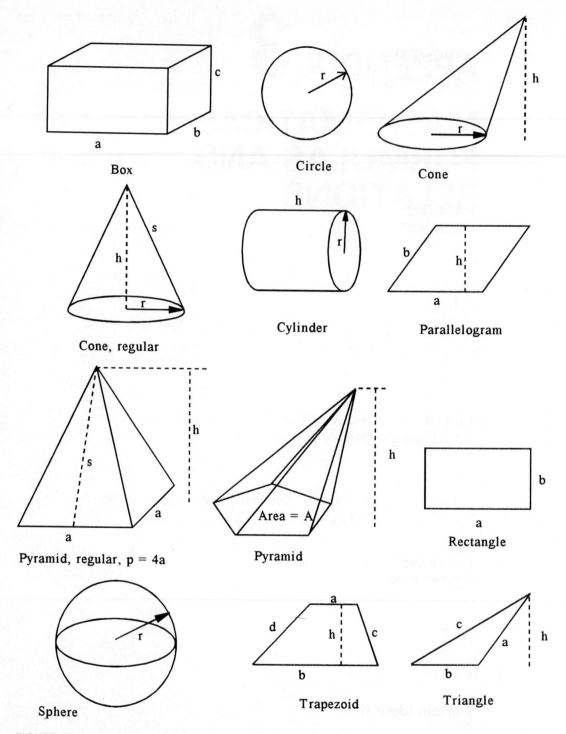

FIGURE A3.1
Geometrical figures.

3.2   TRIGONOMETRY

Right Triangle (Figure A3.2)

$$\sin(\theta) = a/c$$

$$\cos(\theta) = b/c$$

Pythagorean theorem: $c^2 = a^2 + b^2$

$$\tan(\theta) = a/b = \sin(\theta)/\cos(\theta)$$

FIGURE A3.2
A right triangle.

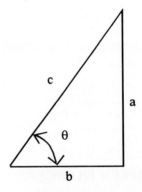

Right Triangle

Any Triangle (Figure A3.3)

$$A + B + C = 180° = \pi \text{ radians}$$

Law of sines: $\dfrac{a}{\sin(A)} = \dfrac{b}{\sin(B)} = \dfrac{c}{\sin(C)}$

Law of cosines: $a^2 = b^2 + c^2 - 2cb \cos(A)$

FIGURE A3.3
A general triangle.

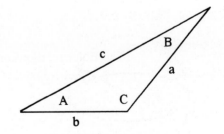

Certain Identities

$$\sin(A \pm B) = \sin(A) \cos(B) \pm \cos(A) \sin(B)$$

$$\cos(A \pm B) = \cos(A) \cos(B) \mp \sin(A) \sin(B)$$

$$\sin(2A) = 2\sin(A)\cos(A)$$
$$\cos(2A) = 2\cos^2(A) - 1$$

3.3 ALGEBRA

Linear Equation

$$y = mx + b$$

where m = slope and b = intercept.

Quadratic Equation

$$y = ax^2 + bx + c$$
$$\text{Roots: } (y = 0)$$
$$x_r = \frac{-b \pm \sqrt{b^2 - 4ac}}{2a}$$

3.4 COMPLEX NUMBERS

$$j \equiv \sqrt{-1}, \text{ thus } j^2 = -1$$

A complex number expressed in rectangular form is

$$z = x + jy$$
$$x = \text{real part of } z, \text{Re}(z)$$
$$y = \text{imaginary part of } z, \text{Im}(z)$$

A complex number expressed in polar form is

$$z = \text{R}e^{j\theta}$$
$$R = \sqrt{x^2 + y^2}$$
$$\theta = \tan^{-1}(y/x)$$

The Euler relations are

$$e^{j\theta} = \cos(\theta) + j\sin(\theta)$$
$$\cos(\theta) = [e^{j\theta} + e^{-j\theta}]/2$$
$$\sin(\theta) = [e^{j\theta} - e^{-j\theta}]/2j$$

INDEX